Abbreviations for Units

A	ampere	keV	kiloelectron volt
Å	angstrom (10^{-10} m)	m	meter
atm	atmosphere	MeV	megaelectron volt (10^6 eV)
Bq	becquerel (sec^{-1})		
C	coulomb	min	minute
°C	degree Celsius	mm	millimeter
cal	calorie	msec	millisecond
Ci	curie	N	newton
cm	centimeter	nm	nanometer (10^{-9} m)
eV	electron volt	R	roentgen
fm	femtometer, fermi (10^{-15} m)	rev	revolution
		sec	second
G	gauss (10^{-4} T)	T	tesla
g	gram	u	unified mass unit
Gy	grey (J/kg)	V	volt
h	hour	W	watt
Hz	hertz (sec^{-1})	y	year
J	joule	μsec	microsecond (10^{-6} sec)
K	kelvin		
kg	kilogram	Ω	ohm
km	kilometer		

Modern Physics

Modern Physics

Paul A. Tipler

Oakland University
Rochester, Michigan

WORTH PUBLISHERS, INC.

Modern Physics

Based on Foundations of Modern Physics

By Paul Tipler, Copyright © 1969

Copyright © 1978 by Worth Publishers, Inc.

Printed in the United States of America

Library of Congress Catalog No. 77-085725

ISBN: 0-87901-088-6

Seventh Printing, June 1985

Designed by Malcolm Grear Designers

Illustrated by Felix Cooper

Worth Publishers, Inc.

444 Park Avenue South

New York, NY 10016

Preface

The revision of this book reflects the many helpful responses and suggestions that came from users of the first edition. In this new edition I have tried to maintain the historical and cultural flavor of the first edition, while reducing the amount and the level of difficulty of the mathematics. In particular, the material formerly in Chapters 6, 7, and 8 of the first edition has been greatly simplified. New topics have been added. A completely new chapter on elementary particles includes material on mesons, resonance particles, the eightfold way, and quarks. Also, there are new sections on general relativity; polyatomic molecules; the structure of solids; impurity semiconductors; semiconductor junctions and devices (including transistors); fission, fusion, and nuclear reactors; and nuclear detectors.

Among the features of the first edition that have been retained are:

1. The logical structure—beginning with an introduction to relativity and quantization; then following with applications.

2. The elementary discussion of kinetic theory in Chapter 2. This topic introduces students to microscopic physics and to the use of distribution functions that they will need later.

3. The qualitative presentation of de Broglie waves and the review of classical waves in Chapter 5, illustrated with many photographs of electron and x-ray diffraction patterns.

4. Many examples with order-of-magnitude estimates and calculations based on simple models.

5. Use of combined quantities such as hc, $\hbar c$, and ke^2 in eV-Å, to simplify numerical calculations.

6. The summaries and the reference lists at the end of every chapter.

7. Separation of the problems into two sets, based on difficulty. The number of easy problems (now called "Exercises") has been greatly increased.

8. Use of real data in figures, numerous photographs of people and apparatus, and quotations from original papers. These features bring to life events in the history of science, and help counter the too-prevalent view among students that physics is a dull and impersonal collection of facts and formulas.

The new edition has been divided into two parts: Part 1, "Introduction to Relativity and Quantum Physics," and Part 2, "Applications." In Chapter 1, the material on special relativity has been shortened and slightly simplified, and a new section on general relativity has been added. Since only the relation $E^2 = p^2c^2 + m^2c^4$ is needed for understanding the later chapters, it is possible to omit this first chapter. Chapters 2 through 5 have been rewritten for simplification and clarification. Nearly all the mathematical derivations have been eliminated, or placed at the ends of sections or chapters and labeled optional. I have simplified Chapter 6 by concentrating on the easiest problems (such as the infinite square well), and by replacing mathematical derivations of the more difficult problems (for example, the harmonic oscillator) with qualitative discussions of their important features. Two new sections have been added to this chapter: Section 6-8 treats a single particle in a three-dimensional infinite square well, and Section 6-9 treats two identical particles in a one-dimensional infinite square well. These two discussions provide a simple framework for introducing many of the important results that arise from the extension of the Schrödinger equation to three dimensions and to systems of more than one particle, without the mathematical complications of Chapters 7 and 8 in the first edition. Part 1 concludes with a chapter on atomic physics. This discussion contains some of the material that was formerly in Chapters 7 and 8, but the presentation has been arranged so that the subject can be covered briefly and qualitatively if desired.

In Part 2, the ideas and methods discussed in Part 1 are applied to the study of molecules, solids, quantum statistics, liquid helium II, nuclei, and elementary particles. Chapter 8 ("Molecular Structure and Spectra") includes material on diatomic molecules (formerly presented in Chapter 9), plus a new section on polyatomic molecules. The material on solids in Chapter 9 has been greatly expanded and rewritten so that the thermal and electrical properties of solids can be understood with only a qualitative discussion of the Fermi energy at $T = 0$. New material on semiconductor junctions and transistors should make this chapter more useful and interesting to both engineering students and physics majors. Quantum statistics and liquid helium II are examined briefly in Chapter 10. Chapter 11 includes a completely rewritten section on radioactivity, and two new sections: one on fission, fusion, and nuclear reactors; the other on nuclear detectors. Those instructors who choose to cover some nuclear physics will probably wish to select only certain sections from this long chapter. The book concludes with a new chapter on elementary particles.

The chapters in Part 2 are completely independent of one another, and can be covered in any order. A one-semester course can probably comprise most of the material in Part 1 and several chapters (or at least parts of chapters) in Part 2.

Some new pedagogical features in this edition include:

1. Questions for discussion and review, given at the ends of various sections.

2. Margin comments highlighting important concepts and equations.

3. Horizontal color rules setting off major equations and underscoring important results in the text.

4. Optional material that is clearly set apart in the right-hand column, and marked by vertical color rules and an "Optional" head in the margin. Much of the material labeled optional in the first edition has been removed. There are now three kinds of optional material: mathematical derivations; more-difficult sections; and extensions of discussions that, while not necessarily difficult, may include more detail than some instructors will wish to cover. Chapters are designed so that optional material usually falls at the end of a section.

5. A short list of learning objectives at the beginning of each chapter, to help students focus on the important information to be presented, and to give them an idea of the depth of understanding that will be required. The objectives also serve as an aid for later review. Additionally, as an enthusiastic believer in discussion questions on examinations, I hope that these learning objectives will prove useful in suggesting such questions.

Acknowledgments

Many people have contributed to one or both editions of this book. I would like to thank everyone who used the first edition, and all those who tried out early drafts of the book. Some of the people whose assistance with the first edition I particularly valued were Marc Ross, Peter Roll, Jerry Griggs, Libor Velinsky, Dorothy Hubert, and Sue Nast. I would like to thank Oakland University for granting me the sabbatical leave during which much of the work on the new edition was done, and the physics department at the University of California at Berkeley for its kind hospitality during my sabbatical. I would like to thank Bob Rogers, who made significant contributions to the planning of this edition; Lynn Hlatky, who helped with the exercises and problems; John McKinley, who suggested many improvements in the final manuscript; and Granvil C. Kyker, who read the galleys and suggested many improvements.

Others who reviewed various parts of the manuscript were: Philip A. Chute, University of Wisconsin, Eau Claire; F. Eugene Dunnam, University of Florida; William Eidson, Drexel University; E. Ni Foo, Drexel University; John Gardner, Oregon State

University; Donald Hall, California State University, Sacramento; Roger Hanson, University of Northern Iowa; Louis V. Holroyd, University of Missouri; Terry Kjeldaas, Brooklyn Polytechnic Institute; Kenneth Krane, Orgeon State University; Edward Saunders, U.S. Military Academy, West Point; John Stewart, University of Virginia; Jay Strieb, Villanova University; Martin Tiersten, City College of New York; and Robert Williamson, Oakland University. Many of these reviewers read several versions of the manuscript, and all offered valuable suggestions for its improvement.

I am grateful to my family, Sue, Becky, and Ruth, for their continuing support and encouragement. Finally, I received much help and encouragement from Worth Publishers, particularly from June Fox, whose tireless efforts in coordinating the project have made its completion both possible and enjoyable.

PAUL A. TIPLER

Rochester, Michigan
December 1977

Contents

PART I

Introduction to Relativity and Quantum Physics

The spectacular successes of the laws of mechanics, electromagnetism, and thermodynamics as expressed by Newton, Maxwell, Carnot, and others led some to believe that there was little left to do in physics except apply these laws to various phenomena. Such optimism (or pessimism, depending on your point of view) was short-lived, for the late nineteenth and early twentieth centuries saw a rapid accumulation of discoveries and experiments which could not be explained in terms of the known laws generally referred to as classical physics. The breakdown in classical physics occurred in many different areas: the Michelson-Morley null result contradicted Newtonian relativity; the blackbody radiation spectrum and the measured heat capacities of solids and gases contradicted predictions of thermodynamics; the photoelectric effect, the Compton effect, and the spectra of atoms could not be explained by electromagnetic theory; and the exciting discoveries of radioactivity and x rays seemed to be outside the framework of classical physics. The development of the theories of relativity and quantum mechanics in the early twentieth century provided answers to all the puzzles listed above and many more. The application of these theories to such microscopic systems as atoms, molecules, nuclei, and elementary particles, and to macroscopic systems of gases, liquids, and solids, has given us a deep understanding of the workings of nature and has revolutionized our way of life.

In Part I we discuss the foundations of modern physics: the theories of relativity and quantum mechanics. The special theory of relativity is discussed at some length in Chapter 1, which also includes a brief discussion of general relativity. In Chapters 2 through 5 the historical development of quantum theory is discussed from the earliest evidences of quantization to the de Broglie postulate of electron waves. An elementary discussion of the Schrödinger equation is provided in Chapter 6, with applications to simple systems. The application of quantum mechanics to atomic physics in Chapter 7 brings out the important new feature of electron spin and extends the treatment to systems of many particles. With the exception of the theory of relativity, which is used only sparingly in later chapters, each chapter in Part I depends on the discussions and developments in the previous chapters.

CHAPTER 1   Relativity

Objectives

After studying this chapter you should:

---

1. Be able to describe the general purpose, method, and result of the Michelson-Morley experiment.

2. Be able to state Einstein's postulates of special relativity and discuss their consequences.

3. Know the meaning of proper time and proper length, and their relation to time and length intervals measured in other reference frames.

4. Know how and why the relativistic and classical Doppler effects differ.

5. Be able to use the Lorentz transformation equations in exercises and problems.

6. Be able to discuss the problems of clock synchronization and simultaneity, and the twin paradox.

7. Know the relativistic expressions for momentum, energy, and kinetic energy, and be able to use them in exercises and problems.

8. Know the relationship between mass and energy and be able to calculate binding energies from given masses.

9. Know the nonrelativistic and extreme relativistic approximations for energy and momentum, and when to use them.

10. Be able to state the equivalence principle and list some of the predictions of the general theory of relativity.

The theory of relativity consists of two rather different theories, the special theory and the general theory. The special theory, developed by Einstein and others in 1905, concerns the comparison of measurements made in different inertial reference frames moving with constant velocity relative to each other. Its consequences, which can be derived with a minimum of mathematics, are applicable in a wide variety of situations encountered in physics and engineering. On the other hand, the general theory, also developed by Einstein and others (around 1916),

Albert Einstein in 1905, at the time of his greatest productivity. (*Courtesy of Lotte Jacobi.*)

is concerned with accelerated reference frames and gravity. A thorough understanding of the general theory requires much sophisticated mathematics (e.g., tensor analysis), and the applications of this theory are chiefly in the area of gravitation. It is of great importance in cosmology but is rarely encountered in other areas of physics or engineering. We shall therefore concentrate on the special theory (often referred to as special relativity), and discuss the general theory only briefly in the last section of this chapter.

The theory of special relativity can be derived from two postulates proposed by Einstein in a paper on the electrodynamics of moving bodies published when he was only 26 years old. These two postulates, simply stated, are [1]

1. Absolute uniform motion cannot be detected.

2. The speed of light is independent of the motion of the source.

Einstein postulates

Although each postulate seems quite reasonable, many implications of the two together are quite surprising and contradict what is often called common sense. For example, a direct consequence is that all observers obtain the same value for the speed of light in a vacuum independent of their relative motion. We shall derive this and other results later. Since we have time

[1] "Zur Electrodynamik bewegter Körper," *Annalen der Physik* (4), **17,** 841 (1905). For a translation from the German, see W. Perrett and G. B. Jeffery (trans.), *The Principle of Relativity: A Collection of Original Memoirs on the Special and General Theory of Relativity* by H. A. Lorentz, A. Einstein, H. Minkowski, and H. Weyl, New York: Dover Publications, Inc., 1952.

here for only a brief introduction, we shall discuss relativistic kinematics in some detail and merely outline relativistic dynamics, omitting derivations. We shall begin by looking at a historically important experiment related to the theory of special relativity, the Michelson-Morley experiment.

1-1   The Michelson-Morley Experiment

All waves except electromagnetic waves require a medium for their propagation. The speed of a wave depends on the properties of the medium. For sound waves, for example, the medium is air, and absolute motion, i.e., motion relative to the still air, can be detected. The Doppler effect for sound depends not only on the relative motion of the source and listener but on the absolute motion of each relative to the air. It was natural to expect that some kind of medium exists which supports the propagation of light and other electromagnetic waves. Such a medium, called the *ether,* was proposed in the nineteenth century. The ether as proposed would have had to possess unusual properties. Although it would require great rigidity to support waves of such high velocity (recall that the velocity of waves on a string depends on the tension of the string), it must introduce no drag force on the planets, as their motion is fully accounted for by the law of gravitation.

It was of considerable interest to determine the velocity of the earth relative to the ether. Maxwell pointed out that in measurements of the speed of light, the earth's speed v relative to the ether appears only in the second order v^2/c^2, an effect then considered too small to measure. Such measurements determine the time for a light pulse to travel to and from a mirror. Figure 1-1 shows a light source and a mirror a distance L apart. If we assume that both are moving with speed v through the ether, classical theory predicts that the light will travel toward the mirror with speed $c - v$ and back with speed $c + v$ (both speeds relative to the mirror and light source). The time for the total trip will be

$$t_1 = \frac{L}{c - v} + \frac{L}{c + v} = \frac{2cL}{c^2 - v^2} = \frac{2L}{c}\left(1 - \frac{v^2}{c^2}\right)^{-1} \qquad 1\text{-}1$$

For v much smaller than c, we can expand this result using the binomial expansion

$$(1 + x)^n \approx 1 + nx + \cdots \qquad \text{for } x \ll 1$$

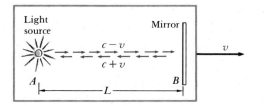

Figure 1-1
Light source and mirror moving with speed v relative to the "ether." According to classical theory, the speed of light relative to the source and mirror would be $c - v$ toward the mirror and $c + v$ away from the mirror.

Then

$$t_1 \approx \frac{2L}{c}\left(1 + \frac{v^2}{c^2} + \cdots\right) \qquad \text{1-2}$$

If we take the orbital speed of the earth about the sun as an estimate of v, we have $v \approx 3 \times 10^4$ m/sec $= 10^{-4}c$ and $v^2/c^2 = 10^{-8}$. Thus the correction for the earth's motion is small indeed.

Albert A. Michelson realized that, although this effect is too small to be measured directly, it should be possible to determine v^2/c^2 by a difference measurement. Figure 1-2a is a diagram of his apparatus, called a *Michelson interferometer*. Light from the source is partially reflected and partially transmitted by mirror A. The transmitted beam travels to mirror B and is reflected back to A. The reflected beam travels to mirror C and is reflected back to A. The two beams recombine and form an interference pattern, which is viewed by an observer at O. Equation 1-2 gives the classical result for the round-trip time t for the transmitted beam. Since the reflected beam travels (relative to the earth) perpendicular to the earth's velocity, the velocity of this beam relative to earth (according to classical theory) is the vector difference $\mathbf{u} = \mathbf{c} - \mathbf{v}$. The magnitude of $\mathbf{u}$ is $\sqrt{c^2 - v^2}$; so the round-trip time for this beam is

$$t_2 = \frac{2L}{\sqrt{c^2 - v^2}} = \frac{2L}{c}\left(1 - \frac{v^2}{c^2}\right)^{-1/2}$$

$$\approx \frac{2L}{c}\left(1 + \frac{1}{2}\frac{v^2}{c^2} + \cdots\right) \qquad \text{1-3}$$

where again the binomial expansion has been used. There is thus a time difference:

$$\Delta t = t_1 - t_2 \approx \frac{2L}{c}\left(1 + \frac{v^2}{c^2}\right) - \frac{2L}{c}\left(1 + \frac{1}{2}\frac{v^2}{c^2}\right)$$

$$\approx \frac{Lv^2}{c^3} \qquad \text{1-4}$$

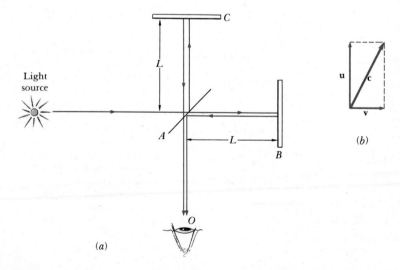

(a)

Figure 1-2
(a) Schematic drawing of the Michelson interferometer. (b) According to classical theory, if the interferometer moves to the right with velocity $\mathbf{v}$ relative to the ether, the light must move with velocity $\mathbf{c}$ in the direction shown (relative to the ether) to strike the upper mirror. Its velocity relative to the interferometer is then $\mathbf{u} = \mathbf{c} - \mathbf{v}$, and its speed is $u = \sqrt{c^2 - v^2}$.

The time difference is to be detected by observing the interference of the two beams of light. Because of the difficulty of making the two paths of equal length to the precision required, the interference pattern of the two beams is observed and then the whole apparatus rotated 90°. The rotation produces a time difference given by Equation 1-4 for each beam. The total time difference of $2 \, \Delta t$ is equivalent to a path difference of $2c \, \Delta t$. The interference fringes observed in the first orientation should thus shift when the apparatus is rotated by a number of fringes ΔN, given by

$$\Delta N = \frac{2c \, \Delta t}{\lambda} = \frac{2L}{\lambda} \frac{v^2}{c^2} \qquad \text{1-5}$$

where λ is the wavelength of the light. In Michelson's first attempt, in 1881, L was about 1.2 m and λ was 590 nm. For $v^2/c^2 = 10^{-8}$, ΔN was expected to be 0.04 fringe.

When no shift was observed, Michelson reported the null result even though the experimental uncertainties were estimated to be about the same order of magnitude as the expected effect. In 1887, when he repeated the experiment with Edward W. Morley, he used an improved system for rotating the apparatus without introducing a fringe shift because of mechanical strains, and he increased the effective path length L to about 11 m by a series of multiple reflections. Figure 1-3 shows the configuration of the Michelson-Morley apparatus. For this attempt, ΔN was expected to be about 0.4 fringe, about 20 to 40 times the minimum shift observable. Once again, no shift was observed. The experiment has since been repeated under various conditions by a number of people, and no shift has ever been found.

A student-type Michelson interferometer. The fringes are produced on a ground-glass screen by light from a laser. (*Courtesy of Libor Velinsky.*)

Figure 1-3
Drawing of Michelson-Morley apparatus used in their 1887 experiment. The optical parts were mounted on a sandstone slab 5 ft square, which was floated in mercury, thereby reducing the strains and vibrations that had affected the earlier experiments. Observations could be made in all directions by rotating the apparatus in the horizontal plane. (*From R. S. Shankland, "The Michelson-Morley Experiment," Copyright © November 1964 by Scientific American, Inc. All rights reserved.*)

Light source
Mirrors
Telescope
Adjustable mirror
Silvered glass plate
Unsilvered glass plate
Mirrors
Mirrors

Albert A. Michelson in his laboratory. (*Courtesy of the Niels Bohr Library, American Institute of Physics.*)

Elmer Taylor

The null result of the Michelson-Morley experiment is easily understood in terms of the Einstein postulates. According to postulate 1, absolute uniform motion cannot be detected. We can consider the whole apparatus and the earth to be at rest. No fringe shift is expected when the apparatus is rotated 90° since all directions are equivalent. It should be pointed out that Einstein did not set out to explain this experiment. His theory arose from his considerations of the theory of electricity and magnetism and the unusual property of electromagnetic waves—namely, that they propagate in a vacuum. In his first paper, which contains the complete theory of special relativity, he made only a passing reference to the Michelson-Morley experiment, and in later years he could not recall whether he was aware of the details of the experiment before he published his theory.

1-2   Consequences of Einstein's Postulates

An immediate consequence of the two Einstein postulates is that

Every observer obtains the same value for the speed of light independent of the relative motion of sources and observers.

Consider a light source S and two observers, R_1 at rest relative to S and R_2 moving toward S with speed v, as shown in Figure 1-4a. The speed of light measured by R_1 is $c = 3 \times 10^8$ m/sec. What is the speed measured by R_2? The answer is not $c + v$. By postulate 1, Figure 1-4a is equivalent to Figure 1-4b, in which R_2 is pictured at rest and the source S and R_1 are moving with speed v. That is, since absolute motion has no meaning, it is not possible to say which is really moving and which is at rest. By postulate 2, the speed of light from a moving source is inde-

(a)   (b)

pendent of the motion of the source. Thus R_2 measures the speed of light to be c, the same as measured by R_1.

This result—that all observers measure the same value for the speed of light—contradicts our intuitive ideas about relative velocities. If a car moves at 30 mi/h away from an observer and another car moves at 70 mi/h in the same direction, the velocity of the second car relative to the first car is 40 mi/h. This result is easily measured and conforms to our intuition. However, according to Einstein's postulates, if a light beam is moving in the direction of the cars, observers in both cars will measure the same speed for the light beam. We shall see later that our intuitive ideas about the combination of velocities are approximations which hold only when the speeds are very small compared with that of light. In practice, even in an airplane moving with the speed of sound, it is not possible to measure the speed of light accurately enough to distinguish the difference between the result c and $c \pm v$, where v is the speed of the airplane. In order to make such a distinction, we must either move with a very great velocity (much greater than that of sound) or make extremely accurate measurements, as in the Michelson-Morley experiment.

Einstein recognized that his postulates have important consequences for measuring time intervals and space intervals as well as relative velocities. He showed that the size of time and space intervals between two events depends on the reference frame in which the events are observed. The changes in such measurements from one reference frame to another are called *time dilation* and *length contraction*. Instead of developing the general formalism of relativity, known as the *Lorentz transformation,* we shall derive these famous relativistic effects directly from the Einstein postulates by considering some simple special cases of measurement of time and space intervals. We shall also study the problems of clock synchronization and simultaneity, which are of central importance in understanding special relativity.

In these discussions we shall be comparing measurements made by observers who are moving relative to each other. We shall use a rectangular coordinate system xyz with origin O, called the *S reference frame,* and another system $x'y'z'$ with origin O', called the *S' frame,* which is moving with constant velocity relative to the *S* frame. For simplicity, we shall consider the *S'* frame to be moving with speed v along the x(or x') axis relative to S. In each frame we assume that there are as many observers as needed, equipped with clocks, metersticks, etc., which are identical when compared at rest (see Figure 1-5).

Figure 1-4
If absolute motion cannot be detected, the reference frame shown in (a), in which the receiver R_1 and source S are stationary and receiver R_2 is moving toward the source with speed v, is equivalent to the reference frame shown in (b), in which R_2 is stationary and the source S and receiver R_1 are moving with speed v. Since the speed of light does not depend on the source speed, receiver R_2 measures the same value for that speed as receiver R_1.

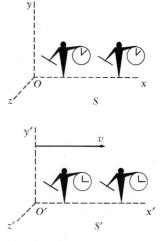

Figure 1-5
Coordinate reference frames S and S'. Each frame moves with speed v relative to the other and has observers with metersticks and clocks.

1-3   Time Dilation and Length Contraction

The results of correct measurements of the time and space intervals between events do not depend on the kind of apparatus used for the measurements or on the events. We are free therefore to choose any events and measuring apparatus which will help us understand the application of the Einstein postulates to the results of the measurement. Convenient events in relativity are those which produce light flashes. A convenient clock is a *light clock,* pictured schematically in Figure 1-6. A photocell detects the light pulse and sends a voltage pulse to an oscilloscope that produces a vertical deflection of the trace on the scope. The phosphorescent material on the face of the oscilloscope tube gives a persistent light that can be observed visually or photographed. The time between two light flashes is determined by measuring the distance between pulses on the scope and knowing the sweep speed of the scope. Such a clock, which can easily be calibrated and compared with other types of clocks, is often used in nuclear-physics experiments.

We first consider an observer A' at rest in frame S' a distance D from a mirror, as shown in Figure 1-7a. He explodes a flash gun and measures the time interval $\Delta t'$ between the original flash and the return flash from the mirror. Since light travels with speed c, this time is

$$\Delta t' = \frac{2D}{c} \qquad\qquad 1\text{-}6$$

We now consider these same two events, the original flash of light and the returning flash, as observed in reference frame S, where observer A' and the mirror are moving to the right with speed v. The events happen at two different places, x_1 and x_2, in frame S because between the original flash and the return flash observer A' has moved a horizontal distance $v\,\Delta t$, where Δt is the

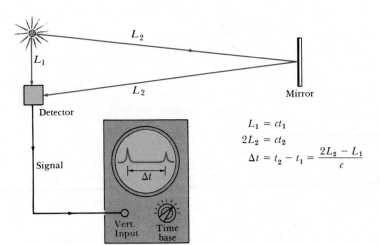

$L_1 = ct_1$

$2L_2 = ct_2$

$\Delta t = t_2 - t_1 = \dfrac{2L_2 - L_1}{c}$

Figure 1-6
Light clock for measuring time intervals. The time is measured by reading the distance between pulses on the oscilloscope after calibrating the sweep speed.

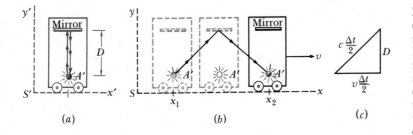

(a)   (b)   (c)

Figure 1-7
(a) A' and mirror are at rest in S'. The time for a light pulse to reach the mirror and return is measured to be $2D/c$ by A'. (b) In frame S, A' and the mirror are moving with speed v. If the speed of light is the same in both reference frames, the time for the light to reach the mirror and return is longer than $2D/c$ in S because the distance traveled is greater than $2D$. (c) Right triangle for computing the time Δt in frame S.

time interval between the events measured in S. In Figure 1-7b we see that the path traveled by the light is longer in S than in S'. However, by Einstein's postulates, light travels with the same speed c in frame S as it does in frame S'. Since it travels farther in S at the same speed, it takes longer in S to reach the mirror and return. The time interval between flashes in S is thus longer than it is in S'. We can easily calculate Δt in terms of $\Delta t'$. From the triangle in Figure 1-7c we see that

$$\left(\frac{c\,\Delta t}{2}\right)^2 = D^2 + \left(\frac{v\,\Delta t}{2}\right)^2$$

or

$$\Delta t = \frac{2D}{\sqrt{c^2 - v^2}} = \frac{2D}{c}\frac{1}{\sqrt{1 - v^2/c^2}}$$

Using $\Delta t' = 2D/c$, we have

$$\Delta t = \frac{\Delta t'}{\sqrt{1 - v^2/c^2}} = \gamma\,\Delta t' \qquad \Delta t = \gamma \Delta t_p \qquad \text{1-7}$$

Time dilation

$t = $ dilated, $t_p = $ proper

where

$$\gamma = \frac{1}{\sqrt{1 - v^2/c^2}} \geq 1 \qquad \text{1-8}$$

Observers in S would say that the clock held by A' runs slow since A' measures a shorter time interval ($\Delta t'$) between these events.

A' measures the times of the light flash and return at the same point in S'; for the observers in S these events happen at two different places. A single clock can be used in S' to measure the time interval, but in S two synchronized clocks are needed, one at x_1 and one at x_2. The time between events that happen at the *same place* in a reference frame (as in S', in this case) is called *proper time*. The time interval measured in any other reference frame is always longer than the proper time. This expansion is called *time dilation*.

Proper time

Time dilation is closely related to another phenomenon, *length contraction*. The length of an object measured in the reference frame in which the object is at rest is called its *proper length*. In a reference frame in which the object is moving, the measured length is shorter (along the direction of relative motion) than its proper length. We can see this from our previous example using the light clocks. Suppose that x_1 and x_2 in that example are at the ends of a measuring rod of length $L_0 = x_2 - x_1$; measured in frame S in which the rod is at rest. Since A' is moving relative to this frame with speed v, the distance moved in time Δt is $v\,\Delta t$. Since A' moves from point x_1 to point x_2 in this time, this distance is $L_0 = x_2 - x_1 = v\,\Delta t$. According to observer A' in frame S' (Figure 1-8), the measuring rod moves with speed v and takes time $\Delta t'$ to move past him. The length of the rod in his frame is thus $L' = v\,\Delta t'$. Since the time interval $\Delta t'$ is less than Δt, the length L' is less than L_0. These lengths are related by

$$L' = v\,\Delta t' = \frac{v\,\Delta t}{\gamma} = \sqrt{1 - \frac{v^2}{c^2}}\,L_0 = \frac{L_0}{\gamma} \qquad \text{1-9} \qquad \textit{Length contraction}$$

Since this contraction is just the amount proposed by Lorentz and FitzGerald to explain the Michelson-Morley experiment, it is often called the *Lorentz-FitzGerald contraction*.

It is important to note that *observers in both reference frames measure the same relative velocity*. Otherwise there would be a lack of symmetry, and postulate 1 would be violated; i.e., we could choose the frame with the smaller or greater relative velocity as a preferred frame.

An interesting example of the observation of these phenomena is afforded by the detection of muons as secondary radiation from cosmic rays. Muons decay according to the statistical law of radioactivity,

$$N(t) = N_0 e^{-t/T} \qquad \text{1-10}$$

where $N(t)$ is the number of muons at time t, N_0 is the number at time $t = 0$, and T is the mean lifetime, which is about 2 μsec for muons at rest. Since they are created (from the decay of π mesons) high in the atmosphere, usually several thousand

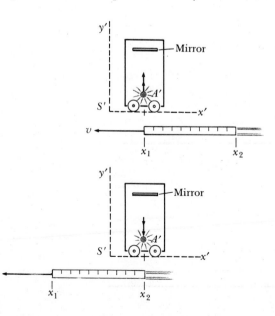

Figure 1-8
Measuring the length of a moving object. The rod has length $L_0 = x_2 - x_1 = v\,\Delta t$ as measured in frame S, in which it is at rest. In frame S' it moves past a fixed point in time $\Delta t' = \sqrt{1 - v^2/c^2}\,\Delta t$, so its length $L' = v\,\Delta t'$ is shorter in this frame.

meters above sea level, we would expect few muons to reach sea level. A typical muon moving with speed of $0.998c$ would travel only about 600 m in 2 μsec. However, the lifetime of the muon measured in the earth's reference frame is increased by the factor $\gamma = 1/\sqrt{1 - v^2/c^2}$, which is 15 for this particular speed. The mean lifetime measured in the earth's reference frame is therefore 30 μsec, and a muon of this speed travels about 9000 m in this time. From the muon's point of view, it lives only 2 μsec, but the atmosphere is rushing past it with a speed of $0.998c$. The distance of 9000 m in the earth's frame is thus contracted to only 600 m, as indicated in Figure 1-9.

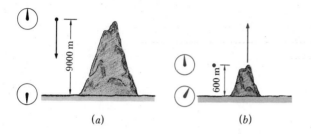

(a)   (b)

Figure 1-9
Although muons are created high above the earth and their mean lifetime is only about 2 μsec when at rest, many appear at the earth's surface. (a) In the earth's reference frame a typical muon moving at $0.998c$ has a mean lifetime of 30 μsec and travels 9000 m in this time. (b) In the reference frame of the muon, the distance traveled by the earth is only 600 m in the muon's proper lifetime of 2 μsec.

It is easy to distinguish experimentally between the classical and relativistic predictions for the number of muons detected at sea level. Suppose that we observe with a muon detector 10^8 muons in some time interval at an altitude of 9000 m. How many would we expect to observe at sea level in the same time interval? According to the nonrelativistic prediction, the time taken for these muons to travel 9000 m is $(9000 \text{ m})/0.998c \approx 30$ μsec, which is 15 lifetimes. Inserting $N_0 = 10^8$ and $t = 15T$ into Equation 1-10, we obtain

$$N = 10^8 e^{-15} = 30.6$$

We would thus expect all but about 31 of the original 100 million muons to decay before reaching sea level.

According to the relativistic prediction, the earth must travel only the contracted distance of 600 m in the rest frame of the muon. This takes only 2 μsec $= T$. Thus the number expected at sea level is

$$N = 10^8 e^{-1} = 3.68 \times 10^7$$

Relativity predicts that we should observe 36.8 million muons. Experiments of this type have confirmed the relativistic predictions.

Question

1. You are standing on a corner and a friend is driving past in an automobile. Both note the times at which the car passes two different intersections. You each determine from your watch readings the time that elapses between these two events. Which of you has determined the proper time interval?

1-4 Clock Synchronization and Simultaneity

At first glance, time dilation and length contraction seem contradictory not only to our intuition but to our ideas of self-consistency. If each reference frame can be considered at rest with the other moving, the clocks in the "other" frame should run slow. How can there be any self-consistency if *each* observer sees the clocks of the other run slow? The answer to this puzzle lies in the problem of clock synchronization and in the concept of simultaneity. We note that the time intervals Δt and $\Delta t'$ considered in Section 1-3 were measured in quite different ways. The events in frame S' happened at the same place, and the times could be measured on a single clock; but in S the two events happened at different places. The time of each event was measured on a different clock and the interval found by subtraction. This procedure requires that the clocks be synchronized. We shall show in this section that

Two clocks synchronized in one reference frame are not synchronized in any other frame moving relative to the first.

A corollary to this result is that

Two events simultaneous in one reference frame are not simultaneous in another frame moving relative to the first.

(unless the events and clocks are in the same plane perpendicular to the relative motion).

Comprehension of these facts usually resolves all relativity paradoxes. Unfortunately, the intuitive (and incorrect) belief that simultaneity is an absolute relation is difficult to get rid of.

How can we synchronize two clocks separated in space? The problem is not difficult but requires some thought. One obvious method is to bring the clocks together and set them to read the same time, then move them back to their original positions. This method has the drawback that, as we have seen, each clock runs slow while it is moving, according to the time-dilation result. Suppose the clocks are at rest at points A and B a distance L apart in frame S. If an observer at A looks at the clock at B and sets his clock to read the same time, the clocks will not be synchronized, because of the time L/c it takes light to travel from one clock to another. To synchronize the clocks, the observer at A must set his clock ahead by the time L/c. Then he will see that the clock at B reads a time which is L/c behind the time on his clock, but he will *calculate* that the clocks are synchronized when he allows for the time L/c it takes the light to reach him. All observers except those midway between the clocks will see the clocks showing different times, but they will also compute that the clocks are synchronized when they correct for the time it takes the light to reach them. An equivalent method for the synchronization of two clocks would be for a third observer C at a point midway between the clocks to send a light signal and for observers at A and B to set their clocks to some prearranged time when they receive the signal.

We now examine the question of simultaneity. Suppose A and B agree to explode bombs at t_0 (having previously synchronized their clocks). Observer C will see the light from the two explosions at the same time, and since he is equidistant from A and B, he will conclude that the explosions were simultaneous. Other observers in S will see light from A or B first, depending on their location, but after correcting for the time the light takes to reach them, they also will conclude that the explosions were simultaneous. We shall thus define two events in a reference frame to be simultaneous if the light signals from the events reach an observer halfway between the events at the same time.

To show that two events which are simultaneous in frame S are not simultaneous in another frame S' moving relative to S, we use an example introduced by Einstein. A train is moving with speed v past the station platform. We have observers A', B', and C' at the front, back, and middle of the train. (We consider the train to be at rest in S' and the platform in S.) We now suppose that the train and platform are struck by lightning at the front and back of the train and that the lightning bolts are simultaneous in the frame of the platform (S) (Figure 1-10). That is, an observer C halfway between the positions A and B, where the lightning strikes, observes the two flashes at the same time. It is convenient to suppose that the lightning scorches both train and platform so that the events can be easily located in each reference frame. Since C' is in the middle of the train, halfway between the places on the train which are scorched, the events are simultaneous in S' only if C' sees the flashes

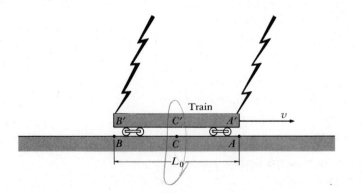

Figure 1-10
Simultaneous lightning bolts strike the ends of a train traveling with speed v in frame S attached to the platform. The light from these simultaneous events reaches observer C midway between the events at the same time. The distance between the bolts is L_0, which is also the length of the train measured in frame S.

at the same time. However, C' sees the flash from the front of the train before he sees the flash from the back. In frame S, when the light from the front flash reaches him, he has moved some distance toward it, so that the flash from the back has not yet reached him, as indicated in Figure 1-11. He must therefore conclude that the events are not simultaneous. The front of the train was struck before the back. As we have discussed above, all observers in S' on the train will agree with C' when they have corrected for the time it takes light to reach them.

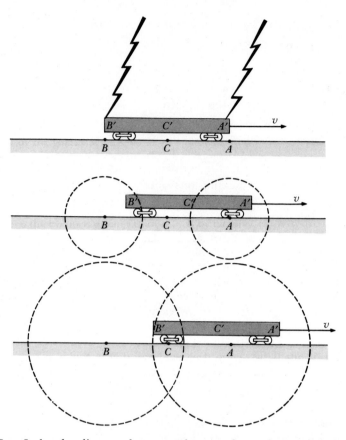

Figure 1-11
In frame S the light from
the bolt at the front of
the train reaches observer
C' in the middle of the
train before light from
the bolt at the rear of the
train because the train is
moving. Since C' is
midway between the
events (which occur at the
front and rear of the
train) these events are not
simultaneous for him.

Let L_0 be the distance between the scorch marks on the plat-form, which is also the length of the train measured in frame S. This distance is smaller than the proper length of the train L'_T because of length contraction. Figure 1-12 shows the situation in frame S', in which the train is at rest and the platform is moving. In this frame the distance between the burns on the platform is contracted and is related to L_0 by

$$L'_P = \frac{L_0}{\gamma} \qquad\qquad 1\text{-}11$$

Similarly the train length is the proper length related to L_0 by

$$L'_T = \gamma L_0 \qquad\qquad 1\text{-}12$$

(The drawing of this figure has been made for $\gamma = 1.5$.) The time interval in S' between these two events is the time it takes the platform to move the distance ΔL, where

$$\Delta L = L'_T - L'_P = \gamma L_0 - \frac{1}{\gamma} L_0$$
$$= \left(1 - \frac{1}{\gamma^2}\right) \gamma L_0 = \frac{v^2}{c^2} \gamma L_0$$

since

$$1 - \frac{1}{\gamma^2} = 1 - \left(1 - \frac{v^2}{c^2}\right) = \frac{v^2}{c^2}$$

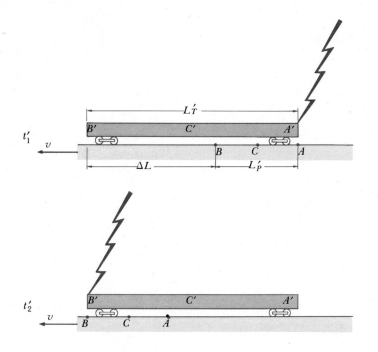

Figure 1-12
Lightning bolts of Figures
1-10 and 1-11 as seen in
reference frame S' fixed
to the train. In this frame
the distance $L'_P = BA$
along the moving plat-
form is contracted, and
the train length
$L'_T = B'A'$ is its proper
length. At time t'_1 lightning strikes the front of
the train when A' and A
are coincident. At a later
time t'_2 the lightning
strikes the rear of the
train when B' and B are
coincident. During the
time between flashes the
platform moves the dis-
tance $L'_T - L'_P$.

Thus

$$t'_2 - t'_1 = \frac{\Delta L}{v} = \frac{\gamma L_0 v}{c^2}$$

During this time (according to observers in S'), the clocks at A
and B ticked off a time interval that is smaller because of time
dilation:

$$\Delta t_s = \frac{1}{\gamma}(t'_2 - t'_1) = \frac{L_0 v}{c^2} \qquad\qquad 1\text{-}13$$

This is the time interval by which the clocks in S are unsynchronized according to observers in S'. That is, the clock at A
is ahead of that at B by the amount $L_0 v / c^2$. This result is worth
remembering.

*If two clocks are synchronized in the frame in which they are at rest,
they will be out of synchronization in another frame. In the frame in
which they are moving, the "chasing clock" leads by an amount
$\Delta t_s = L_0 v / c^2$, where L_0 is the proper distance between the clocks.*

*Lack of synchronization
of moving clocks*

Example 1-1   A numerical example should help clarify time dilation, clock synchronization, and the internal consistency of these
results. Let the light clock used in Section 1-3 be moving with
speed $v = 0.8c$. Then

$$1 - \frac{v^2}{c^2} = 1 - 0.64 = 0.36$$

and $$\gamma = \frac{1}{\sqrt{1 - v^2/c^2}} = \frac{1}{\sqrt{0.36}} = \frac{1}{0.6} = \frac{5}{3}$$

Also, let the distance $x_2 - x_1$ be 40 light-minutes, that is, the distance light travels in 40 min (a convenient notation for this unit is c-min):

$$x_2 - x_1 = 40 \ c\text{-min}$$

The time for the light clock to travel this distance is Δt, given by

$$\Delta t = \frac{x_2 - x_1}{v} = \frac{40 \ c\text{-min}}{0.8c} = 50 \ \text{min}$$

(Note that the c in the denominator cancels the c in the unit c-min, giving the time in minutes.)

The proper time interval in S' is shorter by the factor γ.

$$\Delta t' = \frac{\Delta t}{\gamma} = \frac{50 \ \text{min}}{\frac{5}{3}} = 30 \ \text{min}$$

(Thus we have taken $D = 15 \ c$-min for the distance from the flash gun to the mirror in Figure 1-7.) The distance traveled is

$$\Delta x' = v \ \Delta t' = (0.8c) \ (30 \ \text{min}) = 24 \ c\text{-min}$$

This is just the distance $x_2 - x_1$ which is contracted in S':

$$L' = \frac{L_0}{\gamma} = \frac{40 \ c\text{-min}}{\frac{5}{3}} = 24 \ c\text{-min}$$

Figure 1-13 shows the situation viewed in S'. We assume that the clock at x_1 reads noon at the time of the light flash. The clocks at x_1 and x_2 are synchronized in S but not S'. In S', the clock at x_2, which is chasing the one at x_1, leads by

$$\frac{L_0 v}{c^2} = \frac{(40 \ c\text{-min})(0.8c)}{c^2} = 32 \ \text{min}$$

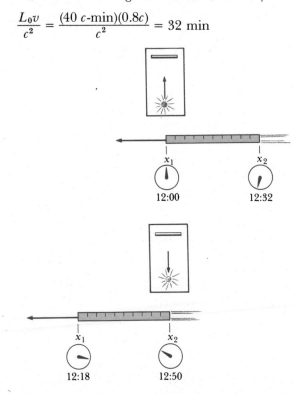

Figure 1-13
Example 1-1. In frame S' the rod passes the light clock in a time $\Delta t' = 30$ min. During this time the clocks at each end of the rod tick off $30/\gamma$ min = 18 min. But the clocks are unsynchronized, with the chasing clock leading by $L_0 v/c^2 =$ 32 min, where L_0 is the proper separation of the clocks. The time interval in S is therefore 32 min + 18 min = 50 min, which is longer than 30 min.

When the light clock coincides with x_2, the clock there reads 50 min past noon. Thus the time between events is 50 min in S. Note that according to observers in S', this clock ticks off $50 - 32 = 18$ min for a trip that takes 30 min in S'. Thus this clock runs slow by the factor $\frac{30}{18} = \frac{5}{3}$.

Thus each observer sees clocks in the other frame run slow. According to observers in S who measure 50 min for the time interval, the time interval in S' is too small (30 min) because the single clock in S' runs too slow by the factor $\frac{5}{3}$. According to the observers in S', the observers in S measure a time which is too *long* despite the fact that their clocks run too slow, because they are out of synchronization. The clocks tick off only 18 min, but the second one leads the first by 32 min, so the time interval found by subtraction is 50 min.

Questions

2. Two observers are in relative motion. In what circumstances can they agree on the simultaneity of two different events?

3. If event A occurs before event B in some frame, is it possible for there to be a reference frame in which event B occurs before event A?

4. Two events are simultaneous in a frame in which they also occur at the same point in space. Are they simultaneous in other reference frames?

1-5 The Doppler Effect

In the Doppler effect for sound waves the change in frequency for a given velocity v depends on whether it is the source or receiver that is moving. Such a distinction is possible for sound because there is a medium (the air) relative to which the motion takes place, and so it is not surprising that the motion of the source or the receiver relative to the still air can be distinguished. Such a distinction between motion of the source or receiver cannot be made for light or other electromagnetic waves in vacuum. The classical expressions for the Doppler effect cannot be correct for light. We now derive the relativistic Doppler-effect equations.

We shall consider a source moving toward a receiver with velocity v and work in the frame of the receiver. Let the source emit N waves. If the source is moving toward the receiver, the first wave will travel a distance $c \, \Delta t_R$, and the source will travel $v \, \Delta t_R$ in the time Δt_R measured in the frame of the receiver. The wavelength will be $\lambda' = (c \, \Delta t_R - v \, \Delta t_R)/N$. The frequency f' observed by the receiver will therefore be

$$f' = \frac{c}{\lambda'} = \frac{c}{c - v} \frac{N}{\Delta t_R} = \frac{1}{1 - v/c} \frac{N}{\Delta t_R}$$

If the frequency of the source is f_0, it will emit $N = f_0\,\Delta t_s$ waves in time Δt_s, measured by the source. Here Δt_s is the proper time interval (the first wave and the Nth wave are emitted at the same place in the source's reference frame). Times Δt_s and Δt_R are related by the usual time-dilating equation $\Delta t_s = \Delta t_R/\gamma$. Thus we obtain for the Doppler effect for a moving source,

$$f' = \frac{1}{1 - v/c}\,\frac{f_0\,\Delta t_s}{\Delta t_R} = \frac{f_0}{1 - v/c}\,\frac{1}{\gamma}$$

$$\boxed{\;= \frac{\sqrt{1 - v^2/c^2}}{1 - v/c}\,f_0\;} \qquad\qquad 1\text{-}14a$$

which differs from the classical equation only in the time-dilation factor.

We now do the calculation in the reference frame of the source. That is, we assume that the source is at rest and the receiver moves with velocity v toward the source. In time Δt_s in the frame of the source, the receiver encounters all the waves in the distance $v\,\Delta t_s$, in addition to the waves in the distance $c\,\Delta t_s$, just as in the classical calculation. The number of waves encountered is thus

$$N = \frac{c\,\Delta t_s + v\,\Delta t_s}{\lambda} = \frac{(c + v)\,\Delta t_s}{c/f_0} = \left(1 + \frac{v}{c}\right) f_0\,\Delta t_s$$

where we have used $\lambda = c/f_0$ for the wavelength. In the classical calculation, we need only divide N by the time interval Δt to obtain the frequency observed by the moving receiver, but here we must be careful to divide by the time interval in the receiver's frame to find the observed frequency. In this case the receiver's time interval Δt_R is proper time, i.e., the time interval between encountering the first wave and the Nth wave. Both events occur at the receiver. Thus $\Delta t_R = \Delta t_s/\gamma$, and the frequency observed is

$$f' = \frac{N}{\Delta t_R} = \left(1 + \frac{v}{c}\right) f_0\,\frac{\Delta t_s}{\Delta t_R} = \gamma\left(1 + \frac{v}{c}\right) f_0$$

$$= \frac{1 + v/c}{\sqrt{1 - v^2/c^2}}\,f_0 \qquad\qquad 1\text{-}14b$$

It is left as an exercise to show that this result is identical to Equation 1-14a.

When the source and receiver move away from each other with relative speed v, the frequency received is given by

$$f' = \frac{1 - v/c}{\sqrt{1 - v^2/c^2}}\,f_0 = \frac{\sqrt{1 - v^2/c^2}}{1 + v/c}\,f_0 \qquad\qquad 1\text{-}15 \qquad \textit{Relativistic Doppler effect}$$

Question

5. How do the relativistic equations for the Doppler effect for sound waves differ from those for light waves?

1-6 The Lorentz Transformation

We now consider the general relation between the coordinates x, y, z, t of an event as seen in reference frame S and the co-ordinates x', y', z', t' of the same event as seen in reference frame S', which is moving with uniform velocity relative to S. We shall consider only the simple special case in which the origins of the two coordinate systems are coincident at time $t = t' = 0$ and S' is moving, relative to S, with speed v along the x (or x') axis. The classical relation, called the *Galilean transformation*, is

$$x = x' + vt' \qquad y = y' \qquad z = z' \qquad t = t' \qquad \text{1-16}$$

with the inverse transformation

$$x' = x - vt \qquad y' = y \qquad z' = z \qquad t' = t$$

(For the rest of this discussion we shall ignore the equations for y and z, which do not change in this special case of motion along the x and x' axes.) These equations are consistent with experiment as long as v is much less than c. They lead to the familiar classical addition law for velocities. If a particle has velocity $u_x = dx/dt$ in frame S, its velocity in frame S' is

$$u'_x = \frac{dx'}{dt'} = \frac{dx}{dt} - v = u_x - v \qquad \text{1-17}$$

If we differentiate this equation again, we find that the acceleration of the particle is the same in both frames: $a_x = du_x/dt = du'_x/dt' = a'_x$.

It should be clear that this transformation is not consistent with the Einstein postulates of special relativity. If light moves along the x axis with speed c in S, these equations imply that the speed in S' is $u'_x = c - v$ rather than $u'_x = c$, which is consistent with Einstein's postulates and with experiment. The classical transformation equations must therefore be modified to be consistent with Einstein's postulates; but they must reduce to the classical equations when v is much less than c. We shall give a brief outline of one method of obtaining the relativistic transformation which is called the *Lorentz transformation*. We assume that the equation for x is of the form

$$x = K(x' + vt')$$

where K is a constant which can depend on v and c but not on the coordinates. If this equation is to reduce to the classical one, K must approach 1 as v/c approaches 0. The inverse transformation must look the same except for the sign of the velocity:

$$x' = K(x - vt)$$

Now let a light pulse start from the origin at $t = 0$. Since we have assumed that the origins coincide at $t = t' = 0$, the pulse also starts out at the origin of S' at $t' = 0$. The equation for the wavefront of the light pulse is $x = ct$ in frame S and $x' = ct'$ in S'. Substituting these in our transformation equations gives

$$ct = K(ct' + vt') = K(c + v)t'$$

and

$$ct' = K(ct - vt) = K(c - v)t$$

We can eliminate either t' or t from these two equations and determine K. The result is $K^2 = (1 - v^2/c^2)^{-1}$, i.e., K is the same as γ:

$$K = \gamma = \frac{1}{\sqrt{1 - v^2/c^2}}$$

The transformation is therefore $x = \gamma(x' + vt')$. We can obtain equations for t and t' by combining this transformation with its inverse

$$x' = \gamma(x - vt)$$

Using this value for x' in the above equation, we get

$$x = \gamma[\gamma(x - vt) + vt']$$

which can be solved for t' in terms of x and t. The complete Lorentz transformation is

---

$$x = \gamma(x' + vt') \qquad y = y'$$

$$t = \gamma\left(t' + \frac{vx'}{c^2}\right) \qquad z = z'$$

1-18   *Lorentz transformation*

and the inverse

$$x' = \gamma(x - vt) \qquad y' = y$$

$$t' = \gamma\left(t - \frac{vx}{c^2}\right) \qquad z' = z$$

1-19

Example 1-2   Two events occur at the same place x_0 at times t_1 and t_2 in S. What is the time interval between them in S'?

$$t_1' = \gamma\left(t_1 - \frac{vx_0}{c^2}\right) \qquad t_2' = \gamma\left(t_2 - \frac{vx_0}{c^2}\right)$$

$$t_2' - t_1' = \gamma(t_2 - t_1)$$

This is our familiar result for time dilation, since $t_2 - t_1$ is the proper time interval.

We can obtain the velocity transformation by differentiating the Lorentz transformation equations or merely by taking differences. We shall do the latter. Suppose a particle moves a distance Δx in time Δt in frame S. Its velocity is $u_x = \Delta x/\Delta t$. Using the transformation equations, we obtain

$$\Delta x' = \gamma(\Delta x - v\,\Delta t) \qquad \text{and} \qquad \Delta t' = \gamma\left(\Delta t - \frac{v\,\Delta x}{c^2}\right)$$

The velocity in S' is

$$u'_x = \frac{\Delta x'}{\Delta t'} = \frac{\gamma(\Delta x - v\,\Delta t)}{\gamma\left(\Delta t - \dfrac{v\,\Delta x}{c^2}\right)} = \frac{\dfrac{\Delta x}{\Delta t} - v}{1 - \left(\dfrac{v}{c^2}\dfrac{\Delta x}{\Delta t}\right)}$$

or

$$u'_x = \frac{u_x - v}{1 - vu_x/c^2} \qquad\qquad 1\text{-}20 \qquad \textit{Velocity addition}$$

Likewise the inverse velocity transformation is

$$u_x = \frac{u'_x + v}{1 + vu'_x/c^2} \qquad\qquad 1\text{-}21$$

If a particle has components of velocity along y and z, it is not difficult to find the components in S'. In these cases, $y' = y$ and $z' = z$; and the Lorentz transformation relates t' and t. Thus

$$u'_y = \frac{\Delta y'}{\Delta t'} = \frac{\Delta y}{\gamma\left(\Delta t - \dfrac{v\,\Delta x}{c^2}\right)} = \frac{\Delta y/\Delta t}{\gamma\left(1 - \dfrac{v\,\Delta x}{c^2\,\Delta t}\right)}$$

or

$$u'_y = \frac{u_y}{\gamma(1 - vu_x/c^2)} \qquad\qquad 1\text{-}22$$

with a similar result for u'_z. The inverse is

$$u_y = \frac{u'_y}{\gamma(1 + vu'_x/c^2)} \qquad\qquad 1\text{-}23$$

Example 1-3   Light moves along the x axis with speed $u_x = c$. What is its speed in S'? From Equation 1-20 with $u_x = c$ we have

$$u'_x = \frac{c - v}{1 - vc/c^2} = \frac{c(1 - v/c)}{1 - v/c} = c$$

as required by the postulates.

Question

6. The Lorentz transformation for y and z is the same as the classical result: $y = y'$ and $z = z'$. Yet the relativistic velocity transformation does not give the classical result $u_y = u'_y$ and $u_z = u'_z$. Explain.

1-7   The Twin Paradox

Homer and Ulysses are identical twins. Ulysses travels at high speed to a planet beyond the solar system and returns while Homer remains at home. When they are together again, which twin is older, or are they the same age? The correct answer is that Homer, the twin who stays at home, is older. This problem, with variations, has been the subject of spirited discussion for decades, though there are now very few who disagree with the answer.[1]

The paradox arises from the seemingly symmetrical roles played by the twins, contrasted with the asymmetric result in their aging. The relativistic result conflicts with common sense based on our strong but incorrect belief in absolute simultaneity. The paradox is resolved when the asymmetry of the twins' roles is noted. We shall consider a particular case with some numerical magnitudes which, though impractical, make the calculations easy.

Let planet P and Homer on earth be fixed in reference frame S a distance L_0 apart, as in Figure 1-14. (Any relative motion of the earth and planet P is ignored.) Reference frames S' and S'' are moving with speed v toward and away from the planet, respectively. Ulysses quickly accelerates to speed v, then coasts in S' until he reaches the planet, where he stops and is momentarily at rest in S. To return, he quickly accelerates to speed v toward earth and coasts in S'' until he reaches earth, where he stops. Let us assume that the acceleration times are negligible compared with the coasting times. (Given the time needed to reach the speed v, we can always formulate the problem with a value of L_0 large enough to meet this condition.) We use the following values for illustration: $L_0 = 8$ light-years and $v = 0.8c$; then $\sqrt{1 - v^2/c^2} = \frac{3}{5}$.

It is easy to analyze the problem from Homer's point of view. According to his clock, Ulysses coasts in S' for a time $L_0/v = 10$ years and in S'' for an equal time. Thus Homer is 20 years older

---

[1] A collection of important papers concerning this paradox can be found in *Special Relativity Theory: Selected Reprints,* New York: American Association of Physics Teachers, 1963.

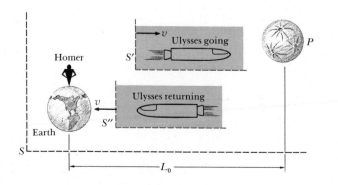

Figure 1-14
Twin paradox. The earth and a distant planet P are fixed in frame S. Ulysses coasts in frame S' to the planet and then coasts in S'' back to earth. His twin Homer stays on earth. When Ulysses returns, he is younger than his twin. The roles played by the twins are not symmetric. Homer remains in one inertial reference frame, but Ulysses must accelerate if he is to return home.

when Ulysses returns. The time interval in S' between leaving earth and arriving at the planet is shorter because it is proper time; the time to reach the planet by Ulysses' clock is

$$\Delta t' = \sqrt{1 - \frac{v^2}{c^2}} \frac{L_0}{v} = \tfrac{3}{5} \times 10 = 6 \text{ years}$$

Since the same time is required for the return trip, Ulysses will have recorded 12 years for the round trip and will thus be 8 years younger than Homer.

From Ulysses' point of view, the calculation of his trip time is not difficult. The distance from earth to planet is contracted and is only

$$L_0 \sqrt{1 - \frac{v^2}{c^2}} = \tfrac{3}{5} \times 8 = 4.8 \text{ light-years}$$

At $v = 0.8c$, it takes only 6 years each way. The real difficulty in this problem is for Ulysses to understand why his twin ages 20 years during his absence. If we consider Ulysses at rest and Homer moving away, doesn't Homer's clock measure proper time? When Ulysses measures 6 years Homer should measure only $\tfrac{3}{5}(6) = 3.6$ years. Then why shouldn't Homer age only 7.2 years during the round trip? This, of course, is the paradox. The difficulty with the analysis from the point of view of Ulysses is that he does not remain in an inertial reference frame. What happens while Ulysses is stopping and starting? To investigate this problem in detail, we should have to treat accelerated reference frames, a topic requiring the general theory of relativity. We shall consider this topic briefly in Section 1-13, but we can gain some insight into this problem from the special theory of relativity by considering the lack of synchronization of moving clocks.

Suppose that there is a clock at the planet P synchronized in S with Homer's clock at earth. In reference frame S', these clocks are unsynchronized by the amount $L_0 v / c^2$. In our example this is 6.4 years. Thus when Ulysses is coasting in S' near the planet, the clock at the planet leads that at the earth by 6.4 years. After he stops, he is in frame S, in which these two clocks are synchronized. Thus in the negligible time (according to Ulysses) it takes him to stop, the clock at earth must gain 6.4 years. Accordingly, his twin on earth ages 6.4 years. This 6.4 years plus the 3.6 years that Homer aged during the coasting makes him 10 years older by the time Ulysses is stopped in frame S. When Ulysses is in frame S'' coasting home, the clock at earth leads that at the planet by 6.4 years, and it will run another 3.6 years before he arrives home. We do not need to know the detailed behavior of the clocks during the acceleration in order to know the cumulative effect; the special theory is enough to show us that if the clocks on earth and at the planet are synchronized in S, the clock on earth lags that at P by $L_0 v / c^2 = 6.4$ years when viewed in S' and the clock on earth leads that at P by this amount when viewed in S''.

The difficulty in understanding the analysis of Ulysses lies in the difficulty of giving up the idea of absolute simultaneity. Suppose Ulysses sends a signal calculated to arrive at earth just as he arrives at P. If the arrival of this signal and the arrival of Ulysses at the planet are simultaneous in S', they are not simultaneous in S; in fact, the signal arrives on earth 6.4 years before Ulysses arrives at the planet, according to observers in S. The roles of the twins are not symmetric because Ulysses does not remain in an inertial reference frame but must accelerate.

It is instructive to have the twins send regular signals to each other so that each can record the other's age continuously. If they arrange to send a signal once a year, the age of the other can be determined merely by counting the signals received. The frequency of arrival of the signals, however, will not be 1 per year because of the Doppler shift. The frequency will be given by Equation 1-14 or 1-15, which for our example is 3 per year or $\frac{1}{3}$ per year, depending on whether the source is approaching or receding.

Consider the situation first from the point of view of Ulysses. During the 6 years it takes him to reach the planet (remember that the distance is contracted in his frame), he receives signals at the rate of $\frac{1}{3}$ per year, or 2 signals. As soon as he turns around and starts back to earth, he receives 3 signals per year; in the 6 years it takes him to return he receives 18 signals, giving a total of 20 for the trip. He accordingly expects his twin to have aged 20 years.

We now consider the situation from Homer's point of view. He receives signals at the rate of $\frac{1}{3}$ per year not only for the 10 years it takes Ulysses to reach the planet, but also for the 8 years it takes for the last signal sent by Ulysses from S' to get back to earth. (He cannot know that Ulysses has turned around until the signals reach him.) During the first 18 years, Homer receives 6 signals. In the final 2 years before Ulysses arrives, Homer receives 6 signals, or 3 per year. (The first signal sent after Ulysses turns around takes 8 years to reach earth, whereas Ulysses, traveling at $0.8c$, takes 10 years to return and therefore arrives just 2 years after Homer begins to receive signals at the faster rate.) Thus Homer expects Ulysses to have aged 12 years. In this analysis, the asymmetry of the twins' roles is apparent. Both twins agree that when they are together again, the one who has been accelerated will be younger than the one who stayed home.

The predictions of the special theory of relativity concerning the twin paradox have been tested many times using small particles which can be accelerated to such large speeds that γ is appreciably greater than 1. Unstable particles can be accelerated and trapped in circular orbits in a magnetic field, for example, and their lifetimes compared with those of identical particles at rest. In all such experiments the accelerated particles live longer on the average than those at rest, as predicted. These predictions have also been confirmed by the results of an experiment using very precise atomic clocks flown around the world in commercial airplanes, but the analysis of this experiment is complicated by the necessity of including gravitational effects treated in the general theory of relativity.[1]

1-8   Relativistic Momentum

In classical mechanics, if Newton's second law $\Sigma \mathbf{F} = m\mathbf{a}$ holds in one reference frame, it holds in any other reference frame moving with constant velocity relative to the first. As discussed in Section 1-6, the Galilean transformation of Equations 1-16 leads to the same accelerations $a'_x = a_x$ in both frames, and forces such as those due to stretching of springs are also the same in both frames. However, according to the Lorentz transformation, accelerations are not the same in two such reference frames. If a particle has acceleration a_x and velocity u_x in frame S, its acceleration in S', obtained by computing du'_x/dt' from Equation 1-20, is

$$a'_x = \frac{a_x}{\gamma^2 (1 - v u_x/c^2)^2} \qquad \qquad 1\text{-}24$$

Either the force must transform in a similar way, or Newton's second law $\Sigma \mathbf{F} = m\mathbf{a}$ does not hold. It is reasonable to expect that $\Sigma \mathbf{F} = m\mathbf{a}$ does not hold at high speeds, for this equation implies that a constant force will accelerate a particle to unlimited velocity if it acts for a long time. However, if a particle's velocity were greater than c in some reference frame S, we could not transform from S to the rest frame of the particle because γ becomes imaginary when $v > c$. We can see from the velocity transformation that if a particle's velocity is less than c in some frame S, it is less than c in all frames moving relative to S with $v < c$. These results lead us to expect that particles never have speeds greater than c.

We can avoid the question of how to transform forces by considering a problem in which the total force is zero, namely, a collision of two masses. In classical mechanics, the total momentum $\Sigma m_i \mathbf{u}_i$ is conserved. We can see by a simple example that this quantity, the classical total momentum, is not conserved relativistically. That is, the conservation of the quantity $\Sigma m_i \mathbf{u}_i$ is an approximation which holds only at low speeds.

Consider an observer in frame S with a ball A and one in S' with a ball B. The balls each have mass m and are identical when compared at rest. Each observer throws his ball along his y-axis with speed u_0 (measured in his own frame) so that the balls collide. Assuming the balls to be perfectly elastic, each observer will see his ball rebound with its original speed u_0. If the total

[1] The details of these tests can be found in J. C. Hafele and R. E. Keating, "Around-the-World Atomic Clocks: Predicted Relativistic Time Gains" and "Around-the-World Atomic Clocks: Observed Relativistic Time Gains," *Science*, **177,** no. 4044, 166 (1972).

momentum is to be conserved, the y component must be zero because the momentum of each ball is merely reversed by the collision. However, if we consider the relativistic velocity transformations, we can see that the quantity mu_y does not have the same magnitude for each ball as seen by either observer.

Let us consider the collision as seen in frame S (Figure 1-15a). In this frame ball A moves along the y axis with velocity $u_{yA} = u_0$. Ball B has an x component of velocity $u_{xB} = v$ and a y component $u_{yB} = u'_{yB}/\gamma = -u_0 \sqrt{1 - v^2/c^2}$. Here we have used the velocity-transformation equations (1-21 and 1-23) and the facts that u'_{yB} is just $-u_0$ and $u'_{xB} = 0$. We see that the y component of velocity of ball B is smaller in magnitude than that of ball A. The factor $\sqrt{1 - v^2/c^2}$ comes from the time-dilation factor. The time taken for ball B to travel a given distance along the y axis in S is greater than the time measured in S' for the ball to travel this same distance. Thus in S the total y component of classical momentum is not zero. Since the velocities are reversed in an elastic collision, momentum as defined by $\mathbf{P} = \Sigma m\mathbf{u}$ is not conserved in S. Analysis of this problem in S' leads to the same conclusion (Figure 1-15b). In the classical limit, $v \ll c$, momentum is conserved, of course, because in that limit $\gamma \approx 1$.

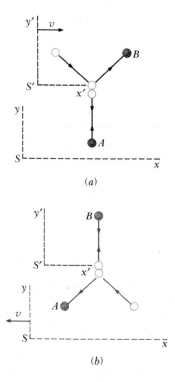

(a)

(b)

Figure 1-15
(a) Elastic collision of two identical balls as seen in frame S. The vertical component of the velocity of ball B is u_0/γ in S if it is u_0 in S'. (b) The same collision as seen in S'. In this frame ball A has vertical component of velocity u_0/γ.

The reason for defining momentum as $\Sigma m\mathbf{u}$ in classical mechanics is that this quantity is conserved when there are no external forces, as in collisions. We now see that this quantity is conserved only in the approximation $v \ll c$. We shall define the *relativistic momentum* $\mathbf{p}$ of a particle to have the following properties:

1. **p** is conserved in collisions.

2. **p** approaches $m\mathbf{u}$ as u/c approaches zero.

We shall state without proof that the quantity meeting these conditions is

$$\mathbf{p} = \frac{m\mathbf{u}}{\sqrt{1 - u^2/c^2}} \qquad\qquad 1\text{-}25$$

Relativistic momentum defined

where u is the speed of the particle. We thus take this equation for the definition of relativistic momentum. It is clear that this definition meets our second criterion because the denominator approaches 1 when u is much less than c. Proof that it also meets the first criterion involves much tedious algebra. From this definition, the momenta of the two balls A and B as seen in S are

$$p_{yA} = \frac{mu_0}{\sqrt{1 - u_0^2/c^2}} \qquad p_{yB} = \frac{mu_{yB}}{\sqrt{1 - (u_{xB}^2 + u_{yB}^2)/c^2}}$$

where $u_{yB} = -u_0\sqrt{1 - v^2/c^2}$ and $u_{xB} = v$. We shall leave as an exercise the details of showing that $p_{yB} = -p_{yA}$. Because of the similarity of the factor $1/\sqrt{1 - u^2/c^2}$ and γ in the Lorentz transformation, Equation 1-25 is often written

$$\mathbf{p} = \gamma m\mathbf{u} \qquad \text{with } \gamma = 1/\sqrt{1 - u^2/c^2} \qquad\qquad 1\text{-}26$$

This use of the symbol γ for two different quantities causes some confusion; the notation is standard, however, and simplifies many of the equations. We shall use this notation except when considering transformations between reference frames. Then, to avoid confusion, we shall write out the factor $\sqrt{1 - u^2/c^2}$ and reserve γ for $1/\sqrt{1 - v^2/c^2}$, where v is the relative speed of the frames.

One interpretation of Equation 1-25 is that the mass of an object increases with speed. The quantity

$$\frac{m}{\sqrt{1 - u^2/c^2}} = \gamma m$$

is sometimes called the *relativistic mass*, written $m(u)$. The mass of the body in its rest frame, written m_0, is then called its *rest mass*. Although this makes the expression $m(u)\mathbf{u}$ for relativistic momentum similar to the nonrelativistic expression, the use of relativistic mass often leads to mistakes. For example, the expression $\frac{1}{2}m(u)u^2$ is *not* the correct relativistic expression for kinetic energy. The measurement of relativistic mass involves measuring the force needed to produce a given change in momentum. The experimental evidence often cited as verification that mass depends on velocity can equally well be interpreted as evidence for the validity of the assumption that the force equals the time rate of change of relativistic momentum. We shall avoid using a symbol for relativistic mass; in this book, m always refers to the rest mass.

1-9   Relativistic Energy

We have seen that the quantity $m\mathbf{u}$ is not conserved in collisions but that $\gamma m\mathbf{u}$ is, with $\gamma = 1/\sqrt{1 - u^2/c^2}$. Evidently Newton's second law in the form $\Sigma\mathbf{F} = m\mathbf{a}$ cannot be correct relativistically, since it leads to the conservation of $m\mathbf{u}$. We can get a hint of the relativistically correct form of the law by writing it $\Sigma\mathbf{F} = d\mathbf{p}/dt$. Let us assume that this equation is correct if relativistic momentum $\mathbf{p}$ is used. The validity of this assumption can be determined only by examining its consequences, since an unbalanced force on a high-speed particle is measured by its effect on momentum and energy. We are in effect defining force by the equation $\Sigma\mathbf{F} = d\mathbf{p}/dt$. As in classical mechanics, we shall define kinetic energy as the work done by an unbalanced force in accelerating a particle from rest to some velocity. Considering motion in one dimension only, we have

$$E_k = \int_{u=0}^{u} \Sigma F ds = \int_0^u \frac{d(\gamma mu)}{dt} ds = \int_0^u u \, d(\gamma mu) \qquad \text{1-27}$$

using $u = ds/dt$. The computation of the integral in Equation 1-27 is not difficult but requires some messy algebra. It is left as a problem to show that

$$d(\gamma mu) = m\left(1 - \frac{u^2}{c^2}\right)^{-3/2} du$$

Substituting this into the integrand in Equation 1-27, we obtain

$$E_k = \int_0^u u \, d(\gamma mu) = \int_0^u m\left(1 - \frac{u^2}{c^2}\right)^{-3/2} u \, du$$

$$= mc^2 \left(\frac{1}{\sqrt{1 - u^2/c^2}} - 1\right)$$

or

$$E_k = \gamma mc^2 - mc^2 \qquad \text{1-28}$$

Relativistic kinetic energy

We can check this expression for low speeds by noting that for $u/c \ll 1$,

$$\gamma = \left(1 - \frac{u^2}{c^2}\right)^{-1/2} \approx 1 + \frac{1}{2}\frac{u^2}{c^2} + \cdots$$

thus

$$E_k \approx mc^2 \left(1 + \frac{1}{2}\frac{u^2}{c^2} + \cdots - 1\right) = \tfrac{1}{2}mu^2$$

The expression for kinetic energy consists of two terms. One term, γmc^2, depends on the speed of the particle (through the factor γ), and the other term, mc^2, is independent of the speed. The quantity mc^2 is called the *rest energy* of the particle. The total energy E is then defined as the sum of the kinetic energy and the rest energy,

Rest energy

$$E = E_k + mc^2 = \gamma mc^2 = \frac{mc^2}{\sqrt{1 - u^2/c^2}} \qquad \text{1-29}$$

Total energy defined

Thus the work done by an unbalanced force increases the energy from the rest energy mc^2 to γmc^2 (or increases the mass from m to γm).

The identification of the term mc^2 as rest energy is not merely a convenience. Whenever additional energy ΔE in any form is stored in an object, the rest mass of the object is increased by $\Delta E/c^2$. This is of particular importance whenever we are concerned with an object that can be broken up into its parts and we want to compare the mass of the parts with that of the object (for example, an atom containing a nucleus and electrons, or a nucleus containing protons and neutrons). In the case of the atom, the mass changes are usually negligibly small (Example 1-6). However, the difference between the mass of a nucleus and that of its constituent parts (protons and neutrons) is often of great importance.

As an example, consider Figure 1-16a in which two particles of mass m are moving toward each other, each with speed u. They collide with a spring that compresses and locks shut. (The spring is merely a device for visualizing the storage of energy.) In the Newtonian mechanics description, the original kinetic energy $E_k = 2(\frac{1}{2}mu^2)$ is converted into potential energy of the spring U. When the spring is unlocked, the potential energy reappears as kinetic energy of the particles. In relativity theory, the internal energy of the system, $E_k = U$ appears as an increase in rest mass of the system. That is, the rest mass of the system M is greater than $2m$ by E_k/c^2. We shall derive this result below. This change in rest mass is too small to be observed for ordinary-sized masses and springs, but it is easily observed in nuclear-energy transformations. For example, in the fission of a ^{235}U nucleus, the energy released as kinetic energy of the fission fragments is an appreciable fraction of the rest energy of the original nucleus. This energy can be calculated from a measurement of the difference in mass of the original system and the total mass of the fragments.

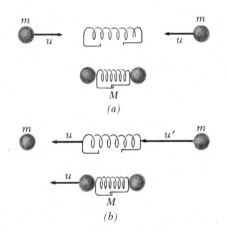

Figure 1-16
Two objects colliding with a massless spring that locks shut. The total rest mass of the system M is greater than that of the parts $2m$ by the amount E_k/c^2 where E_k is the internal energy, which in this case is the original kinetic energy. (*a*) The event as seen in a reference frame S in which the final mass M is at rest. (*b*) The same event as seen in a frame S' moving to the right at speed u relative to S, so that one of the initial masses is at rest.

Derivation of the Increase in Rest Mass of the Two-Particle and Spring System of Figure 1-16

Let m be the rest mass of each particle so that the total mass of the system is $2m$ when the particles are at rest and far apart, and let M be the rest mass of the system when it has internal energy E_k. The original kinetic energy in the reference frame S (Figure 1-16a) is

$$E_k = 2mc^2(\gamma - 1) \qquad \text{1-30}$$

In a perfectly inelastic collision, momentum conservation implies that both particles are at rest after the collision in this frame, which is the center-of-mass frame. The total kinetic energy is therefore lost. We wish to show that if momentum is to be conserved in any reference frame moving with constant velocity relative to S, the total mass of the system must increase by Δm, given by

$$\Delta m = \frac{E_k}{c^2} = 2m(\gamma - 1) \qquad \text{1-31}$$

We therefore wish to show that the total mass of the system with internal energy is M given by

$$M = 2m + \Delta m = 2\gamma m \qquad \text{1-32}$$

To simplify the mathematics, we choose a second reference frame S' moving to the right with speed $v = u$ relative to frame S so that one of the particles is initially at rest, as shown in Figure 1-16b. The initial speed of the other particle in this frame is

$$u' = \frac{u + v}{1 + uv/c^2} = \frac{2u}{1 + u^2/c^2} \qquad \text{1-33}$$

After the collision, the particles move together with speed u (since they are at rest in S). The initial momentum in S' is

$$p_i' = \frac{mu'}{\sqrt{1 - u'^2/c^2}} \qquad \text{to the left}$$

The final momentum is

$$p_f' = \frac{Mu}{\sqrt{1 - u^2/c^2}} \qquad \text{to the left}$$

Using Equation 1-33 for u' and doing some algebra gives

$$1 - \frac{u'^2}{c^2} = 1 - \frac{4u^2/c^2}{(1 + u^2/c^2)^2} = \frac{(1 - u^2/c^2)^2}{(1 + u^2/c^2)^2}$$

Then

$$p_i' = \frac{m[2u/(1 + u^2/c^2)]}{(1 - u^2/c^2)/(1 + u^2/c^2)} = \frac{2mu}{1 - u^2/c^2}$$

Conservation of momentum in frame S' requires $p_f' = p_i'$, or

$$\frac{Mu}{\sqrt{1 - u^2/c^2}} = \frac{2mu}{1 - u^2/c^2}$$

Solving for M we obtain

$$M = \frac{2m}{\sqrt{1 - u^2/c^2}} = 2\gamma m$$

which is Equation 1-32.

1-10 Mass and Binding Energy

When a system of particles is held together by attractive forces, energy is required to break up the system and separate the particles. The magnitude of this energy E_b is called the *binding energy* of the system. As discussed in the previous section, the energy of a system is related to its rest mass. An important result of the theory of special relativity which we shall illustrate by examples in this section is

The rest mass of a bound system is less than that of the separated particles by E_b/c^2 where E_b is the binding energy.

In atomic and nuclear physics, masses and energies are often given in unified mass units and electron volts rather than in the standard SI units of kilograms and joules. The unified mass unit u is defined as one-twelfth the mass of the neutral ^{12}C atom, consisting of the nucleus and six electrons. (This unit replaces older mass units based on the oxygen atom.) The unified mass unit is related to the kilogram by

$$1 \text{ u} = \frac{1 \text{ g}}{6.0220 \times 10^{23}} = 1.6606 \times 10^{-24} \text{ g}$$
$$= 1.6606 \times 10^{-27} \text{ kg} \qquad \text{1-34}$$

The unified mass unit expressed in grams is simply the reciprocal of Avogadro's number $N_A = 6.0220 \times 10^{23}$ (as discussed in Chapter 2). The electron volt (eV) is defined as the energy acquired by a particle of one electronic charge e accelerated through a potential difference of one volt. Since a joule is a coulomb-volt, and the electronic charge is 1.602×10^{-19} C, the eV and joule are related by

$$1 \text{ eV} = 1.602 \times 10^{-19} \text{ C-V} = 1.602 \times 10^{-19} \text{ J} \qquad \text{1-35}$$

Commonly used multiples of the eV are the keV (10^3 eV), the MeV (10^6 eV), and the GeV (10^9 eV). (The GeV was formerly called the BeV.)

The rest energy of a mass of 1 g is

$$(1 \text{ g})c^2 = (10^{-3} \text{ kg})(3 \times 10^8 \text{ m/sec})^2$$
$$= 9 \times 10^{13} \text{ J} = 5.61 \times 10^{32} \text{ eV}$$

The rest energy of a unified mass unit is

$$(1 \text{ u})c^2 = 931.5 \text{ MeV}$$

Table 1-1
Rest energies of some elementary particles and light nuclei

Particle	Symbol	Rest energy, MeV
Photon	γ	0
Neutrino (antineutrino)	$\nu(\bar{\nu})$	0
Electron (positron)	e or e^- (e^+)	0.5110
Muon	$\mu^\pm$	105.7
Pi meson	π^0	135
	$\pi^\pm$	139.6
Proton	p	938.280
Neutron	n	939.573
Deuteron	^{2}H or d	1875.628
Triton	^{3}H or t	2808.944
Alpha	^{4}He or α	3727.409

The rest masses and rest energies of some elementary particles and light nuclei are given in Table 1-1, from which we can see that the mass of a nucleus is not the same as the sum of the masses of its parts.

Example 1-4   The simplest example is that of the deuteron ^{2}H, consisting of a neutron and a proton bound together. Its rest energy is 1875.63 MeV. The sum of the rest energies of the proton and neutron is 938.28 + 939.57 = 1877.85 MeV. Since this is greater than the rest energy of the deuteron, the deuteron cannot spontaneously break up into a neutron and a proton. The binding energy of the deuteron is 1877.85 − 1875.63 = 2.22 MeV. In order to break up the deuteron into a proton and a neutron, at least 2.22 MeV must be added. This can be done by bombarding deuterons with energetic particles or electromagnetic radiation.

If a deuteron is formed by combination of a neutron and proton, energy must be released. When neutrons from a reactor are incident on protons, some neutrons are captured. The nuclear reaction is $n + p \rightarrow d + \gamma$. Most of these reactions occur for low-energy neutrons (kinetic energy less than 1 eV). The energy of the emitted γ (plus the kinetic energy of the deuteron) is 2.22 MeV.

Example 1-5   A free neutron decays into a proton plus an electron plus an antineutrino

$$n \longrightarrow p + e + \bar{\nu}$$

What is the kinetic energy of the decay products? Here rest energy is converted into kinetic energy. Before decay, $(mc^2)_n = 939.57$ MeV. After decay, $(mc^2)_p + (mc^2)_e + (mc^2)_{\bar{\nu}} = 938.28 + 0.511 + 0 = 938.79$ MeV. Thus rest energy of $939.57 - 938.79 = 0.78$ MeV has been converted into kinetic energy of the decay products.

Example 1-6 The binding energy of the hydrogen atom (the energy needed to remove the electron from the atom) is 13.6 eV. How much mass is lost when an electron and a proton form a hydrogen atom?

The mass of a proton plus that of an electron must be greater than that of the hydrogen atom by

$$\frac{13.6 \text{ eV}}{931.5 \text{ MeV/u}} = 1.46 \times 10^{-8} \text{ u}$$

This mass difference is so small that it is usually neglected.

Example 1-7 How much energy is needed to remove one proton from a ^{4}He nucleus? Removal of one proton from ^{4}He leaves ^{3}H. From Table 1-1, the rest energy of ^{3}H plus that of a proton is 2808.94 + 938.28 = 3747.22 MeV. This is about 19.8 MeV greater than the rest energy of ^{4}He.

These examples show that, because atomic binding energies are so small (of the order of 1 eV to 1 keV), the mass changes are negligible in atomic (that is, chemical) reactions, but that nuclear binding energies are quite large and thus nuclear reactions can involve appreciable changes in mass.

1-11 Experimental Determination of Relativistic Momentum

The momentum of a charged particle is usually determined by measuring the radius of curvature of the path of the particle moving in a magnetic field. If the particle has charge q and velocity **u**, it experiences a force in a magnetic field **B** given by

$$\mathbf{F} = q\mathbf{u} \times \mathbf{B}$$

According to our definition of relativistic momentum, this force equals the rate of change of the momentum.

$$\mathbf{F} = \frac{d\mathbf{p}}{dt} = \frac{d(\gamma m \mathbf{u})}{dt} = q\mathbf{u} \times \mathbf{B} \qquad \text{1-36}$$

Since the magnetic force is always perpendicular to the velocity, it does no work on the particle, so the energy of the particle is constant. From Equation 1-29 we see that if the energy is constant, γ must be constant, and therefore the speed u is also constant. Equation 1-36 is then

$$\mathbf{F} = q\mathbf{u} \times \mathbf{B} = \gamma m \frac{d\mathbf{u}}{dt} \qquad \text{1-37}$$

Since the speed is constant, the acceleration must be perpendicular to the velocity. For the case $\mathbf{u} \perp \mathbf{B}$, the particle moves in a circle with centripetal acceleration u^2/R. (If **u** is not perpendicular to **B**, the path is a helix. Since the component of **u** parallel to

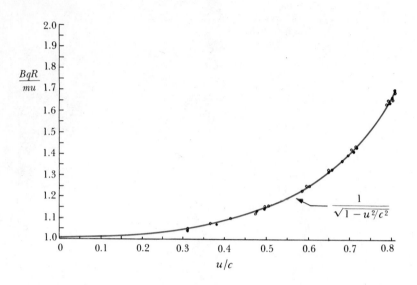

Figure 1-17
A plot of BqR/mu versus u/c for particle of charge q and mass m moving in a circular orbit of radius R in a magnetic field B. The agreement of the data with the curve predicted by relativity theory supports the assumption that the force equals the time rate of change of relativistic momentum. (*Adapted from I. Kaplan, Nuclear Physics, 2d ed., Reading, Mass.: Addison-Wesley Publishing Company, Inc., 1962; by permission.*)

B is unaffected we shall consider only motion in a plane.) We have then

$$quB = m\gamma \left|\frac{d\mathbf{u}}{dt}\right| = m\gamma \left(\frac{u^2}{R}\right)$$

or

$$BqR = m\gamma u = p \qquad\qquad 1\text{-}38$$

Equation 1-38 is the same as the nonrelativistic expression except for the factor of γ. This equation can be tested by measuring the speed of a particle and then finding the magnetic field necessary to bend it in a circle of radius R. The speed can be measured by passing the beam of particles through crossed electric and magnetic fields so that the electric force $q\mathbf{E}$ just cancels the magnetic force $q\mathbf{u} \times \mathbf{B}$ for particles of speed $u = (E/B)$. These particles pass through a collimator, whereas particles traveling at other speeds are deflected away. Figure 1-17 shows a plot of BqR/mu versus u/c. The solid curve is the function $\gamma = (1 - u^2/c^2)^{-1/2}$, predicted by Equation 1-38. As discussed in Section 1-8, these data can also be interpreted as showing that the (relativistic) mass depends on speed.

1-12 Some Useful Equations and Approximations

In practical applications it is often the momentum or energy of a particle that is known rather than speed. Equation 1-25 for the relativistic momentum and Equation 1-28 for the relativistic energy can be combined to eliminate the speed u. The result (see Exercise 24) is

$$E^2 = (pc)^2 + (mc^2)^2 \qquad\qquad 1\text{-}39$$

The triangle shown in Figure 1-18 is sometimes useful in remembering this result. If the energy of a particle is much
greater than its rest energy mc^2, the second term on the right of
Equation 1-39 can be neglected, giving the useful approximation

$$E \approx pc \qquad \text{for } E \gg mc^2 \tag{1-40}$$

This approximation is accurate to about 1 percent or better if E
is greater than about $8mc^2$. (See Example 1-8 below.) Equation
1-40 is the exact relation between energy and momentum for
particles with zero rest mass, such as photons and neutrinos.

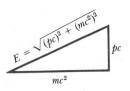

Figure 1-18
Triangle showing the re-
lation between energy,
momentum, and rest
mass in special relativity.

Example 1-8   Compare the exact and approximate expressions
of Equations 1-39 and 1-40 for the case in which the total energy
is eight times the rest energy. We are given $E = 8mc^2$. Equation
1-39   then   gives $(pc)^2 = E^2 - (mc^2)^2 = 64\,(mc^2)^2 - (mc^2)^2 = 63\,(mc^2)^2$. Then

$$\frac{pc}{E} = \frac{\sqrt{63}\ mc^2}{8\ mc^2} = 0.992$$

We see that the approximation $E \approx pc$ is good to slightly better
than 1 percent in this case.

From Equation 1-40 we see that the momentum of a high-
energy particle is simply its total energy divided by c. A convenient unit of momentum is MeV/c. However, the momentum of
a particle is usually found experimentally from Equation 1-38,
where B and R are often known in their SI units of tesla and
meters. It is therefore useful to rewrite Equation 1-38 in terms
of practical but mixed units. It is left as an exercise to show that
Equation 1-38 can be written

$$p = 300BR \left(\frac{q}{e}\right) \tag{1-41}$$

---

where p is in MeV/c, B is in tesla, and R is in meters.

Example 1-9   What is the approximate radius of the path of a
30-MeV electron moving in a magnetic field of 0.05 tesla ($= 500$
gauss)? Since 30 MeV is much greater than the rest energy of the
electron (0.511 MeV), we can use Equation 1-40 for the momentum

$$p \approx \frac{E}{c} = \frac{30\ \text{MeV}}{c}$$

Then from Equation 1-41 we have (with $q = e$),

$$R = \frac{p}{300B} = \frac{30}{300(0.05)} = 2\ \text{m}$$

Since nonrelativistic expressions for energy, momentum, etc.,
are usually easier to use than the relativistic ones, it is important
to know when these expressions are accurate enough. As $\gamma \to 1$,

all the relativistic expressions approach the classical ones. In most situations, the kinetic energy or total energy is given, so that the most convenient expression for calculating γ is, from Equation 1-29,

$$\gamma = \frac{E}{mc^2} = 1 + \frac{E_k}{mc^2} \qquad\qquad 1\text{-}42$$

When the kinetic energy is much less than the rest energy, γ is approximately 1 and nonrelativistic equations can be used. For example, the classical approximation $E_k \approx \frac{1}{2}mu^2 = p^2/2m$ can be used instead of the relativistic expression $E_k = (\gamma - 1)mc^2$ if E_k is much less than mc^2. We can get an idea of the accuracy of these approximations by expanding γ, using the binomial expansion as was done in Section 1-9, and examining the first term that is *neglected* in the classical approximation. We have

$$\gamma = \left(1 - \frac{u^2}{c^2}\right)^{-1/2} \approx 1 + \frac{1}{2}\frac{u^2}{c^2} + \frac{3}{8}\frac{u^4}{c^4} + \cdots$$

and

$$E_k = (\gamma - 1)mc^2 \approx \frac{1}{2}\,mu^2 + \frac{3}{2}\frac{(\frac{1}{2}mu^2)^2}{mc^2}$$

Then

$$\frac{E_k - \frac{1}{2}mu^2}{E_k} \approx \frac{3}{2}\frac{E_k}{mc^2} \qquad\qquad 1\text{-}43$$

We can see from Equation 1-43 that if $E_k/mc^2 \approx 1$ percent, the error in using the approximation $E_k \approx \frac{1}{2}mu^2$ is about 1.5 percent.

At very low energies, the velocity of a particle can be obtained from its kinetic energy $E_k \approx \frac{1}{2}mu^2$ just as in classical mechanics. At very high energies, the velocity of a particle is very near c. The following approximation is sometimes useful (see Exercise 42):

$$\frac{u}{c} \approx 1 - \frac{1}{2\gamma^2} \qquad \text{for } \gamma \gg 1 \qquad\qquad 1\text{-}44$$

An exact expression for the velocity of a particle in terms of its energy and momentum can be obtained by multiplying Equation 1-29 by u, and comparing with Equation 1-26. We have

$$uE = pc^2 \qquad \text{so} \qquad \frac{u}{c} = \frac{pc}{E} \qquad\qquad 1\text{-}45$$

This expression is of course not useful if the approximation $E \approx pc$ has already been made.

Example 1-10   An electron and a proton are each accelerated through 10×10^6 V. Find γ, the momentum, and the speed of each. Since each particle has a charge of magnitude e, each acquires a kinetic energy of 10 MeV. This is much greater than the 0.511-MeV rest energy of the electron and much less than the 938.3-MeV rest energy of the proton. We shall calculate the momentum and speed of each particle exactly, and then by means of the nonrelativistic or the extreme relativistic approximations.

We first consider the electron. From Equation 1-42 we have

$$\gamma = 1 + \frac{E_k}{mc^2} = 1 + \frac{10 \text{ MeV}}{0.511 \text{ MeV}} = 20.57$$

Since the total energy is $E_k + mc^2 = 10.511$ MeV we have, from Equation 1-39,

$$pc = \sqrt{E^2 - (mc^2)^2} = \sqrt{(10.51)^2 - (0.511)^2}$$
$$= 10.50 \text{ MeV}$$

The exact calculation then gives $p = 10.50$ MeV/c. The high-energy or extreme relativistic approximation $p \approx E/c = 10.51$ MeV/c is in good agreement with the exact result. If we use Equation 1-45 we obtain for the speed $u/c = pc/E = 10.50$ MeV/10.51 MeV $= 0.999$. On the other hand, the approximation of Equation 1-44 gives

$$\frac{u}{c} \approx 1 - \frac{1}{2}\left(\frac{1}{\gamma}\right)^2 = 1 - \frac{1}{2}\left(\frac{1}{20.57}\right)^2 = 0.999$$

For the proton, the total energy is $E_k + mc^2 = 10$ MeV $+ 938.3$ MeV $= 948.3$ MeV. From Equation 1-42 we obtain $\gamma = 1 + E_k/mc^2 = 1 + 10/938.3 = 1.01$. Equation 1-39 gives for the momentum

$$pc = \sqrt{E^2 - (mc^2)^2} = \sqrt{(948.3)^2 - (938.3)^2}$$
$$= 137.4 \text{ MeV}$$

(Note that this calculation involves the difference of two large and nearly equal numbers and cannot be done accurately on a slide rule.) The nonrelativistic approximation gives

$$E_k \approx \tfrac{1}{2}mu^2 = \frac{(mu)^2}{2m} \approx \frac{p^2}{2m} = \frac{p^2c^2}{2mc^2}$$

or

$$pc \approx \sqrt{2mc^2E_k} = \sqrt{(2)(938.3)(10)}$$
$$= 137.0 \text{ MeV}$$

The speed can be determined from Equation 1-45 exactly or from $p = mu$ approximately. From Equation 1-45 we obtain

$$\frac{u}{c} = \frac{pc}{E} = \frac{137.4}{948.3} = 0.1449$$

From $p \approx mu$, the nonrelativistic expression for p, we obtain

$$\frac{u}{c} \approx \frac{pc}{mc^2} = \frac{137.0}{938.3} = 0.1460$$

1-13 General Relativity

The generalization of relativity theory to noninertial reference frames by Einstein in 1916 is known as the general theory of relativity. This theory is much more difficult mathematically than the special theory of relativity, and there are fewer situations in which it can be tested. Nevertheless its importance calls for a brief qualitative discussion.

The basis of the general theory of relativity is the principle of equivalence:

A homogeneous gravitational field is completely equivalent to a uniformly accelerated reference frame.

Equivalence principle

This principle arises in Newtonian mechanics because of the apparent identity of gravitational and inertial mass. In a uniform gravitational field, all objects fall with the same acceleration g independent of their mass because the gravitational force is proportional to the (gravitational) mass while the acceleration varies inversely with the (inertial) mass. Consider a compartment in space far from any matter and undergoing uniform acceleration $\mathbf{a}$ as shown in Figure 1-19a. If people in the compartment drop objects, they fall to the "floor" with acceleration $\mathbf{g} = -\mathbf{a}$. If they stand on a spring scale, it will read their "weight" of magnitude ma. No mechanics experiment can be performed *within* the compartment that will distinguish whether the compartment is actually accelerating in space, or it is at rest (or moving with uniform velocity) in the presence of a uniform gravitational field $\mathbf{g} = -\mathbf{a}$.

Einstein assumed that the principle of equivalence applied to all physics and not only to mechanics. In effect, he assumed that there is no experiment of any kind that can distinguish uniformly accelerated motion from the presence of a gravitational field. We shall look qualitatively at only some of the consequences of this assumption.

The first consequence of the principle of equivalence we shall discuss, the deflection of a light beam in a gravitational field, was

Einstein at 53. (*Courtesy of the Archives, California Institute of Technology.*)

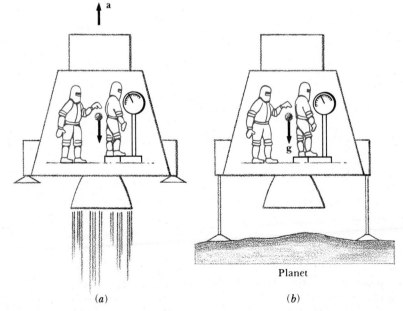

Planet

(*a*)   (*b*)

Figure 1-19
Results from experiments in a uniformly accelerated reference frame (*a*) cannot be distinguished from those in a uniform gravitational field (*b*), if the acceleration **a** and gravitational field **g** have the same magnitude.

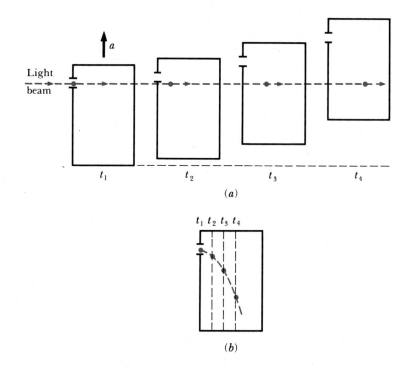

(a)

(b)

Figure 1-20
(a) Light beam moving in a straight line through a compartment that is undergoing uniform acceleration. The position of the light beam is shown at equally spaced times t_1, t_2, t_3, and t_4. (b) In the reference frame of the compartment, the light travels in a parabolic path, as would a ball were it projected horizontally. Note that in both (a) and (b) the vertical displacements are greatly exaggerated for emphasis.

one of the first to be tested experimentally. Figure 1-20 shows a beam of light entering a compartment that is accelerating. Successive positions of the compartment are shown at equal time intervals. Because the compartment is accelerating, the distance it moves in each time interval increases with time. The path of the beam of light, as observed from within the compartment, is therefore a parabola. But according to the equivalence principle, there is no way to distinguish between an accelerating compartment, and one with uniform velocity in a uniform gravitational field. We conclude, therefore, that a beam of light will accelerate in a gravitational field in the same way as do more massive objects. For example, near the surface of the earth, light will fall with acceleration 9.8 m/sec². This is difficult to observe because of the enormous speed of light. For example, in a distance of 3000 km, which takes about 0.01 sec to cover, a beam of light should fall about 0.5 mm. Einstein pointed out that the deflection of a light beam in a gravitational field might be observed when light from a distant star passes close to the sun, as illustrated in Figure 1-21. Because of the brightness of the sun, such a star cannot ordinarily be seen. Such a deflection was first observed in 1919 during an eclipse of the sun.

A second prediction from Einstein's theory of general relativity (which we shall not discuss in detail) is the excess precession of the perihelion of the orbit of Mercury of about 0.01° per century. This effect had been known and unexplained for some time, so in some sense it represented an immediate success of the theory. There is, however, some difficulty in comparing the prediction of general relativity with experiment because of other effects, such as the perturbations due to other planets and the nonspherical shape of the sun.

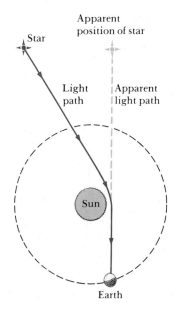

Figure 1-21
Deflection of a beam of light due to the gravitational attraction of the sun. The deflection is greatly exaggerated.

A third prediction of general relativity concerns the change in time intervals and frequencies of light in a gravitational field. If Δt_1 is a time interval measured by a clock where the gravitational potential (potential energy per unit mass) is ϕ_1 and Δt_2 is the same interval measured by another clock where the potential is ϕ_2, general relativity predicts that the fractional difference will be approximately

$$\frac{\Delta t_2 - \Delta t_1}{\Delta t} = \frac{1}{c^2}(\phi_2 - \phi_1) \qquad 1\text{-}46$$

(Since this shift is usually very small it does not matter by which interval we divide on the left side of Equation 1-46.) A clock in a region of low gravitational potential will therefore run more slowly than one in a region of high potential. Since a vibrating atom can be considered to be a clock, the frequency of vibration in a region of low potential (such as near the sun where the gravitational potential, $-GM_s/r$, is negative) will be lower than that of the same atom on earth. This shift toward lower frequency and therefore longer wavelength is called the *gravitational red shift*.

We can understand Equation 1-46 from our knowledge of time dilation in special relativity and the application of the equivalence principle. Consider a turntable rotating with angular velocity ω relative to an inertial reference frame, as shown in Figure 1-22. In the inertial reference frame, the clock at distance r is moving with speed $v = r\omega$. A time interval Δt_r measured on the moving clock is smaller than the corresponding time interval Δt_0 measured on the clock at the center of the table (which is at rest relative to the inertial frame) by the factor γ.

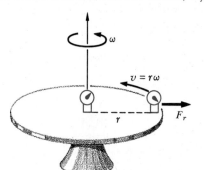

That is

$$\Delta t_r = \frac{1}{\gamma}\Delta t_0 = \sqrt{1 - v^2/c^2}\,\Delta t_0$$
$$\approx \left(1 - \frac{1}{2}\frac{v^2}{c^2}\right)\Delta t_0 \qquad 1\text{-}47$$

where we have used the binomial expansion to approximate $1/\gamma$. Then, using $v = r\omega$, we have

$$\frac{\Delta t_r - \Delta t_0}{\Delta t_0} \approx \frac{1}{c^2}\left(-\frac{1}{2}r^2\omega^2\right)$$

Figure 1-22
Clocks on a rotating turntable. The clock on the edge at a distance r runs slower than that at the center. According to special relativity, this is because it is moving with speed $v = r\omega$. According to general relativity, in the frame of the turntable there is force $F_r = mr\omega^2$ so the apparent gravitational potential ϕ is lower at the edge of the turntable. The slowing down of an atomic clock at low gravitational potential (such as near the surface of a star) leads to the gravitational red shift.

For observers on the rotating table, both clocks are at rest. In their frame they feel a (centrifugal) force outward with magnitude $F = mr\omega^2$. (This frame is of course not undergoing uniform acceleration, but we can approximate the acceleration in any small region as being uniform over short distances and therefore use the equivalence principle.) The relation between this assumed gravitational force and the potential is

$$F_r = -m\,\frac{d\phi}{dr} = mr\omega^2$$

Solving for ϕ we obtain

$$\phi = \phi_0 - \tfrac{1}{2}r^2\omega^2$$

where ϕ_0 is the potential at $r = 0$. Equation 1-47 therefore agrees with Equation 1-46, with $\phi_r - \phi_0 = -\tfrac{1}{2}r^2\omega^2$.

As our final example of the predictions of general relativity, we mention black holes, whose possible existence was first predicted by Oppenheimer and Snyder in 1939. According to the general theory of relativity, if the density of an object such as a star is great enough, the gravitational attraction will be so great that nothing can escape, not even light or other electromagnetic radiation. A remarkable property of such an object is that nothing that happens inside it can be communicated to the outside world. As is often the case in physics, a simple but incorrect calculation gives the correct results for the relation between the mass and critical radius of a black hole. In Newtonian mechanics, the speed needed for the escape of a particle from the surface of a planet or sun is found by requiring the kinetic energy $\tfrac{1}{2}mv^2$ to be equal in magnitude to the potential energy $-GMm/r$ so that the total energy is zero. The resulting escape velocity is

$$v_e = \sqrt{\frac{2GM}{r}}$$

If we set the escape velocity equal to the speed of light, and solve for the radius, we obtain the critical radius R_G, called the *Schwarzschild radius:*

$$R_G = \frac{2GM}{c^2} \qquad\qquad 1\text{-}48$$

For an object of mass equal to that of our sun to be a black hole, its radius must be about 3 km. Since no radiation is emitted from a black hole, and its radius is expected to be small, the detection of such an object is not easy. The best chance of detection would occur if a black hole were a companion to a normal star in a binary star system. Measurements of the Doppler shift of the light from the normal star might then allow a computation of the mass of the unseen companion to determine if it were great enough to be a black hole. At the present time the binary x-ray source Cygnus X-1 (in the constellation Cygnus) appears to be an excellent candidate for a black hole, but the evidence is not conclusive.

Summary

Because of the relative nature of simultaneity implied by the Einstein postulates, time and length intervals between events are not the same in all reference frames. The classical transformation between inertial frames must be replaced by the Lorentz transformation.

When the classical expressions for momentum and energy are replaced by the relativistic expressions, the laws of conservation of momentum and energy hold in all inertial frames. We shall list for future reference some of the more important equations developed in this chapter.

Time Dilation

Proper time is the time between two events that occur at the same space point; thus it can be measured on a single clock. If $\Delta t_0'$ is the proper time interval measured on a clock that moves with speed v in frame S, the time interval measured in S is longer:

$$\Delta t = \gamma \Delta t_0' \quad \text{with } \gamma = \left(1 - \frac{v^2}{c^2}\right)^{-1/2}$$

Length Contraction

The proper length of a rod is the length measured in the rest frame of the rod. If a rod of proper length L_0 moves with speed v in S, its length measured in S is

$$L = \frac{L_0}{\gamma}$$

Clock Synchronization

Two clocks separated by a proper distance, L_0, and synchronized in their rest frame will be unsynchronized in a frame in which they are moving parallel to their separation. If the clocks are moving with speed v in S, with clock b chasing clock a, the time difference at some instant in S will be

$$t_b - t_a = \frac{L_0 v}{c^2}$$

Lorentz Transformation

$$x' = \gamma(x - vt)$$
$$y' = y$$
$$z' = z$$
$$t' = \gamma\left(t - \frac{xv}{c^2}\right)$$

The inverse transformation is obtained by changing v to $-v$ and switching primes.

Velocity Transformation

$$u'_x = \frac{u_x - v}{1 - u_x v/c^2}$$

$$u'_y = \frac{u_y}{\gamma(1 - u_x v/c^2)}$$

$$u'_z = \frac{u_z}{\gamma(1 - u_x v/c^2)}$$

Momentum

$$\mathbf{p} = \frac{m\mathbf{u}}{\sqrt{1 - u^2/c^2}} = \gamma m\mathbf{u}$$

Energy

$$E = \frac{mc^2}{\sqrt{1 - u^2/c^2}} = \gamma mc^2$$

$$E^2 = (pc)^2 + (mc^2)^2$$

Kinetic Energy

$$E_k = E - mc^2 = (\gamma - 1)mc^2$$

For $u/c \ll 1$, $E_k \approx \frac{1}{2}mu^2$; for $u/c \approx 1$, $E_k \approx pc - mc^2$.

The general theory of relativity is based on the principle of equivalence between a uniform gravitational field and an accelerated reference frame. Some predictions from this theory are the deflection of a light beam in a gravitational field, the gravitational red shift and time dilation, the precession of the perihelion of Mercury, and the possible existence of black holes.

References

1.   T. M. Helliwell, *Introduction to Special Relativity,* Boston: Allyn and Bacon, Inc., 1966.

2.   C. Kacser, *Introduction to the Special Theory of Relativity,* Englewood Cliffs, N.J.: Prentice-Hall, Inc., 1967.

3.   E. P. Ney, *Electromagnetism and Relativity,* New York: Harper & Row Publishers, Inc., 1962.

4.   R. Resnick, *Introduction to Special Relativity,* New York: John Wiley & Sons, Inc., 1968.

5.   E. F. Taylor and J. A. Wheeler, *Spacetime Physics,* San Francisco: W. H. Freeman and Company, 1966.

Each of the above references is a short book written for students.

6.   A. Einstein et al., *The Principle of Relativity,* New York: Dover Publications, Inc., 1923. A collection of original papers pertaining to special and general relativity.

7.   "Resource Letter SRT-1 on Special Relativity Theory," *American Journal of Physics,* **30,** 462 (1962). A list of references.

8.   *Special Relativity Theory: Selected Reprints,* New York: American Association of Physics Teachers, 1963. This booklet is a collection of some of the papers listed in "Resource Letter SRT-1."

9.   A. P. French, *Special Relativity,* New York: W. W. Norton, Inc., 1968. An excellent text with a particularly good discussion of the historical basis of special relativity.

10. F. K. Richtmyer, E. H. Kennard, and J. N. Cooper, *Introduction to Modern Physics,* New York: McGraw-Hill Book Company, 1969. The sixth edition of an excellent text originally published in 1928 and intended as a survey course for graduate students. A standard source for reference material pertaining to modern physics.

11. G. Gamow, *Mr. Tompkins in Paper Back,* New York: Cambridge University Press, 1965. Contains the delightful "Mr. Tompkins in Wonderland," "Mr. Tompkins Explores the Atom," and other stories. In one of the stories, Mr. Tompkins visits a dream world in which the speed of light is 10 mi/h and relativistic effects are quite noticeable.

Exercises

Section 1-1, The Michelson-Morley Experiment

1.   In one series of measurements of the speed of light, Michelson used a path length L of 35.4 km (22 mi). (*a*) What is the time needed for light to make the round-trip of distance $2L$? (*b*) What is the classical correction term in seconds in Equation 1-2, assuming the earth's speed is $v = 10^{-4}c$? (*c*) From about 1600 measurements, Michelson arrived at a result for the speed of light of $299,796 \pm 4$ km/sec. Is this experimental value accurate enough to be sensitive to the correction term in Equation 1-2?

2.   An airplane flies with speed c relative to still air from point A to point B and returns. Compare the time required for the round trip when the wind blows from A to B with speed v with that when the wind blows perpendicularly to the line AB with speed v.

Section 1-2, Consequences of Einstein's Postulates

There are no exercises for this section.

Section 1-3, Time Dilation and Length Contraction

✓3.   Derive the following results for values of v much less than c and use when applicable in the exercises and problems that follow.

(*a*) $\gamma \approx 1 + \dfrac{1}{2}\dfrac{v^2}{c^2}$

(*b*) $\dfrac{1}{\gamma} \approx 1 - \dfrac{1}{2}\dfrac{v^2}{c^2}$

(*c*) $\gamma - 1 \approx 1 - \dfrac{1}{\gamma} \approx \dfrac{1}{2}\dfrac{v^2}{c^2}$

4.   How great must the relative speed of two observers be for their time-interval measurements to differ by 1 percent (see Exercise 3)?

$\gamma = 1.01$
$v = .14c$

5.   The proper mean lifetime of π mesons is 2.6×10^{-8} sec. If a beam of such particles has speed $0.9c$, (a) what would their mean life be as measured in the laboratory? (b) How far would they travel (on the average) before they decay? (c) What would your answer be to part (b) if you neglected time dilation?

$\gamma = 2.29$
a) 5.95×10^{-8} s
b.) 16.1 m
c.) 7.0 m

6.   (a) In the reference frame of the π mesons in Exercise 5, how far does the laboratory travel in a typical lifetime of 2.6×10^{-8} sec? (b) What is this distance in the laboratory frame?

a.) $L = L'/\gamma = 7.0$ m
b.) 16.1 m

7.   A meterstick moves parallel to its length with speed $v = 0.6c$ relative to you. (a) Find the length of the stick measured by you. (b) How long does it take for the stick to pass you?

$\gamma = 1.25$
a.) 80 cm
b.) 4.4 n s

8.   Supersonic jets achieve maximum speeds of about $3 \times 10^{-6}c$. (2000 mi/hr) (a) By what percentage would you observe such a jet to be contracted in length? (b) During a time of 1 y $= 3.16 \times 10^7$ sec on your clock, how much time would elapse on the pilot's clock? How many minutes are lost by the pilot's clock in 1 y of your time?

$\gamma = 1.00..043$, 12 zeros
a) 4.5×10^{-11} % o
b.) 1.42×10^{-3} s loss

Section 1-4, Clock Synchronization and Simultaneity

Exercises 9 to 13 refer to the following situation: an observer in S' lays out a distance L' = 100 c-min between points A' and B' and places a flashbulb at the midpoint C'. He arranges for the bulb to flash and for clocks at A' and B' to be started at zero when the light from the flash reaches the clocks (see Figure 1-23). Frame S' is moving to the right and speed 0.6c relative to an observer C in S who is at the midpoint between A' and B' when the bulb flashes and sets his clock to zero at that time.

$\gamma = (1 - (.6)^2)^{-1/2} = 1.25$

9.   What is the distance between clocks A' and B', according to the observer in S? $L = \frac{L'}{\gamma} = \frac{100 \text{ c-min}}{1.25} = 80 \text{ c-min}$

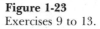

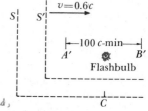

10. As the light pulse from the flashbulb travels toward A' and B' with speed c, A' travels toward C with speed $0.6c$. Show that the clock in S reads 25 min when the flash reaches A'. $\Delta t' = 50$ min in s', but, in S, A' is unsync'd by 30 min w/c', $\Delta t' = 20$ min, $\Delta t = \gamma \Delta t' = 25$ min

11. Show that the clock in S reads 100 min when the light flash reaches B', which is traveling away from C with speed $0.6c$. $\Delta t' = 50$ min in s', but, in S, B' is unsync'd by -30 min w/c', $\Delta t' = 80$ min, $\Delta t = \gamma \Delta t' = 100$ min

12. The time interval between reception of the flashes is 75 min according to the observer in S. How much time does he expect to have elapsed on the clock at A' during this 75 min? from #10 $\Delta t' = 20$ min, from #11, $\Delta t' = 80$ min, A' = $80 - 20 = 60$ min

13. The time interval calculated in Exercise 12 is the amount that the clock at A' leads that at B', according to observers in S. Compare this result with L_0v/c^2. $L_0 v/c^2 = \frac{(100 \text{ c-min})(.6c)}{c^2} = 60$ min

Figure 1-23
Exercises 9 to 13.

Section 1-5, The Doppler Effect

14. (a) Show that Equations 1-14a and b are identical and can both be written

$$f' = f_0 \sqrt{\frac{1 + \beta}{1 - \beta}} \qquad \text{where } \beta = \frac{v}{c}$$

(b) Show that Equation 1-15 can be written

$$f' = f_0 \sqrt{\frac{1 - \beta}{1 + \beta}}$$

✓15. How fast must you move toward a red light (λ = 650 nm) for it to appear green (λ = 525 nm)? $v = \dfrac{((f/f_0)^2 - 1)c}{(f/f_0)^2 + 1} = v$, $\dfrac{f'}{f_0} = 1.238$ $v = .21c$

16. A distant galaxy is moving away from us at speed 2×10^7 m/sec. Calculate the fractional red shift $(\lambda' - \lambda_0)/\lambda_0$ of the light from this galaxy.

17. A distant galaxy is moving away from the earth such that each wavelength is shifted by a factor of 2; that is, $\lambda' = 2\lambda_0$. What is the speed of the galaxy relative to us (see Exercise 14)?

Section 1-6, The Lorentz Transformation

18. Two events happen at the same point x'_0 in frame S' at times t'_1 and t'_2. (a) Use Equations 1-18 to show that in frame S the time interval between the events is greater than $t'_2 - t'_1$ by factor γ. (b) Why are Equations 1-19 less convenient than Equations 1-18 for this problem?

✓19. You measure the length of an object moving in your frame of reference by recording the positions of each end x_2 and x_1 at the same time t_0. (a) Use Equations 1-19 to show that the result is smaller than the proper length $x'_2 - x'_1$ measured in the frame in which the object is at rest. (b) Why are Equations 1-19 more convenient for this problem than Equations 1-18? $x'_2 - x'_1 = \gamma(x_2 - vt) - \gamma(x_1 - vt) = \gamma(x_2 - x_1)$, $\gamma \geq 1$

b.) - only necessary to substitute

20. Two spaceships are approaching each other. (a) If the speed of each is $0.9c$ relative to the earth, what is the speed of one relative to the other? (b) If the speed of each relative to the earth is 30,000 m/sec (about 100 times the speed of sound), what is the speed of one relative to the other?

21. A light beam moves along the y' axis with speed c in frame S', which is moving to the right with speed v relative to frame S. (a) Find u_x and u_y, the x and y components of the velocity of the light beam in frame S. (b) Show that the magnitude of the velocity of the light beam in S is c.

22. A particle moves with speed $0.9c$ along the x'' axis of frame S'', which moves with speed $0.9c$ in the positive x' direction relative to frame S'. Frame S' moves with speed $0.9c$ in the positive x direction relative to frame S. (a) Find the speed of the particle relative to frame S'. (b) Find the speed of the particle relative to frame S.

Section 1-7, The Twin Paradox

There are no exercises for this section.

Section 1-8, Relativistic Momentum, and Section 1-9, Relativistic Energy

23. Show that $p_{yA} = -p_{yB}$, where p_{yA} and p_{yB} are the relativistic momenta of the balls in Figure 1-15, given by

$$p_{yA} = \frac{mu_0}{\sqrt{1 - u_0^2/c^2}} \qquad p_{yB} = \frac{mu_{yB}}{\sqrt{1 - (u_{xB}^2 + u_{yB}^2)/c^2}}$$

$$u_{yB} = -u_0 \sqrt{1 - \frac{v^2}{c^2}} \qquad u_{xB} = v$$

24. Combine Equations 1-25 and 1-29 to derive Equation 1-39: $E^2 = (pc)^2 + (mc^2)^2$.

✓25. An electron of rest energy $mc^2 = 0.511$ MeV moves with speed $u = 0.6c$. Find (a) γ, (b) p in units of MeV/c, (c), E, (d) E_k.

$\gamma = 1.25$

$p = \dfrac{(.511 \text{ MeV}/c^2)(.6c)}{1.25} = .245 \dfrac{\text{MeV}}{c}$

$E = .638$ MeV

$E_k = E - rest = .638 - .511 = .128$ MeV

26. Work Exercise 25 with $u = 0.8c$.

27. Work Exercise 25 with $u = 0.99c$.

28. How much energy would be required to accelerate a particle of mass m from rest to a speed of (a) $0.5c$, (b) $0.9c$, (c) $0.99c$? Express your answers as multiples of the rest energy.

29. Two 1-kg masses are separated by a spring of negligible mass. They are pushed together, compressing the spring. If the work done in compressing the spring is 10 J, find the change in mass of the system in kilograms. Does the mass increase or decrease?

Section 1-10, Mass and Binding Energy

30. The energy released in the fission of a ^{235}U nucleus is about 200 MeV. How much rest mass is converted to energy in this fission?

31. How much energy is required to remove one of the neutrons from ^{3}H to yield ^{2}H plus a neutron?

32. The rest mass of ^{3}He (a nucleus consisting of two protons and one neutron) is 3.01440 u. (a) What is the rest energy of ^{3}He in MeV? (b) How much energy is needed to remove a proton to make ^{2}H plus a proton?

33. The energy released when sodium and chlorine combine to form NaCl is 4.2 eV. (a) What is the increase in mass (in unified mass units) when a molecule of NaCl is dissociated into an atom of Na and an atom of Cl? (b) What percentage error is made in neglecting this mass difference? (The mass of Na is about 23 u, and that of Cl is about 35.5 u.)

34. In a nuclear fusion reaction two ^{2}H atoms are combined to produce ^{4}He. (a) Calculate the decrease in rest mass in unified mass units. (b) How much energy is released in this reaction? (c) How many such reactions must take place per second to produce 1 W of power?

✓35. Calculate the rate of conversion of rest mass to energy (in kg/h) needed to produce 100 MW.

Section 1-11, Experimental Determination of Relativistic Momentum

There are no exercises for this section.

Section 1-12, Some Useful Equations and Approximations

36. Show that Equation 1-39 can be written $E = mc^2(1 + p^2/m^2c^2)^{1/2}$ and use the binomial expansion to show that, when pc is much less than mc^2, $E \approx mc^2 + p^2/2m$.

✓37. An electron of rest energy 0.511 MeV has a total energy of 5 MeV. (a) Find its momentum in units of MeV/c. (b) Find u/c.

38. Make a sketch of the total energy of an electron E as a function of its momentum p. (See Equation 1-40 and Exercise 36 for the behavior of E at large and small values of p.)

39. Find the momentum of a 20-MeV (E_k) electron using (a) the exact Equation 1-39 and (b) the approximation $E \approx pc$.

40. What magnetic field is needed to bend a 20-MeV electron in a circular path of radius 1 m?

41. A proton is bent into a circular path of radius 2m by a magnetic field of 0.5 T. (a) What is the momentum of the proton? (b) What is its kinetic energy?

42. Write u/c in terms of γ and use the binomial expansion for the case $1/\gamma \ll 1$ to derive Equation 1-44.

Section 1-13, General Relativity

There are no exercises for this section.

Problems   447 m/s

✓1. An airplane flies at a speed of 1000 mi/h. How long must it fly before its clock loses 1 sec because of time dilation? $\Delta t = \gamma \Delta t'$
9×10¹¹ SEC OR   286 centurys

2. (a) Show that the speed u of a particle of mass m and total energy E is given by $u/c = [1 - (mc^2/E)^2]^{1/2}$ and that, if E is much greater than mc^2, the approximation $u/c \approx 1 - \frac{1}{2}(mc^2/E)^2$ holds. (b) Find the speed of an electron of kinetic energy 0.511 MeV and that of an electron of kinetic energy 10 MeV.

3. What percentage error is made in using $\frac{1}{2}mu^2$ for the kinetic energy of a particle if its speed is (a) $u = 0.1c$, (b) $u = 0.9c$?

4. Two spaceships each 100 m long when measured at rest travel toward each other, each with speed $0.8c$ relative to earth. (a) How long is each ship as measured by someone on earth? (b) How fast is each ship traveling as measured by the other? (c) How long is one ship when measured by the other? (d) At some time $t = 0$ (on earth clocks) the fronts of the ships are together as they begin to pass each other. At what time (on earth clocks) are their backs together? (e) Sketch diagrams in the frame of one of the ships showing the passing of the other ship.

5. For the special case of a particle moving with speed u along the y axis in S, show that the momentum and energy in frame S' are related to the momentum and energy in S by the transformation equations

$$p'_x = \gamma\left(p_x - \frac{vE}{c^2}\right) \qquad p'_y = p_y$$

$$p'_z = p_z \qquad \frac{E'}{c} = \gamma\left(\frac{E}{c} - \frac{vp_x}{c}\right)$$

Compare these equations with the Lorentz transformation for x', y', z', and t'. These show that (in this special case) the quantities p_x, p_y, p_z, and E/c transform in the same way as x, y, z, and ct.

6. The equation for a spherical wavefront of a light pulse which begins at the origin at time $t = 0$ is $x^2 + y^2 + z^2 - (ct)^2 = 0$. Using the Lorentz transformation equations, show that such a light pulse also has a spherical wavefront in frame S' by showing that $x'^2 + y'^2 + z'^2 - (ct')^2 = 0$.

✓7. In Problem 6 you showed that the quantity $x^2 + y^2 + z^2 - (ct)^2$ has the same value (0) in both S and S'. Such a quantity is called an *invariant*. From the results of Problem 5, the quantity $p_x{}^2 + p_y{}^2 + p_z{}^2 - (E/c)^2$ must also be invariant. Show that this quantity has the value $-m^2c^2$ in both S and S' reference frames.

8.   Two events in S are separated by a distance $D = x_2 - x_1$ and time $T = t_2 - t_1$. (*a*) Use the time-transformation equation to show that in frame S' (moving with speed v relative to S) the time separation is $t_2' - t_1' = \gamma(T - vD/c^2)$. (*b*) Show that the events can be simultaneous in frame S' only if D is greater than cT. (*c*) If one of the events is the *cause* of the other, the separation D must be less than cT since D/c is the smallest time a signal can take to travel from x_1 to x_2 in frame S. Show that if D is less than cT, t_2' is greater than t_1' in all reference frames. This shows that the cause must precede the effect in all reference frames (assuming that it does in one frame). (*d*) Suppose that a signal could be sent with speed c' *greater* than c so that in frame S the cause precedes the effect by the time $T = D/c'$, which is less than D/c. Show that there is then a reference frame moving at speed v less than c in which the effect precedes the cause.

9.   Two observers agree to test time dilation. They use identical clocks, and one observer in frame S' moves with speed $v = 0.6c$ relative to the other observer in frame S. When their origins coincide, they start their clocks. They agree to send a signal when their clocks read 60 min and to send a confirmation signal when each receives the other's signal. (*a*) When does the observer in S receive the first signal from the observer in S'? (*b*) When does he receive the confirmation signal? (*c*) Make a table showing the times in S when the observer sent the first signal, received the first signal, and received the confirmation signal. How does this table compare with one constructed by the observer in S'?

10. Show that if v is much less than c, the Doppler frequency shift is approximately given by $\Delta f/f = \pm v/c$, both classically and relativistically. A radar transmitter-receiver bounces a signal off an aircraft and observes a fractional increase in the frequency of $\Delta f/f = 8 \times 10^{-7}$. What is the speed of the aircraft? (Assume the aircraft to be moving directly toward the transmitter.)

52

CHAPTER 2 The Kinetic Theory of Matter

Objectives

After studying this chapter you should:

---

1. Know the value of Avogadro's number and its relation to the unified mass unit.

2. Know the relation between the kinetic energy of the molecules of a gas and the pressure and temperature of the gas.

3. Know the approximate value of kT in eV at room temperature.

4. Be able to state the equipartition theorem and discuss its applications.

5. Know what a distribution function is and how to use it to calculate average values of quantities.

6. Be able to sketch the Maxwell-Boltzmann velocity and speed distribution functions.

7. Be able to calculate average values of various quantities, such as v_x, v, v^2, E_k, etc., from the Maxwell-Boltzmann distribution function.

8. Be able to discuss the connection between viscosity, heat conduction, and diffusion based on simple transport theory.

In this chapter we shall study some aspects of kinetic theory, the first successful microscopic model of matter. We shall see how the first estimates and measurements of the size and mass of molecules were made. The introduction to the statistical methods of distribution functions and averages in this chapter should prove valuable in Chapter 6, where the same methods are used in quantum mechanics.

The idea that all matter is composed of tiny particles, or atoms, dates back to the speculations of the Greek philosopher Democritus and his teacher Leucippus about 450 B.C. However, there was little attempt to correlate such speculations with observations of the physical world until the seventeenth century. Pierre Gassendi, in the middle of the seventeenth century, and somewhat later Robert Hooke attempted to explain states of matter and the transitions between them with a model of tiny indestructible solid objects flying in all directions.

In 1662, Robert Boyle published results of his experiments showing that the product of the pressure and volume of a gas remains constant at constant temperature. Isaac Newton in his *Principia* (1687) showed that Boyle's law could be derived by assuming the gas to consist of hard *static* particles that repel each other with a force varying inversely with their separation. The first mathematical derivation of Boyle's law using a *kinetic* model was done by Daniel Bernoulli in 1738.

Little more was done along these lines for nearly a hundred years. The nineteenth century saw a rapid development of the kinetic theory of matter by many people, notably Herapath, Waterston, Joule, Clausius, Maxwell, and Boltzmann. A parallel development of the theory of atoms emerged in the beginning of the nineteenth century from attempts to understand the laws of chemistry. To explain the law of definite proportions postulated by J. L. Proust (1754–1826), which states that the elements that make up a chemical compound always combine in the same definite proportions by weight, John Dalton in 1808 assumed that an element consisted of identical indestructible atoms. In the same year, J. L. Gay-Lussac announced the law of combining volumes; when two gases combine to form a third, the ratios of the volumes are ratios of integers. He showed, for example, that whenever hydrogen combined with oxygen to form water vapor, the ratio of the volume of hydrogen to that of oxygen was 2 to 1 within 0.1 percent accuracy. (It is interesting to note that Dalton did not accept Gay-Lussac's law because it did not agree with his static atomic model, a model which he thought had been proved by Newton's derivation of Boyle's law. Dalton also had data less accurate than Gay-Lussac's, which showed deviations from ratios of integers.) In 1811 an Italian physicist, Amedeo Avogadro, proposed a remarkable hypothesis which, though not accepted for some time, eventually paved the way for the atomic theory of chemistry. Avogadro assumed that:

1. Particles of a gas were small compared with the distances between them.   *Avogadro's hypotheses*

2. The particles of elements sometimes consisted of two or more atoms stuck together. These particles he called "molecules" to distinguish them from atoms.

3. Equal volumes of gases at constant temperature and pressure contained equal numbers of molecules.

Using these hypotheses along with the work of Gay-Lussac, Dalton, Proust, and others, Avogadro worked out the composition of molecules and, in particular, found that it was necessary to assume that the molecules of a gas such as hydrogen and oxygen contained two atoms. At first, few scientists believed these hypotheses, primarily because of the difficulty of understanding why, if two oxygen atoms attracted each other to form the molecule O_2, three or four atoms did not likewise bind together to form O_3 or O_4. (This was not completely understood until the development of quantum mechanics.)

Avogadro's hypotheses were not really accepted until the latter half of the nineteenth century. It is interesting to note that he had no knowledge of the magnitude of the number of molecules in a given volume of gas, only that the number was very large. The first calculation of the number was done by J. Loschmidt in 1865 from the kinetic theory of gases. We do not have space in this brief introduction to go into more detail concerning the fascinating history of the discovery of the atomic theory of chemistry. The interested reader is referred to the excellent discussion in Reference 1, on which much of this introduction is based.

2-1   Avogadro's Number

From the characteristic ratios of weights in which different elements combined, a scale of relative *atomic weights* was established by Avogadro and his contemporaries. Hydrogen, the lightest element, was assigned the weight of unity, and other elements were assigned weights relative to hydrogen. Avogadro's number was defined as the number of atoms needed to make up an amount of an element equal to its atomic weight[1] in grams.

For example, using this scheme, Avogadro's number, N_A, of hydrogen atoms had a mass of 1 g. The basis for assigning relative weights was subsequently changed from hydrogen to oxygen and more recently to carbon. Avogadro's number of ^{12}C atoms is now defined as having a mass of exactly 12.000 g, and the mass of one ^{12}C atom is exactly 12.000 unified mass units (u). (Unified mass unit replaces the older atomic mass unit, amu.) On this basis the relative weight of the hydrogen atom (i.e., the atomic weight) is 1.0078. Avogadro's number of hydrogen atoms therefore has a mass of 1.0078 g. N_A atoms or molecules of a substance are called a *mole* of that substance. One mole of H atoms thus has a mass of 1.0078 g, while one mole of H_2 molecules has a mass of 2.0158 g. This mass is called the gram-molecular weight of H_2. The present value of Avogadro's number is

$$N_A = 6.0220 \times 10^{23} \qquad 2\text{-}1$$

Avogadro's number

The mass of a mole of atoms (or molecules) divided by Avogadro's number is the mass of a single atom (or molecule). For example, the mass of the hydrogen atom is

$$m_H = \frac{1.0078 \text{ g/mole}}{6.022 \times 10^{23} \text{ atoms/mole}}$$
$$= 1.674 \times 10^{-24} \text{ g/atom}$$

Since Avogadro's number of unified mass units has a mass of exactly 1 g, the unified mass unit is simply the reciprocal of N_A.

$$1 \text{ u} = \frac{1}{N_A} = \frac{1 \text{ g}}{6.022 \times 10^{23}} = 1.661 \times 10^{-24} \text{ g} \qquad 2\text{-}2$$

Unified mass unit

[1] Strictly speaking, the word "weight" should be replaced with "mass" but we shall follow common usage and refer to relative atomic mass as atomic weight.

2-2   The Pressure of a Gas

Kinetic theory attempts to describe the properties of gases in terms of a microscopic picture of the gas as a collection of particles in motion. The pressure exerted by a gas on the walls of its container is an example of a property that is readily calculated by kinetic theory. The gas exerts a pressure on its container because as molecules of the gas collide with the walls of the container, they must transfer momentum to the walls. The total change in momentum per second is the force exerted on the walls by the gas. We start by making the following assumptions:

1. The gas consists of a large number, N, of molecules that make elastic collisions with each other and with the walls of the container.

2. The molecules are separated by distances that are large compared with their diameters, and they exert no forces on each other except when they collide.

3. In the absence of external forces (we can neglect gravity), there is no preferred position for a molecule in the container, and there is no preferred direction for the velocity vector.

For the moment we shall ignore the collisions the molecules make with each other. This is not a serious flaw in our calculation for, since momentum is conserved, collisions of molecules with one another will not affect the total momentum in any given direction. Let m be the mass of each molecule. If we take the x axis to be perpendicular to the wall, the x component of momentum of a molecule is $+mv_x$ before it hits the wall and $-mv_x$ afterward. The magnitude of the change in momentum of the molecule due to its collision with the wall is $2mv_x$. <u>The total change in the momentum of all the molecules in some time interval Δt is $2mv_x$ times the number that hit the wall during this interval</u>.

Let us consider a rectangular container of volume V with a right wall of area A (Figure 2-1). Let N_i be the number of gas molecules whose x component of velocity is v_{xi}. The number of molecules hitting the right wall in time Δt is the number within a distance $v_{xi} \Delta t$ and traveling to the right. Since there are N_i such molecules in volume V, the number in the volume $Av_{xi} \Delta t$ is $(N_i/V)(Av_{xi} \Delta t)$. If we assume for the moment that v_{xi} is positive, the number that hit the right wall in time Δt is

$$\frac{N_i}{V} A v_{xi} \, \Delta t$$

The impulse exerted by the wall on these molecules equals the total change in momentum of these molecules, which is $2mv_{xi}$, times the number that hit:

$$I_i = \left(\frac{N_i v_{xi} A \, \Delta t}{V} \right) 2mv_{xi} = \frac{2N_i m v_{xi}{}^2 A \, \Delta t}{V}$$

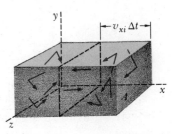

Figure 2-1
Gas molecules in a rectangular container. A molecule with velocity component v_{xi} will hit the right side within time Δt if it is within the distance $v_{xi} \Delta t$ and if v_{xi} is positive.

This also equals the magnitude of the impulse exerted by these molecules *on* the wall. We obtain the average force exerted by these molecules by dividing the impulse by the time interval Δt. The pressure is this average force divided by the area A. The pressure exerted by these molecules is thus

$$P_i = \frac{I_i}{\Delta t\,A} = \frac{2N_i m v_{xi}^2}{V}$$

The total pressure exerted by all the molecules is obtained by summing over all the x components of velocity v_{xi} that are positive. Since, on the average, half the molecules will be moving to the right (positive v_{xi}) and half to the left (negative v_{xi}) at any time, we can sum over all the molecules and multiply by $\frac{1}{2}$:

$$P = \frac{1}{2}\,\Sigma P_i = \frac{1}{2}\,\Sigma\,\frac{2N_i m v_{xi}^2}{V} = \frac{m}{V}\,\Sigma N_i v_{xi}^2$$

We can write this in terms of the average value of v_x^2, defined as

$$(v_x^2)_{av} = \frac{1}{N}\,\Sigma N_i v_{xi}^2$$

where $N = \Sigma N_i$ is the total number of molecules. Thus we can write for the pressure (on the wall perpendicular to the x axis),

$$P = \frac{Nm}{V}\,(v_x^2)_{av} \qquad\qquad 2\text{-}3$$

If there is no preferred direction of motion of the molecules, $(v_x^2)_{av}$ must be the same as $(v_y^2)_{av}$ and $(v_z^2)_{av}$. The square of the speed is

$$v^2 = v_x^2 + v_y^2 + v_z^2$$

Hence

$$(v^2)_{av} = (v_x^2)_{av} + (v_y^2)_{av} + (v_z^2)_{av} = 3(v_x^2)_{av}$$

Thus we can write the pressure in terms of the average square speed:

$$P = \frac{1}{3}\frac{N}{V}\,m(v^2)_{av} = \frac{2}{3}\frac{N}{V}\left(\frac{1}{2}\,mv^2\right)_{av}$$
$$= \tfrac{2}{3}n\,(\tfrac{1}{2}mv^2)_{av} \qquad\qquad 2\text{-}4$$

where $n = N/V$ is called the number density. This result shows that the pressure is proportional to the number of molecules per unit volume and to their average kinetic energy.

If we write $\overline{E}_k = (\tfrac{1}{2}mv^2)_{av}$ for the average kinetic energy we have

$$PV = \tfrac{2}{3}N\overline{E}_k \qquad\qquad 2\text{-}5$$

Let us compare this result with the ideal gas relation

$$PV = \nu RT$$

where ν is the number of moles, which is the total number of molecules divided by Avogadro's number

$$\nu = \frac{N}{N_A}$$

and R is the gas constant

$$R = 8.31 \text{ J/K-mole} = 1.99 \text{ cal/K-mole} \qquad \text{2-6}$$

We then have

$$\nu RT = \tfrac{2}{3} N \bar{E}_k = \tfrac{2}{3} \nu N_A \bar{E}_k$$

or

$$\bar{E}_k = \frac{3}{2} \frac{R}{N_A} T = \frac{3}{2} kT \qquad \text{2-7}$$

Kinetic energy and temperature

where $k = R/N_A$ is called Boltzmann's constant:

$$k = 1.381 \times 10^{-23} \text{ J/K} = 8.617 \times 10^{-5} \text{ eV/K} \qquad \text{2-8}$$

The absolute temperature is therefore a measure of the average translational kinetic energy of the molecules. (We include the word "translational" because a molecule may have other kinds of kinetic energy, e.g., rotational or vibrational. Only the translational kinetic energy enters the calculation of the pressure exerted on the walls of the container.) The total translational kinetic energy of ν moles of a gas containing N molecules is

$$E_k = N \bar{E}_k = \tfrac{3}{2} NkT = \tfrac{3}{2} \nu RT \qquad \text{2-9}$$

The translational kinetic energy is $\tfrac{3}{2} kT$ per molecule or $\tfrac{3}{2} RT$ per mole. At a typical temperature of $T = 300$ K ($=81°$F), the quantity kT has the value

$$kT = 2.585 \times 10^{-2} \text{ eV} \approx \tfrac{1}{40} \text{ eV} \qquad \text{2-10}$$

kT at 300 K

The mean translational kinetic energy of a gas molecule at room temperature is only a few hundredths of an electron volt. Two important results are obtained from this simple calculation:

1. *Speed of a molecule in a gas.* We do not expect all the molecules in a gas to have the same speed. The distribution of molecular speeds will be discussed in Section 2-5. However, even without knowing this distribution, we can calculate the average square speed $(v^2)_{av}$ and the root-mean-square (rms) speed $v_{rms} = \sqrt{(v^2)_{av}}$. We have

$$(v^2)_{av} = \frac{2\bar{E}_k}{m} = \frac{3RT}{N_A m} = \frac{3RT}{\mathcal{M}} \qquad \text{2-11}$$

where $\mathcal{M}$ is the *molecular weight*. Then

$$v_{rms} = \sqrt{\frac{3RT}{\mathcal{M}}} \qquad \text{2-12}$$

It is not hard to remember the order of magnitude of molecular speeds if we recall that the speed of sound in a gas is given by

$$v_{\text{sound}} = \sqrt{\frac{\gamma RT}{\mathcal{M}}}$$

where γ is the ratio of the heat capacity at constant pressure to that at constant volume. (For air, $\gamma = C_p/C_v = 1.4$.) Thus the rms speed of a gas molecule is of the same order of magnitude as the speed of sound in the gas.

2. *Heat capacities* The molar heat capacity at constant volume is defined by

$$C_v = \lim_{\Delta T \to 0} \frac{\Delta Q}{\Delta T}$$

where ΔQ is the heat input and ΔT is the temperature rise for 1 mole of a substance. Since no work is done if the volume is constant, the heat input equals the change in internal energy U (from the first law of thermodynamics). Thus

$$C_v = \left(\frac{\partial U}{\partial T}\right)_v$$

If we assume that the total internal energy is *translational* kinetic energy, we have from Equation 2-9 for 1 mole,

$$U = N_A \bar{E}_k = \tfrac{3}{2} RT \qquad\qquad 2\text{-}13$$

and

$$C_v = \tfrac{3}{2} R = 2.98 \text{ cal/mole}$$

This result agrees well with experiments for monatomic gases such as argon and helium (see Table 2-1, page 61). For other gases, the measured molar heat capacity is greater than this, indicating that some of the heat input goes into forms of internal energy other than translational kinetic energy, such as energy of molecular rotation or vibration.

Example 2-1   Calculate the root-mean-square speed of nitrogen molecules at $T = 300$ K. We have

$$\mathcal{M} = 28 \text{ g/mole} = 28 \times 10^{-3} \text{ kg/mole}$$

$$v_{\text{rms}} = \left(\frac{3 \times 8.31 \text{ J-K}^{-1} \text{ mole}^{-1} \times 300 \text{ K}}{28 \times 10^{-3} \text{ kg/mole}}\right)^{1/2}$$

$$= 517 \text{ m/sec}$$

Questions

1. Why can we neglect collisions of the molecules with the top, bottom, and sides of the box when calculating the pressure exerted on the end of the box?

2. How does $\bar{E}_k$ for H_2 molecules compare with $\bar{E}_k$ for O_2 molecules under standard conditions?

3. How does v_{rms} for H_2 molecules compare with v_{rms} for O_2 molecules under standard conditions?

2-3 Equipartition Theorem and Heat Capacities of Gases and Solids

Equation 2-7 can be written

$$\overline{E}_k = (\tfrac{1}{2}mv^2)_{av} = (\tfrac{1}{2}mv_x{}^2)_{av} + (\tfrac{1}{2}mv_y{}^2)_{av} + (\tfrac{1}{2}mv_z{}^2)_{av}$$
$$= \tfrac{3}{2}kT$$

But since $(v_x{}^2)_{av} = (v_y{}^2)_{av} = (v_z{}^2)_{av}$, we have

$$(\tfrac{1}{2}mv_x{}^2)_{av} = (\tfrac{1}{2}mv_y{}^2)_{av} = (\tfrac{1}{2}mv_z{}^2)_{av} = \tfrac{1}{2}kT \qquad 2\text{-}14$$

In equilibrium, the kinetic energy is shared equally among the three terms $\tfrac{1}{2}mv_x{}^2$, $\tfrac{1}{2}mv_y{}^2$, and $\tfrac{1}{2}mv_z{}^2$. This sharing is the natural consequence of collisions between molecules. Suppose we tried to increase the energy associated with motion in the x direction, $\tfrac{1}{2}mv_x{}^2$, without affecting the energy associated with the motion in the y or z direction. We could do this momentarily by replacing the wall perpendicular to the x direction with a movable piston. If we compress the gas by moving the piston, a molecule moving toward the piston with x component of velocity v_x will rebound with v_x' of greater magnitude so that the energy associated with motion in the x direction will be increased, with no change in that associated with motion in the y or z directions. But immediately after colliding with the piston, molecules collide with other nearby molecules and a new equilibrium is established with $\tfrac{1}{2}mv_x{}^2$, $\tfrac{1}{2}mv_y{}^2$, and $\tfrac{1}{2}mv_z{}^2$ each having the same average value $\tfrac{1}{2}kT$. (This average value is greater than before—thus the work done on the gas by the moving piston has increased the temperature of the gas.) This sharing of energy is a special case of the *equipartition theorem* which can be derived from statistical mechanics.

We call each coordinate, velocity component, angular-velocity component, etc., that appears squared in the expression for the energy of a molecule a *degree of freedom*. The equipartition theorem states that

In equilibrium, there is associated with each degree of freedom an average energy of $\tfrac{1}{2}kT$ per molecule.

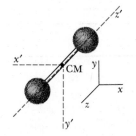

Figure 2-2
Rigid-dumbbell model of a diatomic gas molecule that can translate along the x, y, or z axis, and rotate about the x' or y' axis. If the spheres are smooth or are points, rotation about the z' axis can be neglected.

Equipartition theorem

As an example of the use of the equipartition theorem, consider a rigid-dumbbell model of a diatomic molecule (Figure 2-2) that can translate in the x, y, and z directions and can rotate about axes x' and y' through the center of mass and perpendicular to the z' axis along the line joining the two atoms.[1] The kinetic energy for this rigid-dumbbell-model molecule is then

$$E_k = \tfrac{1}{2}mv_x{}^2 + \tfrac{1}{2}mv_y{}^2 + \tfrac{1}{2}mv_z{}^2 + \tfrac{1}{2}I_{x'}\omega_{x'}{}^2 + \tfrac{1}{2}I_{y'}\omega_{y'}{}^2$$

where $I_{x'}$ and $I_{y'}$ are the moments of inertia about the x' and y'

---

[1] We rule out rotation about the z' axis of the dumbbell by assuming either that the atoms are points and the moment of inertia about this axis is therefore zero, or that the atoms are hard smooth spheres, in which case rotation about this axis cannot be changed by collisions and therefore does not participate in the exchange of energy. Either of these assumptions also rules out the possibility of rotation of a monatomic molecule.

axes. Since this molecule has 5 degrees of freedom, 3 translational and 2 rotational, the equipartition theorem predicts the
average energy to be $\frac{5}{2}kT$ per molecule. The energy per mole is
then $\frac{5}{2}N_A kT = \frac{5}{2}RT$ and the molar heat capacity at constant volume is $\frac{5}{2}R$. From the observation that C_v for both nitrogen and
oxygen is about $\frac{5}{2}R$, Clausius speculated (about 1880) that these
gases must be diatomic gases which can rotate about two axes as
well as translate.

If a diatomic molecule is not rigid, the atoms can vibrate
along the line of separation (Figure 2-3). Then, in addition to
the translational energy of the center of mass and rotational energy, there can be vibrational energy. The vibration adds two
more squared terms to the energy, one for the potential energy,
which is proportional to $(r - r_0)^2$, and one for kinetic energy
proportional to $(dr/dt)^2$, where r is the separation of the atoms
which has the value r_0 at equilibrium. For a diatomic molecule
that is translating, rotating, and vibrating, the equipartition
theorem predicts a molar heat capacity of $(3 + 2 + 2)\frac{1}{2}R$, or $\frac{7}{2}R$
(Figure 2-4).

Table 2-1 lists experimental values of C_v for several gases. For
all the diatomic molecules except Cl_2, these data are consistent
with the equipartition-theorem prediction assuming a rigid non-
vibrating molecule. The value for Cl_2 is about halfway between
that predicted for a rigid molecule and that predicted for a vibrating molecule. The situation for molecules with three or
more atoms is more complicated and will not be examined
in detail here.

It is difficult to understand why the equipartition theorem in
conjunction with the point-atom, rigid-dumbbell model is so successful in predicting the molar heat capacity for most diatomic
molecules but not for all of them. Why should not diatomic gas
molecules vibrate? If the atoms are not points, the moment of inertia about the line joining the atoms is not zero, and there are
three terms for rotational energy rather than two. Assuming no
vibration, C_v should be $\frac{6}{2}R$. This agrees with the measured value
for Cl_2 but not for the other diatomic gases. Furthermore, monatomic molecules would have three terms for rotational energy

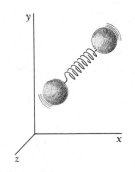

Figure 2-3
Nonrigid-dumbbell model
of a diatomic gas mole-
cule that can translate, rotate, and vibrate.

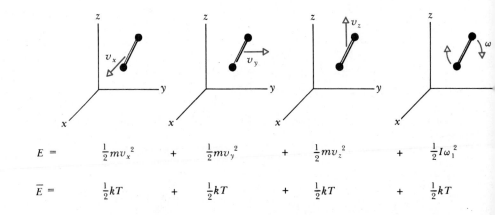

$$E = \quad \frac{1}{2}mv_x^2 \quad + \quad \frac{1}{2}mv_y^2 \quad + \quad \frac{1}{2}mv_z^2 \quad + \quad \frac{1}{2}I\omega_1^2$$

$$\overline{E} = \quad \frac{1}{2}kT \quad + \quad \frac{1}{2}kT \quad + \quad \frac{1}{2}kT \quad + \quad \frac{1}{2}kT$$

Table 2-1
Molar heat capacities of some gases at 15°C and 1 atm

Gas	C_v (cal/mole-deg)	C_v/R
Ar	2.98	1.50
He	2.98	1.50
CO	4.94	2.49
H_2	4.87	2.45
HCl	5.11	2.57
N_2	4.93	2.49
NO	5.00	2.51
O_2	5.04	2.54
Cl_2	5.93	2.98
CO_2	6.75	3.40
CS_2	9.77	4.92
H_2S	6.08	3.06
N_2O	6.81	3.42
SO_2	7.49	3.76

$$R = 1.987 \text{ cal/mole-deg}$$

From J. R. Partington and W. G. Shilling, *The Specific Heats of Gases,* London: Ernest Benn, Ltd., 1924.

if the atoms were not points, and C_v should be $\frac{6}{2}R$ for these atoms also. Since the average energy is calculated by *counting* terms, it should not matter how small the atoms are as long as they are not merely points.

In addition to these difficulties, the molar heat capacity is found to depend on temperature, contrary to the predictions from the equipartition theorem. The most spectacular case is that of H_2, shown in Figure 2-5. It seems as if at very low temperatures, H_2 behaves like a monatomic molecule and does not rotate. At very high temperatures H_2 begins to vibrate, but the molecule dissociates before C_v reaches $3.5R$. Other diatomic gases show similar behavior except that at low temperatures they liquefy before C_v reaches $1.5R$.

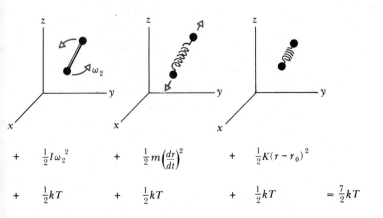

$$+ \quad \frac{1}{2}I\omega_2^2 \qquad\qquad + \quad \frac{1}{2}m\left(\frac{dr}{dt}\right)^2 \qquad\qquad + \quad \frac{1}{2}K(r - r_0)^2$$

$$+ \quad \frac{1}{2}kT \qquad\qquad + \quad \frac{1}{2}kT \qquad\qquad + \quad \frac{1}{2}kT \qquad\qquad = \frac{7}{2}kT$$

Figure 2-4
Energy modes of a diatomic molecule. With each of the seven possible motions there is associated an average energy $\frac{1}{2}kT$, giving a total energy of $\frac{7}{2}kT$ per molecule.

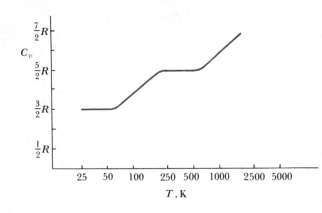

Figure 2-5
Temperature dependence of molar heat capacity of H_2. Between about 250 and 1000 K, C_v is $\frac{5}{2}R$, as predicted by the rigid-dumbbell model. At low temperatures, C_v is only $\frac{3}{2}R$, as predicted for a nonrotating molecule. At high temperatures C_v seems to be approaching $\frac{7}{2}R$, as predicted for a dumbbell model that rotates and vibrates, but the molecule dissociates before this plateau is reached.

The equipartition theorem is also useful in understanding the heat capacity of solids. In 1819, Dulong and Petit pointed out that the molar heat capacity of most solids was very nearly equal to 6 cal/K-mole $\approx 3R$. This result was used by them to obtain unknown molecular weights from the experimentally determined heat capacities. The Dulong-Petit law is easily derived from the equipartition theorem by assuming that the internal energy of a solid consists of the vibrational energy of the molecules (Figure 2-6). If the force constants in the x, y, and z directions are K_1, K_2, and K_3, the vibrational energy of each molecule is

Dulong-Petit law

$$E = \tfrac{1}{2}mv_x{}^2 + \tfrac{1}{2}mv_y{}^2 + \tfrac{1}{2}mv_z{}^2 + \tfrac{1}{2}K_1x^2 + \tfrac{1}{2}K_2y^2 + \tfrac{1}{2}K_3z^2$$

Since there are six squared terms, the average energy per molecule is $6(\tfrac{1}{2}kT)$, and the total energy of 1 mole is $3N_AkT = 3RT$, giving $C_v = 3R$.

At high temperatures, all solids obey the Dulong-Petit law. For temperatures below some critical value, C_v drops appreciably below the value of $3R$ and approaches zero as T approaches zero. The critical temperature is characteristic of the solid. It is lower for soft solids such as lead than for hard solids such as diamond. The general temperature dependence of C_v for solids is shown in Figure 2-7.

The fact that C_v for metals is not appreciably different from that for insulators is somewhat puzzling. A model of a metal that is moderately successful in describing electrical and heat conduction assumes that approximately one electron per atom is free to move about the metal, colliding with the atoms much as molecules do in a gas. According to the equipartition theorem, this "electron gas" should have an average energy of $\frac{3}{2}kT$ per electron; thus the molar heat capacity should be about $\frac{3}{2}R$ greater for a conductor than for an insulator. Although the molar heat capacity for metals is slightly greater than $3R$ at very high temperatures, the difference is much less than the $1.5R$ predicted for the contribution of the electron gas.

The failure of the kinetic theory in predicting heat capacities of gases and solids is not a failure of the model but rather a failure of classical mechanics. The search for an understanding of

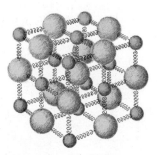

Figure 2-6
Simple model of a solid consisting of atoms connected to each other by springs. The internal energy of the solid then consists of kinetic and potential vibrational energy.

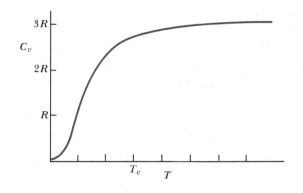

Figure 2-7
Temperature dependence of molar heat capacity of solids. At high temperatures C_v is $3R$, as predicted by the equipartition theorem. At low temperatures C_v approaches zero. The critical temperature at which C_v becomes nearly $3R$ is different for different solids.

specific heats was instrumental in the discovery of energy quantization in the beginning of the twentieth century. We shall see in the next chapter how energy quantization provides a basis for the complete understanding of the problems discussed in this section.

Question

4. Discuss the effect of molecular collisions on a gas which initially had half its molecules moving to the right and half moving to the left with the same speed v. Assume a random placement of molecules in a box with no initial v_y or v_z.

2-4 Distribution Functions

The calculation of the pressure of a gas gives us interesting information about the average square speed, and therefore the average energies, of the molecules in a gas, but it does not yield any details about the *distribution* of molecular velocities. Before we consider this problem in Section 2-5, we shall discuss the idea of distribution functions in general, with some elementary examples from common experience. This discussion should prove useful not only in the next section, but also in Chapter 6 where the same methods are applied to the more abstract probability distribution functions of quantum mechanics.

Suppose a teacher gave a 25-point quiz to a large number, N, of students. In order to describe the results of the quiz, he might give the average score or the median score, but this would not be a complete description. For example, if all N students received 12.5, this is quite a different result than if $N/2$ students received 25 and $N/2$ received 0, though both results have the same average. A complete description would be to give the number n_i who received the score s_i for all scores s_i between 0 and 25. An alternative would be to divide n_i by the total number of students, N, to give the fraction of the students, $f_i = n_i/N$, receiving the score s_i. Both n_i and f_i (which depend on the variable s) are called *distribution functions*. The fractional distribution, f_i, is slightly more convenient to use. The probability that one of the N students selected at random received the score s_i equals the number of students that received that score, $n_i = Nf_i$, divided by the total number N; thus this probability equals the distribution function f_i. Note that

$$\sum_i f_i = \sum_i \frac{n_i}{N} = \frac{1}{N} \sum_i n_i$$

and since

$$\sum_i n_i = N$$

we have

$$\sum_i f_i = 1 \qquad\qquad 2\text{-}15$$

Normalization condition

Equation 2-15 is called the *normalization condition* for fractional-distribution functions. A possible distribution function for a 25-point quiz is shown in Figure 2-8.

To find the average score, all the scores are added and the result is divided by N. Since each score s_i was obtained by $n_i = Nf_i$ students, this procedure is equivalent to

$$\bar{s} = \frac{1}{N} \sum_i s_i n_i = \sum_i s_i f_i \qquad\qquad 2\text{-}16$$

Mean or average value

We shall take Equation 2-16 as the definition of the *mean* (or *average*) score $\bar{s}$. Similarly, the average of any function $g(s)$ is defined by

$$\overline{g(s)} = \sum_i g(s_i) f_i \qquad\qquad 2\text{-}17$$

In particular, the mean square score is often useful:

$$\overline{s^2} = \sum_i s_i^2 f_i$$

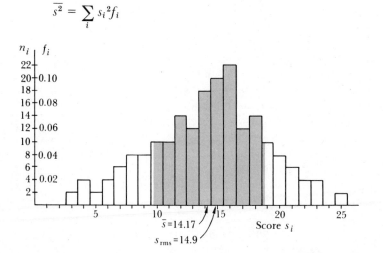

Figure 2-8
Grade distribution for a 25-point quiz given to 200 students; n_i is the number, and $f_i = n_i/N$ is the fraction, of students receiving the score s_i. The average score $\bar{s}$ and the root-mean-square score s_{rms} are indicated. The shaded area indicates the scores within 1 standard deviation of the mean.

A useful quantity characterizing a distribution is the *standard deviation*, σ, defined by

$$\sigma = \left[\sum_i (s_i - \bar{s})^2 f_i \right]^{1/2} \qquad \text{2-18}$$

Standard deviation

Note that

$$\sum_i (s_i - \bar{s})^2 f_i = \sum_i s_i^2 f_i + \bar{s}^2 \sum_i f_i - 2\bar{s} \sum_i s_i f_i = \overline{s^2} - \bar{s}^2$$

Therefore,

$$\sigma = (\overline{s^2} - \bar{s}^2)^{1/2} \qquad \text{2-19}$$

The standard deviation measures the spread of the values s_i about the mean. For most distributions there will be few values that differ from $\bar{s}$ by more than a few multiples of σ. In the case of the normal or Gaussian distribution, common in the theory of errors, about two-thirds of the values will lie within $\pm\sigma$ of the mean value. A Gaussian distribution is shown in Figure 2-9.

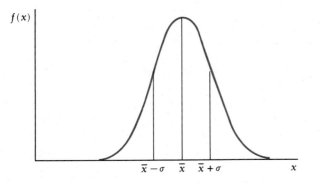

Figure 2-9
Gaussian or normal distribution curve. The curve is symmetrical about the mean value $\bar{x}$ which is also the most probable value. 68% of the area under the curve is within 1 standard deviation of the mean. This curve describes the distribution of random errors in many experimental situations.

If a student were selected at random from the class and one had to guess his score, the best guess would be the score obtained by the greatest number of students, called the most probable score, s_m. For the distribution in Figure 2-8, s_m is 16 and the average score, $\bar{s}$, is 14.17. The root-mean-square score, $s_{rms} = (\overline{s^2})^{1/2}$, is 14.9, and the standard deviation, σ, is 4.6. Note that 66 percent of the scores for this distribution lie within $\bar{s} \pm \sigma = 14.17 \pm 4.6$.

Now consider the case of a continuous distribution. Suppose we wanted to know the distribution of heights of a large number of people. For a finite number N, the number of persons *exactly* 6 ft tall would be zero. If we assume that height can be determined to any desired accuracy, there is an infinite number of possible heights, and the chance that anybody has a particular exact height is zero. We would therefore divide the heights into intervals Δh (for example, Δh could be 0.1 ft) and ask what fraction of people have heights that fall in any particular interval.

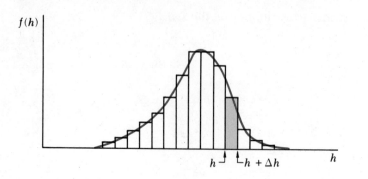

Figure 2-10
A possible height distribution. The fraction of the number of heights between h and $h + \Delta h$ is proportional to the shaded area. The histogram can be approximated by a continuous curve as shown.

This number depends on the size of the interval. We define the distribution function $f(h)$ as the fraction of the number of people with heights in a particular interval, divided by the size of the interval. Thus for N people, $Nf(h) \Delta h$ is the number of people whose height is in the interval between h and $h + \Delta h$. A possible height-distribution function is plotted in Figure 2-10. The fraction of people with heights in a particular interval is the area of the rectangle $\Delta h \times f(h)$. The total area represents the sum of all fractions; thus it must equal 1. If N is very large, we can choose Δh very small and still have $f(h)$ vary only slightly between intervals. The histogram $f(h)$ versus h approaches a smooth curve as $N \to \infty$ and $\Delta h \to 0$. In many cases of importance, the number of objects N is extremely large and the intervals can be taken as small as measurement allows. The distribution functions $f(h)$ are usually considered to be continuous functions, intervals are written dh, and the sums are replaced by integrals. For example, if $f(h)$ is a continuous function, the average height h is[1]

$$\bar{h} = \int hf(h)\, dh \qquad \qquad 2\text{-}20$$

Mean or average value

and the normalization condition expressing the fact that the sum of all fractions is 1 is

$$\int f(h)\, dh = 1 \qquad \qquad 2\text{-}21$$

Normalization condition

Example 2-2   The distribution function for lifetimes of radioactive nuclei is given by

$$f(t) = Ce^{-\lambda t} \qquad \qquad 2\text{-}22$$

where λ, called the *decay constant,* depends on the particular kind of nucleus (and the type of radioactivity). Assuming λ is known, find the constant C and the mean lifetime.

---

[1] The limits on the integration depend on the range of the variable. For this case, h ranges from 0 to ∞. We shall often omit explicit indication of the limits when the range of the variable is clear.

The fraction of lifetimes between t and $t + dt$ is $f(t)\,dt$. The fraction of lifetimes between $t = 0$ and $t = \infty$ must be 1; thus the normalization condition is

$$\int_0^\infty f(t)\,dt = \int_0^\infty Ce^{-\lambda t}\,dt = 1$$

The integral $\int_0^\infty e^{-\lambda t}\,dt$ has the value λ^{-1}. Then $C = \lambda$. Because the constant C is determined by the normalization condition, it is called the *normalization constant*. The mean lifetime is calculated by

$$\bar{t} = \int_0^\infty tf(t)\,dt = \lambda \int_0^\infty te^{-\lambda t}\,dt = \lambda^{-1}$$

The mean lifetime is the reciprocal of the decay constant.

Question

5. If the distribution function $f(x)$ is symmetric about the origin, i.e., if $f(-x) = f(x)$, the mean value of x is zero. Must the most probable value x_m also be zero?

2-5 The Maxwell-Boltzmann Distribution

The distribution function for molecular velocities was first obtained by James Clerk Maxwell in 1859. The problem can be stated as follows: Consider a gas consisting of N molecules confined to some volume V and in thermal equilibrium at temperature T. We wish to know how many molecules have their x component of velocity between v_x and $v_x + dv_x$, their y component between v_y and $v_y + dv_y$, and their z component between v_z and $v_z + dv_z$. We write this number as $NF(v_x,v_y,v_z)\,dv_x dv_y dv_z$. The problem is then to find the form of the distribution function $F(v_x,v_y,v_z)$.

Some insight into this problem can be gained by examining some simple distributions to see if they are possible solutions. Suppose first that all molecules are moving with the same speed, one-sixth of them in the positive x direction, one-sixth in the negative x direction, one-sixth in the positive y direction, etc. Place the molecules at random positions in the box at time zero. It is obvious that the molecules will collide and that many of the collisions will not be head on; thus their velocities will change and the original distribution will not persist. If we assume some model such as hard spheres for the molecules, we can calculate (statistically) what collisions will take place, and how the distribution changes as a result, from a knowledge of the original distribution. The *equilibrium* distribution is the one that remains unchanged by the collisions determined by the distribution.

Culver Pictures, Inc.

James Clerk Maxwell. (*Courtesy of Trinity College Library, Cambridge.*)

Maxwell assumed that the components v_x, v_y, and v_z were independent and that, therefore, the probabilities of a molecule having a certain v_x, v_y, v_z could be factored into the product of the probability of having v_x times the probability of having v_y times the probability of having v_z. He also assumed that the distribution could depend only on the speed, i.e., the velocity components could occur only in the combination $v_x^2 + v_y^2 + v_z^2$. He thus wrote

$$F(v_x, v_y, v_z) = f(v_x)f(v_y)f(v_z) \qquad\qquad 2\text{-}23$$

where $f(v_x)$ is the distribution function for v_x only, i.e., $f(v_x)\, dv_x$ is the fraction of the total number of molecules which have their x component of velocity between v_x and $v_x + dv_x$.[1]

We shall omit Maxwell's derivation and merely state his result. The form of $f(v_x)$ is

$$f(v_x) = Ce^{-mv_x^2/2kT} \qquad\qquad 2\text{-}24$$

with similar expressions for $f(v_y)$ and $f(v_z)$, where C is a constant determined by the normalization condition

$$\int_{-\infty}^{+\infty} f(v_x)\, dv_x = \int_{-\infty}^{+\infty} Ce^{-mv_x^2/2kT}\, dv_x = 1 \qquad\qquad 2\text{-}25$$

We shall need to evaluate integrals of the form

$$I_n = \int_0^\infty x^n e^{-\lambda x^2}\, dx \qquad\qquad 2\text{-}26$$

several times in this chapter. Table 2-2, derived in Appendix B, lists I_n for values of n from 0 to 5. Using this table to evaluate Equation 2-25 with $\lambda = m/2kT$ we find

$$C = \left(\frac{\lambda}{\pi}\right)^{1/2} = \left(\frac{m}{2\pi kT}\right)^{1/2} \qquad\qquad 2\text{-}27$$

---

[1] To avoid having to repeat this rather long phrase, we shall hereafter use the expression "the number in dv_x at v_x," or simply "the number in dv_x."

Table 2-2
Values of the integral $I_n = \displaystyle\int_0^\infty x^n e^{-\lambda x^2} dx$ for values of $n = 0$ to $n = 5$

n	I_n
0	$\frac{1}{2}\pi^{1/2}\lambda^{-1/2}$
1	$\frac{1}{2}\lambda^{-1}$
2	$\frac{1}{4}\pi^{1/2}\lambda^{-3/2}$
3	$\frac{1}{2}\lambda^{-2}$
4	$\frac{3}{8}\pi^{1/2}\lambda^{-5/2}$
5	λ^{-3}
If n is even:	$\displaystyle\int_{-\infty}^{+\infty} x^n e^{-\lambda x^2} dx = 2I_n$
If n is odd:	$\displaystyle\int_{-\infty}^{+\infty} x^n e^{-\lambda x^2} dx = 0$

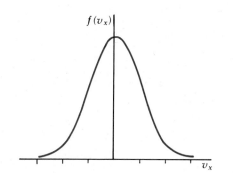

Figure 2-11
The distribution function
$f(v_x)$ for the x component of velocity. This is a
Gaussian curve symmetric
about the origin.

Substituting this result for C into Equation 2-24 we have

$$f(v_x) = \left(\frac{m}{2\pi kT}\right)^{1/2} e^{-mv_x{}^2/2kT} \qquad\qquad \text{2-28}$$

Figure 2-11 shows a sketch of $f(v_x)$ versus v_x. Of course, $f(v_x)$ is symmetric about the origin, $f(v_x) = f(-v_x)$, so the average of v_x is zero. As can be seen from the figure, the most probable v_x is also zero. The complete velocity distribution is

$$
\begin{aligned}
F(v_x, v_y, v_z) &= f(v_x)f(v_y)f(v_z) \\
&= \left(\frac{m}{2\pi kT}\right)^{3/2} e^{-m(v_x{}^2 + v_y{}^2 + v_z{}^2)/2kT} \qquad \text{2-29}
\end{aligned}
$$

*Maxwell velocity
distribution*

Example 2-3   Find the mean value of $v_x{}^2$. We have

$$(v_x{}^2)_{\text{av}} = \int_{-\infty}^{+\infty} v_x{}^2 f(v_x)\, dv_x$$

where $f(v_x)$ is given by Equation 2-24. Writing $\lambda = m/2kT$ we have

$$(v_x{}^2)_{\text{av}} = \int_{-\infty}^{+\infty} C v_x{}^2 e^{-\lambda v_x{}^2}\, dv_x$$

$$= 2C \int_0^{\infty} v_x{}^2 e^{-\lambda v_x{}^2}\, dv_x$$

with C given by Equation 2-27. This integral is of the form I_n given in Table 2-2 with $n = 2$. Then

$$(v_x{}^2)_{\text{av}} = 2CI_2 = 2\left(\frac{\lambda}{\pi}\right)^{1/2} \tfrac{1}{4}\pi^{1/2}\lambda^{-3/2}$$

$$= \tfrac{1}{2}\lambda^{-1} = \frac{kT}{m}$$

Note that this agrees with the equipartition theorem

$$(\tfrac{1}{2}mv_x{}^2)_{\text{av}} = \tfrac{1}{2}kT$$

The velocity distribution can be represented pictorially in what is called *velocity space*. Imagine the velocity vector of each molecule placed with its tail at the origin of a coordinate system v_x, v_y, v_z as in Figure 2-12. If we have N molecules, the

Velocity space

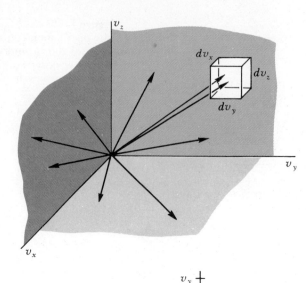

Figure 2-12
Velocity vectors in velocity space. The velocity distribution function gives the fraction of molecular velocities whose vectors end in a cell $dv_x \, dv_y \, dv_z$.

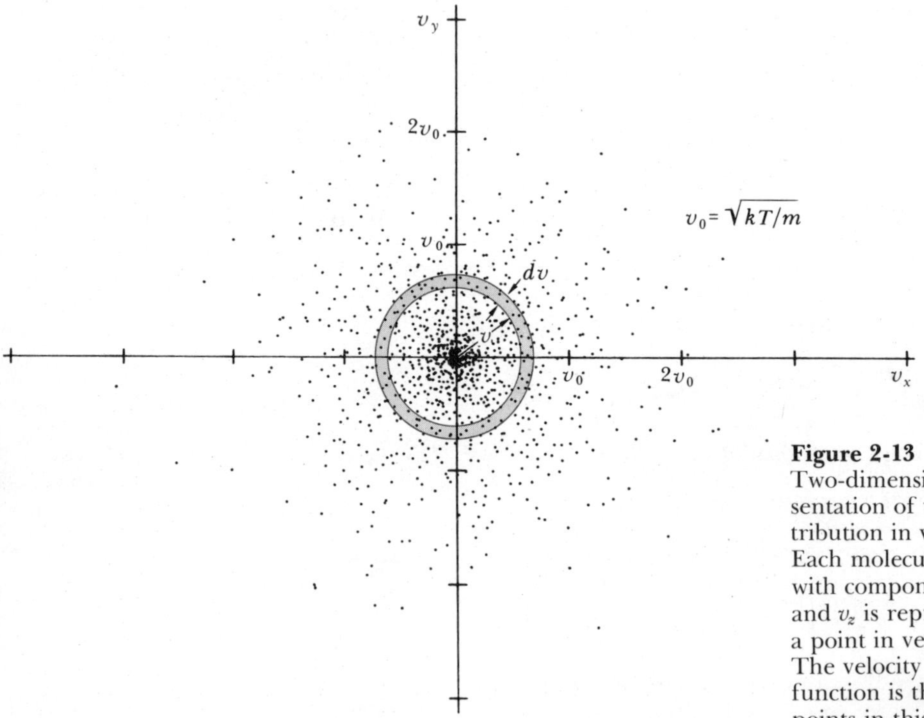

$$v_0 = \sqrt{kT/m}$$

Figure 2-13
Two-dimensional representation of velocity distribution in velocity space. Each molecular velocity with components v_x, v_y, and v_z is represented by a point in velocity space. The velocity distribution function is the density of points in this space. The density is maximum at the origin. The speed distribution is found by multiplying the density times the volume of the spherical shell $4\pi v^2 dv$. (*This computer-generated plot courtesy of Paul Doherty, Oakland University.*)

number of these vectors whose tips end in the "volume" element $dv_x \, dv_y \, dv_z$ is $NF(v_x,v_y,v_z) \, dv_x \, dv_y \, dv_z$. A simpler representation is shown in Figure 2-13. Here each molecular velocity with components v_x, v_y, and v_z is represented by a point in velocity space. The quantity $NF(v_x,v_y,v_z)$ is then the number of points per unit volume, i.e., the density in velocity space. The density is maximum at the origin since $F(v_x,v_y,v_z)$ has its maximum value there. Note that this density is spherically symmetric in velocity space; i.e., it depends only on the "distance" $v = (v_x{}^2 + v_y{}^2 + v_z{}^2)^{1/2}$, which is the molecular speed.

We can now calculate the *speed* distribution from the velocity distribution. Let $Ng(v) \, dv$ be the number of molecules with speeds between v and $v + dv$. In Figure 2-13 this is just the number of points in a spherical shell between v and $v + dv$. This number is the density $NF(v_x, v_y, v_z)$ times the volume of the shell $4\pi v^2\ dv$. Thus

$$Ng(v)\ dv = NF(v_x, v_y, v_z)4\pi v^2\ dv$$

or

$$g(v) = 4\pi \left(\frac{m}{2\pi kT}\right)^{3/2} v^2 e^{-mv^2/2kT} \qquad 2\text{-}30 \qquad \textit{Speed distribution}$$

The speed distribution is sketched in Figure 2-14. The most probable speed v_m, the mean speed $\bar{v}$, and the rms speed v_{rms} are indicated in the figure. Although the density function F is a maximum at the origin ($v = 0$), the speed distribution function $g(v)$ approaches zero as $v \to 0$ because the volume of a spherical shell, $4\pi v^2\ dv$, approaches zero. The factor $4\pi v^2$ thus shifts the maximum in the function $g(v)$ away from the origin to the value $v = v_m$ indicated in Figure 2-14. At very high speeds, the speed distribution function approaches zero because of the exponential factor $e^{-mv^2/2kT}$.

Example 2-4   Calculate the mean speed $\bar{v}$. We have

$$\bar{v} = \int_0^\infty vg(v)\ dv = \int_0^\infty Av^3 e^{-\lambda v^2}\ dv = AI_3$$

with $\lambda = m/2kT$ as before and $A = 4\pi(m/2\pi kT)^{3/2}$. Note that the integration ranges from 0 to ∞ rather than from $-\infty$ to $+\infty$ since speed is always positive. Using Table 2-2 for I_3 we have

$$\bar{v} = A\tfrac{1}{2}\lambda^{-2} = \frac{1}{2}\, 4\pi \left(\frac{m}{2\pi kT}\right)^{3/2}\left(\frac{2kT}{m}\right)^2$$

$$= \left(\frac{8kT}{\pi m}\right)^{1/2} \qquad 2\text{-}31$$

The mean speed is slightly less than $v_{rms} = (3kT/m)^{1/2}$, as indicated in Figure 2-14. The rms speed can be calculated directly from the speed distribution, or from the equipartition theorem

$$(\tfrac{1}{2}mv^2)_{av} = (\tfrac{1}{2}mv_x{}^2)_{av} + (\tfrac{1}{2}mv_y{}^2)_{av} + (\tfrac{1}{2}mv_z{}^2)_{av}$$
$$= \tfrac{3}{2}kT$$

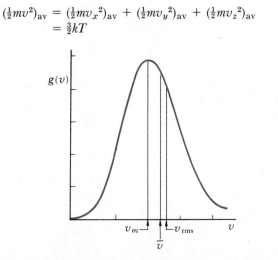

Figure 2-14
Maxwell speed distribution function $g(v)$. The most probable speed v_m, the mean speed $\bar{v}$, and the rms speed v_{rms} are indicated.

It is left as an exercise to show that the most probable speed is

$$v_m = \left(\frac{2kT}{m}\right)^{1/2} \qquad\qquad 2\text{-}32$$

The energy distribution function, $F(E)$, is the fraction of molecules with (kinetic) energies between E and $E + dE$. We can calculate the energy distribution by noting that

$$F(E)\,dE = g(v)\,dv$$

with $E = \tfrac{1}{2}mv^2$ and $dE = mv\,dv$. Thus

$$v^2\,dv = \frac{v\,dE}{m} = \left(\frac{2E}{m}\right)^{1/2}\frac{dE}{m}$$

The energy distribution is thus

$$F(E) \propto E^{1/2}e^{-E/kT} \qquad\qquad 2\text{-}33$$

Energy distribution

The proportionality constant can be determined by the normalization condition.

The first direct measurement of the speed distribution of molecules was made by O. Stern in 1926. Since then, measurements have been made by Zartman and Ko (1930); I. Estermann, O. C. Simpson, and O. Stern (1946); and Miller and Kusch (1955). These experiments employed various methods of selecting a range of speeds of molecules escaping from a small hole in an oven and determining the number of molecules in this range. Zartman and Ko, for example, allowed the beam to pass through a slit in a rotating cylinder and measured the intensity versus position on the collecting plate. In the more recent experiment of Miller and Kusch, illustrated in Figure 2-15, a collimated beam from the oven is aimed at a fixed detector. Most of the beam is stopped by a rotating cylinder. Small helical slits in

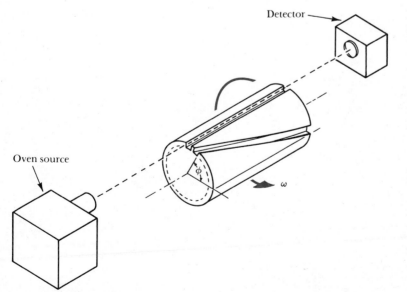

Detector

ω

Oven source

Figure 2-15
Schematic sketch of apparatus of Miller and Kusch for measuring the speed distribution of molecules. Only one of the 720 helical slits in the cylinder is shown. For a given angular velocity ω, only molecules of a certain speed from the oven pass through the helical slit to the detector. [*From R. C. Miller and P. Kusch,* Physical Review, **99**, *1314* (*1955*).]

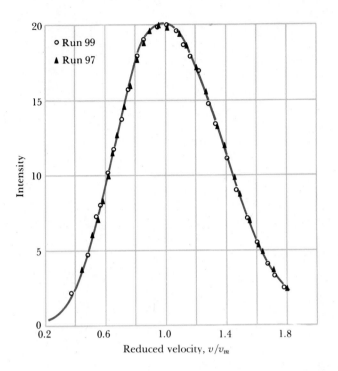

Figure 2-16
Data of Miller and Kusch showing the distribution of speeds of thallium atoms from an oven. The data have been corrected to give the distribution inside the oven. The solid curve is that predicted by the Maxwell speed distribution. [*From R. C. Miller and P. Kusch,* Physical Review, **99,** *1314 (1955).*]

the cylinder allow passage of those molecules in a narrow speed range determined by the angular velocity of the cylinder. The Miller and Kusch results are shown in Figure 2-16.

The velocity distribution for molecules in a gas (Equation 2-29) is a special case of the general Maxwell-Boltzmann distribution, which can be derived by the methods of statistical mechanics. Since this distribution is applicable to a wide variety of systems other than gases (for example, atoms in a solid), it will be stated in a more general form here. We consider a system of particles for which the energy E can be a function of the coordinates x, y, z and momenta p_x, p_y, and p_z.[1] The six-dimensional space x, y, z, p_x, p_y, p_z is called *phase space*. The probability of a particle being in a cell in phase space of "volume" $d\tau = dx\, dy\, dz\, dp_x\, dp_y\, dp_z$ is

$$f(x,y,z,p_x,p_y,p_z)\, d\tau = Ce^{-E/kT}\, d\tau \qquad \text{2-34}$$

Generalized Maxwell-Boltzmann distribution

where the constant C is determined by the normalization condition

$$\int Ce^{-E/kT}\, d\tau = 1 \qquad \text{2-35}$$

We can apply this to the case of an ideal gas by writing the energy

$$E = \frac{p_x^{\,2}}{2m} + \frac{p_y^{\,2}}{2m} + \frac{p_z^{\,2}}{2m}$$

[1] This distribution also holds if we interpret **p** to be angular momentum and the coordinates to be the corresponding angles. For example, p_z can be the z component of angular momentum, in which case z is the angle of rotation ϕ.

Since the energy does not depend on x, y, or z, we can integrate over these coordinates to obtain the total volume of the container V. (This factor will cancel a factor $1/V$ in the normalization constant C given by Equation 2-35.) We are left with a momentum distribution function which is essentially the same as the velocity distribution of Equation 2-29 except for the normalization constants.

We shall see in the next chapters that, in general, the energy of a system is not a continuous variable but takes on only a discrete set of values. (The energy often appears to be continuous because these discrete values are very close together.) It is useful therefore to have a statement of the Maxwell-Boltzmann distribution for the case of discrete energy states.

Given a system of particles for which the energy has a discrete set of values, the probability of a particle having energy E_i is

$$f_i = Cg_i e^{-E_i/kT} \qquad\qquad 2\text{-}36 \qquad \textit{Discrete distribution}$$

where the constant C is determined by the normalization condition and g_i is called the *statistical weight*. The statistical weight is the number of quantum states having the same energy value E_i. In our study of quantum mechanics and atomic physics in Chapters 6 and 7 we shall see how g_i is determined for various systems.

Example 2-5   The Law of Atmospheres   Consider an ideal gas in a uniform gravitational field. Find how the density of the gas depends on height above the ground. Let the force of gravity be in the negative z direction and consider a column of gas of cross-sectional area A. The energy of a gas molecule is then

$$E = \frac{p_x^2}{2m} + \frac{p_y^2}{2m} + \frac{p_z^2}{2m} + mgz = \frac{p^2}{2m} + mgz$$

where $p^2 = p_x^2 + p_y^2 + p_z^2$ and mgz is the potential energy of a molecule at height z above the ground.

From Equation 2-34 we have

$$f(p_x,p_y,p_z,x,y,z) = Ce^{-p^2/2mkT}\, e^{-mgz/kT}$$

Since we are interested only in the dependence on z, we can integrate over the other variables dx, dy, dp_x, dp_y, and dp_z. The integration merely gives a new normalization constant C', i.e., the result is equivalent to ignoring these variables. The fraction of the molecules between z and $z + dz$ is then

$$f(z)\, dz = C'e^{-mgz/kT}\, dz \qquad\qquad 2\text{-}37 \qquad \textit{Law of atmospheres}$$

The constant C' is obtained from the normalization condition $\int_0^\infty f(z)\, dz = 1$. The result is $C' = kT/mg$. We see that the density decreases exponentially with distance above the ground. This is known as the *law of atmospheres*.

Example 2-6   The first excited state of the hydrogen atom is 10.2 eV above the ground state. What is the ratio of the number of atoms in the first excited state to the number in the ground state at $T = 300$ K? We shall see in Chapter 7 that the statistical weights for these states are $g_1 = 2$ and $g_2 = 8$. The ratio is f_2/f_1, given by Equation 2-36, using $kT \approx 0.026$ eV:

$$\frac{f_2}{f_1} = \frac{g_2 e^{-E_2/kT}}{g_1 e^{-E_1/kT}} = \frac{g_2}{g_1} e^{-(E_2-E_1)/kT}$$
$$= 4e^{-(10.2/0.026)} = 4e^{-393} \approx 10^{-171}$$

We see that because of the great energy difference compared with kT, very few atoms are in the excited state.

Questions

6. If we derive the energy distribution for an ideal gas from the general Maxwell-Boltzmann distribution, we obtain a factor $E^{1/2}$ multiplying $e^{-E/kT}$. Where does this factor come from?

7. H_2 molecules can escape so freely from the earth's gravitational field that H_2 is not found in the earth's atmosphere. Yet the mean speed of H_2 molecules at ordinary atmospheric temperatures is much less than the escape speed. How then can H_2 molecules escape?

2-6 Transport Phenomena

In the calculation of the pressure exerted by a gas on its container the size of the molecules was not involved, and we could neglect the collision of the molecules with each other. We shall now consider the phenomena of viscosity, heat conduction, and diffusion, which depend directly on the size of gas molecules and on molecular collisions. The success of the application of kinetic theory to these phenomena provided one of the first convincing demonstrations of its essential validity, and consequently of the existence of molecules.

In the kinetic theory, viscosity involves the transport of momentum, heat conduction involves the transport of kinetic energy, and diffusion involves the transport of the density of the molecules. Molecular collisions play an important role in the transport of these quantities, and the frequency of collisions depends directly on the size of the molecules and the number of molecules per unit volume.

It is not difficult to see that if either Avogadro's number or the size of a molecule is known, the other can be estimated. Consider, for example, a solid in which the molecules are close together. If we assume each molecule to occupy a cube of side d, where d is the diameter of the molecule and also the distance between the centers of the molecules, the volume of 1 mole of the solid is $N_A d^3$. If $\mathcal{M}$ is its molecular weight, its density is

$$\rho = \frac{\mathcal{M}}{N_A d^3} \qquad\qquad 2\text{-}38$$

Either N_A or d can be found from a simple measurement of density if the other is known.[1] If neither is known, they can both be obtained if a second relationship can be found.

The comparison of the predictions of kinetic theory with macroscopic measurements of viscosity and heat conduction provided one of the first estimates of molecular sizes and of Avogadro's number. We shall consider here only the most elementary treatment of kinetic theory of transport phenomena.

An important quantity characterizing molecular collisions is the average distance a molecule travels between collisions. This distance is called the *mean free path*, ℓ. We should expect ℓ to depend inversely on the molecular size and the density of the gas. We can relate ℓ to the number density n and the diameter d as follows.

Mean free path

Consider one molecule moving with speed v through a region of stationary molecules of number density n (Figure 2-17). It will collide with another molecule if the centers are a distance $2r$, or d, apart. In time t the molecule moves a distance vt and collides with every molecule in the cylindrical volume $\pi d^2 vt$. The number of molecules in this volume is $n\pi d^2 vt$. (After each collision, the direction of the molecule changes; thus the path is really a zigzag one.) The total path length divided by the number of collisions is the mean free path:[2]

$$\ell \approx \frac{vt}{n\pi d^2 vt} = \frac{1}{n\pi d^2} \qquad \text{2-39}$$

The quantity πd^2 is the effective area presented by one molecule of diameter d to another of the same size. This area is called the *collision cross section*, σ. If the molecules are of different size the collision cross section would be $\pi(r_1 + r_2)^2$, where r_1 and r_2 are the radii. In terms of the collision cross section,

Cross section

$$\ell = \frac{1}{n\sigma} \qquad \text{2-40}$$

Let us now examine the phenomenon of viscosity. Consider a gas between two plates; the upper plate is pulled to the right

[1] The quantity d can now be accurately determined from x-ray diffraction measurements.

[2] Of course, the other molecules are not stationary. If we assume a Maxwell-Boltzmann distribution of velocities, the calculation is considerably more involved, with the result $\ell = 0.707/n\pi d^2$. For our purposes, we may neglect this and other corrections.

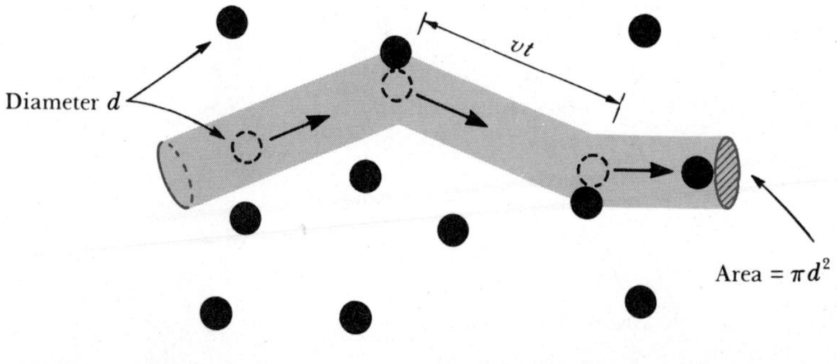

Diameter d

vt

Area $= \pi d^2$

Figure 2-17
Model of a molecule moving in a gas. In time t the molecule with diameter d will collide with any similar molecule whose center is in a cylinder of volume $\pi d^2 vt$, where v is the molecular speed. In this picture all the molecules but one are assumed to be at rest.

with a speed u_0, while the bottom plate is held stationary. It is found that the gas has a net flow to the right, the speed varying with height z from the bottom plate. Essentially, the gas near the top tends to follow the upper plate with speed u_0, whereas that near the bottom tends to remain at rest. The flow velocity $u(z)$ is superimposed on the random, or thermal, velocity of the molecules. A force is necessary to keep the upper plate moving with constant speed and to hold the bottom plate at rest. Evidently there is a drag force, called a *viscous force*, exerted by the gas. The coefficient of viscosity is defined as follows.

Consider a hypothetical plane surface of area A parallel to the plates at a height z_1 above the lower plate, as in Figure 2-18. The gas above this surface exerts a force to the right on the gas below, and of course the gas below exerts an equal but opposite force to the left on the gas above. This force is tangential to the plane and proportional to the area A and to the velocity gradient du/dz. The force per unit area is called the *viscous stress, S.*

$$S = \eta \frac{du}{dz} \qquad 2\text{-}41$$

This equation defines the coefficient of viscosity, η.

We shall consider qualitatively the kinetic-theory explanation of the force exerted on the gas below z_1 (shaded region in Figure 2-18). In this theory we consider each molecule to have an average "drift" velocity u to the right (superimposed on its thermal velocity) which is characteristic of the gas velocity u at the point of the last collision of the molecule. Molecules crossing the plane from above bring in x momentum mu_1 and those crossing from below carry away x momentum mu_2. Since the average momentum of those from above is greater than that from below, there is a net transfer of momentum across the plane. The net transfer of momentum across the plane per second per unit area equals the stress exerted on the lower gas. From this model we expect the net rate of transfer of momentum to be proportional to the rate at which molecules cross the plane, which is proportional to the number density n and the mean speed $\bar{v}$. It should also be proportional to the mean free path. For example, for molecules from above, the greater the distance to its last collision, the greater its average momentum mu. A detailed calculation gives for the viscous stress

$$S = \frac{1}{3}\, n\bar{v}\ell\, \frac{d(mu)}{dz} = \frac{1}{3}\, n\bar{v}\ell m\, \frac{du}{dz} \qquad 2\text{-}42$$

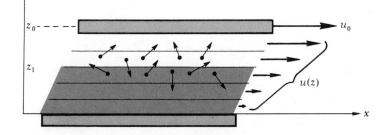

Figure 2-18
Viscous flow of a gas. Because of the relative motion of the plates, the gas between them has a flow velocity that varies from 0 at the bottom plate to u_0 at the top plate. Momentum is transferred from one layer to the other by molecules that cross the boundary.

so the coefficient of viscosity is

$$\eta = \tfrac{1}{3}n\bar{v}\ell m \qquad\qquad\qquad 2\text{-}43$$

where m is the mass of the molecule. Substituting $\ell = 1/n\pi d^2$ from Equation 2-39 we have

$$\eta = \frac{1}{3}\frac{m\bar{v}}{\pi d^2} \qquad\qquad\qquad 2\text{-}44$$

Note that this expression is independent of density. This surprising result was first pointed out by Maxwell and verified by him experimentally over a wide range of densities. (At extremely low densities, this theory breaks down when the mean free path becomes of the order of the size of the container.) Equation 2-44 also implies that η depends on temperature only through $\bar{v}$, which increases as $T^{1/2}$. The experimental results, so easily understood by a simple kinetic-theory model, that the viscosity of gas is independent of density (and therefore independent of pressure at constant T) and increases with temperature (rather than decreasing, as is the case for liquids), were an important factor in the general acceptance of the kinetic theory in the nineteenth century.

If we write $m = \mathcal{M}/N_A$ for the mass of a molecule, and use the result $\bar{v} = (8RT/\pi\mathcal{M})^{1/2}$ from the Maxwell distribution, we can write for the coefficient of viscosity

$$\eta = \frac{\mathcal{M}}{3\pi N_A d^2}\left(\frac{8RT}{\pi\mathcal{M}}\right)^{1/2} \qquad\qquad 2\text{-}45$$

Loschmidt in 1885 used this result along with Equation 2-38 and measurements of the viscosities of gases and of the densities of solids to obtain the first reliable estimate of N_A and d. (The number N_A is often called "Loschmidt's number" in Europe.) He obtained $d \approx 10^{-10}$ m and $N_A \approx 10 \times 10^{23}$. This is reasonably close to the modern value $N_A = 6.022 \times 10^{23}$.

Table 2-3
Some values of molecular mean free paths, and molecular radii computed from viscosity measurements

Gas	$\eta(15°C)$ (newton-sec/m)	$\ell(15°C, 1\ \text{atm})$ (Å)	r (Å)
He	1.94×10^{-6}	1860	1.09
Ne	31.0	1320	1.30
Ar	22.0	666	1.82
H_2	8.71	1180	1.37
N_2	17.3	628	1.88
O_2	20.0	679	1.80
CO_2	14.5	419	2.30
NH_3	9.7	451	2.22
CH_4	10.8	516	2.07

From J. F. Lee, F. W. Sears, and D. L. Turcotte, *Statistical Thermodynamics*, Reading, Mass.: Addison-Wesley Publishing Company, Inc., 1963.

If we use the modern value of N_A, we can compute the molecular radius[1] from viscosity measurements using Equation 2-44 and the mean free path using Equation 2-39. Table 2-3 lists the results for several gases. From this table we see that molecular radii are about 1 or 2 Å (1 Å = 10^{-10} m), and at normal densities the mean free paths are several hundred times this.

Example 2-7 What is the order of magnitude of the time between collisions for N_A molecules in a gas at standard conditions? Let τ be the time between collisions. The average distance traveled by a molecule in this time is $\ell = v\tau$. Using $\ell \approx$ 600 Å = 600×10^{-10} m from Table 2-3, and $v \approx 500$ m/sec from the calculation of v_{rms} in Example 2-1, we have

$$\tau = \frac{\ell}{v} \approx \frac{600 \times 10^{-10} \text{ m}}{500 \text{ m/sec}} \approx 10^{-10} \text{ sec}$$

The collision frequency is of the order of $1/\tau \approx 10^{10}$ collisions per second.

The treatment of heat conduction is similar to that of viscosity except that we consider the transport of molecular energy rather than of momentum. Consider the plates shown in Figure 2-18 to be at rest and at different temperatures. If ΔQ is the heat conducted across area A in time Δt, it is found that ΔQ is proportional to A, Δt, and the temperature gradient dT/dz. The coefficient of heat conduction, K, is defined by

$$\frac{\Delta Q}{A\,\Delta t} = K\,\frac{dT}{dz} \qquad\qquad 2\text{-}46$$

We can use the same analysis that we used for viscosity if we replace the momentum mu by the average energy per molecule $\overline{E}$. Molecules crossing the plane from above transport more energy than those from below if the upper plate is at a higher temperature. Equation 2-42 then becomes for the case of heat conduction

$$\frac{\Delta Q}{A\,\Delta t} = \frac{1}{3}\,n\bar{v}\ell\,\frac{d\overline{E}}{dz} \qquad\qquad 2\text{-}47$$

If we multiply the average energy per molecule by N_A, we obtain the energy per mole. Thus $\overline{E}N_A = C_v T$ and

$$\frac{\Delta Q}{A\,\Delta t} = \frac{1}{3}\,\frac{n\bar{v}\ell C_v}{N_A}\,\frac{dT}{dz} \qquad\qquad 2\text{-}48$$

The coefficient of heat conduction is therefore

$$K = \frac{1}{3}\,\frac{n\bar{v}\ell C_v}{N_A} \qquad\qquad 2\text{-}49$$

---

[1] It should be pointed out that we are not implying that molecules are spherical. It is the collision cross section that is determined from Equation 2-44. By radius, we mean the quantity related to the collision cross section by $\sigma = \pi d^2 = 4\pi r^2$.

Comparing with Equation 2-43 for η, we have

$$\frac{K}{\eta} = \frac{C_v}{N_A m} = \frac{C_v}{\mathcal{M}}$$

or

$$\frac{K \mathcal{M}}{\eta C_v} = 1 \qquad\qquad 2\text{-}50$$

The experimental determination of this ratio yields numbers between about 1.5 and 2.5 for most gases. The agreement within a factor of 3 of theory and experiment is another success of the kinetic model, for there is little reason from the macroscopic point of view to suspect heat conduction and viscosity to be simply related. The discrepancy is due to the oversimplification of the theory.

The coefficient of self-diffusion[1] is defined by

$$\frac{\Delta n}{A \, \Delta t} = D \frac{dn}{dz} \qquad\qquad 2\text{-}51$$

where Δn is the number of molecules crossing the plane of area A in time Δt. In this case it is the number of molecules that varies, leading to the transport of molecules. The simple theory gives

$$D = \tfrac{1}{3}\ell v \qquad\qquad 2\text{-}52$$

We should note that the numerical factors such as the factor $\tfrac{1}{3}$ in the result for the viscosity, heat conduction, and diffusion coefficients come from the simplest kinetic-theory calculations and are often modified by more detailed treatment. In the case of mutual diffusion, or diffusion of large objects through a gas or liquid, this simple mean-free-path treatment is not even an adequate starting point.

Question

8. If we double the number density n, twice as many molecules cross a given area per second. Does this double the rate of heat conduction? Why or why not?

---

[1] Self-diffusion is the diffusion of molecules into others of the same kind because of a density difference. Restricting the problem to like molecules simplifies the calculations because all the collision cross sections are the same. Experimentally, self-diffusion can be observed by using radioactive-tracer methods to tag certain molecules without changing their collision cross sections.

Optional

2-7   Brownian Motion and the Random-Walk Problem

In 1828 a botanist, Robert Brown, observed an irregular zigzag motion of pollen grains suspended in water. After much experimentation, he concluded that the cause of the motion was not organic, for he observed it in a wide variety of materials. This motion, now called *Brownian motion*, went unexplained for nearly half a century, until the kinetic theory was developed. (Many thought that the motion was due to convection currents or vibrations transmitted through the liquid.) The true cause of Brownian motion, the irregular bombardment of the grains by the molecules of the suspending fluid, was finally understood at the beginning of the twentieth century. The first complete theory was given by Einstein in 1905. In 1908, Jean Perrin made exhaustive quantitative observations of the paths of many suspended particles of different sizes. From these observations, which were in good agreement with Einstein's theory, Perrin calculated Avogadro's number. Perrin's monumental work finally laid aside all doubts as to the validity of the kinetic theory of matter.

The Brownian motion of suspended particles is similar to the diffusion of molecules, except that the particles are much larger than the molecules. We can get some insight into a number of statistical processes, such as diffusion, Brownian motion, and the combination of errors, by considering a simple statistical problem called the *random-walk problem*. In the one-dimensional version of this problem, a man flips a coin and takes one step forward if the result is heads or one step backward if the result is tails. We are interested in determining how far the man gets from the starting point on the average. Suppose the man takes N steps, each of size unity. After N steps, the man has a displacement x_N from the origin. Since the probabilities of a forward step and a backward step are equal, the average displacement $\bar{x}_N$ will be zero. We shall now show that the rms distance from the origin after N unit steps equals $N^{1/2}$.

After one step the displacement is

$$x_1 = \pm 1$$

Squaring, we obtain $x_1^2 = +1$. After two steps, the displacement is

$$x_2 = x_1 \pm 1$$

Squaring, we obtain

$$x_2^2 = x_1^2 \pm 2x_1 + 1$$

When we take the average, the middle term drops out because $(x_1)_{av} = 0$. Then

$$(x_2^2)_{av} = (x_1^2)_{av} + 1 = 2$$

If we continue, we find that $(x_3^2)_{av} = (x_2^2)_{av} + 1 = 3$, and so on. Therefore, after N steps we have

$$(x_N^2)_{av} = N \qquad\qquad 2\text{-}53$$

If the step size is ℓ, the above argument gives

$$(x_N^2)_{av} = N\ell^2 \qquad\qquad 2\text{-}54$$

and

$$x_{rms} = N^{1/2}\ell \qquad\qquad 2\text{-}55$$

This result can be applied to error theory. The probable resulting error due to the combination of a large number, N, of small random errors of size ℓ is given by Equation 2-55. We can also relate this to the problem of self-diffusion by taking N to be the number of collisions made and ℓ to be the mean free path. If the mean speed of the molecules is $\bar{v}$, the number of collisions made in time t is $N = \bar{v}t/\ell$; so

$$(x^2)_{\text{av}} = \bar{v}\ell t$$

Thus the mean-square distance is proportional to the time.

Since Brownian motion of a suspended particle is the result of many small irregular movements due to random molecular bombardment, the mean-square distance for this motion is also proportional to the number of collisions made by the particle and therefore to the time.

In 1905 Einstein applied kinetic theory to the calculation of the rms displacement of a large sphere of radius a undergoing Brownian motion in a gas of viscosity η. His result was

$$(x^2)_{\text{av}} = \frac{RT}{3\pi\eta a N_A}\, t \qquad\qquad 2\text{-}56$$

Einstein pointed out that this result could be used to obtain Avogadro's number.

In 1908 Jean Perrin made a series of remarkable measurements of Avogadro's number. In order to use Equation 2-56, he needed a large number of small but visible particles of equal radius a. He found that he could make emulsions of gamboge (prepared from a dried vegetable latex) and mastic which, after several months of separation by centrifuging, contained grains of nearly equal size.[1] In one series of measurements, he watched individual particles as they moved about and recorded their positions at equal time intervals. He verified that the mean-square displacement was proportional to the time and determined N_A. Figure 2-19 is a diagram of the horizontal projections of the positions of a grain with radius 0.53×10^{-6} m observed at intervals of 30 sec. The following quotation is taken from Perrin's Nobel Prize address in 1926.

[1] An interesting account of Perrin's experiments can be found in Jean Perrin, *Atoms*, New York: D. Van Nostrand Company, Inc., 1923.

Figure 2-19
Brownian motion. Points indicate successive positions of a particle observed at 30-sec intervals. The lines between the points are added to indicate the sequence of positions; the particle does not move in straight lines between observations because it is struck by millions of molecules each second. (*From Jean Perrin*, Atoms, *trans. D. Hammick, New York: D. Van Nostrand Company, Inc., 1923.*)

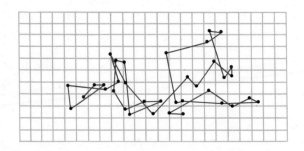

Wide World Photos

Jean-Baptiste Perrin.

These theories can be judged by experiment if we know how to *prepare spherules of a measurable radius.* I was, therefore, in a position to attempt this check as soon as I knew, thanks to Langevin, of the work of Einstein.

I must say that, right at the beginning, Einstein and Smoluchovski had pointed out that the order of magnitude of the Brownian movement seemed to correspond to their predictions. And this approximate agreement gave already much force to the kinetic theory of the phenomenon, at least in broad outline.

It was impossible to say anything more precise so long as spherules of known size had not been prepared. Having such grains, I was able to check Einstein's formulae by seeing whether they led always to the same value for Avogadro's number and whether it was appreciably equal to the value already found.

This is obtained for the displacements by noting on the camera lucida (magnification known) the horizontal projections of the same grain at the beginning and at the end of an interval of time equal to the duration chosen, in such a manner as to measure a large number of displacements, for example, in one minute.

In several series of measurements I varied, with the aid of several collaborators, the size of the grains (in the ratio of 1 to 70,000) as well as the nature of the liquid (water, solutions of sugar or urea, glycerol) and its viscosity (in the ratio of 1 to 125). They gave values between 55×10^{22} and 72×10^{22}, with differences which could be explained by experimental errors. The agreement is such that it is impossible to doubt the correctness of the kinetic theory of the translational Brownian movement.[1]

In another series of measurements, Perrin determined N_A by measuring the density of particles suspended in an emulsion at different heights. If a fluid is in a uniform gravitational field in the negative z direction, the number density is given by the law of atmospheres (Equation 2-37)

$$n(z) = n_0 e^{-mgz/kT} = n_0 e^{-Mgz/RT} \qquad \text{2-57}$$

where n_0 is the number at $z = 0$. If a visible particle is suspended in the fluid, its tendency to sink because of gravity is counteracted by a tendency to rise because it is struck by more molecules from below than from above (due to the greater density below, as shown by Equation 2-57). The equilibrium distribution of the visible particles is given by Equation 2-57 with $M = N_A m$, where m is the mass of the particle (Figure 2-20). Thus N_A can be determined by measuring the mass of the particles and the number versus height.

Perrin also measured the rotation of particles in a fluid due to bombardment by molecules, and calculated N_A from the theory of rotational Brownian motion given by Einstein.

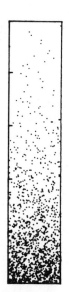

Figure 2-20
Equilibrium height distribution of particles in a gravitational field. The distribution is the same as that given by the law of atmospheres,
$n(z) = n_0 e^{-mgz/kT}$. *(This computer-generated plot courtesy of Paul Doherty, Oakland University.)*

[1] From *Nobel Prize Lectures: Physics,* Amsterdam and New York: Elsevier Publishing Company, 1964.

Summary

The number of molecules in a mole is Avogadro's number $N_A = 6.022 \times 10^{23}$, which is the reciprocal of the unified mass unit expressed in grams.

A simple model assuming that the pressure of a gas is due to collisions of molecules with the walls of a container implies that the mean kinetic energy of the molecules is proportional to the temperature of the gas.

The probability of occurrence of a value of x in the range dx is given by $f(x)\,dx$, where $f(x)$ is the distribution function. Such a function obeys the normalization condition

$$\int f(x)\,dx = 1$$

The Maxwell-Boltzmann velocity distribution is

$$F(v_x, v_y, v_z) = \left(\frac{m}{2\pi kT}\right)^{3/2} e^{-m(v_x^2 + v_y^2 + v_z^2)/2kT}$$

and the speed distribution is

$$g(v) = 4\pi v^2 \left(\frac{m}{2\pi kT}\right)^{3/2} e^{-mv^2/2kT}$$

The Maxwell-Boltzmann energy distribution is

$$F(E) = C E^{1/2} e^{-E/kT}$$

where C is determined by normalization.

The mean translational kinetic energy of gas molecules is $\frac{3}{2}kT$, independent of any characteristics of the molecules. This is an example of the equipartition theorem: there is a mean energy of $\frac{1}{2}kT$ associated with each squared coordinate or momentum in the expression for the energy of a molecule. Molecular speeds are of the order of magnitude of the speed of sound.

An elementary theory of transport yields similar expressions for the coefficients of viscosity, heat conduction, and diffusion in terms of the molecular density, mean speed, and mean free path between collisions. The mean free path varies inversely with density and with the square of the molecular diameter. In particular, this theory predicts that the coefficient of viscosity is independent of density and is proportional to the square root of the temperature, in agreement with experiment. From this theory, and macroscopic measurements of viscosity, the first estimates of molecular size and Avogadro's number were made.

In the random-walk problem, the rms distance is proportional to the square root of the number of steps. This problem is useful in vizualizing the processes of diffusion and Brownian motion. Since the number of molecular collisions is proportional to time, the rms distance for diffusion or Brownian motion is proportional to the square root of the time. Detailed observations of the position of particles in colloidal suspension allowed Perrin to make the first accurate measurements of Avogadro's number and to verify directly and quantitatively the predictions of kinetic theory.

References

1. G. Holton, *Introduction to Concepts and Theories in Physical Science*, Reading, Mass.: Addison-Wesley Publishing Company, Inc., 1952.

2. H. Boorse and L. Motz (eds.), *The World of the Atom*, New York: Basic Books, Inc., Publishers, 1966. This two-volume, 1873-page work is a collection of original papers translated and edited. Much of the work referred to in this chapter and throughout this text can be found in these volumes. Of particular interest for this chapter are the papers by Boyle, Hooke, Bernoulli, Dalton, Gay-Lussac, Avogadro, Herapath, Brown, Waterston, Joule, and Maxwell.

3. W. Niven (ed.), *Scientific Papers of James Clerk Maxwell*, New York: Dover Publications, Inc., 1965.

4. J. Perrin, *Atoms*, trans. D. Hammick, New York: D. Van Nostrand Company, Inc., 1923. In this short book, Perrin describes in detail his experiments with suspended particles.

5. A. Einstein, *Investigations on the Theory of the Brownian Movement*, ed. R. Furth, trans. A. D. Cowper, New York: Dover Publications, Inc., 1965. This book contains five papers on Brownian motion by Einstein.

6. L. L. Whyte, *Essay on Atomism from Democritus to 1960*, Middletown, Conn.: Wesleyan University Press, 1961. This book contains a chronological table with comments on the development of atomistic ideas.

7. F. Friedman and L. Sartori, *The Classical Atom*, Reading, Mass.: Addison-Wesley Publishing Company, Inc., 1965. A concise, sophisticated (but using elementary mathematics) treatment of the topics discussed in this and the next two chapters.

8. J. F. Lee, F. W. Sears, and D. L. Turcotte, *Statistical Thermodynamics*, Reading, Mass.: Addison-Wesley Publishing Company, Inc., 1963.

9. R. Present, *Kinetic Theory of Gases*, New York: McGraw-Hill Book Company, 1958.

References 8 and 9 are intermediate-level textbooks.

Exercises

Section 2-1, Avogadro's Number

1. The molecular weight of O_2 is 32.0 g/mole. Find the mass of an O_2 molecule.

2. The quantity $N_A e$ is called the faraday, where $e = 1.60 \times 10^{-19}$ coulomb is the electron charge. Find the number of coulombs in a faraday.

Section 2-2, The Pressure of a Gas

3. Show that the SI units of $(3RT/\mathcal{M})^{1/2}$ are m/sec.

4. (a) Find the total kinetic energy of translation of 1 mole of N_2 molecules at $T = 273$ K. (b) Would your answer be the same, greater, or less for 1 mole of He atoms at the same temperature?

✓5.   (a) Calculate v_{rms} for H_2 at $T = 300$ K. (b) Calculate the temperature T for which v_{rms} for H_2 equals the escape speed of 11.2 km/sec.

6.   (a) The ionization energy for hydrogen atoms is 13.6 eV. At what temperature is the average kinetic energy of translation equal to 13.6 eV? (b) What is the average kinetic energy of translation of hydrogen atoms at $T = 10^7$ K, a typical temperature in the interior of the sun?

Section 2-3, Equipartition Theorem and Heat Capacities of Gases and Solids

7.   For an ideal gas, the molar heat capacity at constant pressure C_p is related to that at constant volume C_v by $C_p = C_v + R$. A quantity that is often measured is the ratio $\gamma = C_p/C_v$. Find C_p and γ for an ideal gas which consists of (a) spherical atoms that do not rotate, (b) rigid-dumbbell-shaped molecules, (c) nonrigid-dumbbell-shaped molecules that vibrate as well as rotate.

8.   The measured value of γ for air is 1.4. Which of the models in Exercise 7 best describes air molecules?

9.   A monatomic gas is confined to move in two dimensions so that the energy of a molecule is $E_k = \frac{1}{2}mv_x^2 + \frac{1}{2}mv_y^2$. What are C_v, C_p, and γ for this gas? (See Exercise 7.)

10.   Use the Dulong-Petit law that $C_v = 3R$ for solids to calculate the specific heat $c_v = C_v/\mathcal{M}$ in cal/mole for (a) aluminum, $\mathcal{M} = 27.0$ g/mole, (b) copper, $\mathcal{M} = 63.5$ g/mole, and (c) lead, $\mathcal{M} = 207$ g/mole, and compare your results with values given in a handbook.

Section 2-4, Distribution Functions

✓11.   A class of 50 students was given a 10-point quiz. The grade distribution was:

Score s_i: 10 9 8 7 6   5 4 3 2 1 0
Number n_i: 1 4 8 7 6 15 6 0 3 0 0

Find the mean grade and the standard deviation. Indicate these on a histogram plot of this distribution.

12.   The speed of sound is measured in an elementary laboratory by students A and B. Each takes nine readings. The data are:

A: 345, 350, 338, 340, 335, 334, 346, 342, 330 m/sec
B: 340, 350, 345, 330, 325, 360, 320, 355, 335 m/sec

Compute the mean value and the standard deviation for each set.

13.   The distribution function for some quantity x is given by

$f(x) = C$ for $|x| < a$ where C is a constant

$f(x) = 0$ for $|x| > a$

Find C, $\bar{x}$, x_{rms}, and σ, and indicate these quantities on a sketch of $f(x)$.

✓14.   The distribution function for some positive quantity x is

$f(x) = Ax$ for $x < x_0$

$f(x) = 0$ for $x > x_0$

Find A, $\bar{x}$, x_{rms}, and σ. Indicate $\bar{x}$, x_{rms}, and σ on a sketch of $f(x)$.

Section 2-5, The Maxwell-Boltzmann Distribution

15. Neutrons in a reactor have a Maxwell-Boltzmann velocity distribution when they are in thermal equilibrium. Find $\bar{v}$ and v_m for neutrons in thermal equilibrium at $T = 300$ K.

16. Show that $g(v)$ (Equation 2-30) has its maximum value at $v = v_m = (2kT/m)^{1/2}$.

17. (a) Show that Equation 2-28 can be written

$$f(v_x) = (2\pi)^{-1/2} v_0^{-1} e^{-v_x^2/2v_0^2}$$

where $v_0 = v_{x,\text{rms}} = (kT/m)^{1/2}$. Consider 1 mole of gas and approximate dv_x by $\Delta v_x = 0.01 v_0$. Find the number of molecules in Δv_x at (b) $v_x = 0$, (c) $v_x = v_0$, (d) $v_x = 2v_0$, (e) $v_x = 8v_0$.

18. Calculate the mean value of the reciprocal speed $(\overline{1/v})$ from the speed distribution $g(v)$.

19. (a) Show that the velocity distribution function (Equation 2-29) can be written

$$F(v_x, v_y, v_z) = \pi^{-3/2} v_m^{-3} e^{-(v/v_m)^2}$$

where $v_m = (2kT/m)^{1/2}$ is the most probable speed and $v^2 = v_x^2 + v_y^2 + v_z^2$. Consider 1 mole of molecules and approximate dv_x, dv_y, and dv_z by $\Delta v_x = \Delta v_y = \Delta v_z = 0.01 v_m$. Find the number of molecules in a cell centered at (b) $v_x = v_y = v_z = 0$, (c) $v_x = v_y = v_z = v_m$.

20. Calculate the most probable energy E_m from the energy distribution (Equation 2-33) and indicate it on a sketch of $F(E)$ versus E.

21. (a) Show that the speed distribution function can be written $g(v) = 4\pi^{-1/2} (v/v_m)^2 v_m^{-1} e^{-(v/vm)^2}$, where v_m is the most probable speed. Consider 1 mole of molecules and approximate dv by $\Delta v = 0.01\, v_m$. Find the number of molecules with speeds in dv at (b) $v = 0$, (c) $v = v_m$, (d) $v = 2v_m$, (e) $v = 8v_m$.

22. From the absorption spectrum it is determined that about one out of 10^6 hydrogen atoms in a certain star is in the first excited state, 10.2 eV above the ground state (other excited states can be neglected). What is the temperature of the star? (Take the ratio of statistical weights to be 4, as in Example 2-6.)

23. The first rotational energy state of the H_2 molecule ($g_2 = 3$) is about 4×10^{-3} eV above the lowest energy state ($g_1 = 1$). What is the ratio of the numbers of molecules in these two states at room temperature (300 K)?

Section 2-6, Transport Phenomena

24. One mole of a gas occupies a volume of 22.4 liters = 22.4×10^{-3} m³ under standard conditions. Using $N_A L^3$ for this volume, find the order of magnitude of the molecular separation of the molecules L. Compare this with the size of the molecules and with their mean free path.

25. Use the known value of N_A to estimate d from Equation 2-38 for (a) aluminum with density 2.7 g/cm³ and molecular weight 27 g/mole, (b) lead with density 11.2 g/cm³ and molecular weight 207 g/mole, (c) copper with density 8.96 g/cm³ and molecular weight 63.5 g/mole.

26. In the classical theory of electrical conduction, the free electrons in a metal are similar to molecules in a gas except that the electrons are treated as points which collide with the lattice ions of the metal. (*a*) Calculate the number density of copper atoms from $n = N_A \rho / M$, where $\rho = 8.96$ g/cm^3 and $M = 63.5$ g/mole. (*b*) Estimate the mean free path for electrons using πr^2 for the collision cross section and reasonable values of r for copper from Exercise 25.

Section 2-7, Brownian Motion and the Random-Walk Problem

✓27. A man has $20 and bets $1 at a time on either red or black on the roulette wheel, for which his odds of winning are approximately even (we neglect the slight advantage of the house resulting from the green 0 and 00, which do not pay off). He decides to play until he has either lost the $20 or won an additional $20. (*a*) About how many times can he play? (*b*) What is his average expected winning?

Problems

✓1. Given the distribution function

$$f(x) = A \sin^2(\pi x/L) \qquad \text{for } 0 < x < L$$

$$f(x) = 0 \qquad\qquad\quad \text{for other } x$$

find $A, \bar{x}, x_m$, and x_{rms} and sketch $f(x)$ versus x.

2. Given the distribution function

$$f(x) = A \sin^2(2\pi x/L) \qquad \text{for } 0 < x < L$$

$$f(x) = 0 \qquad\qquad\qquad \text{for other } x$$

find $A, \bar{x}, x_m$, and x_{rms} and sketch $f(x)$ versus x. What is the chance of x having the value $x = \bar{x}$?

The distribution functions in Problems 1 and 2 arise in the quantum-mechanical problem of a particle confined to a one-dimensional box of length L. The quantity $f(x)dx$ gives the probability of finding the particle at x in the range dx when the particle is in the ground state (Problem 1) and first excited state (Problem 2).

3. Calculate the average value of $|v_x|$ from the Maxwell-Boltzmann distribution.

✓4. The average deviation AD from the mean

$$\text{AD} = \frac{1}{N} \sum_i \left| \bar{x} - x_i \right|$$

is sometimes used because it is easier to calculate than σ (however, it is not as significant in error theory). For a normal (Gaussian) distribution, $\text{AD} = \sqrt{2/\pi}\, \sigma$. Show this for the special case $\bar{x} = 0$ by comparing your result for the average value of $|v_x|$ from Problem 3 with $v_{x,rms}$.

5. Calculate the *order of magnitude* of the fractional Doppler shift due to thermal motion by assuming the light source is an oxygen atom moving relative to the observer with speed v. Do you need to use the relativistic formula for the Doppler effect for this calculation?

6. Consider a layer of air of thickness dz and area A. Show that for equilibrium in the earth's gravitational field, the pressure difference

must be $-dP = \rho g\,dz = (N/V)\,mg\,dz$, where N/V is the number of molecules per unit volume and m is the mass of each molecule. Using the ideal gas law and assuming constant temperature, derive the law of atmospheres $P = P_0 e^{-mgz/kT}$. Relate this result to Equation 2-37.

7.   In this problem you are to use a simple but artificial model of a gas to derive Equation 2-4. Assume that there are N molecules in a box of length L (along the x axis) and end area A all moving with the same speed v, and that $\frac{1}{3}$ of them are moving parallel to the x axis, $\frac{1}{3}$ parallel to the y axis, and $\frac{1}{3}$ parallel to the z axis. (a) Show that in time $t_1 = 2L/v$ each molecule moving parallel to the x axis will hit the right side once, so that in time Δt the number that hit the right face is $\frac{1}{3}N(\Delta t/t_1) = \frac{1}{6}Nv\,\Delta t/L$. (b) Using the fact that each molecule delivers an impulse of $2mv$ to the right wall, find the total impulse in time Δt. (c) Divide by the time Δt and by the area A to obtain the pressure exerted on the right face.

8.   A gas consists of N point particles in an infinite space. Each particle is attracted to the origin by a force proportional to the distance from the origin, $\mathbf{F} = -C\mathbf{r}$, so in addition to its translational kinetic energy, each particle has a potential energy $U = \frac{1}{2}Cr^2 = \frac{1}{2}C(x^2 + y^2 + z^2)$. (a) What is the Maxwell-Boltzmann distribution function $f(x,y,z,v_x,v_y,v_z)$? (b) Show that the probability of finding a particle in dr at distance r is $Ar^2 e^{-Cr^2/2kT}\,dr$, and evaluate the constant A. (c) Find the average values of x, x^2, r, and r^2. (Hint: The equipartition theorem can be used to find the average values of x^2 and r^2.) (d) What is the average energy per particle? (e) What is the molar heat capacity C_v for this gas?

9.   Consider the problem of a two-dimensional gas confined to a rectangle of area A. Derive the expression $P = \frac{1}{2}nm\overline{v^2}$, where the "pressure" P is the force per unit length on the edge of the rectangle.

10. Assuming that the velocity distribution for the gas of Problem 9 is $f(v_x,v_y) = Ae^{-m(v_x^2+v_y^2)/2kT}$, find A, the speed distribution, and $\overline{v}$.

11. The speed distribution of molecules in a container is the Maxwell distribution $f(v) \propto v^2 e^{-mv^2/2kT}$. The number with speed v that hit the wall in a given time is proportional to the speed v and to $f(v)$. Thus, if there is a very small hole in the wall (too small to have much effect on the distribution inside), the speed distribution of those that escape is $F(v) \propto vf(v) \propto v^3 e^{-mv^2/2kT}$. Show that the mean energy of those that escape is $2kT$.

12. This problem is related to the equipartition theorem. Consider a system in which the energy of a particle is given by $E = Au^2$, where A is a constant and u is any coordinate or momentum which can vary from $-\infty$ to $+\infty$. (a) Write the probability of the particle having u in the range du and calculate the normalization constant C in terms of A. (b) Calculate the average energy $\overline{E} = \overline{Au^2}$ and show that $\overline{E} = \frac{1}{2}kT$.

13. In this problem you derive the distribution function for free paths of a molecule in a gas. Let N be the number of molecules that travel a distance x without making a collision. This number is decreased in the interval dx by the number of collisions made in dx. Assuming that the number of collisions in dx is proportional to N and to dx, we can write $-dN = \alpha N(x)\,dx$, where α is a constant. (a) Solve this equation for $N(x)$, assuming N_0 molecules at $x = 0$. (b) The distribution function for free paths $f(x)\,dx$ is the fraction that make their first collision in dx, which is $-dN/N_0$. Show that $f(x) = \alpha e^{-\alpha x}$. (c) Show that the mean free path is $1/\alpha$. (d) What is the most probable free path?

CHAPTER 3 The Quantization of Electricity, Light, and Energy

Objectives

After studying this chapter you should:

---

1. Know the relationships among the faraday, the electronic charge, and Avogadro's number.

2. Be able to describe the J. J. Thomson measurement of e/m and the Millikan measurement of e, and know the value of e in coulombs.

3. Know the Stefan-Boltzmann law and be able to sketch the spectral distribution of blackbody radiation.

4. Know what is meant by the "ultraviolet catastrophe."

5. Know the assumptions made by Planck to derive the correct formula for the spectral distribution of blackbody radiation.

6. Be able to discuss which features of the photoelectric effect are in accord with the predictions of classical physics and which are not.

7. Know the Einstein equation for the photoelectric effect.

8. Be able to discuss how the photon concept explains (a) all features of the photoelectric effect, (b) the Duane-Hunt rule for the cutoff wavelength of x rays, and (c) Compton scattering of x rays.

9. Be able to discuss how the Einstein model of a solid explains the failure of the Dulong-Petit law for the heat capacity of solids.

10. Be able to discuss how the idea of energy quantization explains the observed temperature dependence of the heat capacity of solids and gases.

The great success of Avogadro's hypothesis in interpreting chemical reactions and of the kinetic theory in the late nineteenth and early twentieth centuries led to general (though not unanimous) acceptance of the molecular theory of matter. Apparently matter is not continuous, as it appears, but is *quantized* (i.e., discrete) on the microscopic scale. It is because of the enormous size of Avogadro's number that the discreteness of matter is not readily observable. In this chapter we shall study how three great discoveries were made: the quantization of (1) electric charge, (2) light energy, and (3) energy states of mechanical systems. The quantization of electric charge was not particularly surprising to scientists in 1900; it was quite analogous to the quantization of mass as implied by Dalton, Avogadro, and others. However, the quantization of light energy and mechanical energy were revolutionary ideas.

3-1   Early Estimates of e and e/m

The first estimates of the order of magnitude of the electric charges found in atoms were obtained from Faraday's law. The work of Michael Faraday (1791–1867) in the early- to mid-1800s stands out even today for its vision, experimental ingenuity, and thoroughness. The story of this self-educated blacksmith's son who rose from errand boy and bookbinder's apprentice to become the director of the distinguished Royal Institution of London and the foremost experimental investigator of his time is a fascinating one. One aspect of his work concerned the study of the conduction of electricity in liquids. His results and his subsequent statement of the law of electrolysis (1833) were of great importance at the time for the evidence they gave of the electrical nature of atomic forces. The phenomenon still has interest, for, besides underlying the study of the field of electrochemistry, electrolysis is used in the most precise modern determinations of the electronic charge.

In his experiments, Faraday passed a current through weakly conducting solutions and observed the subsequent liberation of the components of the solution. For instance, one important experiment consisted of passing a current through zinc and platinum electrodes immersed in a weak solution of sulfuric acid and water. He collected and measured the hydrogen and oxygen gas that evolved, and also measured the loss in weight of the zinc electrode. He found that upon repeated trials, the ratio of water decomposed to zinc oxidized was always just 1 to 3.59. He correctly interpreted this as giving the ratio of molecular weights of the water to zinc; his value is within 1 percent of what is accepted today. He was thus able to conclude the existence of a definite unit of charge associated with each atom.

Quantitatively, Faraday's law of electrolysis states that the same quantity of electricity, F, called the faraday and equal to about 96,500 C, always decomposes 1 gram-ionic weight of monovalent ions. For example, if 96,500 C pass through a solution of NaCl, 23 g of Na appear at the cathode and 35.5 g of Cl at the anode. For ions of valence 2, such as Cu or SO_4, it takes 2 faradays to decompose 1 gram-ionic weight. Since a gram-ionic weight is just Avogadro's number of ions, it is reasonable to assume that each monovalent ion contains the same charge, e, and

$$F = N_A e$$ 3-1   *Faraday's law*

Since the faraday could be measured quite accurately, N_A or e could each be determined if the other were known. Faraday was aware of this but could not determine either quantity. In 1874, G. J. Stoney estimated e to be about 10^{-20} C, using estimates of N_A from kinetic theory. Helmholtz pointed out in 1880 that it is apparently impossible to obtain a subunit of this charge. The first direct measurement of this smallest unit of charge was made by Townsend in 1897, by an ingenious method that was the forerunner of the famous Millikan oil-drop experiment.

The first evidence for the existence of atomic particles with a specific charge-to-mass ratio was obtained by P. Zeeman in 1896, by looking at the light emitted by atoms placed in a strong magnetic field. When viewed through a spectroscope, this light appears as a discrete set of lines called *spectral lines.* According to classical electromagnetic theory, a charge oscillating in simple harmonic motion will emit electromagnetic radiation at the frequency of oscillation. If the charge is placed in a magnetic field, there will be an additional force on the charge which, to a first approximation, merely changes the frequency of oscillation. The frequency is either slightly increased, slightly decreased, or unchanged, depending on the orientation of the line of oscillation relative to the field. Then, according to classical theory, if a spectral line from an atom is due to the oscillation of a charged particle in the atom, that line will be split into three lines when the atom is placed in a magnetic field. The magnitude of the splitting depends on the charge-to-mass ratio q/m of the oscillating particle. Zeeman measured such a splitting and calculated q/m to be about 1.6×10^{11} C/kg, which compares favorably with the presently accepted value for the electron of 1.76×10^{11} C/kg. From the polarization of the spectral lines, Zeeman concluded that the oscillating particles were negatively charged.

3-2   The J. J. Thomson Experiment

The direct measurement of the charge-to-mass ratio e/m of electrons by J. J. Thomson in 1897 can justly be considered the beginning of our understanding of atomic structure. Thomson's classic experiment evolved from the study of electrical discharges in gases.

Many studies of this phenomenon were done in the late nineteenth century, and interest was greatly increased by the discovery of x rays, which ionized the gases, thus permitting the control of the conductivity of the gases. It was found that the ions responsible for gaseous conduction carried the same charge as did those in electrolysis. The cathode-ray tube used by J. J. Thomson (Figure 3-1) is typical of those used. At sufficiently low pressure, the space near the cathode becomes dark, and as the pressure is lowered, this dark space extends across the tube until finally it reaches the glass, which then glows. When apertures are placed at A and B, the glow is limited to a well-defined spot

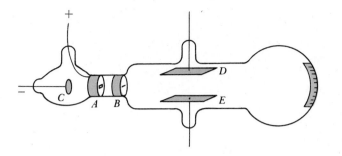

Figure 3-1
J. J. Thomson's tube for
measuring e/m. Electrons
from the cathode C pass
through the slits at A and
B and strike a phospho-
rescent screen. The beam
can be deflected by an
electric field between the
plates D and E or by a
magnetic field (not
shown). From measure-
ments of the deflections,
e/m can be determined.
[*From J. J. Thomson,*
Philosophical Magazine
(5), **44,** *293 (1897).*]

on the glass. This spot can be deflected by electrostatic or magnetic fields.[1]

In 1895, J. Perrin collected these "cathode rays" on an
electrometer and found them to carry a negative electric
charge. In 1897, Thomson measured the ratio of the mass to
charge of these rays. Figure 3-1, taken from his paper on
"Cathode Rays," *Philosophical Magazine* (5), **44,** 293 (1897),
shows his apparatus. We quote from this paper:

> The experiments discussed in this paper were undertaken in
> the hope of gaining some information as to the nature of the
> Cathode Rays. The most diverse opinions are held as to these
> rays; according to the almost unanimous opinion of German
> physicists they are due to some process in the aether to
> which—inasmuch as in a uniform magnetic field their course
> is circular and not rectilinear—no phenomenon hitherto ob-
> served is analogous: another view of these rays is that, so far
> from being wholly aetherial, they are in fact wholly material,
> and that they mark the paths of particles of matter charged
> with negative electricity. It would seem at first sight that it
> ought not to be difficult to discriminate between views so dif-
> ferent, yet experience shows that this is not the case, as
> amongst the physicists who have most deeply studied the sub-
> ject can be found supporters of either theory.
>
> The electrified-particle theory has for purposes of research
> a great advantage over the aetherial theory, since it is definite
> and its consequences can be predicted; with the aetherial
> theory it is impossible to predict what will happen under any
> given circumstances, as in this theory we are dealing with hith-
> erto unobserved phenomena in the aether, of whose laws we
> are ignorant.
>
> The following experiments were made to test some of the
> consequences of the electrified-particle theory. . . .
>
> The rays from the cathode C pass through a slit in the
> anode A [see Figure 3-1], which is a metal plug fitting tightly
> into the tube and connected with the earth; after passing

---

[1] Much of the early confusion about the nature of cathode rays was due to the
failure of Heinrich Hertz in 1883 to observe any deflection of the rays in an electric field. This failure was later found to be the result of ionization of the gas in
the tube; the ions quickly neutralized the charges on the deflecting plates so that
there was actually no electric field between the plates. With better vacuum technology in 1897, Thomson was able to work at lower pressure and observe electrostatic deflection.

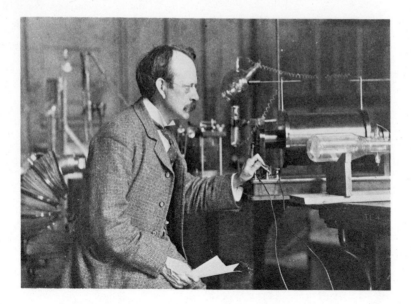

Sir. J. J. Thomson in his
laboratory. (*Courtesy of
Cavendish Laboratory.*)

through a second slit in another earth-connected metal plug B,
they travel between two parallel aluminium plates about 5 cm
long by 2 broad and at a distance of 1.5 cm apart; they then fall
on the end of the tube and produce a narrow well-defined
phosphorescent patch. A scale pasted on the outside of the
tube serves to measure the deflexion of this patch. At high ex-
haustions the rays were reflected when the two aluminium
plates were connected with the terminals of a battery of small
storage cells; the rays were depressed when the upper plate
was connected with the negative pole of the battery, the lower
with the positive, and raised when the upper plate was connected with the positive, the lower with the negative pole. The
deflexion was proportional to the difference of potential be
tween the plates, and I could detect the deflexion when the
potential-difference was as small as 2 volts.

When a magnetic field of strength B is placed perpendicular to
the original path, the particles move in a circular path. The
radius R of the path can be obtained from Newton's second law,
by setting the magnetic force qvB equal to the mass m times the
acceleration v^2/R.

$$qvB = \frac{mv^2}{R}$$

or

$$R = \frac{mv}{qB} \qquad\qquad 3\text{-}2$$

In his first measurement, Thomson determined the velocity
from measurements of the total charge and the temperature
change occurring when the beam struck an insulated collector.

For N particles, the total charge is $Q = Ne$, while the temperature rise is proportional to the energy loss $W = N(\frac{1}{2}mv^2)$. Eliminating N and v from these equations, we obtain

$$\frac{e}{m} = \frac{2W}{B^2 R^2 Q} \qquad\qquad 3\text{-}3$$

In his second method, which came to be known as the *J. J. Thomson experiment,* he adjusted perpendicular B and $\mathscr{E}$ fields so that the particles were *undeflected.* This allowed him to determine the speed by equating the magnitudes of the magnetic and electric forces.

$$qvB = q\mathscr{E}$$

or

$$v = \frac{\mathscr{E}}{B} \qquad\qquad 3\text{-}4$$

He then turned off the B field and measured the deflection of the particles on the screen. This deflection is made up of two parts (Figure 3-2). While the particles are between the plates they undergo a vertical deflection y_1, given by

$$y_1 = \frac{1}{2} a t_1^2 = \frac{1}{2} \frac{e\mathscr{E}}{m} \left(\frac{x_1}{v_x}\right)^2 \qquad\qquad 3\text{-}5$$

where x_1 is the horizontal distance traveled. After they leave the plates they undergo additional deflection y_2, given by

$$y_2 = v_y t_2 = a t_1 \left(\frac{x_2}{v_x}\right) = \frac{e\mathscr{E}}{m} \left(\frac{x_1}{v_x}\right)\left(\frac{x_2}{v_x}\right)$$

$$= \frac{e\mathscr{E}}{m} \frac{x_1 x_2}{v_x^2} \qquad\qquad 3\text{-}6$$

where x_2 is the horizontal distance traveled beyond the deflection plates. The total deflection $y_1 + y_2$ is proportional to e/m. It is interesting to note that his original values of e/m from his first method, about 2×10^{11} C/kg, were closer to the present value, 1.76×10^{11}, than those from his second method, 0.7×10^{11}. (The inaccuracy of the results obtained by his second method was due to his having neglected the magnetic field outside the region of the deflecting plates. Despite this inaccuracy, however, the second method had the advantage of reproducibility, i.e., there was considerably less scatter in the data in this method.) Thomson repeated the experiment with different gases in

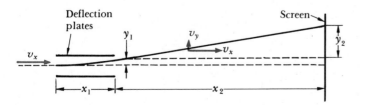

Figure 3-2
Deflection of the electron beam in Thomson's apparatus.

tubes and different metals for cathodes and always obtained the same value of e/m within his experimental accuracy, thus showing that these particles were common to all metals. The agreement of these results with Zeeman's led to the unmistakable conclusion that these particles, called *corpuscles* by Thomson and later called *electrons,* having one unit of negative charge e and mass about 2000 times less than the lightest known atom, were constituents of all atoms.

Questions

1. One advantage of Thomson's evidence over others (such as that of Faraday or Zeeman) was its directness. Another was that it was not just a statistical inference. How is it shown in the Thomson experiment that e/m is the same for each particle rather than being just an average value for a large number of particles?

2. Thomson noted that his values for e/m were about 2000 times larger than those for the lightest known ion, that of hydrogen. Could he distinguish from his data between the possibilities that this was a result of the electron having either a greater charge or a smaller mass than the hydrogen ion?

3-3 Quantization of Electric Charge

After his measurement of e/m, Thomson initiated a series of measurements to determine the electronic charge e. These experiments were carried out by his student J. S. Townsend.

Townsend used electrolysis to produce charged gaseous ions, which formed a cloud when bubbled through water. He measured the mass of the cloud by passing it through drying tubes and determining the increased weight of these tubes. He measured the total charge with an electrometer and determined the average radius of the individual water droplets in the cloud by observing the rate of fall of the cloud due to gravity. Newton's law $\Sigma \mathbf{F} = m\mathbf{a}$ for a drop falling in a medium with a retarding force proportional to the velocity is

$$mg - bv = m \frac{dv}{dt} \qquad\qquad 3\text{-}7$$

The drop quickly reaches its terminal velocity, which we can find by setting $dv/dt = 0$:

$$v_t = \frac{mg}{b}$$

The quantity b is related to the radius of the drop (a) and the coefficient of viscosity of the fluid (η) by a result from fluid mechanics known as Stokes' law:

$$b = 6\pi\eta a \qquad\qquad 3\text{-}8$$

If we write the mass in terms of the density, ρ,

$$m = \tfrac{4}{3}\pi a^3 \rho$$

we obtain for the terminal velocity

$$v_t = \frac{2}{9}\,ga^2\,\frac{\rho}{\eta}$$ 3-9   *Terminal velocity of a drop*

where η is the coefficient of viscosity of air in this case. Equation 3-9 was used to determine a. Knowing the average size of the drops and the total mass, Townsend could compute the number of drops in the cloud. Dividing the total charge of the cloud by the number of drops (he assumed each ion formed one water drop), he estimated the charge on each ion as about 1.0×10^{-19} C, the same order of magnitude as determined from both Faraday's laws of electrolysis and kinetic-theory estimates of N_A. Variations on this technique were made by J. J. Thomson and H. A. Wilson, with little improvement in accuracy. Wilson produced clouds between the plates of a capacitor and observed their fall due to gravity alone and due to the combination of gravity and an electric field produced between the plates. Table 3-1, giving final results of 11 different trials, illustrates the accuracy obtained.

The accuracy of Thomson's method was limited by the uncertain rate of evaporation of the drop, and the assumption that each droplet contained a single charge could not be verified. R. M. Millikan tried to eliminate the evaporation problem by using a field strong enough to hold the top surface of the cloud stationary so that he could observe the rate of evaporation, and correct for it. The results are described in his paper "A New Modification of the Cloud Method of Determining the Elementary Charge and the Most Probable Value of That Charge," which appeared in the *Philosophical Magazine* (6), **19,** 209 (1910). The following quotation is taken from this paper:

The Balancing of Individual Charged Drops by an Electrostatic Field

My original plan for eliminating the evaporation error was to obtain, if possible, an electric field strong enough exactly to balance the force of gravity upon the cloud and then by means of a sliding contact to vary the strength of this field so as to hold the cloud balanced throughout its entire life. In this way it was thought that the whole evaporation history of the cloud might be recorded, and that suitable allowances might then be made in the observations on the rate of fall to eliminate entirely the error due to evaporation. It was not found possible to balance the cloud as had been originally planned, but it was found possible to do something much better: namely, to hold individual charged drops suspended by the field for periods varying from 30 to 60 seconds. I have never actually timed drops which lasted more than 45 seconds, although I have several times observed drops which in my judgment lasted considerably longer than this. The drops which it was found possible to balance by an electric field always carried multiple charges,

Table 3-1
Results of Wilson's determination of e†

e (C)
0.77×10^{-19}
0.87×10^{-19}
$1.5 \ \times 10^{-19}$
0.90×10^{-19}
$1.1 \ \times 10^{-19}$
$1.3 \ \times 10^{-19}$
$1.3 \ \times 10^{-19}$
$1.0 \ \times 10^{-19}$
$1.2 \ \times 10^{-19}$
0.67×10^{-19}
0.77×10^{-19}

† Mean value:
1.03×10^{-19} C.
From *Philosophical Magazine* (6), **5,** 439 (1903).

Millikan's original oil-drop apparatus for determining the fundamental unit of electric charge *e*. [*From R. A. Millikan,* Electrons (+ and −), Protons, Neutrons, Mesotrons, and Cosmic Rays, *Chicago: University of Chicago Press, 1947; By permission of the publishers.* © *1947 by University of Chicago. All rights reserved.*]

and the difficulty experienced in balancing such drops was less than had been anticipated.

The procedure is simply to form a cloud and throw on the field immediately thereafter. The drops which have charges of the same sign rapidly fall, while those which are charged with too many multiples of the sign opposite to that of the upper plate are jerked up against gravity to this plate. The result is that after a lapse of seven or eight seconds the field of view has become quite clear save for a relatively small number of drops, which have just the right ratio of charge to mass to be suspended by the electric field. These appear as perfectly distinct bright points. I have on several occasions obtained but one single such "star" in the whole field and held it there for nearly a minute. For the most part, however, the observations recorded below were made with a considerable number of such points in view. Thin, flocculent clouds, the production of which seemed to be facilitated by keeping the water jackets . . . a degree or two above the temperature of the room, were found to be particularly favorable to observations of this kind.

Furthermore, it was found possible so to vary the mass of a drop by varying the ionization, that drops carrying in some cases two, in some three, in some four, in some five, and in some six, multiples could be held suspended by nearly the same field. The means of gradually varying the field which had been planned were therefore found to be unnecessary. If a given field would not hold any drops suspended it was varied by steps of 100 or 200 volts until drops were held stationary, or nearly stationary. When the P. D. [potential difference] was thrown off, it was often possible to see different drops move down under gravity with greatly different speeds, thus showing that these drops had different masses and correspondingly different charges.

During this experiment, he noticed that balanced drops sometimes suddenly moved upward or downward, evidently because they had picked up a positive or negative ion. This led to the possibility of observing the charge of a single ion. In 1909, Millikan began a series of experiments which not only showed that charges occurred in multiples of an elementary unit, e, but measured the value of e to about 1 part in 1000. To eliminate evaporation, he used oil drops sprayed into dry air between the plates of a capacitor. These drops were already charged by the spraying process, and during the course of observation they picked up or lost additional charges. By switching the field between the plates, a drop could be moved up or down and observed for several hours. When the charge on a drop changed, the velocity of the drop changed. Assuming only that the terminal velocity of the drop was proportional to the force acting on it (this "assumption" was carefully checked experimentally), Millikan's experiment gave conclusive evidence that charges always occur in multiples of a fundamental unit, e.

Optional

Figure 3-3
Millikan oil-drop apparatus. The drops are sprayed from the atomizer and pick up a static charge. Their fall due to gravity and their rise due to the electric field between the capacitor plates can be observed with the telescope. From measurements of the rise and fall times, the electric charge on a drop can be calculated.

Let us examine this experiment in some detail. Figure 3-3 shows a sketch of Millikan's apparatus. With no electric field, the downward force is mg and the upward force is bv, where b is given by Stokes' law (Equation 3-8). The equation of motion is Equation 3-7, and the terminal velocity of the falling drop is

$$v_f = \frac{mg}{b} \qquad \text{3-10}$$

When an electric field $\mathscr{E}$ is applied, the upward motion of a charge q_n is given by

$$q_n \mathscr{E} - mg - bv = m \frac{dv}{dt}$$

Thus the terminal velocity in the presence of the electric field is

$$v_r = \frac{q_n \mathscr{E} - mg}{b} \qquad \text{3-11}$$

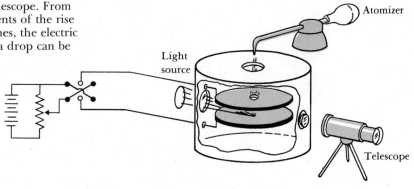

Atomizer

Light
source

Telescope

In this experiment, the terminal speeds were reached almost immediately, and the drops drifted a distance L upward or downward at a constant speed. Solving Equations 3-10 and 3-11 for q_n, we have

$$q_n = \frac{mg}{\mathscr{E}v_f}(v_f + v_r) = \frac{mgT_f}{\mathscr{E}}\left(\frac{1}{T_f} + \frac{1}{T_r}\right) \qquad 3\text{-}12$$

where $T_f = L/v_f$ is the fall time and $T_r = L/v_r$ is the rise time.

If an additional charge is picked up, the terminal velocity becomes v_r', which is related to the new charge q_n' by Equation 3-11:

$$v_r' = \frac{q_n'\mathscr{E} - mg}{b}$$

The amount of charge gained is thus

$$q_n' - q_n = \frac{mg}{\mathscr{E}v_f}(v_r' - v_r) = \frac{mgT_f}{\mathscr{E}}\left(\frac{1}{T_r'} - \frac{1}{T_r}\right) \qquad 3\text{-}13$$

The velocities, v_f, v_r, and v_r' are determined by measuring the time taken to fall or rise the distance L between the capacitor plates.

If we write $q_n = ne$ and $q_n' - q_n = n'e$, Equations 3-12 and 3-13 can be written

$$\frac{1}{n}\left(\frac{1}{T_f} + \frac{1}{T_r}\right) = \frac{\mathscr{E}e}{mgT_f} \qquad 3\text{-}14$$

and

$$\frac{1}{n'}\left(\frac{1}{T_r'} - \frac{1}{T_r}\right) = \frac{\mathscr{E}e}{mgT_f} \qquad 3\text{-}15$$

Table 3-2, taken from Millikan's book, is typical of his early data on a single drop. The fall times in column 1 are all the same within experimental error. There are no numbers in column 3 until there is a change in the rise time in column 2. The numbers in column 6 are proportional to the total charge on the drop (Equation 3-12). When these numbers are divided by the appropriate choice of n (column 7), the resulting numbers in column 8 are the same for all trials. The numbers in column 3 are proportional to the changes in the charge indicated by changes in the rise time. When these are divided by the appropriate choice of n' (column 4), the numbers in column 5 are the same for all trials and the same as those in column 8 (Equations 3-14 and 3-15). This drop began with a charge of $18e$. It then picked up $6e$, giving it a charge of $24e$ (row 2). It then lost $7e$, gained $1e$, etc.

Millikan did experiments like these with thousands of drops, some of nonconducting oil, some of semiconductors like glycerine, and some of conductors like mercury (see Reference 1). In no case was a fractional charge found.

To obtain a value of e from these data, one needs the mass of

Table 3-2
Rise and fall times of a single oil drop with calculated number of elementary charges on drop†

1 T_f	2 T_r	3 $\dfrac{1}{T'_r} - \dfrac{1}{T_r}$	4 n'	5 $\dfrac{1}{n'}\left(\dfrac{1}{T'_r} - \dfrac{1}{T_r}\right)$	6 $\dfrac{1}{T_f} + \dfrac{1}{T_r}$	7 n	8 $\dfrac{1}{n}\left(\dfrac{1}{T_r} + \dfrac{1}{T_f}\right)$
11.848	80.708				0.09655	18	0.005366
11.890	22.366	0.03234	6	0.005390	0.12887	24	0.005371
11.908	22.390						
11.904	22.368						
11.882	140.566	0.03751	7	0.005358	0.09138	17	0.005375
11.906	79.600	0.005348	1	0.005348	0.09673	18	0.005374
11.838	34.748	0.01616	3	0.005387	0.11289	21	0.005376
11.816	34.762						
11.776	34.846						
11.840	29.286				0.11833	22	0.005379
11.904	29.236						
11.870	137.308	0.026872	5	0.005375	0.09146	17	0.005380
11.952	34.638	0.021572	4	0.005393	0.11303	21	0.005382
11.860							
11.846	22.104	0.01623	3	0.005410	0.12926	24	0.005386
11.912	22.268						
11.910	500.1	0.04307	8	0.005384	0.08619	16	0.005387
11.918	19.704	0.04879	9	0.005421	0.13498	25	0.005399
11.870	19.668						
11.888	77.630	0.03794	7	0.005420	0.09704	18	0.005390
11.894	77.806						
11.878	42.302	0.01079	2	0.005395	0.10783	20	0.005392
11.880		Means		0.005389	Means		0.005384

Duration of experiment	45 min		Pressure	75.62 cm Hg
Plate distance	16 mm		Oil viscosity	0.9199 poise
Fall distance	10.21 mm		Air viscosity	1.824×10^{-4} poise
Initial volts	5088.8		Radius (a)	0.000276 cm
Final volts	5081.2		Speed of fall	0.08584 cm/sec
Temperature	22.82°C			

† $e = 4.991 \times 10^{-10}$. The value of 4.991×10^{-10} esu = 1.664×10^{-19} C.

the drop (or its radius, since the density is known). The radius is obtained from Stokes' law using Equation 3-9. (For high precision, the buoyant force of the air on the drop must be taken into account. In the course of these experiments, Millikan found also that Stokes' law did not hold for the smallest of his drops, because of density fluctuations of the medium on a scale comparable with the size of the drops. He found, by successive approximations, an experimental correction to Stokes' law.)

The value of e determined by Millikan was 1.591×10^{-19} C. This value was accepted for about 20 years, until it was discovered that x-ray diffraction measurements of N_A gave values of e that differed from Millikan's by about 0.4 percent. The discrepancy was traced to the value of the coefficient of viscosity η used by Millikan, which was too low. Improved measurements of η gave a value about 0.5 percent higher, thus changing the value of e resulting from the oil-drop experiments to 1.601×10^{-19} C, in good agreement with the x-ray diffraction data. For the modern determination of the "best" value of e and other atomic constants, the reader is referred to Reference 5.

Questions

3. If the value of n in column 7 (Table 3-2) were chosen to be 3 rather than 18, the first value of n' to be 1, and the second value of n to be 4, the first two numbers in column 8 and the first in column 5 would still be equal. Why then is the first n chosen to be 18 rather than 3?

4. The quantization of charge could be shown and the value of e calculated using only Equation 3-12, without bothering with Equation 3-13 and columns 3, 4, and 5 in Table 3-2. Why are the numbers in these columns important?

3-4   Blackbody Radiation

The first clue as to the quantum nature of radiation came from the study of thermal radiation emitted by opaque bodies. When radiation falls on an opaque body, part of it is reflected and the rest absorbed. Light-colored bodies reflect most of the radiation incident on them, whereas dark bodies absorb most of it. If an opaque body is in thermal equilibrium with its surroundings, it must emit and absorb radiation at the same rate. If not, the body would spontaneously become warmer or cooler than its surroundings, contrary to the assumption of thermal equilibrium. A good absorber of radiation is therefore a good emitter.

The radiation emitted under these circumstances is called thermal radiation. At ordinary temperatures (below about 600°C) the thermal radiation emitted by a body is not visible; most of the energy is concentrated in wavelengths much longer than those of visible light. As a body is heated the quantity of thermal radiation emitted increases, and the energy radiated extends to shorter and shorter wavelengths. At about 600–700°C there is enough energy in the visible spectrum so that the body glows and becomes a dull red, and at higher temperatures it becomes bright red or even "white hot."

A body that absorbs *all* radiation incident on it is called an *ideal blackbody*. In 1879 Josef Stefan found an empirical relation between the power per unit area radiated by a blackbody and the temperature:

$$R = \sigma T^4 \qquad \qquad 3\text{-}16$$

Stefan-Boltzmann law

where R is the power per unit area, T is the absolute temperature, and $\sigma = 5.6703 \times 10^{-8}$ W/m²-K⁴ is a constant called Stefan's constant. This result was also derived by Ludwig Boltzmann about five years later and is now called the Stefan-Boltzmann law. Note that the power per unit area depends only on the temperature, and not on any other characteristic of the object. The same is true of the *spectral distribution* of the radiation emitted by a blackbody. Let $R(\lambda)\ d\lambda$ be the power emitted per unit area with wavelength between λ and $\lambda + d\lambda$. Figure 3-4 shows the spectral distribution function $R(\lambda)$ versus λ for several values of T. The wavelength at which the distribution is a maximum varies inversely with the temperature:

$$\lambda_m \propto \frac{1}{T}$$

Wien's law

or

$$\lambda_m T = \text{constant} = 2.898 \times 10^{-3} \text{ m-K} \qquad 3\text{-}17$$

This result is known as Wien's displacement law.

The calculation of the distribution function $R(\lambda)$ involves the calculation of the energy density of electromagnetic waves in a cavity. Materials such as black velvet or lampblack come close to being ideal blackbodies, but the best practical realization of an ideal blackbody is a small hole leading into a cavity (Figure 3-5) (such as the keyhole in a closet door). Radiation incident on the hole has little chance of being reflected back out the hole before it is absorbed by the walls of the cavity. The power radiated out of the hole is proportional to the total energy density U (energy per unit volume of the radiation in the cavity). The proportionality constant can be shown to be $c/4$, where c is the speed of light.[1]

$$R = \tfrac{1}{4}cU \qquad 3\text{-}18$$

Similarly, the spectral distribution of the power emitted from the hole is proportional to the spectral distribution of the energy in the cavity. If $u(\lambda)\ d\lambda$ is the fraction of the energy per unit volume in the cavity in the range $d\lambda$, $u(\lambda)$ and $R(\lambda)$ are related by

$$R(\lambda) = \tfrac{1}{4}cu(\lambda) \qquad 3\text{-}19$$

The energy-density distribution function $u(\lambda)$ can be calculated from classical physics in a straightforward way. The method involves finding the number of modes of oscillation of the electromagnetic field in the cavity with wavelengths in the interval $d\lambda$ and multiplying by the average energy per mode. We shall not go into the details of the calculation here. The number of modes of oscillation per unit volume, $n(\lambda)$, is independent of the shape of the cavity and is given by

$$n(\lambda) = 8\pi\lambda^{-4} \qquad 3\text{-}20$$

According to classical theory, the average energy per mode of oscillation is kT, the same as for a one-dimensional oscillator.

[1] See, for example, pp. 135–137 of Reference 7.

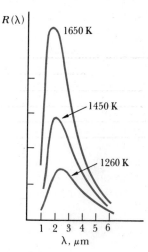

Figure 3-4
Spectral distribution of radiation from a blackbody for three different temperatures.

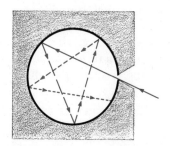

Figure 3-5
Cavity approximating an ideal blackbody. Radiation entering the cavity has little chance of leaving before it is completely absorbed.

The classical theory thus predicts for the spectral distribution function

$$u(\lambda) = kTn(\lambda) = 8\pi kT\lambda^{-4} \qquad \text{3-21}$$

Rayleigh-Jeans law

This prediction is called the *Rayleigh-Jeans law*.

At very long wavelengths the Rayleigh-Jeans law agrees with the experimentally determined spectral distribution, but at short wavelengths this law predicts that $u(\lambda)$ becomes infinite whereas experiment shows that the distribution approaches zero. This result was called the *ultraviolet catastrophe.*

In 1900 the German physicist Max Planck announced that by making somewhat strange assumptions, he could derive a function $u(\lambda)$ which agreed with the experimental data. He first found an empirical function that fit the data, then searched for a way to modify the usual calculation so as to predict his empirical formula. We can see the type of modification needed if we note that, for a given size cavity, the shorter the wavelength, the more standing-wave modes there will be. As $\lambda \to 0$ the number of modes of oscillation approaches infinity, as evidenced in Equation 3-20. In order for the energy-density distribution function $u(\lambda)$ to approach zero as λ approaches zero, the average energy per mode must depend on the wavelength λ and approach zero as λ approaches zero. That is, the equipartition theorem result that $\bar{E} = kT$ must be modified.

The average energy for a one-dimensional simple harmonic oscillator is calculated classically from the energy distribution function, which in turn is found from the Maxwell-Boltzmann distribution function. The energy distribution function has the form[1]

$$f(E) = Ce^{-E/kT} \qquad \text{3-22}$$

where C is determined by normalization. The average energy is then found from

$$\bar{E} = \int_0^\infty Ef(E)\, dE = \int_0^\infty ECe^{-E/kT}\, dE \qquad \text{3-23}$$

with the result $\bar{E} = kT$.

Planck found that he could derive his empirical function by calculating the average energy $\bar{E}$ *assuming the energy to be a discrete variable:* i.e., that it could take on only the values $0, \epsilon, 2\epsilon, \ldots,$ $n\epsilon$ where n is an integer; and further, that ϵ is proportional to the frequency of the radiation. Planck therefore wrote the energy as

$$E_n = n\epsilon = nhf \qquad n = 0, 1, 2, \ldots \qquad \text{3-24}$$

Planck's hypothesis

---

[1] The derivation of this result from the Maxwell-Boltzmann distribution of Equation 2-34 involves computing the number of cells in phase space $dx\, dp$ in the energy interval dE. For an ideal gas, this calculation introduces a factor $E^{1/2}$ into the energy distribution. For a simple harmonic oscillator of energy $E = p^2/2m + \frac{1}{2}Kx^2$, the calculation is somewhat more difficult but the result is simple, and given by Equation 3-22.

where h is a constant now called *Planck's constant.* The Maxwell-Boltzmann distribution law then becomes

$$f_n = Ce^{-E_n/kT} = Ce^{-n\epsilon/kT} \qquad\qquad 3\text{-}25$$

where C is determined by the normalization condition

$$\sum_{n=0}^{\infty} f_n = C \sum_{n=0}^{\infty} e^{-n\epsilon/kT} = 1 \qquad\qquad 3\text{-}26$$

The average energy of an oscillator is then given by

$$\overline{E} = \sum_{n=0}^{\infty} E_n f_n = \sum_{n=0}^{\infty} E_n Ce^{-E_n/kT} \qquad\qquad 3\text{-}27$$

The calculation of the sums in Equations 3-26 and 3-27 is given in Section 3-8. The result is

$$\overline{E} = \frac{\epsilon}{e^{\epsilon/kT} - 1} = \frac{hf}{e^{hf/kT} - 1} = \frac{hc/\lambda}{e^{hc/\lambda kT} - 1} \qquad\qquad 3\text{-}28$$

Multiplying this result by the number of oscillators in the interval $d\lambda$ given by Equation 3-20, we obtain for the distribution function

$$u(\lambda) = \frac{8\pi hc\lambda^{-5}}{e^{hc/\lambda kT} - 1} \qquad\qquad 3\text{-}29 \qquad \textit{Planck's law}$$

This function is sketched in Figure 3-6. It is clear from the figure that the result fits the data quite well.

For very large λ, the exponential in Equation 3-29 can be expanded using $e^x \approx 1 + x + \cdot\cdot\cdot$ for $x \ll 1$, where $x = hc/\lambda kT$. Then

$$e^{hc/\lambda kT} - 1 \approx \frac{hc}{\lambda kT}$$

and

$$u(\lambda) \longrightarrow 8\pi\lambda^{-4}kT$$

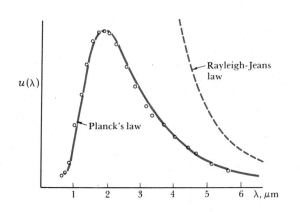

Figure 3-6
Comparison of Planck's law and the Rayleigh-Jeans law with experimental data at $T = 1600$ K. (*Adapted from F. K. Richtmyer, E. H. Kennard, and J. N. Cooper,* Introduction to Modern Physics, *6th ed., New York: McGraw-Hill Book Company, 1969, by permission.*)

which is the Rayleigh-Jeans formula. For short wavelengths, we can neglect the 1 in the denominator of Equation 3-29, and we have

$$u(\lambda) \longrightarrow 8\pi hc \lambda^{-5} e^{-hc/\lambda kT} \longrightarrow 0$$

as $\lambda \to 0$.

The value of Planck's constant, h, can be determined by fitting the function given by Equation 3-29 to the experimental data. The presently accepted value is

$$\begin{aligned} h &= 6.626 \times 10^{-34} \text{ J-sec} \\ &= 4.136 \times 10^{-15} \text{ eV-sec} \end{aligned} \qquad 3\text{-}30$$

Planck tried at length to reconcile his treatment with classical physics but was unable to do so. The fundamental importance of the quantization assumption implied by Equation 3-24 was suspected by Planck and others but was not generally appreciated until 1905. In that year Einstein applied the same ideas to explain the photoelectric effect and suggested that, rather than being merely a mysterious property of blackbody radiation, quantization is a fundamental characteristic of light energy.

Example 3-1   The surface temperature of the sun is about 5000 K. If the sun is assumed to be a blackbody radiator, at what wavelength λ_m would its spectrum peak? From the Wien displacement law (Equation 3-17) we have

$$\lambda_m = \frac{2.898 \times 10^{-3} \text{ m-K}}{5000 \text{ K}} = \frac{2.898 \times 10^6 \text{ nm-K}}{5000 \text{ K}}$$

$$= 579.6 \text{ nm}$$

where 1 nm = 10^{-9} m. Note that this is in the middle of the visible spectrum.

Example 3-2   What is the mean energy $\bar{E}$ for an oscillator that has a frequency $f = kT/h$ according to Planck's calculation? What is it according to the equipartition theorem? From Equation 3-29 with $\epsilon = hf = kT$ we have

$$\bar{E} = \frac{\epsilon}{e^{\epsilon/kT} - 1} = \frac{kT}{e^1 - 1} = 0.582 kT$$

According to the equipartition theorem $\bar{E} = kT$ regardless of the frequency.

Example 3-3   Show that the total energy density in a blackbody cavity is proportional to T^4 in accordance with the Stefan-Boltzmann law.

The total energy density is obtained from the distribution function (Equation 3-29) by integrating over all wavelengths:

$$U = \int_0^\infty u(\lambda) \, d\lambda = \int_0^\infty \frac{8\pi hc \lambda^{-5}}{e^{hc/\lambda kT} - 1} \, d\lambda$$

Define the dimensionless variable $x = hc/\lambda kT$. Then $dx = -hc \, d\lambda/\lambda^2 kT$ or $d\lambda = -\lambda^2(kT/hc) \, dx$. Then

$$U = -\int_\infty^0 \frac{8\pi hc\lambda^{-3}}{e^x - 1}\left(\frac{kT}{hc}\right) dx$$

$$= 8\pi hc \left(\frac{kT}{hc}\right)^4 \int_0^\infty \frac{x^3}{e^x - 1} \, dx$$

Since the integral is now dimensionless, this shows that U is proportional to T^4. The value of the integral can be obtained from tables; it is $\pi^4/15$. Then $U = (8\pi^5 k^4/15h^3 c^3)T^4$. This result can be combined with Equations 3-16 and 3-18 to express Stefan's constant σ in terms of π, k, h, and c (see Exercise 12).

3-5   The Photoelectric Effect

It is one of the ironies in the history of science that in the famous experiment of Heinrich Hertz in 1887 in which he produced and detected electromagnetic waves, thus confirming Maxwell's theory, he also discovered the photoelectric effect that led directly to the particle description of light.

Hertz was using a spark gap in a tuned circuit to generate the waves and another similar circuit to detect them. He noticed accidentally that when the light from the generating gap was shielded from the receiving gap, the receiving gap had to be made shorter to allow the sparks to pass. Light from any spark that fell on the terminals of the gap facilitated the passage of the sparks.

Hertz did not pursue this investigation, but others did. It was found that negative particles were emitted from a clean surface when exposed to light. P. Lenard in 1900 deflected these "rays" in a magnetic field and found that they had a charge-to-mass ratio of the same magnitude as that measured by Thomson for cathode rays.

Figure 3-7 shows a schematic diagram of the basic apparatus. When light is incident on a clean metal surface (cathode C), electrons are emitted. If some of these electrons strike the anode A, there is a current in the external circuit. The number of the emitted electrons reaching the anode can be increased or decreased by making the anode positive or negative with respect to the cathode. Let V be the potential difference between cathode and anode. Figure 3-8 shows the current versus V for two values of the intensity of light incident on the cathode. When V is positive, the electrons are attracted to the anode. At sufficiently large V all the emitted electrons reach the anode and the current reaches its maximum value. Lenard observed that the maximum current is proportional to the light intensity, an expected result since doubling the energy per unit time incident on the cathode should double the number of electrons emitted. When V is negative, the electrons are repelled from the anode. Then only electrons with initial kinetic energy $\frac{1}{2}mv^2$ greater than $e|V|$ can reach the anode. From Figure 3-8 we see that if V is less than $-V_0$ no

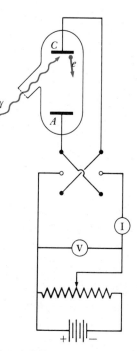

Figure 3-7
Schematic drawing of photoelectric-effect apparatus. Light strikes the cathode C and ejects electrons. The number of electrons which reach the anode A is measured by the current in the ammeter. The anode can be made positive or negative with respect to the cathode to attract or repel the electrons.

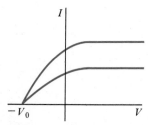

Figure 3-8
Photoelectric current versus voltage V for two values of intensity of the incident light. There is no current when V is less than $-V_0$. The saturation current observed for large V is proportional to the light intensity. The stopping potential V_0 is independent of the light intensity.

electrons reach the anode. The potential V_0 is called the *stopping potential*. It is related to the maximum kinetic energy of the emitted electrons by

$$(\tfrac{1}{2}mv^2)_{\max} = eV_0 \qquad\qquad 3\text{-}31$$

The experimental result that V_0 is independent of the incident light intensity was surprising. Apparently, increasing the rate of energy falling on the cathode does not increase the maximum kinetic energy of the emitted electrons, contrary to classical expectations. In 1905, Einstein offered an explanation of this result in a remarkable paper in the same volume of *Annalen der Physik* that contains his papers on special relativity and Brownian motion.

Einstein assumed that energy quantization used by Planck in the blackbody problem was a universal characteristic of light. Rather than being distributed evenly in the space through which it is propagated, light energy consists of discrete quanta of energy hf. When one of these quanta, called a *photon,* penetrates the surface of the cathode, all of its energy is given completely to an electron. If ϕ is the energy necessary to remove an electron from the surface (ϕ is called the *work function* and is characteristic of the metal), the maximum energy of the electrons leaving the surface will be $hf - \phi$. (Some electrons will have less than this amount because of energy loss in traversing the metal.) Thus the stopping potential V_0 should be given by

$$eV_0 = \tfrac{1}{2}mv^2 = hf - \phi \qquad\qquad 3\text{-}32$$

Einstein equation for the photoelectric effect

Equation 3-32 is called the *Einstein equation.*

> If the derived formula is correct, then V_0, when represented in cartesian coordinates as a function of the frequency of the incident light, must be a straight line whose slope is independent of the nature of the emitting substance.[1]

[1] From Einstein's original paper, as translated by A. B. Arons and M. B. Peppard in the *American Journal of Physics,* **33,** 367 (1965).

Albert A. Michelson, Albert Einstein, and Robert A. Millikan at a meeting in Pasadena, California, in 1931.

As can be seen from Equation 3-32, the slope of V_0 versus f should equal h/e. At the time of this prediction, there was no evidence that Planck's constant had anything to do with the photoelectric effect. There was also no evidence for the dependence of the stopping potential V_0 on frequency. In fact, this turned out to be a difficult experiment, and as late as 1913[1] it was not clear whether the data could be fitted better by f proportional to V_0 or to $V_0^{1/2}$. Careful experiments by Millikan, reported in 1914 and in more detail in 1916, showed that Equation 3-32 was correct, and measurements of h from it agreed with the value obtained by Planck. A plot taken from this work is shown in Figure 3-9.

The threshold frequency, labeled f_t in this plot, and the corresponding threshold wavelength, λ_t, are related to the work function ϕ by setting $V_0 = 0$ in Equation 3-32:

$$\phi = hf_t = \frac{hc}{\lambda_t} \qquad \text{3-33}$$

Work function and threshold frequency and wavelength

Photons of frequency lower than f_t (and therefore having wavelengths greater than λ_t) do not have enough energy to eject an electron from the metal. Work functions for metals are typically on the order of a few electron volts. Since wavelengths are usually given in nanometers or angstroms and energies in electron volts, it is useful to have the value of hc in electron volt–angstroms or electron volt–nanometers. We have

$$hc = (4.14 \times 10^{-15} \text{ eV-sec})(3 \times 10^8 \text{ m/sec})$$
$$= 1.24 \times 10^{-6} \text{ eV-m}$$

or

$$hc = 1240 \text{ eV-nm} = 12,400 \text{ eV-Å} \qquad \text{3-34}$$

[1] See Reference 1 for a complete discussion of the difficulties in the experimental verification of the Einstein equation.

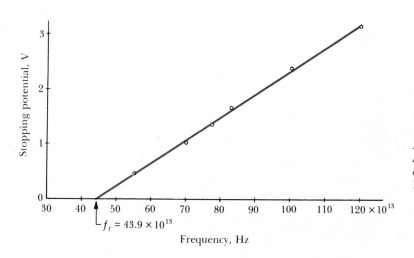

Figure 3-9
Millikan's data for stopping potential versus frequency for the photoelectric effect. The data fall on a straight line of slope h/e, as predicted by Einstein a decade before the experiment. [*From R. A. Millikan, "A Direct Photoelectric Determination of Planck's h,"* Physical Review, **7,** *362 (1916).*]

Example 3-4   The threshold wavelength for potassium is 558 nm. What is the work function for potassium? What is the stopping potential when light of wavelength 400 nm is used?

$$\phi = hf_t = \frac{hc}{\lambda_t} = \frac{1240 \text{ eV-nm}}{558 \text{ nm}} = 2.22 \text{ eV}$$

The energy of a photon of wavelength 400 nm is

$$E = \frac{hc}{\lambda} = \frac{1240 \text{ eV-nm}}{400 \text{ nm}} = 3.10 \text{ eV}$$

The maximum kinetic energy of the emitted electrons is then

$$(\tfrac{1}{2}mv^2)_{max} = hf - \phi = 3.10 \text{ eV} - 2.22 \text{ eV}$$
$$= 0.88 \text{ eV}$$

The stopping potential is therefore 0.88 V.

Another interesting feature of the photoelectric effect which is contrary to classical physics but easily explained by the photon hypothesis is the lack of any time lag between the turning on of the light source and the appearance of electrons. Classically, the incident energy is distributed uniformly over the illuminated surface; the time required for an area the size of an atom to acquire enough energy to allow the emission of an electron can be calculated from the intensity (power per unit area) of the incident radiation. Experimentally, the incident intensity can be adjusted so that this calculated time lag should be several minutes or even hours. But no time lag is ever observed; electrons are apparently ejected immediately. The photon explanation of this result is that although the rate at which photons are incident upon the metal is very small when the intensity is low, *each* photon has enough energy to eject an electron, and there is some chance that a photon will be absorbed immediately. The classical calculation gives the correct *average* number of photons absorbed per unit time.

Example 3-5   Light of wavelength 400 nm and intensity 10^{-2} W/m² is incident on potassium. Estimate the time lag expected classically. According to the previous example, the work function for potassium is 2.22 eV. If we take $r = 10^{-10}$ m as a typical radius of an atom, the total energy falling on the atom in time t is

$$E = (10^{-2} \text{ W/m}^2)(\pi r^2)t = (10^{-2} \text{ W/m}^2)(\pi 10^{-20} \text{ m}^2)t$$
$$= (3.14 \times 10^{-22} \text{ J/sec})t$$

Setting this energy equal to 2.22 eV (= $2.22 \times 1.6 \times 10^{-19}$ J) gives

$$(3.14 \times 10^{-22} \text{ J/sec})t = (2.22)(1.6 \times 10^{-19} \text{ J})$$

$$t = 1.13 \times 10^3 \text{ sec} = 18.8 \text{ min}$$

According to classical prediction, no atom could emit an electron until 18.8 min after the light source was turned on. According to the photon model of light, each photon has enough energy to eject an electron immediately. Because of the low intensity, there are few photons incident per second so that the chance of any particular atom absorbing a photon and emitting an electron in any given time interval is small. However, there are so many atoms in the cathode that some emit electrons immediately.

Example 3-6   In the previous example, how many photons are incident per second per square meter? The energy of each photon is $E = hf = hc/\lambda = (1240$ eV-nm$)/400$ nm $= 3.1$ eV $= (3.1$ eV$)(1.6 \times 10^{-19}$ J/eV$) = 4.96 \times 10^{-19}$ J. Since the incident intensity is 10^{-2} W/m^2 = 10^{-2} J/sec-m^2, the number of photons per second per square meter is

$$N = \frac{10^{-2} \text{ J/sec-m}^2}{4.96 \times 10^{-19} \text{ J/photon}}$$

$$= 2.02 \times 10^{16} \text{ photons/sec-m}^2$$

Questions

5. How is the result that the maximum current is proportional to the intensity explained in the photon model of light?

6. What experimental features of the photoelectric effect can be explained by classical physics? What features cannot?

3-6   X Rays and the Compton Effect

Further evidence of the correctness of the photon concept was furnished by Arthur H. Compton, who measured the scattering of x rays by free electrons. Before we examine Compton scattering in detail, we shall briefly describe some of the early work with x rays.

X rays were discovered in 1895 by W. Roentgen when he was working with a cathode-ray tube. He found that "rays" originating from the point where cathode rays (electrons) hit the glass tube, or a target within the tube, could pass through materials opaque to light and activate a fluorescent screen or photographic film. He investigated this phenomenon extensively and found that all materials were transparent to these rays to some degree and that the transparency decreased with increasing density. This fact led to the medical use of x rays within months after Roentgen's first paper.[1]

Roentgen was unable to deflect these rays in a magnetic field, nor was he able to observe refraction or the interference phenomena associated with waves. He thus gave the rays the somewhat mysterious name of x rays. Since classical electromagnetic theory predicts that charges will radiate electromagnetic waves

---

[1] A translation of this paper can be found in E. C. Watson, "The Discovery of X Rays," *American Journal of Physics,* **13,** 284 (1945), in Reference 2 of Chapter 2, and in Reference 2 of this chapter.

Early x-ray tube. (*Courtesy of Cavendish Laboratory.*)

THE NEW PHOTOGRAPHIC DISCOVERY.

Thanks to the discovery of Professor Röntgen, the German Emperor will now be able to obtain an exact Photograph of a " Backbone " of unsuspecied size and strength !

Cartoon that appeared in *Punch,* January 25, 1896, one month after Roentgen announced his discovery of x rays. (© Punch; *by permission of the publishers.*)

when accelerated, it is natural to expect that x rays are electromagnetic waves produced by the acceleration of electrons when they are stopped by a target. In 1899 H. Haga and C. H. Wind[1] observed a slight broadening of an x-ray beam after passing through slits a few thousandths of a millimeter wide; assuming that this was due to diffraction, they estimated the wavelength to be of the order of 10^{-10} m = 1 Å. In 1912 Laue suggested that since the wavelengths of x rays were of the same order of magnitude as the spacing of atoms in a crystal, the regular array of atoms in a crystal might act as a three-dimensional grating for the diffraction of x rays. Acting on this suggestion, W. Friedrich and P. Knipping allowed a collimated beam of x rays to pass through a crystal, behind which was a photographic plate. In addition to the central beam, they observed a regular array of spots (see Figure 3-10). From an analysis of the positions of the spots, they were able to calculate that their x-ray beam contained wavelengths ranging from about 0.1 to 0.5 Å. This experiment confirmed two important assumptions: that x rays are a form of electromagnetic radiation, and that atoms in crystals are arranged in a regular array.

A simple and convenient way of analyzing the diffraction of x rays by crystals was proposed by William Lawrence Bragg in 1912. He considered the interference of x rays due to scattering from various sets of parallel planes of atoms, now called *Bragg planes.* Two sets of Bragg planes are illustrated in Figure 3-11 for NaCl, which has a simple crystal structure called *face-centered cubic.* Consider Figure 3-12. Waves scattered from two successive atoms within a plane will be in phase and thus interfere constructively, independent of the wavelength, if the scattering angle equals the incident angle. (This condition is the same as for reflection.) Waves scattered at equal angles from atoms in

---

[1] "Die Beugung, der Röntgen strahlen," *Annalen der Physik,* **68,** 884 (1899).

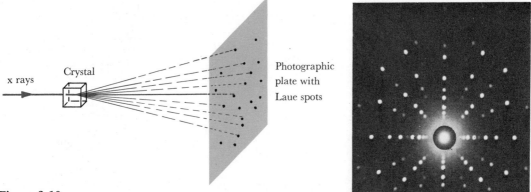

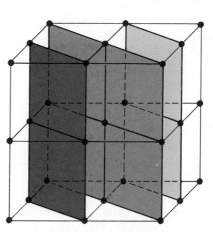

General Electric Company

Figure 3-10
(*Left*) Schematic sketch of Laue experiment. The crystal acts as a three-dimensional grating, which diffracts the x-ray beam and produces a regular array of spots, called a *Laue pattern,* on a photographic plate. (*Right*) Modern Laue-type x-ray diffraction pattern using a niobium diboride crystal and 20-kV molybdenum x rays.

Figure 3-11
A crystal of NaCl showing two sets of Bragg planes.

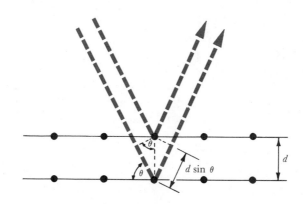

Figure 3-12
Bragg scattering from two successive planes. The waves from the two atoms shown have a path difference of $2d \sin \theta$. They will be in phase if the Bragg condition $2d \sin \theta = m\lambda$ is met.

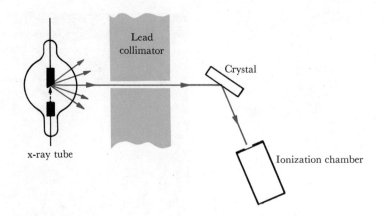

x-ray tube

Figure 3-13
Schematic diagram of
Bragg crystal spectrom-
eter. A collimated x-ray
beam is incident on a
crystal and scattered into
an ionization chamber.
The crystal and ionization
chamber can be rotated
to keep the angles of incidence and scattering
equal as both are varied.
By measuring the ioniza-
tion in the chamber as a
function of angle, the
spectrum of the x rays
can be determined using
the Bragg condition
$2d \sin \theta = m\lambda$, where d is
the separation of the
Bragg planes in the
crystal. If the wavelength
λ is known, the spacing d
can be determined.

two different planes will be in phase if the path length is an integral number of wavelengths. From Figure 3-12 we see that this
condition is satisfied if

$$2d \sin \theta = m\lambda \qquad \qquad 3\text{-}35$$

where m = an integer.

Equation 3-35 is called the *Bragg condition*. At angles meeting
this condition, waves will be strongly scattered because the
waves scattered from many atoms interfere constructively. Figure 3-13 shows the main features of a crystal spectrometer first
built by William Henry Bragg, the father of W. L. Bragg. X rays
with wavelengths satisfying the Bragg condition are scattered at
an angle θ equal to the incident angle. A measurement of the
scattered intensity versus angle gives the distribution of wavelengths in the incident x-ray beam if the spacing d is known. For
a simple crystal such as NaCl, the spacing of Bragg planes can be
easily calculated because the distance between the Na^+ and Cl^-
ions can be obtained from a measurement of the density and a
knowledge of Avogadro's number N_A. The scattering of x rays
of known wavelength from crystals can be used to obtain information about the structure of crystals. W. H. Bragg and W. L.
Bragg were awarded the Nobel Prize in 1915 for their contributions to crystal analysis.

Figure 3-14 shows a typical x-ray spectrum produced by bombarding a molybdenum target with electrons. In this figure,
$I(\lambda)\, d\lambda$ is the intensity of the x-ray beam in the wavelength interval $d\lambda$. The spectrum consists of a series of sharp lines, called
the *characteristic spectrum*, superimposed on the continuous *bremsstrahlung* spectrum (from the German for "braking radiation").
The line spectrum is characteristic of the target material and
varies from element to element. The continuous spectrum has a
sharp cutoff wavelength, λ_m, which is independent of the target
material but depends on the energy of the bombarding electrons. If the voltage of the x-ray tube is V_0 in volts, the cutoff
wavelength is given by

$$\lambda_m = \frac{1.24 \times 10^4}{V_0} \text{ Å} \qquad \qquad 3\text{-}36$$

Bragg's law

Duane-Hunt rule

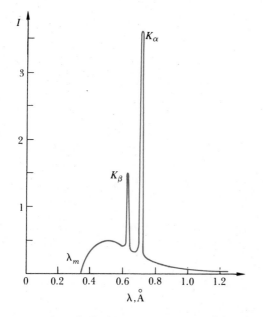

Figure 3-14
X-ray spectrum of molyb-
denum showing the K_α
and K_β lines superimposed on a continuous
spectrum. The cutoff
wavelength λ_m is independent of the target element and is related to the
voltage of the x-ray tube
V_0 by $\lambda_m = hc/eV_0$. The
wavelengths of the lines
are characteristic of the
target element.

Equation 3-36 is called the *Duane-Hunt rule.* It is easily explained by the photon theory of x rays. The kinetic energy of
each electron accelerated through a potential difference of V_0 is
eV_0. (We can neglect the work function and initial kinetic energy
of the electrons at the cathode because these are only a few electron volts, whereas V_0 is typically on the order of several thousand volts.) If an electron loses all its energy to a single photon,
the wavelength of the photon will be given by[1]

$$\frac{hc}{\lambda_m} = eV_0$$

This is the same as Equation 3-36 since hc is 1.24×10^4 eV-Å.
We can see that a measurement of the cutoff wavelength can be
used to determine h/e.

Let us now consider the scattering of x rays by electrons.
Compton pointed out that if the scattering process were considered to be a "collision" between a photon of energy hf_1 (and momentum hf_1/c) and an electron, the recoiling electron would absorb part of the total energy, and the scattered photon would
therefore have less energy and thus a lower frequency. [The fact
that electromagnetic radiation of energy E carries momentum
E/c was known from classical theory and from experiments of
Nichols and Hull in 1903. This relation is also consistent with the
relativistic expression $E^2 = p^2c^2 + (mc^2)^2$ for a particle with zero
rest mass.] Compton applied the laws of conservation of momentum and energy to the collision of a photon with an isolated
electron to obtain the change in the wavelength of the photon as
a function of the scattering angle. His result is

$$\lambda_2 - \lambda_1 = \frac{h}{mc}(1 - \cos\theta) \qquad\qquad 3\text{-}37 \qquad \textit{Compton scattering}$$

[1] Most electrons lose their energy by emitting a number of photons of randomly
varied energies, hence the continuous bremsstrahlung spectrum.

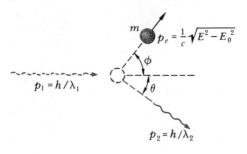

$$p_e = \frac{1}{c}\sqrt{E^2 - E_0^{\,2}}$$

$$p_1 = h/\lambda_1$$

$$p_2 = h/\lambda_2$$

Figure 3-15
The scattering of x rays can be treated as a collision of a photon of initial momentum h/λ_1 and a free electron. Using conservation of momentum and energy, the momentum of the scattered photon h/λ_2 can be related to the initial momentum, the electron mass, and the scattering angle. The resulting Compton equation for the change in the wavelength of the x ray is Equation 3-37.

Compton wavelength of electron

where θ is the scattering angle for the photon, as shown in Figure 3-15. The change in wavelength is independent of the original wavelength. The quantity $h/mc = hc/mc^2$ depends only on the mass of the electron. It has the dimensions of length and is called the *Compton wavelength of the electron*. Its value is

$$\lambda_c = \frac{hc}{mc^2} = \frac{1.24 \times 10^4 \text{ eV-Å}}{5.11 \times 10^5 \text{ eV}} = 0.0243 \text{ Å} \qquad \text{3-38}$$

Because $\lambda_2 - \lambda_1$ is small, it is difficult to observe unless λ_1 is very small so that the fractional change $(\lambda_2 - \lambda_1)/\lambda_1$ is appreciable.

Compton verified his result experimentally using the characteristic x-ray line of wavelength 0.711 Å from molybdenum for the incident monochromatic photons, and scattering these photons from electrons in graphite. The wavelength of the scattered photons was measured using a Bragg crystal spectrometer. His experimental arrangement is shown in Figure 3-16; Figure 3-17 shows his results. The first peak corresponds to scattering with no shift in the wavelength due to scattering by the inner electrons of carbon. Since these are tightly bound to the atom, it is the whole atom that recoils rather than the individual electron. The expected shift for this case is given by Equation 3-37, with m being the mass of the atom, which is about 10^4 times that of the electron; thus this shift is negligible. The variation of $\Delta\lambda$ with θ was found to be that predicted by Equation 3-37.

We have seen in this and the preceding two sections that the interaction of electromagnetic radiation with matter is a quantum interaction. It is perhaps curious that after so many years of debate about the nature of light, we now find that we must have both a particle theory to describe in detail the energy

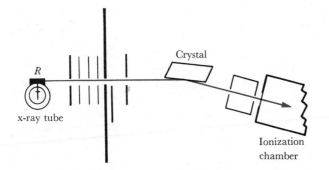

R

Crystal

x-ray tube

Ionization
chamber

Figure 3-16
Schematic sketch of Compton apparatus. X rays from the tube strike the carbon block R and are scattered into a Bragg-type crystal spectrometer. In this diagram, the scattering angle is 90°.

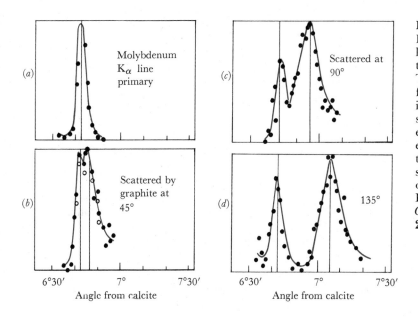

(a) Molybdenum K_α line primary

(b) Scattered by graphite at 45°

(c) Scattered at 90°

(d) 135°

6°30' 7° 7°30'
Angle from calcite

6°30' 7° 7°30'
Angle from calcite

Figure 3-17
Intensity versus wavelength for Compton scattering at several angles. The first peak results from photons of the original wavelength that are scattered by tightly bound electrons, which have an effective mass equal to that of the atom. The separation in wavelength of the peaks is given by Equation 3-37. [*From A. Compton*, Physical Review, **22,** *411 (1932)*.]

exchange between electromagnetic radiation and matter, and a wave theory to describe the interference and diffraction of electromagnetic radiation. We shall discuss this so-called "particle-wave duality" in more detail in Chapter 5.

Optional

Derivation of Equation 3-37

Let λ_1 and λ_2 be the wavelengths of the incident and scattered x rays, respectively, as shown in Figure 3-15. The corresponding momenta are

$$p_1 = \frac{E_1}{c} = \frac{hf_1}{c} = \frac{h}{\lambda_1}$$

and

$$p_2 = \frac{E_2}{c} = \frac{h}{\lambda_2}$$

using $f\lambda = c$. Since the energy of the incident x ray (17.4 keV for $\lambda = 0.711$ Å) is much greater than the binding energy of the valence electrons in carbon (about 11 eV), the electron can be considered to be free.

Conservation of momentum gives

$$\mathbf{p}_1 = \mathbf{p}_2 + \mathbf{p}_e$$

or

$$
\begin{aligned}
p_e^2 &= p_1^2 + p_2^2 - 2\mathbf{p}_1\cdot\mathbf{p}_2 \\
&= p_1^2 + p_2^2 - 2p_1 p_2 \cos\theta
\end{aligned}
\qquad 3\text{-}39
$$

where $\mathbf{p}_e$ is the momentum of the electron after the collision. The energy of the electron before the collision is simply its rest energy $E_0 = mc^2$. After the collision, the energy of the electron is $(E_0^2 + p_e^2 c^2)^{1/2}$.

Conservation of energy gives

$$p_1 c + E_0 = p_2 c + (E_0^2 + p_e^2 c^2)^{1/2}$$

Transposing the term $p_2 c$ and squaring, we obtain

$$E_0^2 + c^2(p_1 - p_2)^2 + 2cE_0(p_1 - p_2) = E_0^2 + p_e^2 c^2$$

or

$$p_e^2 = p_1^2 + p_2^2 - 2p_1 p_2 + \frac{2E_0(p_1 - p_2)}{c} \qquad 3\text{-}40$$

If we eliminate p_e^2 from Equations 3-39 and 3-40, we obtain

$$\frac{E_0(p_1 - p_2)}{c} = p_1 p_2 (1 - \cos \theta)$$

Multiplying each term by $hc/p_1 p_2 E_0$ and using $\lambda = h/p$, we obtain Compton's equation:

$$\lambda_2 - \lambda_1 = \frac{hc}{E_0}(1 - \cos \theta) = \frac{hc}{mc^2}(1 - \cos \theta)$$

Question

7. Why is it practically impossible to observe the Compton effect using visible light?

3-7 Quantization of Energy States of Matter

In 1908 Einstein showed that the failure of the equipartition theorem in predicting the specific heats of solids at low temperatures could be understood if it were assumed that the atoms of the solid could have only certain discrete energy values. Einstein's calculation is closely related to Planck's calculation of the average energy of a harmonic oscillator assuming the oscillator can take on only a discrete set of energies. We shall see in this section how the idea of quantized energy states also explains the puzzling behavior of the heat capacities of diatomic gases. In particular we shall be able to understand why the H_2 molecule seems to have only 3 degrees of freedom (corresponding to translation) at low temperatures, 5 degrees of freedom at intermediate temperatures (corresponding to translation and rotation), and 7 degrees of freedom at high temperatures (corresponding to translation, rotation, and vibration).

Consider 1 mole of a solid consisting of N_A molecules, each free to vibrate in three dimensions about a fixed center. For simplicity, Einstein assumed that all the molecules oscillate at the same frequency f in each direction. The problem is then equivalent to $3N_A$ one-dimensional oscillators, each with frequency f. The classical distribution function for the energy of a set of one-dimensional oscillators is given by Equation 3-22:

$$f(E) = Ce^{-E/kT}$$

The average energy turns out (from Equation 3-23) to be kT, which agrees with the equipartition theorem for a system with 2 degrees of freedom (the kinetic energy and potential energy of vibration). The classical calculation thus gives for the total energy of $3N_A$ one-dimensional oscillators

$$U = 3N_A kT = 3RT$$

and the molar heat capacity is

$$C_v = \frac{dU}{dT} = 3R$$

which is the Dulong-Petit law.

Following Planck, Einstein assumed that the energy of each oscillator could take on only the values given by

$$E_n = n\epsilon = nhf \qquad\qquad\qquad 3\text{-}41$$

where h is Planck's constant and $n = 0, 1, 2, \ldots$. He then used the discrete form of the Maxwell-Boltzmann distribution to calculate the average energy from

$$\overline{E} = \sum_{n=0}^{\infty} E_n f_n = \sum_{n=0}^{\infty} E_n C e^{-E_n/kT}$$

where C is the normalization constant determined by

$$\sum_{n=0}^{\infty} f_n = \sum_{n=0}^{\infty} C e^{-E_n/kT} = 1$$

These equations are the same as Equations 3-26 and 3-27. These sums are worked out in Section 3-8. The result is

$$\overline{E} = \frac{hf}{e^{hf/kT} - 1} \qquad\qquad\qquad 3\text{-}42$$

Modification of equipartition theorem

(This result is of course the same as Equation 3-28.) At high temperatures the quantity hf/kT is small and we can expand the exponential, using $e^x \approx 1 + x + \cdots$ for $x \ll 1$, where $x = hf/kT$. Then

$$e^{hf/kT} - 1 \approx \left(1 + \frac{hf}{kT} + \cdots\right) - 1 \approx \frac{hf}{kT}$$

and $\overline{E}$ approaches kT, in agreement with the equipartition theorem.

The total energy for $3N_A$ oscillators is now

$$U = 3N_A \overline{E} = \frac{3N_A hf}{e^{hf/kT} - 1} \qquad\qquad\qquad 3\text{-}43$$

and the heat capacity is

$$C_v = \frac{dU}{dT} = 3N_A k \left(\frac{hf}{kT}\right)^2 \frac{e^{hf/kT}}{(e^{hf/kT} - 1)^2} \qquad\qquad 3\text{-}44$$

Einstein heat capacity

It is left as an exercise to show directly from Equation 3-44 that $C_v \to 0$ as $T \to 0$ and $C_v \to 3N_A k = 3R$ as $T \to \infty$.

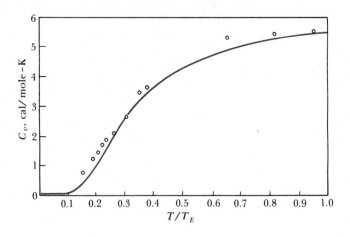

Figure 3-18
Molar heat capacity of diamond versus reduced temperature. The solid curve is that predicted by Einstein. [*From Einstein's original paper, Annalen der Physik (4), 22, 180 (1907).*]

Figure 3-18 shows a comparison of this calculation with experiments. As can be seen, the curve fits the experimental points quite well except at very low temperatures, where the data fall slightly above the curve. The lack of detailed agreement of the curve with the data at low T is due to the oversimplification of the model. A refinement of this model was made by P. Debye, who gave up the assumption that all the molecules vibrate at the same frequency. He allowed for the possibility that the motion of one molecule could be affected by that of the others and treated the solid as a system of coupled oscillators. Calculations with the Debye model are somewhat involved and will not be considered here. The improvement of the Debye model over the Einstein model is shown by Figure 3-19.

By comparing the Einstein calculation of the average energy per molecule with the classical one, we can gain some insight into the problem of when the classical theory will work and when it will fail. Let us define the critical temperature,

$$T_E = \frac{hf}{k} \qquad\qquad 3\text{-}45 \qquad \textit{Einstein temperature}$$

called the *Einstein temperature*. The energy distribution in terms of this temperature is $f_n(E) = Ce^{-nT_E/T}$. For temperatures T much higher than T_E, small changes in n have little effect on the expo-

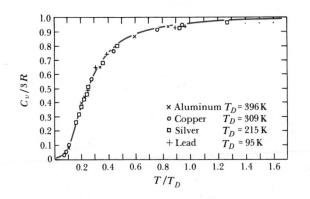

Figure 3-19
Molar heat capacity of several solids versus reduced temperature. The solid curve is that predicted by Debye. The data are taken from Debye's original paper. [*From* Annalen der Physik (4), **39**, 789 (1912), *as adapted by David MacDonald,* Introductory Statistical Mechanics for Physicists, *New York: John Wiley & Sons, Inc., 1963; by permission.*]

nential in the distribution, that is, $f_n \approx f_{n+1}$. Then E might as well be treated as a continuous variable and the sum in $\bar{E}$ be replaced by an integral. However, for temperatures much lower than T_E, even the smallest possible change in n, $\Delta n = 1$, results in a significant change in $e^{-nT_E/T}$, and we would expect that the discontinuity of possible energy values becomes significant, and the result is quite different. Since hard solids have stronger binding forces than soft ones, their frequencies of molecular oscillation and therefore their Einstein temperatures are higher. For lead and gold, T_E is of the order of 50 to 100 K; ordinary temperatures of around 300 K are "high" for these metals, and they obey the classical Dulong-Petit law at these temperatures. For diamond, T_E is well over 1000 K; in this case 300 K is a "low" temperature, and C_v is much less than the Dulong-Petit value of $3R$ at this temperature.

Let us now see if we can understand the specific heat of diatomic gases on the basis of discrete, or quantized, energies. In Chapter 2 we wrote the energy of a diatomic molecule as the sum of translational, rotational, and vibrational energies. If f is the frequency of vibration, and the vibrational energy is quantized $E_{\text{vib}} = nhf$, as we assumed for solids, we know from the previous calculation that for low temperatures the average energy of vibration approaches zero and vibration will not contribute to C_v. We can define a critical temperature for vibration of a diatomic gas molecule by

$$T_v = \frac{hf}{k} \qquad\qquad 3\text{-}46$$

where f is the frequency of vibration. Apparently $T_v > 15°C$ for all the diatomic gases listed in Table 2-1 except Cl_2. From Figure 2-5 we can see that T_v is of the order of 1000 to 5000 K for H_2.

The rotational energy of a diatomic molecule is

$$E_R = \tfrac{1}{2}I\omega^2$$

where I is the moment of inertia and ω is the angular velocity of rotation. It is not obvious how the rotational energy is quantized, or if it is; however, let us borrow a result from Bohr's theory of the hydrogen atom, to be discussed in detail in the next chapter. Bohr found that he could derive the radiation spectrum of the hydrogen atom if he assumed that the angular momentum of the atom took on only the discrete values $nh/2\pi$, where n is an integer and h is Planck's constant. If L is the angular momentum of a diatomic molecule, $L = I\omega$, and we can write the energy as

$$E_R = \frac{L^2}{2I}$$

and using the Bohr quantum condition $L = n(h/2\pi)$,

$$E_R = n^2 \frac{h^2}{8\pi^2 I} \qquad\qquad 3\text{-}47$$

The energy distribution will contain the factor

$$e^{-E_R/kT} = e^{-n^2(h^2/8\pi^2 IkT)}$$

and we can define a critical temperature for rotation

$$T_R = \frac{h^2}{8\pi^2 I k}$$ 3-48

If this procedure is correct, we expect that for temperatures $T \gg T_R$, the equipartition theorem will hold for rotation and the average energy of rotation will approach $\frac{1}{2}kT$ for each axis of rotation, while for low temperatures, $T \ll T_R$, the average energy of rotation will approach 0. Let us estimate T_R for some cases of interest:

1. H_2, for rotation about the x or y axis, taking the z axis as the line joining the atoms. The moments of inertia I_x and I_y through the center of mass are

$$I_x = I_y = \frac{1}{2}MR^2$$

The separation of the atoms is about $R \approx 0.8$ Å. The mass of the H atom is about $M \approx 940 \times 10^6$ eV/c^2. We first calculate kT_R:

$$kT_R = \frac{h^2}{8\pi^2 I} = \frac{(hc)^2}{4\pi^2 Mc^2 R^2} = \frac{(1.24 \times 10^4 \text{ eV-Å})^2}{4\pi^2 (940 \times 10^6 \text{ eV})(0.8 \text{ Å})^2}$$
$$\approx 6.4 \times 10^{-3} \text{ eV}$$

Using $k \approx 2.6 \times 10^{-2}$ eV/300 K, we obtain

$$T_R = \frac{6.4 \times 10^{-3}}{2.6 \times 10^{-2}} \, 300 \text{ K} \approx 74 \text{ K}$$

As can be seen from Figure 2-5, this is indeed the temperature region below which the rotational energy does not contribute to the heat capacity.

2. O_2. Since the mass of the oxygen atom is 16 times that of the hydrogen atom and the separation is roughly the same, the critical temperature for rotation will be $T_R \approx (\frac{74}{16}) \approx 4.6$ K. For all temperatures at which O_2 exists as a gas, $T \gg T_R$.

3. A monatomic gas, or rotation of diatomic gas about the z axis. We shall take the H atom for calculation. The moment of inertia of the atom is due mainly to the electron since the radius of the nucleus is extremely small (about 10^{-15} m). The distance from the nucleus to the electron is about the same as the separation of atoms in the H_2 molecule. Since the mass of the electron is about 2000 times smaller than that of the atom, we have

$$I_H \approx \frac{1}{2000} I_{H_2}$$

and

$$T_R \approx 2000 \times 74 \text{ K} \approx 1.5 \times 10^5 \text{ K}$$

This is always much higher than the dissociation temperature for any gas. Thus $\bar{E}_R \approx 0$ for monatomic gases, and for rotation of diatomic gases about the line joining the atoms for all attainable temperatures.

We see that energy quantization explains, at least qualitatively, the temperature dependence of the specific heats of gases and solids.

Example 3-7   What is the average energy of vibration of the molecules in a solid if the temperature is (a) $T = hf/2k$, (b) $T = 4hf/k$?

(a) This is lower than the critical temperature hf/k, so we expect a result considerably lower than the high-temperature limit of kT given by the equipartition theorem. From Equation 3-42 we have

$$\overline{E} = \frac{hf}{e^{hf/kT} - 1} = \frac{2kT}{e^2 - 1} = 0.31\, kT$$

(b) This temperature is four times the critical temperature so we expect a result near the high-temperature limit of kT. Using $hf/kT = \frac{1}{4}$ in Equation 3-42 we have

$$\overline{E} = \frac{0.25kT}{e^{0.25} - 1} = 0.880kT$$

Example 3-8   At the "low" and "high" temperatures of Example 3-7, find the ratio of the number of oscillators with energy $E_1 = hf$ to the number with $E_0 = 0$. At any temperature T, the Maxwell-Boltzmann distribution for the fraction of oscillators with energy $E_n = nhf$ is $f_n = Ce^{-E_n/kT} = Ce^{-nhf/kT}$. For $n = 0$ this gives $f_0 = Ce^0 = C$. The ratio f_n/f_0 is then $f_n/f_0 = e^{-nhf/kT}$.

(a) For $n = 1$ and $kT = \frac{1}{2}hf$ we have $f_1/f_0 = e^{-hf/kT} = e^{-2} = 0.135$. Most of the oscillators are in the lowest energy state $E_0 = 0$.

(b) For the higher temperature $kT = 4hf$ we get $f_1/f_0 = e^{-hf/kT} = e^{-0.25} = 0.779$.

At the higher temperature the states are more nearly equally populated and the average energy is larger.

Optional

3-8   Calculation of the Sums in Equations 3-26 and 3-27

Let $x = \epsilon/kT = hf/kT$. The sum in Equation 3-26 is then

$$\sum_{n=0}^{\infty} f_n = \sum_{n=0}^{\infty} Ce^{-nx} = C[e^0 + e^{-x} + (e^{-x})^2 + (e^{-x})^3$$

$$+ \cdots] = C(1 + y + y^2 + y^3 + \cdots)$$

where $y = e^{-x}$. This sum is the series expansion for the function $(1 - y)^{-1}$, that is,

$$(1 - y)^{-1} = 1 + y + y^2 + y^3 + \cdots$$

Then $\Sigma f_n = C(1 - y)^{-1} = 1$ gives $C = 1 - y$. In terms of x and y, Equation 3-27 is

$$\overline{E} = \sum_{n=0}^{\infty} E_n\, Ce^{-E_n/kT} = C \sum_{n=0}^{\infty} nhfe^{-nhf/kT}$$

$$= Chf \sum_{n=0}^{\infty} ne^{-nx}$$

Now note that Σne^{-nx} is the negative derivative of Σe^{-nx} with respect to x. But $\Sigma e^{-nx} = (1 - y)^{-1}$. We thus have

$$\sum ne^{-nx} = -\frac{d}{dx} \sum e^{-nx} = -\frac{d}{dx}(1 - y)^{-1}$$

$$= (1 - y)^{-2}\left(-\frac{dy}{dx}\right) = y(1 - y)^{-2}$$

since $dy/dx = d(e^{-x})/dx = -e^{-x} = -y$. Multiplying this sum by hf and by $C = 1 - y$, we obtain for the average energy

$$\overline{E} = hfC \sum_{n=0}^{\infty} ne^{-nx} = hf(1 - y)y(1 - y)^{-2} = \frac{hfy}{1 - y}$$

$$= \frac{hfe^{-x}}{1 - e^{-x}}$$

If we multiply both numerator and denominator by e^x, we obtain

$$\overline{E} = \frac{hf}{e^x - 1} = \frac{hf}{e^{hf/kT} - 1}$$

which is Equation 3-28.

Summary

Avogadro's number, the faraday, and e are related by $F = N_A e$. J. J. Thomson's measurements with cathode rays showed that the same particle (the electron), with e/m about 2000 times that of ionized hydrogen, exists in all elements. The fundamental electric charge e can be measured directly by the Millikan oil-drop experiment. There is overwhelming evidence that charges always occur in multiples of e.

   The power per unit area radiated by a blackbody is related to the absolute temperature of the body by the Stefan-Boltzmann law, $R = \sigma T^4$. The spectral distribution of blackbody radiation cannot be derived by classical physics, but the correct formula can be obtained by the assumption of energy quantization. The photoelectric effect and the Compton effect show that light energy interacts with matter in quantized amounts called photons. The energy and momentum of a photon is proportional to its frequency. The Einstein equation for the photoelectric effect is

$$hf = eV_0 + \phi$$

The wavelength shift in the Compton effect is given by

$$\lambda_2 - \lambda_1 = \frac{h}{mc}(1 - \cos\theta)$$

The assumption of discrete energy states for atoms and molecules leads to a temperature dependence of the specific heat which is in good agreement with experiment. At high temperatures, kT is much greater than the energy-level spacing, and the quantum calculation gives the same result as the classical calculation. When the temperature is low enough so that kT is of the same order as the energy-level spacing, the quantum and classical calculations differ. Because of the finite energy gap between the lowest and first excited states, the average energy approaches the energy of the lowest state as T approaches zero, and C_v approaches zero.

Planck's constant has the value

$$h = 6.626 \times 10^{-34} \text{ J-sec} = 4.136 \times 10^{-15} \text{ eV-sec}$$

A useful combination of constants is

$$hc = 1.24 \times 10^4 \text{ eV-Å} = 1.24 \times 10^3 \text{ eV-nm}$$

References

1. R. A. Millikan, *Electrons (+ and −), Protons, Photons, Neutrons, Mesotrons, and Cosmic Rays, 2d edition,* Chicago: The University of Chicago Press, 1947. This book on modern physics by one of the great experimentalists of his time contains fascinating, detailed descriptions of Millikan's oil-drop experiment and his verification of the Einstein photoelectric equation. (There is a new edition of this book under the title *The Electron,* but with different page numbers. It is a facsimile of the first edition without the corrections made for the second edition. All page references given in the text refer to the second edition.)

2. M. H. Shamos (ed.), *Great Experiments in Physics,* New York: Holt, Rinehart and Winston, Inc., 1962. This book contains 25 original papers and extensive editorial comment. Those of particular interest for this chapter are by Faraday, Hertz, Roentgen, J. J. Thomson, Einstein (photoelectric effect), Millikan, Planck, and Compton.

3. A. Goble and D. Baker, *Elements of Modern Physics,* New York: The Ronald Press Company, 1962. Chapters 6 and 8 of this textbook discuss determinations of e, N_A, F, and h.

4. E. Cohen, K. Crowe, and J. Dumond, *Fundamental Constants of Physics,* New York: Interscience Publishers, Inc., 1957.

5. J. H. Sanders, *Fundamental Atomic Constants,* Fairlawn, N.J.: Oxford University Press, 1961. This and Reference 4 discuss present methods, as well as earlier methods, of determining the values of fundamental constants.

6. G. P. Thomson, *J. J. Thomson, Discoverer of the Electron,* Garden City, N.Y.: Anchor Books, Doubleday & Company, Inc., 1964. An interesting study of J. J. Thomson by his son, G. P. Thomson, also a physicist.

7.   F. K. Richtmyer, E. H. Kennard, and J. N. Cooper, *Introduction to Modern Physics*, New York: McGraw-Hill Book Company, 1969. The sixth edition of an excellent text originally published in 1928, intended as a survey course for graduate students.

Exercises

Section 3-1, Early Estimates of e and e/m

1.   Use the known values of the faraday and Avogadro's number in the table in Appendix D to calculate the electron charge e.

2.   Using a current of 3 A, how long does it take to obtain 0.1 g of copper from the electrolysis of $CuSO_4$?

Section 3-2, The J. J. Thomson Experiment

3.   Using the known values of e and m, find e/m for (a) an electron, (b) a proton.

4.   If electrons have kinetic energy of 2000 eV, find (a) their speed, (b) the time needed to traverse a distance of 5 cm between plates D and E in Figure 3-1, (c) the vertical component of their velocity after passing between the plates if the electric field is 3.33×10^3 V/m.

5.   In Figure 3-1, the plates D and E are 1.5 cm apart and 5 cm long and are kept at 50 V potential difference. (a) If the electrons have kinetic energy of 2000 eV, find the deflection produced in the 5-cm path between the plates. (b) What is the total deflection of the spot on the screen, assuming the beam travels an additional 30 cm in the field-free region before striking the screen? (c) What strength magnetic field would be needed between the plates for no deflection?

√6.   In J. J. Thomson's first method, the heat capacity of the beam stopper was about 5×10^{-3} cal/°C and the temperature increase was about 2°C. How many 2000-eV electrons struck the beam stopper?

Section 3-3, Quantization of Electric Charge

7.   Calculate the standard deviation σ for Wilson's data in Table 3-1. What percentage of the mean value is σ?

8.   Show that the electric field needed to make the rise time of the oil drop equal to its field-free fall time is $\mathscr{E} = 2mg/q$.

9.   (a) Find the terminal fall velocity v_f from Table 3-2 using the mean fall time and the distance given (10.21 mm). (b) Use the density of oil $\rho = 0.943$ g/cm³ $= 943$ kg/m³, the viscosity of air $\eta = 1.824 \times 10^{-5}$ N-sec/m², and $g = 9.81$ m/sec² to calculate the radius a of the oil drop from Stokes' law as expressed in Equation 3-9.

Section 3-4, Blackbody Radiation

√10.   Show that Planck's constant h has units of angular momentum.

11.   Find λ_m for blackbody radiation at (a) $T = 3$ K, (b) $T = 300$ K, (c) $T = 3000$ K.

12.   Use the result of Example 3-3 and Equations 3-16 and 3-18 to express Stefan's constant in terms of h, c, and k. Using the known values of these constants, calculate Stefan's constant.

13. Find the temperature of a blackbody if its spectrum has its peak at (*a*) $\lambda_m = 700$ nm, (*b*) $\lambda_m = 3$ cm (microwave region), (*c*) $\lambda_m = 3$ m (FM radio waves).

14. The derivation of the one-dimensional analogue of Equation 3-20 for the number of modes of radiation in a cavity is not difficult. The number of standing waves n of a particular polarization in a one-dimensional box of length L is given by the standing-wave condition $n\lambda/2 = L$. Show that the number of standing waves of a given polarization in the region $d\lambda$ is $2L\lambda^{-2}\, d\lambda$. This result multiplied by two for the two polarizations and divided by the length L is the one-dimensional analogue of Equation 3-20.

15. If the absolute temperature of a blackbody is doubled, by what factor is the total emitted power increased?

16. What is the total power per unit area emitted by a blackbody at a temperature of (*a*) 300 K, (*b*) 3000 K?

17. Calculate the average energy $\bar{E}$ per mode of oscillation for (*a*) a long wavelength $\lambda = 10\ hc/kT$, (*b*) a short wavelength $\lambda = 0.1\ hc/kT$, and compare your results with that from the equipartition theorem.

Section 3-5, The Photoelectric Effect

18. Calculate the range of photon energies in the visible spectrum from about 400 to 700 nm wavelength.

19. Find the photon energy corresponding to (*a*) a wavelength of 1 Å (about 1 atomic diameter), (*b*) a wavelength of 1 fm (= 10^{-15} m, about 1 nuclear diameter), (*c*) a frequency of 100 MHz in the FM radio band.

20. The work function for tungsten is 4.58 eV. (*a*) Find the threshold frequency and wavelength for the photoelectric effect on tungsten. Find the stopping potential if the wavelength of the incident light is (*b*) 200 nm, (*c*) 250 nm.

21. When light of wavelength 300 nm is incident on potassium, the emitted electrons have maximum kinetic energy of 1.91 eV. (*a*) What is the energy of the incident photons? (*b*) What is the work function for potassium? (*c*) What would be the stopping potential if the incident light had a wavelength of 350 nm?

22. The threshold wavelength for the photoelectric effect for silver is 262 nm. (*a*) Find the work function for silver. (*b*) Find the stopping potential if the incident radiation has a wavelength of 200 nm.

23. What is the threshold wavelength and the work function for the metal used by Millikan to obtain Figure 3-9?

24. The work function for cesium is 1.9 eV. (*a*) Find the threshold frequency and wavelength for the photoelectric effect. Find the stopping potential if the wavelength of the incident light is (*b*) 300 nm, (*c*) 400 nm.

25. What number of photons per second of frequency 100 MHz is needed by an FM radio that can just detect a signal of 10^{-12} W?

26. (*a*) If 5 percent of the power of a 100-W bulb is radiated in the visible spectrum, how many visible photons are radiated per second? (*b*) If the bulb is a point source radiating equally in all directions, what is the

flux of photons (number per unit time per unit area) at a distance of 2 m?

27. Under optimum conditions, the eye will perceive a flash if about 60 photons arrive at the cornea. How much energy is this in joules if the wavelength is 550 nm?

Section 3-6, X Rays and the Compton Effect

28. An x-ray tube operates at a potential of 40,000 V. What is the cut-off wavelength of the continuous x-ray spectrum from this tube?

29. The minimum wavelength in the continuous x-ray spectrum from a TV tube is 0.124 nm. What is the voltage of the tube?

30. Find the momentum of a photon in eV/c and in kg-m/sec if the wavelength is (a) 400 nm, (b) 1 Å = 0.1 nm, (c) 3 cm.

√31. The wavelength of Compton-scattered photons is measured at $\theta = 90°$. If $\Delta\lambda/\lambda$ is to be 1 percent, what should the wavelength of the incident photons be?

32. Compton used photons of wavelength 0.0711 nm. (a) What is the energy of these photons? (b) What is the wavelength of the photons scattered at $\theta = 180°$? (c) What is the energy of the photons scattered at $\theta = 180°$? (d) What is the recoil energy of the electrons if $\theta = 180°$?

Section 3-7, Quantization of Energy States of Matter

√33. Find the average energy of an oscillator at (a) $T = 10\ hf/k$, (b) $T = hf/k$, (c) $T = 0.1\ hf/k$, and compare your results with that from the equipartition theorem.

34. Calculate the ratio f_n/f_0 as in Example 3-8 for the $n = 2$ state for (a) $kT = \frac{1}{2}hf$, (b) $kT = 4hf$.

35. Use Equation 3-44 to calculate the value of C_v for a solid at the Einstein temperature $T_E = hf/k$.

36. Repeat Example 3-8 for (a) $kT = 0.1hf$, (b) $kT = 10hf$.

Section 3-8, Calculation of the Sums in Equations 3-26 and 3-27

There are no exercises for this section.

Problems

1. (a) Calculate the standard deviation σ for the numbers in column 5, Table 3-2. What percentage is σ of the mean value? (b) For a Gaussian distribution, the chance of a value differing from the mean by more than 3σ is less than 0.3 percent. The number corresponding to $n' = 9$ differs from the mean by more than any other (except $n' = 1$). To investigate the possibility of this being due to the existence of fractional charges, recalculate the number in column 5, assuming n' to be $9\frac{1}{3}$. By how many standard deviations does this number differ from the mean? Are these data good evidence for the existence of "quarks," particles with charge $\frac{1}{3}e$ or $\frac{2}{3}e$?

2. Data for stopping potential versus wavelength for the photoelectric effect using sodium are

λ, Å	2000	3000	4000	5000	6000
V_0, V	4.20	2.06	1.05	0.41	0.03

Plot these data in such a way as to be able to obtain (*a*) the work function, (*b*) the threshold frequency, (*c*) the ratio h/e.

3.   This problem is one of *estimating* the time lag (expected classically but not observed) for the photoelectric effect. Assume that a point light source gives 1 W = 1 joule/sec of light energy. (*a*) Assuming uniform radiation in all directions, find the light intensity in eV/m^2-sec at a distance of 1 m from the source. (*b*) Assuming some reasonable size for an atom, find the energy per unit time incident on the atom for this intensity. (*c*) If the work function is 2 eV, how long does it take for this much energy to be absorbed, assuming that all the energy hitting the atom is absorbed?

4.   The energy from the sun falling on a unit area at normal incidence at the earth, called the *solar constant,* is equal to 1370 W/m^2. (*a*) What is the total power radiated by the sun? (The radius of the earth is 6.37×10^6 m, and the distance from sun to earth is 1.49×10^{11} m.) (*b*) Assuming the sun to be a blackbody of radius 6.96×10^8 m, calculate its temperature T from the Stefan-Boltzmann law.

5.   What should the energy of the photons be so that the maximum change in wavelength due to Compton scattering by electrons is 1 percent? Using these photons, what is $\Delta\lambda$ in angstroms for Compton-scattered photons at 60° from the incident beam? What is the energy of the recoil electrons in this case?

6.   A photon can be absorbed by a system that can have internal energy. Assume that a 15-MeV photon is absorbed by a carbon nucleus initially at rest. The momentum of the carbon nucleus must be 15 MeV/c. (*a*) Calculate the kinetic energy of the carbon nucleus. What is the internal energy of this nucleus? (*b*) The carbon nucleus comes to rest and then loses its internal energy by emitting a photon. What is the energy of the photon?

7.   Consider the distribution function $f(x) = C(6 - x)$ for $0 \le x \le 6$; $f(x) = 0$ for $x > 6$, where x is a dimensionless variable proportional to the energy (such as E/kT, for example). (*a*) Sketch this function and compute the normalization constant C and $\bar{x}$ for this distribution. (*b*) Assume that x is now a discrete variable given by $x_n = n$, $n = 0, 1, 2, \ldots, 5$ and the distribution function is $f_n = C(6 - x_n)$. Sketch this function and compute C and $\bar{x}$. (*c*) Repeat (*b*) for $x_n = 2n, n = 0, 1, 2$.

8.   Prove that the photoelectric effect cannot occur with a free electron. (*Hint:* Consider the reference frame in which the total momentum of the electron and incident photon is zero.)

9.   The molar heat capacity data given in Table 3-3 are taken from *AIP Handbook,* 2d ed., New York: McGraw-Hill Book Company, 1963. Plot the data for these solids all on one graph and sketch in the curves C_v versus T. Estimate the Einstein temperature for each of the solids using the result of Exercise 35.

Table 3-3
Heat capacities in cal/mole-K for Au, Diamond, Al, and Be

T, K	20	50	70	100	150	200	250	300	400	500	600	800	1000
Au	0.77	3.41	4.39	5.12	5.62	5.84	5.96	6.07	6.18	6.28	6.40	6.65	6.90
Diamond	0.00	0.005	0.016	0.059	0.24	0.56	0.99	1.46	2.45	3.24	3.85	4.66	5.16
Al	0.05	0.91	1.85	3.12	4.43	5.16	5.56	5.82	6.13	6.42	6.72	7.31	7.00
Be	0.003	0.04	0.12	0.43	1.36	2.41	3.30	3.93	4.77	5.26	5.59	6.07	6.51

10. Sketch the histogram of f_n versus n, where

$$f_n = e^{-E_n/kT} = e^{-n(hf/kT)}$$

(a) for the temperature $T = 10hf/k = 10T_E$, (b) for the temperature $T = hf/k = T_E$. On each of these histograms, sketch the continuous function

$$f(E) = e^{-E/kT}$$

11. From the Einstein temperatures found in Problem 9, calculate the frequency of vibration $f = kT_E/h$ of the atoms.

12. Show that the expression for C_v given by Equation 3-44 approaches zero as T approaches zero and approaches $3N_A k$ as T approaches infinity.

13. Consider a system of N particles which has only two possible energy states, $E_1 = 0$ and $E_2 = \epsilon$. The distribution function is $f_i = Ce^{-E_i/kT}$. (a) What is C for this case? (b) Compute the average energy $\bar{E}$ and show that $\bar{E} \to 0$ as $T \to 0$ and $\bar{E} \to \epsilon/2$ as $T \to \infty$. (c) Show that the heat capacity is

$$C_v = Nk \left(\frac{\epsilon}{kT}\right)^2 \frac{e^{-\epsilon/kT}}{(1 + e^{-\epsilon/kT})^2}$$

(d) Sketch C_v versus T.

14. Estimate the frequency and energy of vibration of the H_2 molecule from the graph of C_v versus T in Figure 2-5.

15. This problem is to derive the Wien displacement law of Equation 3-17. (a) Show that the energy-density distribution function can be written $u = C\lambda^{-5}(e^{a/\lambda} - 1)^{-1}$, where C is a constant and $a = hc/kT$. (b) Show that the value of λ for which $du/d\lambda = 0$ satisfies the equation $5\lambda(1 - e^{-a/\lambda}) = a$. (c) This equation can be solved with a calculator by the trial-and-error method. Try $\lambda = \alpha a$ for various values of α until λ/a is determined to four significant figures. (d) Show that your solution in (c) implies $\lambda_m T = $ constant and calculate the value of the constant.

16. In some cases the sums in Equations 3-26 and 3-27 can be done numerically with a calculator to reasonable accuracy using only a few terms. Calculate C and $\bar{E}$ numerically (a) using only three terms for the case $hf = 2kT$, and (b) using five terms for $hf = kT$. (c) Compare your results for $\bar{E}$ with those obtained exactly from Equation 3-28 for these cases.

CHAPTER 4   The Nuclear Atom

4-1   Empirical Spectral Formulas

The study of the characteristic radiation emitted by atoms in a gas excited by an electrical discharge, or by atoms in a flame, was vigorously pursued during the late nineteenth century. When viewed through a spectroscope, this radiation appears as a discrete set of lines, each of a particular color or wavelength; the positions and intensities of the lines are characteristic of the element. (The light appears as lines because the source aperture is a narrow slit.) The wavelengths of these lines could be determined with great precision, and much effort went into finding regularities in the spectra. A major breakthrough was made in 1885 by a Swiss schoolteacher, Johann Balmer, who found that the lines in the spectrum of hydrogen could be represented by the formula

$$\lambda = 364.6 \, \frac{m^2}{m^2 - 4} \quad \text{nm}$$

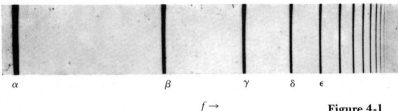

α   β   γ   δ   ε

$f \rightarrow$

where m is a variable integer which takes on the values $m = 3, 4,$ 5, ... Figure 4-1 shows the set of spectral lines of hydrogen (now known as the *Balmer series*) whose wavelengths are given by Balmer's formula. Balmer suggested that his formula might be a special case of a more general expression applicable to the spectra of other elements. Such an expression, found by J. R. Rydberg and W. Ritz, gives the reciprocal wavelength as

Figure 4-1
The Balmer series of hydrogen. [*From G. Herzberg,* Annalen der Physik, **84,** *565 (1927).*]

$$\frac{1}{\lambda} = R \left(\frac{1}{m^2} - \frac{1}{n^2} \right) \qquad n > m \qquad\qquad 4\text{-}1$$

Rydberg-Ritz formula

where m and n are integers, and R, called the *Rydberg constant* or the *Rydberg*, is the same for all series of the same element and varies only slightly, in a regular way, from element to element. For hydrogen, the value of R is $R_{\mathrm{H}} = 1.096776 \times 10^7$ m^{-1}. For very heavy elements, R approaches the value $R_\infty = 1.097373 \times 10^7$ m^{-1}. (The quantity $1/\lambda$ was used instead of the frequency $f = c/\lambda$ because λ could be measured much more accurately than the speed of light c.) Such empirical expressions were successful in predicting other spectra, e.g., other hydrogen lines outside the visible spectrum were predicted and found.

Rydberg constant

Many attempts were made to construct a model of the atom that yields these formulas for its radiation spectrum. It was known that an atom was about 10^{-10} m in diameter, that it contained electrons much lighter than the atom, and that it was electrically neutral. Electromagnetic theory showed that charges radiate when accelerated. In particular, when vibrating, they radiate at the vibration frequency. It was natural to assume that the atom was analogous to an acoustic system whose general motion was made up of a set of vibrations with a discrete frequency spectrum. The atom would thus emit all frequencies simultaneously.

The most popular model was that of J. J. Thomson. He considered various arrangements of electrons embedded in some kind of fluid that contained most of the mass of the atom and had enough positive charge to make the atom electrically neutral. He then searched for configurations that were stable and had normal modes of vibration corresponding to the known frequency spectrum. One difficulty with all such models was that electric forces alone could not produce stable equilibrium. Despite rather elaborate mathematical calculations, Thomson was unable to obtain from his model a set of frequencies of vibration that corresponded with the frequencies of observed spectra.

In Section 4-3 we shall see how a completely different model of the atom, the Bohr planetary model, leads to Equation 4-1 for the wavelengths of the line spectra of hydrogen, and predicts the correct value for the Rydberg constant. Before we consider this model we shall digress from our discussion of spectra to describe the historic Rutherford scattering experiment. This experiment led to our present picture of the nuclear atom, with its positive charge concentrated in a small region of space (the nucleus) that is much smaller than the region occupied by the negatively charged electrons.

4-2   Rutherford Scattering

The Thomson model of the atom was essentially ruled out by the results of a set of experiments conducted by Ernest Rutherford and his students H. W. Geiger and E. Marsden. Rutherford was investigating radioactivity and had shown that the radiations from uranium consisted of at least two types, which he labeled α and β. He showed, by an experiment similar to that of J. J. Thomson, that q/m for the α particle was half that of the proton. Suspecting that the α particles were doubly ionized helium, Rutherford and his coworkers let a radioactive substance decay in a previously evacuated chamber; then by spectroscopy they detected ordinary helium gas in the chamber. Realizing that this energetic, massive particle would make an excellent probe for investigating other atoms, Rutherford began a series of experiments for this purpose.

In these experiments, a narrow beam of α particles fell on a zinc sulfide screen, which gave off visible light scintillations when struck (Figure 4-2). The distribution of scintillations on the screen was observed when various thin metal foils were placed between it and the source. Most of the particles were either undeflected, or deflected through very small angles of the

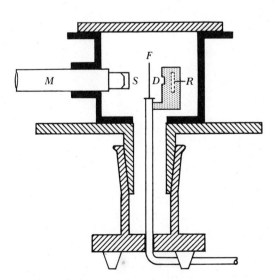

Figure 4-2
Apparatus used by Geiger and Marsden. The α particles from a fixed source R strike a fixed foil F after passing through a collimating diaphragm D. Scintillations on screen S, resulting from scattered particles, are observed with a microscope M. The chamber is evacuated and can be rotated about the foil. [*From H. Geiger and E. Marsden*, Philosophical Magazine (6), **25**, *507 (1913)*.]

Figure 4-3
Head-on (elastic) collision
between an α particle
and a light electron. The
heavy particle is hardly
affected by the collision.
Since the relative velocity
of separation after the
collision equals the relative velocity of approach
before the collision, the
electron's speed after the
collision must be
about $2v$.

order of $1°$. Entirely unexpectedly, however, some were
deflected through angles as large as $90°$ or more. If the atom
consisted of a positively charged sphere of radius 10^{-10} m,
containing electrons as in the Thomson model, only a very
small deflection could result from a single encounter between
a particle and an atom, even if the α particle penetrated into
the atom.

Let us *estimate* the order of magnitude of the maximum deflection of an α particle in such an encounter. We first consider
the collision of an α particle with a single electron. Because the
mass of the α particle is about 8000 times that of an electron, the
electron can have little effect on the momentum of the α particle. Figure 4-3 shows a head-on collision between a large particle of mass m_α and speed v and a small particle of mass m_e initially at rest. This is like a collision between a bowling ball and a
BB shot. The heavy particle is hardly affected, and continues
after the collision with nearly the same speed v. The lighter particle acquires a speed of approximately $2v$ since the relative
speed of separation after an elastic collision is the same as the
relative speed of approach before the collision. The loss of momentum of the α particle equals the gain in momentum of the
electron, which is approximately $2m_e v \approx m_\alpha v/4000$. We can get
an upper-limit estimate on the angle of deflection by taking this
maximum momentum change $\Delta\mathbf{p}$ to be perpendicular to the
original momentum $\mathbf{p}$ of the α particle, as in Figure 4-4. (Of
course, $\Delta\mathbf{p}$ could be perpendicular to $\mathbf{p}$ only for a glancing collision, in which $\Delta\mathbf{p}$ would be less than $2m_e v$; however, we are
interested only in the order of magnitude of the deflection
angle.) Then $\Delta p/p \approx \theta \approx \frac{1}{4000}$ rad $\approx 0.01°$.

*Estimate of deflection due
to electrons*

We now consider the possible effect of the positive charge.
The electric force on a point charge due to a uniformly charged
sphere is shown as a function of r in Figure 4-5. The force is
strongest at $r = R$. We can estimate the change in momentum
of the α particle, Δp, due to this charge by assuming that the
maximum force acts on it for the time it takes to pass the atom at
speed v; this time is $\Delta t \approx 2R/v$. The force on the particle at a
distance R from a positive charge Q is given by Coulomb's
law, $F = kq_\alpha Q/R^2$, where q_α is the charge of the α particle and
$k = 1/4\pi\epsilon_0 \approx 9 \times 10^9$ N-m^2/C^2 is the Coulomb constant. The
change in momentum is then of the order of

$$\Delta p \approx F \, \Delta t = \frac{kq_\alpha Q}{R^2} \frac{2R}{v}$$

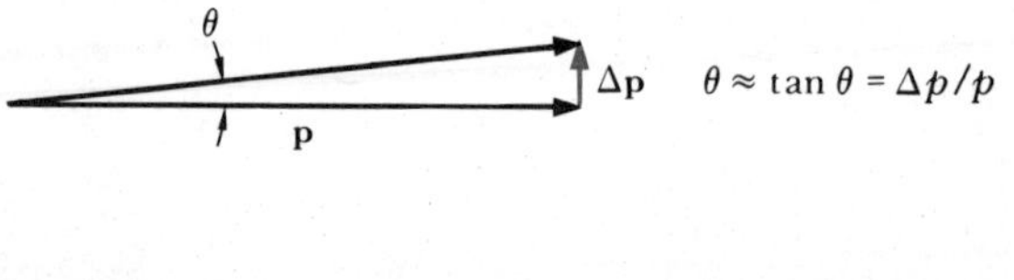

Figure 4-4
The maximum deflection
can be estimated by
taking the maximum mo-
mentum change to be
perpendicular to the original momentum.

$\theta \approx \tan\theta = \Delta p/p$

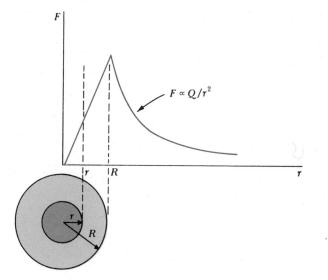

Figure 4-5
Force on a point charge
versus distance r from
the center of a uniformly
charged sphere of radius
R. Outside the sphere
the force is proportional
to Q/r^2, where Q is the
total charge. Inside the
sphere, the force is pro-
portional to
$q'/r^2 = Qr/R^3$, where
$q' = Q(r/R)^3$ is the
charge within a sphere
of radius r. The max-
imum force occurs at
$r = R$.

Again taking this change to be at right angles to the original momentum $m_\alpha v$, we get for the maximum deflection angle:

$$\tan \theta \approx \frac{\Delta p}{p} = \frac{2kq_\alpha Q}{Rm_\alpha v^2} = \frac{kq_\alpha Q}{R(\frac{1}{2}m_\alpha v^2)} \qquad \text{4-2}$$

*Estimate of deflection due
to charged sphere of
radius R*

Let us evaluate this expression for a typical case of an α particle of charge $q = 2e$, with energy of 5 MeV, incident on a gold
atom of charge $Q = 79e$. For this calculation and others, it is
convenient to express the quantity ke^2, which has dimensions of
energy $\times$ length, in units of eV-Å or eV-nm. We have

$$ke^2 = (9 \times 10^9 \text{ N-m}^2/\text{C}^2)(1.6 \times 10^{-19} \text{ C})^2$$

$$\times \frac{1 \text{ eV}}{1.6 \times 10^{-19} \text{ J}}$$

$$= 1.44 \times 10^{-9} \text{ eV-m}$$

or

$$ke^2 = 1.44 \text{ eV-nm} = 14.4 \text{ eV-Å} \qquad \text{4-3}$$

ke^2

For our example, Equation 4-2 then gives, with $R = 1$ Å,

$$\frac{\Delta p}{p} = \frac{(2)(79)(14.4 \text{ eV-Å})}{(1 \text{ Å})(5 \times 10^6 \text{ eV})} = 4.55 \times 10^{-4}$$

Then

$$\tan \theta \approx \theta \approx 4.55 \times 10^{-4} \text{ rad} \approx 0.026°$$

We can see from these crude estimates that a deflection even
as small as 1° must be the result of many collisions, according to
the Thomson model of the atom. If this is the case, the number
of particles scattered through large angles can be predicted
from the statistical theorem of multiple scattering. For example,

a result of this theory (which we shall not derive) is that the fraction scattered through angles greater than some angle θ is $e^{-(\theta/\theta_{rms})^2}$, where θ_{rms} is the rms scattering angle. If we take $\theta_{rms} \approx 1°$, as observed by Geiger and Marsden, we obtain for the fraction scattered through 90° or more, $e^{-(90/1)^2} = e^{-8100} \approx 10^{-3500}$. The fraction observed was about $\frac{1}{8000}$, which is tremendously large compared with 10^{-3500}.

The question is, then, how can one obtain large-angle scattering? The trouble with the Thomson atom is that it is too "soft"—the maximum force experienced by the α particle is too weak to give a large deflection. If the positive charge of the atom is concentrated in a more compact region, however, a much larger force will occur at near impacts. For example, we can take the result of our crude estimate in Equation 4-2 and ask what value of R will give $\Delta p \approx p$. Setting $\tan \theta = 1$ in Equation 4-2 and solving for R, we obtain

$$R = \frac{kq_\alpha Q}{\frac{1}{2}m_\alpha v^2} = \frac{(2)(79)(14.4 \text{ eV-Å})}{5 \times 10^6 \text{ eV}} = 4.6 \times 10^{-4} \text{ Å}$$

Rutherford concluded that such large-angle scattering could result only from a single encounter of the α particle with a massive charge confined to a volume much smaller than that of the whole atom. Assuming this "nucleus" to be a point charge, he calculated the expected angular distribution for the scattered α particles. His predictions as to the dependence of scattering probability on angle, nuclear charge, and kinetic energy were completely verified in a series of experiments carried out in his laboratory by Geiger and Marsden.

We shall not go through Rutherford's derivation in detail, but merely outline the assumptions and conclusions. Figure 4-6 shows the geometry of an α particle being scattered by a nucleus which we take to be a point charge Q at the origin. Initially, the α particle approaches with speed v along a line a distance b from a parallel line COA through the origin. The force on the α particle is $F = kq_\alpha Q/r^2$, given by Coulomb's law. After scattering, when the α particle is again far from the nucleus, it is moving

Figure 4-6
Rutherford scattering geometry. The nucleus is assumed to be a point charge Q at the origin O. At any distance r the α particle experiences a repulsive force $kq_\alpha Q/r^2$. The α particle travels along a hyperbolic path that is initially parallel to line OA a distance b from it and finally parallel to line OB, which makes an angle θ with OA. The scattering angle θ can be related to the impact parameter b by classical mechanics.

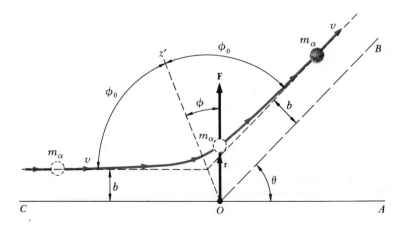

with the same speed v parallel to the line *OB*, which makes an angle θ with line *COA*. (Since the potential energy is again zero, the final speed must equal the initial speed by conservation of energy.) The distance b is called the *impact parameter* and the angle θ the *scattering angle*. The path of the α particle can be shown to be a hyperbola and the scattering angle θ can be related to the impact parameter b from the laws of classical mechanics. The result is[1]

Impact parameter

$$b = \frac{kq_\alpha Q}{m_\alpha v^2} \cot \tfrac{1}{2}\theta \qquad\qquad 4\text{-}4$$

Impact parameter and scattering angle

Of course, it is not possible to choose or to know the impact parameter for any α particle; but all such particles with impact parameters less than, or equal to, a particular b will be scattered through an angle θ greater than or equal to that given by Equation 4-4 (Figure 4-7). Let the intensity of the incident α particle beam be I_0 particles per second per unit area. The number per second scattered by one nucleus through angles greater than θ equals the number per second that have impact parameters less than $b(\theta)$. This number is $\pi b^2 I_0$.

The quantity πb^2, which has the dimensions of an area, is called the *cross section* for scattering through angles greater than θ. The cross section is thus defined to be the number scattered per nucleus per unit time divided by the incident intensity. The total number of particles scattered per second is obtained by multiplying $\pi b^2 I_0$ by the number of nuclei in the scattering foil (this assumes the foil to be thin enough to make the chance of overlap negligible). Let n be the number of nuclei per unit volume:

Cross section

$$n = \frac{\rho(\text{g/cm}^3)N_A \text{ (atoms/mole)}}{\mathcal{M} \text{ (g/mole)}} = \frac{\rho N_A}{\mathcal{M}} \frac{\text{atoms}}{\text{cm}^3} \qquad 4\text{-}5$$

[1] A derivation of this result is given at the end of this section.

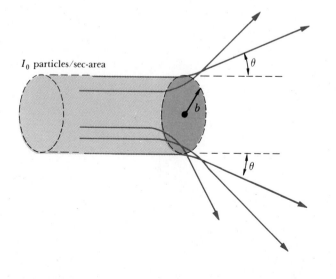

I_0 particles/sec-area

Figure 4-7
Particles with impact parameters less than or equal to b are scattered through angles greater than or equal to θ, related to b by Equation 4-4. The area πb^2 is called the cross section for scattering through angles greater than θ.

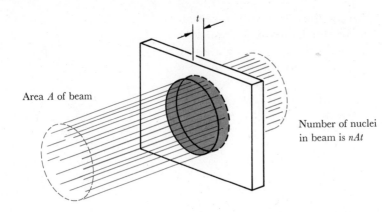

For a foil of thickness t the total number of nuclei is nAt, where A is the area of the beam (Figure 4-8). The total number scattered per second through angles greater than θ is thus $\pi b^2 I_0 ntA$. If we divide this by the number of α particles incident per second, $I_0 A$, we get the fraction scattered through angles greater than θ:

$$f = \pi b^2 nt \qquad\qquad 4\text{-}6$$

Let us evaluate this fraction for the gold foil 10^{-4} cm thick, used by Geiger and Marsden, for $\theta = 90°$. Using $\cot\left(\frac{90}{2}\right) = 1$ and taking $\frac{1}{2}m_\alpha v^2 = 5$ MeV for a typical α-particle energy, we have, from Equation 4-4,

$$b = \frac{(2)(79)ke^2}{m_\alpha v^2} = \frac{(2)(79)(14.4 \text{ eV-Å})}{(2)(5 \times 10^6 \text{ eV})} \approx 2.3 \times 10^{-4} \text{ Å}$$

and from Equation 4-5,

$$n = \frac{(19.3 \text{ g/cm}^3)(6.02 \times 10^{23} \text{ atoms/mole})}{197 \text{ g/mole}}$$

$$= 5.9 \times 10^{22} \text{ atoms/cm}^3$$

Then

$$f = \pi(2.3 \times 10^{-12})^2(5.9 \times 10^{22})(10^{-4}) \approx 10^{-4}$$

This is in good agreement with their observation of about 1 in 8000 in their first trial.

Geiger and Marsden did a series of experiments in which they measured:

1. The number of particles per unit area on the screen, scattered through angles between θ and $\theta + d\theta$

2. The variation in the number scattered with foil thickness

3. The variation in the number scattered with the atomic weight of the foil

4. The variation in the number scattered with incident velocity v, which they varied by placing thin absorbers in the incident beam to slow down the α particles

The number scattered by one nucleus at angles between θ and $\theta + d\theta$ is the number incident with impact parameters between b and $b + db$ (Figure 4-9). This number equals the product of the incident intensity I_0 and the area $2\pi b\, db$ shown in Figure 4-9. We shall omit the algebraic details of the calculation of this number from Equation 4-4. The result can be written

$$I_0 2\pi b\, db = I_0 2\pi \left(\frac{kZe^2}{m_\alpha v^2}\right)^2 \frac{\sin\theta\, d\theta}{\sin^4 \frac{1}{2}\theta} \qquad 4\text{-}7$$

The area of the screen, from Figure 4-10, is $(2\pi r \sin\theta)\,(r\, d\theta)$. The number scattered by one nucleus per unit area on the screen is therefore proportional to

$$\frac{N}{\text{Area}} \propto I_0 \left(\frac{kZe^2}{m_\alpha v^2}\right)^2 \frac{1}{\sin^4 \frac{1}{2}\theta} \qquad 4\text{-}8$$

The point-nucleus model thus predicts that the observed number of scattered particles per unit area on the screen is pro-

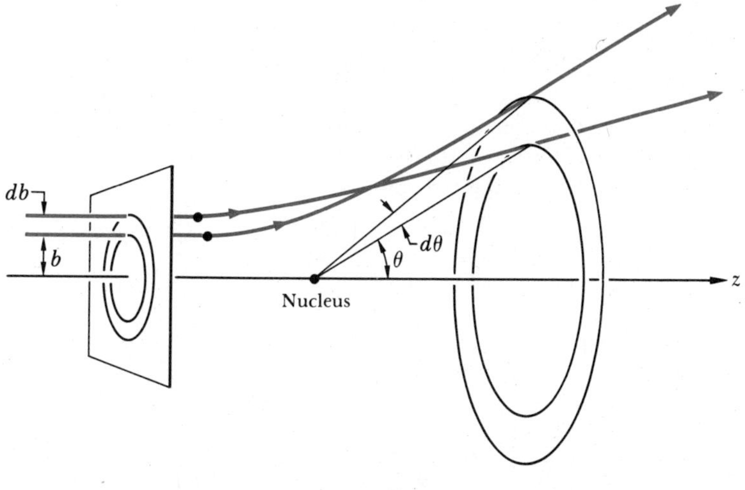

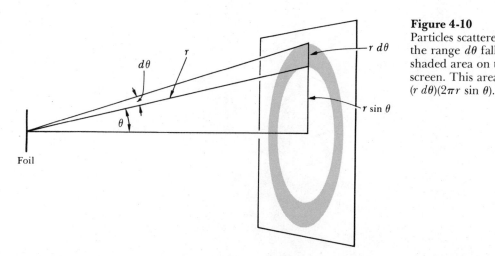

Figure 4-9
The number of particles with impact parameters between b and $b + db$ is proportional to the area $2\pi b\, db$. These particles are scattered into the range $d\theta$.

Figure 4-10
Particles scattered into the range $d\theta$ fall on the shaded area on the screen. This area is $(r\, d\theta)(2\pi r \sin\theta)$.

portional to $\sin^{-4}(\theta/2)$, to Z^2, and to v^{-4}. Since the number of scattering nuclei is proportional to the foil thickness, the number of particles scattered at a given angle should be proportional to the foil thickness if the scattering is due to a single encounter between the α particle and a nucleus.

We quote the summary from the paper "Deflection of α Particles through Large Angles," by Geiger and Marsden, *Philosophical Magazine* (6), **25,** 605 (1913).

> The experiments described in the foregoing paper were carried out to test a theory of the atom proposed by Prof. Rutherford, the main feature of which is that there exists at the center of the atom an intense, highly concentrated electrical charge. The verification is based on the laws of scattering which were deduced from this theory. The following relations have been verified experimentally:
>
> 1. The number of α particles emerging from a scattering foil at an angle θ with the original beam varies as $1/\sin^4(\theta/2)$, when the α particles were counted on a definite area at a constant distance from the foil. This relation has been tested for angles varying from 5° to 150°, and over this range the number of α particles varied from 1 to 250,000 in a good agreement with the theory.
>
> 2. The number of α particles scattered in a definite direction is directly proportional to the thickness of the scattering foil for small thicknesses. For large thicknesses the decrease of velocity of the α particles in the foil causes a somewhat more rapid increase in the amount of scattering.
>
> 3. The scattering per atom of foils of different materials varies approximately as the square of the atomic weight. This relation was tested for foils of atomic weight from that of carbon to that of gold.

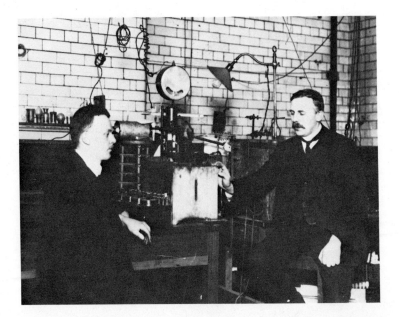

Hans Geiger (left) and Ernest Rutherford in their Manchester Laboratory. (*Courtesy of University of Manchester.*)

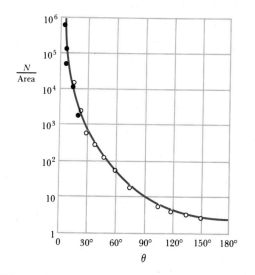

Figure 4-11
The number of scattered
α particles as a function
of θ. The curve is
$\sin^{-4}\frac{1}{2}\theta$. The data are
from Geiger and Mars-
den. (*From R. D. Evans,*
The Atomic Nucleus,
*New York: McGraw-Hill
Book Company, 1955.*)

4. The amount of scattering by a given foil is approximately
proportional to the inverse fourth power of the velocity of the
incident α particles. This relation was tested over a range of velocities such that the number of scattered particles varied as
$1:10$.

5. Quantitative experiments show that the fraction of particles
of Ra C, which is scattered through an angle of 45° by a gold
foil of 1 mm air equivalent (2.1×10^{-5} cm), is 3.7×10^{-7} when
scattered particles are counted on a screen of 1-sq mm area
placed at a distance of 1 cm from the scattering foil. From this
figure and the foregoing results, it can be calculated that the
number of elementary charges composing the center of the
atom is equal to half the atomic weight.

Figure 4-11 is a plot of their data, showing the angular
dependence of the scattering using 7.7-MeV α particles. The
excellent agreement of their data with the $\sin^{-4}\frac{1}{2}\theta$ prediction
of Equation 4-8 indicates that the force law $F = kq_\alpha Q/r^2$ used
to derive Equation 4-8 is correct. This does not imply that the
nucleus is a mathematical point charge, however; the force
law would be the same even if the nucleus were a ball of charge
of some radius R_0, as long as the α particle did not penetrate
the ball (Figure 4-12). For a given scattering angle, the distance
of closest approach of the α particle to the nucleus can be
calculated from the geometry of the collision. For the largest
angle, near 180°, the collision is nearly "head on." The corresponding distance of closest approach thus is an experimental
upper limit on the size of the target nucleus.

We can calculate the distance of closest approach for a
head-on collision r_d by setting the potential energy at this distance equal to the original kinetic energy:

$$\frac{kq_\alpha Q}{r_d} = \frac{1}{2}m_\alpha v^2$$

or

$$r_d = \frac{kq_\alpha Q}{\frac{1}{2}m_\alpha v^2}$$

4-9

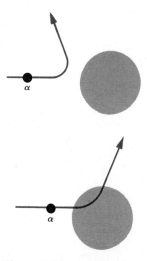

Figure 4-12
If the α particle does not
penetrate the nuclear
charge, the nucleus can
be considered a point
charge. If the particle has
enough energy to pene-
trate the nucleus, as in
the figure on the bottom,
the Rutherford scattering
law does not hold.

For the case of 7.7-MeV α particles, the distance of closest approach for a head-on collision is

$$r_d = \frac{(2)(79)(14.4 \text{ eV-Å})}{7.7 \times 10^6 \text{ eV}} \approx 3 \times 10^{-4} \text{ Å} = 3 \times 10^{-14} \text{ m}$$

For other collisions, the distance of closest approach is somewhat greater than this, but for those scattered at large angles it is of the same order of magnitude. The excellent agreement of the data of Geiger and Marsden at large angles with the prediction of Equation 4-8 thus indicates that the radius of the gold nucleus is less than about 3×10^{-14} m. If higher-energy particles could be used, the distance of closest approach would be smaller, and as the energy of the α particles increased, we might expect that eventually the particles would penetrate the nucleus. Since, for this case, the force law is no longer $F = kq_\alpha Q/r^2$, the data would not agree with the point-nucleus calculation. Rutherford did not have higher-energy α particles available, but he could reduce the distance of closest approach by using targets of lower atomic numbers. For the case of aluminum with $Z = 13$, the most energetic α particles scattered at large angles did not follow the predictions of Equation 4-8. From these data, Rutherford estimated the radius of the aluminum nucleus to be about 10^{-14} m.

A unit of length convenient for describing nuclear sizes is the fermi, or femtometer (fm), defined by 1 fm $= 10^{-15}$ m. As we shall see in Chapter 11, the nuclear radius varies from about 1 to 10 fm from the lightest to the heaviest atoms.

Optional

Derivation of Equation 4-4

We can derive the relation between the impact parameter b and the scattering angle θ as given by Equation 4-4 without going into the details of finding the path followed by the α particle.

Let $\mathbf{p}_1$ and $\mathbf{p}_2$ be the initial and final momentum vectors of the α particle. From Figure 4-13 it is evident that the total change in momentum $\Delta \mathbf{p} = \mathbf{p}_2 - \mathbf{p}_1$ is along the z' axis of Figure 4-6. The

Figure 4-13
Momentum diagram for Rutherford scattering. The magnitude of the momentum change Δp is related to the scattering angle θ by $\Delta p = 2m_\alpha v \sin \frac{1}{2}\theta$.

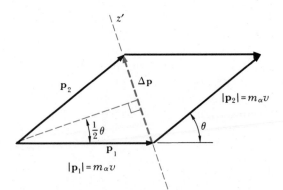

magnitude of either $\mathbf{p}_1$ or $\mathbf{p}_2$ is $m_\alpha v$. From the isosceles triangle formed by $\mathbf{p}_1$, $\mathbf{p}_2$, and $\Delta \mathbf{p}$ shown in Figure 4-13, we have

$$\frac{\frac{1}{2}\Delta p}{m_\alpha v} = \sin \tfrac{1}{2}\theta$$

or

$$\Delta p = 2m_\alpha v \sin \tfrac{1}{2}\theta \qquad\qquad 4\text{-}10$$

We now write Newton's law for the α particle: $\mathbf{F} = d\mathbf{p}/dt$, or

$$d\mathbf{p} = \mathbf{F}\, dt$$

The force F is given by Coulomb's law, $F = kq_\alpha Q/r^2$, and is in the radial direction. Taking components along the z' direction, and integrating to obtain Δp, we have

$$\Delta p = \int (dp)_{z'} = \int F \cos \phi \, dt = \int F \cos \phi \, \frac{dt}{d\phi}\, d\phi \quad 4\text{-}11$$

where we have changed the variable of integration from t to the angle ϕ. We can relate $dt/d\phi$ to the angular momentum of the α particle about the origin. Since the force is central (i.e., acts along the line joining the α particle and the nucleus at the origin), there is no torque about the origin, and the angular momentum of the α particle is conserved. Initially, the angular momentum has the magnitude $m_\alpha vb$. At a later time, it is $m_\alpha r^2\, d\phi/dt$. Conservation of angular momentum thus gives

$$m_\alpha r^2\, \frac{d\phi}{dt} = m_\alpha vb$$

or

$$\frac{dt}{d\phi} = \frac{r^2}{vb}$$

Substituting this result and $F = kq_\alpha Q/r^2$ for the force into Equation 4-11, we obtain

$$\Delta p = \int \frac{kq_\alpha Q}{r^2} \cos \phi \, \frac{r^2}{vb}\, d\phi = \frac{kq_\alpha Q}{vb} \int \cos \phi \, d\phi$$

or

$$\Delta p = \frac{kq_\alpha Q}{vb}\, (\sin \phi_2 - \sin \phi_1) \qquad\qquad 4\text{-}12$$

From Figure 4-6 we see that $\phi_1 = -\phi_0$ and $\phi_2 = +\phi_0$, where $2\phi_0 + \theta = 180°$. Then $\sin \phi_2 - \sin \phi_1 = 2 \sin (90° - \tfrac{1}{2}\theta) = 2 \cos \tfrac{1}{2}\theta$. Combining Equations 4-10 and 4-12 for Δp we have

$$\Delta p = 2m_\alpha v \sin \tfrac{1}{2}\theta = \frac{kq_\alpha Q}{vb}\, 2 \cos \tfrac{1}{2}\theta$$

or

$$b = \frac{kq_\alpha Q}{m_\alpha v^2} \cot \tfrac{1}{2}\theta$$

which is Equation 4-4.

Questions

1. Why can't the impact parameter for a particular α particle be chosen?

2. Why is it necessary to use a very thin target foil?

3. Why could Rutherford place a lower limit on the radius of the Al nucleus but not on the Au nucleus?

4-3 The Bohr Model of the Hydrogen Atom

In 1913, the Danish physicist Niels Bohr proposed a model of the hydrogen atom which combined the work of Planck, Einstein, and Rutherford, and was remarkably successful in predicting the observed spectrum of hydrogen. Bohr, who had been working in the Rutherford laboratory during the experiments of Geiger and Marsden, made the assumption that the electron in the hydrogen atom moved in an orbit about the positive nucleus, bound by the electrostatic attraction of the nucleus. Classical mechanics allows circular or elliptical electron orbits in this system, just as in the case of the planets orbiting the sun. For simplicity, Bohr chose to consider circular orbits.

Such a model is mechanically stable, because the attractive Coulomb force provides the centripetal force necessary for the electron to move in a circle; but it is electrically unstable because the electron is always accelerating toward the center of the circle. The laws of electrodynamics predict that such an accelerating charge will radiate light of frequency equal to that of the periodic motion, which in this case is the frequency of revolution. As energy is lost to the radiation, the electron's orbit will become smaller and smaller. The time required for the electron to spiral into the nucleus can be calculated from classical mechanics and electrodynamics; it turns out to be less than a microsecond. Thus, at first sight, this model predicts that the atom will radiate a continuous spectrum (since the frequency of revolution changes continuously as the electron spirals in) and will collapse after a very short time, a result that fortunately does not occur.

Bohr "solved" this difficulty by *postulating* that the electron could move in certain orbits without radiating. He called these stable orbits *stationary states*. He then assumed that the atom radiates when the electron somehow makes a transition from one stationary state to another (Figure 4-14), and that the frequency of radiation is not the frequency of motion in either stable orbit but is related to the energies of the orbits by

First Bohr postulate: nonradiating orbits

$$hf = E_i - E_f \qquad \text{4-13}$$

Second Bohr postulate: photon frequency from energy conservation

where h is Planck's constant and E_i and E_f are the total energies in the initial and final orbits. This assumption, which is equivalent to that of energy conservation with the emission of a

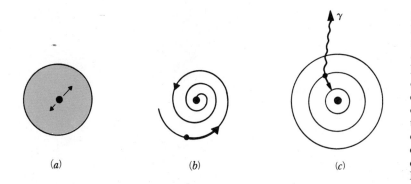

(a)   (b)   (c)

Figure 4-14
Models of an atom. (*a*) In the Thomson model, the electron is embedded in a large positive charge and oscillates about the center. (*b*) In the classical orbital model, the electron orbits about the nucleus and spirals into the center because of the energy radiated. (*c*) In the Bohr model, the electron orbits without radiating until it jumps to another radius, at which time radiation is emitted.

photon, is crucial because it deviates from classical theory, which requires the frequency of radiation to be that of the motion of the charged particle. (Planck made a similar assumption in his blackbody-radiation theory, but in that case the energy difference between two adjacent states is hf so that when the oscillator emitted or absorbed radiation, the frequency of the radiation was the same as that of the oscillator.)

In order to determine the radii of the allowed, nonradiating orbits, Bohr made an additional assumption, which is now known as the *correspondence principle*:

In the limit of large orbits and large energies, quantum calculations must agree with classical calculations.

Correspondence principle

Thus the correspondence principle says that, whatever modifications of classical physics are made to describe matter at the submicroscopic level, when the results are extended to the macroscopic world they must agree with those from the classical laws of physics that have been so abundantly verified in the everyday world. While Bohr's detailed model of the hydrogen atom has been supplanted by modern quantum theory, which we shall discuss in later chapters, his radiation hypothesis (Equation 4-13) and the correspondence principle remain as essential features of the new theory.

In his first paper, in 1913, Bohr pointed out that his results imply that the angular momentum of the electron in the hydrogen atom can only take on values which are integral multiples of Planck's constant divided by 2π. That is, angular momentum is quantized; it can assume only the values $nh/2\pi$, where n is an integer. Rather than follow Bohr's derivation based on the correspondence principle, we shall use the fundamental conclusion of angular-momentum quantization to find his expression for the observed spectra.

If the nuclear charge is $+Ze$ and the electron charge $-e$, the potential energy at a separation r is

$$V = -\frac{kZe^2}{r}$$

where $k = 1/4\pi\epsilon_0$ is the Coulomb constant. (For hydrogen

Culver Pictures

(a)

Mark Oliphant

(b)

(c)

Mark Oliphant

(d)

(a) Niels Bohr in 1922. (b) The Rutherfords and Bohrs in the Rutherfords' garden in 1930. (*Courtesy of the Niels Bohr Library, American Institute of Physics, Margrethe Bohr Collection.*) (c) The Bohrs take a ride on George Gamow's motorcycle at their country house Tisvilde in the summer of 1930. (*Courtesy of George Gamow.*) (d) Niels Bohr explains a point in front of the blackboard (1956). (*Courtesy of the Niels Bohr Library, American Institute of Physics. Margrethe Bohr Collection.*)

$Z = 1$, but it is convenient not to specify Z at this time, so that the results can be applied to other atoms.) The total energy of the electron moving in a circular orbit with speed v is then

$$E = \tfrac{1}{2}mv^2 + V = \tfrac{1}{2}mv^2 - \frac{kZe^2}{r} \qquad\qquad 4\text{-}14$$

The kinetic energy can be obtained as a function of r by using Newton's law $\Sigma \mathbf{F} = m\mathbf{a}$. Setting the Coulomb attractive force equal to the mass times the centripetal acceleration, we have

$$\frac{kZe^2}{r^2} = m\,\frac{v^2}{r}$$

or

$$\tfrac{1}{2}mv^2 = \frac{1}{2}\frac{kZe^2}{r} \qquad\qquad 4\text{-}15$$

For circular orbits, the kinetic energy is equal to half the magnitude of the potential energy, a result which holds for circular motion in any inverse-square force field. The total energy is then

$$E = -\frac{1}{2}\frac{kZe^2}{r} \qquad\qquad 4\text{-}16$$

Total energy in circular orbit

Using Equation 4-13 for the frequency of radiation when the electron changes from an initial orbit of radius r_i to a final orbit of radius r_f, we obtain

$$f = \frac{E_i - E_f}{h} = -\frac{1}{2}\frac{kZe^2}{h}\left(\frac{1}{r_i} - \frac{1}{r_f}\right) \qquad 4\text{-}17$$

To obtain the Balmer-Ritz formula $f = c/\lambda = cR(1/m^2 - 1/n^2)$ (Equation 4-1), it is evident that the radii of the stable orbits must be proportional to the squares of integers. The postulate of the quantization of angular momentum leads to this result. The angular momentum of a particle moving in a circular path is mvr. Setting this equal to an integer times $h/2\pi$, we obtain

Third Bohr postulate: quantized angular momentum

$$mvr = n\,\frac{h}{2\pi} = n\hbar \qquad\qquad 4\text{-}18$$

where n is an integer called a *quantum number* and $\hbar = h/2\pi$. (The constant $\hbar$, read "h bar," is often more convenient than h itself, just as angular frequency $\omega = 2\pi f$ is often more convenient than the frequency f.) Using this postulate, we can obtain a quantum condition on r by eliminating v from Equations 4-15 and 4-18. We have

Quantum number

$$v^2 = n^2\,\frac{\hbar^2}{m^2 r^2} = \frac{kZe^2}{mr}$$

or

$$r = n^2\,\frac{\hbar^2}{mkZe^2} = n^2\,\frac{a_0}{Z} \qquad\qquad 4\text{-}19$$

where

$$a_0 = \frac{\hbar^2}{mke^2} = 0.529 \text{ Å} \qquad\qquad \text{4-20} \qquad \textit{Bohr radius}$$

is called the first Bohr radius. Combining Equations 4-17 and 4-19 we obtain an expression for $1/\lambda$ similar to the Rydberg-Ritz formula of Equation 4-1. We have

$$\frac{hf}{hc} = Z^2 \frac{mk^2e^4}{4\pi c\hbar^3}\left(\frac{1}{n_f^2} - \frac{1}{n_i^2}\right)$$

or

$$\frac{1}{\lambda} = Z^2 R \left(\frac{1}{n_f^2} - \frac{1}{n_i^2}\right) \qquad\qquad \text{4-21}$$

where

$$R = \frac{mk^2e^4}{4\pi c\hbar^3} \qquad\qquad \text{4-22} \qquad \textit{Rydberg constant}$$

is Bohr's prediction for the value of the Rydberg constant.

Using the values of m, e, c, and $\hbar$ known in 1913, Bohr calculated R and found his result to agree (within the limits of the uncertainties of the constants) with the value obtained from spectroscopy. Bohr noted in his original paper that this equation might be valuable in determining best values for the constant e, m, and $\hbar$ because of the extreme precision possible in measuring R. This has indeed turned out to be the case.

The possible values of the energy of the hydrogen atom predicted by the Bohr model are given by Equation 4-16, with r given by Equation 4-19. They are

$$E_n = -\frac{mk^2e^4}{2\hbar^2}\frac{Z^2}{n^2} = -Z^2\frac{E_0}{n^2} \qquad\qquad \text{4-23} \qquad \textit{Energy levels}$$

where

$$E_0 = \frac{mk^2e^4}{2\hbar^2} = 13.6 \text{ eV} \qquad\qquad \text{4-24}$$

It is convenient to plot these energies versus n, as in Figure 4-15. Such a plot is called an *energy-level diagram*. Various series of transitions are indicated in this diagram by vertical arrows between the levels. The frequency of the light emitted in one of these transitions is the energy difference divided by h according to Bohr's equation (4-13). The energy required to remove the electron from the atom, 13.6 eV, is called the *ionization energy*, or *binding energy*, of the electron.

At the time of Bohr's paper there were two series known for hydrogen: the Balmer series, corresponding to $n_f = 2$, $n_i = 3$, 4, 5, ... and another, named after its discoverer Paschen (1908), corresponding to $n_f = 3$, $n_i = 4$, 5, 6, ... Equation

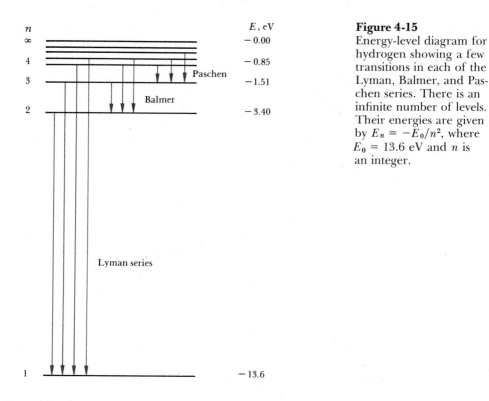

n
∞
4
3
2

1

E, eV
-0.00
-0.85
-1.51
-3.40

-13.6

Paschen

Balmer

Lyman series

Figure 4-15
Energy-level diagram for hydrogen showing a few transitions in each of the Lyman, Balmer, and Paschen series. There is an infinite number of levels. Their energies are given by $E_n = -E_0/n^2$, where $E_0 = 13.6$ eV and n is an integer.

4-21 indicates that other series should exist for different values of n_f. In 1916 Lyman found the series corresponding to $n_f = 1$, and in 1922 and 1924 Brackett and Pfund, respectively, found series corresponding to $n_f = 4$ and $n_f = 5$. As can be easily determined by computing the wavelengths for these series, only the Balmer series lies in the visible portion of the electromagnetic spectrum. The Lyman series is in the ultraviolet, the others in the infrared.

The assumption that the nucleus is fixed is equivalent to the assumption that it has infinite mass. If the nucleus has mass M its kinetic energy will be $\frac{1}{2}Mv^2 = p^2/2M$, where $p = Mv$ is the momentum. If we assume that the total momentum of the atom is zero, the momenta of nucleus and electron must be equal in magnitude. The total kinetic energy is then

$$E_k = \frac{p^2}{2M} + \frac{p^2}{2m} = \frac{M + m}{2mM}p^2 = \frac{p^2}{2\mu}$$

where

$$\mu = \frac{mM}{m + M} = \frac{m}{1 + m/M} \qquad\qquad 4\text{-}25 \qquad \textit{Reduced mass}$$

This is slightly different from the kinetic energy of the electron because μ, called the *reduced mass*, is slightly different from the electron mass. The results derived above for a nucleus of infinite mass can be applied directly for the case of a nucleus of mass M if we replace the electron mass in the equations by reduced mass μ, defined by Equation 4-25. (The validity of this procedure is proved in most intermediate and advanced mechanics books.)

This correction amounts to only 1 part in 2000 for the case of hydrogen and to even less for other nuclei; however, the predicted variation in the Rydberg constant from atom to atom is precisely that which is observed. For example, the spectrum of a singly ionized helium atom, which has one remaining electron, is just that predicted by Equations 4-21 and 4-22 with $Z = 2$ and the proper reduced mass.

According to the correspondence principle, when the energy levels are closely spaced, quantization should have little effect, and classical and quantum calculations should give the same results. From the energy-level diagram of Figure 4-15, we see that the energy levels are close together when the quantum number n is large. This leads us to a slightly different statement of Bohr's correspondence principle: In the region of very large quantum numbers (n in this case) classical calculation and quantum calculation must yield the same results. To see that the Bohr model of the hydrogen atom does indeed obey the correspondence principle, let us compare the frequency of a transition between level $n_i = n$ and level $n_f = n - 1$ for large n, with the classical frequency, which is the frequency of revolution of the electron. From Equation 4-21 we have

$$f = \frac{c}{\lambda} = \frac{Z^2 m k^2 e^4}{4\pi\hbar^3} \left[\frac{1}{(n-1)^2} - \frac{1}{n^2} \right]$$

$$= \frac{Z^2 m k^2 e^4}{4\pi\hbar^3} \frac{2n-1}{n^2(n-1)^2}$$

For large n we can neglect the 1s compared with n and $2n$ to obtain

$$f \simeq \frac{Z^2 m k^2 e^4}{4\pi\hbar^3} \frac{2}{n^3} = \frac{Z^2 m k^2 e^4}{2\pi\hbar^3 n^3} \qquad 4\text{-}26$$

The frequency of revolution of the electron is

$$f_{\text{rev}} = \frac{v}{2\pi r}$$

Using $v = n\hbar/mr$ from Equation 4-18 and $r = n^2\hbar^2/mkZe^2$ from Equation 4-19, we obtain

$$f_{\text{rev}} = \frac{(n\hbar/mr)}{2\pi r} = \frac{n\hbar}{2\pi m r^2} = \frac{n\hbar}{2\pi m (n^2\hbar^2/mkZe^2)^2}$$

$$= \frac{m^2 k^2 Z^2 e^4 n\hbar}{2\pi m n^4 \hbar^4} = \frac{m k^2 Z^2 e^4}{2\pi\hbar^3 n^3} \qquad 4\text{-}27$$

which is the same as Equation 4-26.

A natural extension of the Bohr model is the treatment of elliptical orbits. A result of Newtonian mechanics, familiar from planetary motion, is that in an inverse-square force field, the energy of an orbiting particle depends only on the major axis of the ellipse and not on its eccentricity. There is consequently no change at all unless the force differs from inverse square or unless Newtonian mechanics is modified. A. Sommerfeld considered the effect of special relativity on the Bohr model. Since the relativistic corrections should be of the order of v^2/c^2, it is likely that a highly eccentric orbit would have a larger correction, because v becomes greater as the electron moves nearer the nucleus. The Sommerfeld calculations are quite complicated, but we can estimate the order of magnitude of the effect of special relativity by calculating v/c for the first Bohr orbit in hydrogen. For $n = 1$, we have

$$mvr = \hbar$$

Then using $r = a_0 = \hbar^2/mke^2$ we have

$$v = \frac{\hbar}{mr} = \frac{\hbar}{m(\hbar^2/mke^2)} = \frac{ke^2}{\hbar}$$

and

$$\frac{v}{c} = \frac{ke^2}{\hbar c} = \frac{14.4 \text{ eV-Å}}{1973 \text{ eV-Å}} = \frac{1}{137} \qquad\qquad 4\text{-}28$$

where we have used another convenient combination

$$\hbar c = \frac{1.24 \times 10^4 \text{ eV-Å}}{2\pi} = 1973 \text{ eV-Å}$$

$$= 197.3 \text{ eV-nm} \qquad\qquad 4\text{-}29$$

Though v^2/c^2 is very small, an effect of this magnitude is observable. Upon examination with high resolution, some spectral lines of hydrogen are seen to consist of several closely spaced lines. In Sommerfeld's theory, this is explained in the following way. For each allowed circular orbit of radius r_n and energy E_n, a set of n elliptical orbits is possible of equal major axis but different eccentricities, and therefore slightly different energies. Thus, the energy radiated when the electron changes orbit depends slightly on the eccentricities of the initial and final orbits as well as on their major axes. The splitting of the energy levels is called *fine-structure splitting*, and the dimensionless constant

$$\alpha = \frac{ke^2}{\hbar c} = \frac{1}{137} \qquad\qquad 4\text{-}30 \qquad \textit{Fine-structure constant}$$

is called the *fine-structure constant*. As we shall see in Chapter 7, fine structure is associated with a completely nonclassical property of the electron called *spin*. Though Sommerfeld's explanation does not provide the correct picture, it is remarkable because the result of his calculation agrees perfectly with experiment and also with a detailed and complex calculation based on the Dirac relativistic wave equation, which includes spin.

Questions

4. If the electron moves to an orbit of greater radius, does its total energy increase or decrease? Does its kinetic energy increase or decrease?

5. What is the energy of the shortest-wavelength photon that can be emitted by the hydrogen atom?

4-4 X-Ray Spectra

The extension of the Bohr theory to atoms more complicated than hydrogen proved difficult. Quantitative calculations of the energy levels of atoms of more than one electron could not be made from the model. However, experiments by H. Moseley in 1913 and J. Franck and G. Hertz in 1914 strongly supported the general Bohr-Rutherford picture of the atom as a positively charged core surrounded by electrons that moved in quantized energy states relatively far from the core. Moseley's analysis of x-ray spectra will be discussed in this section, and the Franck-Hertz measurement of the transmission of electrons through gases will be discussed in the next section.

Using the methods of crystal spectrometry that had just been developed by W. H. and W. L. Bragg, Moseley measured the wavelengths of the characteristic x-ray line spectrum for about 40 different target elements. (A typical x-ray spectrum is shown in Figure 3-14.) He noted that the x-ray line spectrum varied in a regular way from element to element, unlike the irregular variations of optical spectra. He surmised that this regular variation occurred because characteristic x-ray spectra were due to transitions involving the innermost electrons of the atoms. Because of the shielding of the other electrons, the inner-electron energies do not depend on the complex interactions of the outer electrons, which are responsible for the complicated optical spectra. Furthermore, the inner electrons are well shielded from the interatomic forces which are responsible for the binding of atoms in solids.

According to the Bohr theory (published earlier the same year, 1913), the energy of an electron in the first Bohr orbit is proportional to the square of the nuclear charge. Moseley reasoned that the energy, and therefore the frequency, of a characteristic x-ray photon should vary as the square of the atomic number of the target element. He therefore plotted the square root of the frequency of a particular x-ray line versus the atomic number Z of the element. Such a plot, now called a *Moseley plot*, is shown in Figure 4-16. These curves can be fitted by the equation

$$f^{1/2} = A_n(Z - b) \qquad\qquad 4\text{-}31$$

where A_n and b are constants for each x-ray line. One family of lines, called the *K series*, has $b = 1$ and different values for A_n. The other family shown in Figure 4-16, called the *L series*, could be fitted by Equation 4-31 with $b = 7.4$.

If the bombarding electron in the x-ray tube knocks an electron from the inner orbit $n = 1$ in a target atom completely out of the atom, photons will be emitted corresponding to transitions of other electrons to the vacancy in the $n = 1$ orbit. (This orbit was called the *K shell*; thus the name "*K* series" for these lines.) The lowest-frequency line corresponds to the lowest-energy transition ($n = 2$ to $n = 1$). This line is called the K_α *line.*

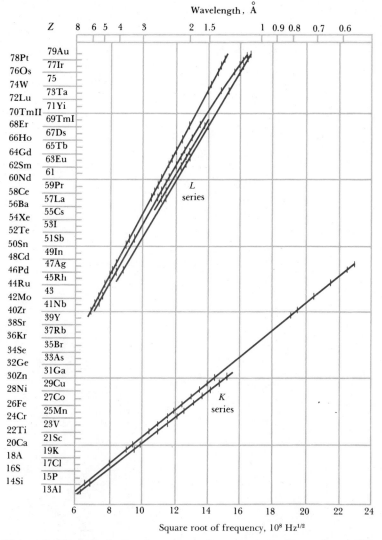

Figure 4-16
Moseley's plots of the
square root of frequency
versus Z for characteristic
x rays. When an atom is
bombarded by high-
energy electrons, an inner
atomic electron is some-
times knocked out,
leaving a vacancy in the
inner shell. The K series
x rays are produced by
atomic transitions to va-
cancies in the $n = 1$ (K)
shell, whereas the L series
is produced by transitions
to the vacancies in the
$n = 2$ (L) shell. [*From H.
Moseley,* Philosophical
Magazine *(6),* **27***, 713
(1914*).]

Using the Bohr relation for a one-electron atom (Equation 4-21)
with $n_f = 1$, and using $(Z - 1)$ in place of Z, we obtain for the
frequencies of the K series

$$f = \frac{mk^2e^4}{4\pi\hbar^3}(Z - 1)^2 \left(\frac{1}{1^2} - \frac{1}{n^2}\right)$$

$$= cR(Z - 1)^2 \left(1 - \frac{1}{n^2}\right) \qquad \text{4-32}$$

where R is the Rydberg. Comparing this with Equation 4-31 we
see that A_n is given by

$$A_n^2 = cR\left(1 - \frac{1}{n^2}\right) \qquad \text{4-33}$$

The wavelengths of the lines in the K series are then given by

$$\lambda = \frac{c}{f} = \frac{c}{A_n^2(Z - 1)^2} = \frac{1}{R(Z - 1)^2(1 - 1/n^2)} \qquad \text{4-34}$$

Example 4-1   Calculate the wavelength of the K_α line of molybdenum, $Z = 42$, and compare with the value $\lambda = 0.721$ Å measured by Moseley. Using $n = 2$, $R = 1.097 \times 10^7$ m^{-1}, and $Z = 42$ we obtain

$$\lambda = [(1.097 \times 10^7 \text{ m}^{-1})(41)^2(1 - \tfrac{1}{4})]^{-1}$$
$$= 7.23 \times 10^{-11} \text{ m}$$
$$= 0.723 \text{ Å}$$

The fact that f is proportional to $(Z - 1)^2$ rather than to Z^2 is explained by the shielding of the nuclear charge by another electron remaining in the K shell. Using this reasoning, Moseley concluded that, since $b = 7.4$ for the L series, these lines involved electrons farther from the nucleus, which were thus shielded by more inner electrons. Assuming that the L series was due to transitions to the $n = 2$ shell, the frequencies for this series are given by

$$f = cR \left(\frac{1}{2^2} - \frac{1}{n^2} \right) (Z - 7.4)^2 \qquad \text{4-35}$$

where $n = 3, 4, 5, \ldots$.

Before Moseley's work the atomic number was merely the place number of the element in Mendeleev's periodic table of the elements arranged by weight. It was known to be approximately half the atomic weight. The experiments of Geiger and Marsden showed that the nuclear charge was approximately $A/2$, while x-ray scattering experiments by Barkla showed that the number of electrons in an atom was approximately $A/2$. These two experiments are consistent, since the atom as a whole must be electrically neutral. However, several discrepancies were found in the periodic table as arranged by weight. For example, the 18th element in order of weight is potassium (39.102) and the 19th is argon (39.948). Arrangement by weight, however, puts potassium in the column with the inert gases and argon with the active metals, the reverse of their known chemical properties. Moseley showed that for these elements to fall on the line $f^{1/2}$ versus Z, argon had to have $Z = 18$ and potassium $Z = 19$. Arranging the elements by the Z number obtained from the Moseley plot, rather than by weight, gave a periodic chart in complete agreement with the chemical properties. We quote from the summary of Moseley's second paper, "The High-Frequency Spectra of the Elements":[1]

Henry G. J. Moseley. (*Courtesy of University of Manchester.*)

  1. Every element from aluminium to gold is characterized by an integer Z which determines its x-ray spectrum. Every detail in the spectrum of an element can therefore be predicted from the spectra of its neighbors.

  2. This integer Z, the atomic number of the element, is identi-

[1] *Philosophical Magazine* (6), **27,** 713 (1914). In his paper Moseley used N for the atomic number. We have changed this to Z to conform with our notation, which is standard today.

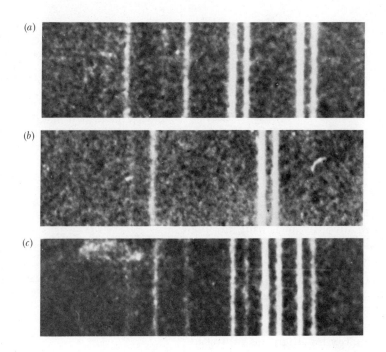

(a)

(b)

(c)

Characteristic x-ray spectra. (*a*) Part of the spectra of neodymium ($Z = 60$) and samarium ($Z = 62$). (*b*) Part of the spectrum of the synthetic element promethium ($Z = 61$). This element was first positively identified in 1945 at the Clinton Laboratory (now Oak Ridge). (*c*) Part of the spectra of the three elements neodymium, promethium, and samarium. (*Courtesy of Dr. J. A. Swartout, Oak Ridge National Laboratory.*)

fied with the number of positive units of electricity contained in the atomic nucleus.

3. The atomic numbers for all elements from Al to Au have been tabulated on the assumption that Z for Al is 13.

4. The order of the atomic numbers is the same as that of the atomic weights, except where the latter disagrees with the order of the chemical properties.

5. Known elements correspond with all the numbers between 13 and 79 except three. There are here three possible elements still undiscovered.

6. The frequency of any line in the x-ray spectrum is approximately proportional to $A(Z - b)^2$, where A and b are constants.

Question

6. Why did Moseley plot $f^{1/2}$ versus Z rather than f versus Z?

4-5 The Franck-Hertz Experiment

An important experiment confirming Bohr's hypothesis of energy quantization in atoms was done in 1914 by J. Franck and G. Hertz. It is now a standard undergraduate laboratory experiment. Figure 4-17 is a schematic diagram of the apparatus. Electrons are ejected from a heated cathode and accelerated toward a grid, which is at a potential V_0 relative to the cathode. Some electrons pass through the grid and reach the plate P, which is at

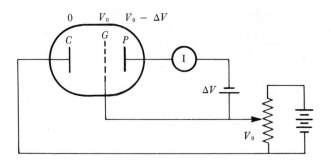

a slightly lower potential $V_p = V_0 - \Delta V$. The tube is filled with
mercury vapor. The experiment involves measuring the plate
current as a function of V_0. As V_0 is increased from 0, the current increases until a critical value (about 4.9 V for Hg) is
reached, at which point the current suddenly decreases. As V_0 is
increased further, the current rises again. The explanation of
this result is that the first excited state of Hg (i.e., the next-to-
the-lowest energy level) is about 4.9 eV above the lowest level
(ground state). Electrons with energy less than this cannot lose
energy to the Hg atoms, but electrons with energy greater than
4.9 eV can make inelastic collisions and lose 4.9 eV. If this
happens near the grid, these electrons cannot gain enough energy to overcome the small back voltage ΔV and reach the plate;
the current therefore decreases. If this explanation is correct,
the Hg atoms that are excited to an energy level of 4.9 eV above
the ground state should return to the ground state by emitting
light of wavelength

$$\lambda = \frac{c}{f} = \frac{hc}{eV_0} = \frac{1.24 \times 10^4 \text{ eV-Å}}{4.9 \text{ eV}} = 2530 \text{ Å}$$

There is indeed a line of this wavelength in the mercury spectrum. When the tube is viewed with a spectroscope, this line is
seen when V_0 is greater than 4.9 eV, while no lines are seen
when V_0 is less than this amount. For further increases in V_0,
additional sharp decreases in the current are observed, corresponding either to excitation of other levels in Hg or to multiple
excitation of the first excited state, i.e., due to an electron losing
4.9 eV more than once. (In the usual setup, multiple excitations
of the first level are observed and dips are seen every 4.9 eV.
The probability of observing such multiple first-level excitations,
or excitation of other levels, depends on the detailed variation of
the potential in the tube. The reader is referred to Reference 1,
page 8, for details of this experiment.) A plot of the data of
Franck and Hertz is shown in Figure 4-18.

The Franck-Hertz experiment was an important confirmation of the idea that discrete optical spectra were due to the
existence in atoms of discrete energy levels which could be excited by nonoptical methods. It is particularly gratifying to be
able to detect the existence of discrete energy levels by measurements using only voltmeters and ammeters.

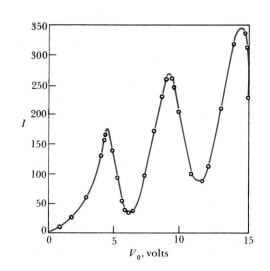

V_0, volts

Figure 4-18
Current versus accelerating voltage in the Franck-Hertz experiment. The current decreases because many electrons lose energy due to inelastic collisions with mercury atoms in the tube, and therefore cannot overcome the small back potential indicated in Figure 4-17. The regular spacing of the peaks in this curve indicates that only a certain quantity of energy, 4.9 eV, can be lost to the mercury atoms. This interpretation is confirmed by the observation of radiation of photon energy 4.9 eV emitted by the mercury atoms, when V_0 is greater than this energy. [*From J. Franck and G. Hertz,* Verband Deutscher Physikalischer Gesellschaften, **16**, *457 (1914).*]

Wilson-Sommerfeld rule

Optional

4-6   The Wilson-Sommerfeld Quantization Rule

It was surprising that Bohr could derive the hydrogen spectrum by postulating the quantization of angular momentum $L = n\hbar$, whereas Einstein and Planck had used the quantization of energy $E = nhf$ to obtain the specific heats of solids and the spectral distribution of blackbody radiation. Certainly there had to be some connection between these quantum postulates; yet the connection remained a mystery for some time. In 1916, W. Wilson and A. Sommerfeld announced a general rule for the quantization of periodic systems. Their rule is

$$\oint P \, dq = nh \qquad\qquad 4\text{-}36$$

P can be a component of linear momentum such as p_x, in which case q is the corresponding coordinate x; or P can be a component of angular momentum such as L_z, in which case q is the angle ϕ associated with rotation about the z axis. The symbol $\oint$ indicates that the integral is to be taken over one complete cycle of the system. In many cases, the evaluation of Equation 4-36 is somewhat involved. We give three examples in this section.

Example 4-2   Particle Moving in a Circle in a Central Field   The Wilson-Sommerfeld condition for this problem is

$$\oint L \, d\phi = nh$$

where L is the angular momentum.

Since the angular momentum is constant, this becomes

$$L \oint d\phi = nh$$

The integration of $d\phi$ for one cycle gives 2π; thus

$$L = \frac{nh}{2\pi} = n\hbar$$

which is the Bohr quantum condition.

Example 4-3   Simple Harmonic Motion   Newton's law of motion for a mass m on a spring of force constant K is

$$-Kx = m \, \frac{d^2x}{dt^2}$$

A solution of this equation is

$$x = A \sin \omega t \qquad\qquad 4\text{-}37$$

where A is the amplitude and ω is the angular frequency $\omega = 2\pi f = \sqrt{K/m}$. The sum of the potential energy $\frac{1}{2}Kx^2$ and the kinetic energy $p^2/2m$ is constant and equal to the maximum value of either:

$$E = \tfrac{1}{2}KA^2 = \tfrac{1}{2}m\omega^2 A^2 \qquad\qquad 4\text{-}38$$

Using Equation 4-37 we can calculate dx and the momentum p. We have

$$dx = \omega A \cos \omega t \, dt$$

$$p = m \, \frac{dx}{dt} = m\omega A \cos \omega t$$

Then

$$\oint p \, dx = \oint m\omega^2 A^2 \cos^2 \omega t \, dt = nh$$

or

$$2E \oint \cos^2 \omega t \, dt = nh$$

where we have used $m\omega^2 A^2 = 2E$. If we let $\theta = \omega t$, the integration over one cycle corresponds to integrating θ from 0 to 2π. We have

$$\frac{2E}{\omega} \int_0^{2\pi} \cos^2 \theta \, d\theta = \frac{2E}{\omega}\, \pi = nh$$

or

$$E = \frac{nh\omega}{2\pi} = nhf = n\hbar\omega \qquad\qquad 4\text{-}39$$

This is the condition used by Planck and Einstein for the quantization of oscillators.

Example 4-4   A particle of mass m moves back and forth in a one-dimensional box of length L (Figure 4-19). No forces act on the particle except when it hits the walls and is reflected elasti-

Figure 4-19
Particle moving with constant speed in a "box" of length L. The Wilson-Sommerfeld quantization rule predicts that the momentum is quantized to the values $p_n = nh/2L$, where n is an integer.

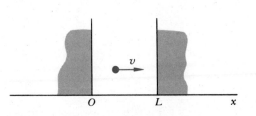

cally. If we consider the particle at the left wall at the beginning of the cycle, we have

$$\oint p \, dx = \int_0^L (+mv) \, dx + \int_L^0 (-mv) \, dx = 2mvL = nh$$

or

$$p = mv = \frac{nh}{2L} \qquad\qquad 4\text{-}40$$

Assuming the particle to be nonrelativistic, the energy is

$$E = \tfrac{1}{2}mv^2 = \frac{p^2}{2m} = n^2 \frac{h^2}{8mL^2} \qquad\qquad 4\text{-}41$$

4-7 Critique of the Bohr Theory and of "Old Quantum Mechanics"

We have seen in this and the preceding chapters that many phenomena—blackbody radiation, the photoelectric effect, Compton scattering, specific heats, optical spectra of hydrogen, and the x-ray spectra of many elements—could be "explained" by various quantum assumptions, which could be summarized by the Wilson-Sommerfeld quantum rule. This "theory" is now usually referred to as "old quantum mechanics." It is a strange mixture of classical physics and quantum assumptions. The application of this quantum mechanics in the early years of the twentieth century was more of an art than a science, for no one knew exactly what the rules were. It must be said, however, that the successes of the Bohr theory were substantial and spectacular. The existence of unknown spectral lines was predicted and later observed. Not only was the Rydberg constant given in terms of known constants, but its slight variation from atom to atom was accurately predicted by the slight variation in the reduced mass. The radius of the first Bohr orbit in hydrogen, 0.53 Å, corresponded well with the known diameter of the hydrogen molecule, about 2.2 Å. The wavelengths of the characteristic x-ray spectra could be calculated from the Bohr theory.

The failures of the Bohr theory and the "old quantum mechanics" were mainly those of omission. There was no way of predicting the relative intensities of spectral lines. There was little success in applying the theory to the optical spectra of more complex atoms. Finally, there was the considerable philosophical problem that its assumptions lacked foundation. There were no *a priori* reasons to expect that Coulomb's law would work but that the laws of radiation would not, or that Newton's laws could be used even though only certain values of angular momentum were allowed. Though the Wilson-Sommerfeld quantization rule worked well for periodic systems, it was not known why, and there was no theory at all for nonperiodic systems. In the

1920s scientists struggled with these difficulties, and a systematic theory, now known as *quantum mechanics* or *wave mechanics*, was formulated by de Broglie, Schrödinger, Heisenberg, Pauli, Dirac, and others. We shall study some aspects of this theory in the next two chapters and apply it to the study of atoms, nuclei, and solids in the remaining chapters of this book. We shall see that, though this theory is much more satisfying from a philosophical point of view, it is somewhat abstract and difficult to apply in detail to problems. In spite of its shortcomings, the Bohr theory provides a model that is easy to visualize, gives the correct energy levels in hydrogen, and is often useful in describing a quantum-mechanical calculation.

Summary

The scattering of α particles indicates that the positive charge in an atom is concentrated in a small nucleus, which for gold has a radius of about 10^{-14} m. The charge on the nucleus is Ze, where the atomic number Z is the place number in the periodic table.

If the incoming intensity is I_0 (particles/sec-area), the number per second scattered by one atom at angles greater than θ is related to the impact parameter b by

$$N_{>\theta} = I_0 \pi b^2(\theta)$$

The cross section σ is defined as the number of particles scattered per atom divided by the incident intensity. The cross section has dimensions of area and represents the effective area of an atom for scattering. The cross section for scattering through angles greater than θ is

$$\sigma_{>\theta} = \frac{N_{>\theta}}{I_0} = \pi b^2(\theta)$$

The impact parameter for Rutherford scattering is related to the scattering angle by

$$b = \tfrac{1}{2} r_d \cot \tfrac{1}{2}\theta$$

where the distance of closest approach for a head-on collision is

$$r_d = \frac{k q_\alpha Q}{\tfrac{1}{2} m v^2}$$

The fraction of α particles scattered per unit area between θ and $\theta + d\theta$ is proportional to $\sin^{-4} \tfrac{1}{2}\theta$.

The Bohr postulates are:

1. Electrons move only in certain nonradiating, stable, circular orbits consistent with Coulomb's law and Newton's law, and specified by the quantization of angular momentum

$$L = mvr = \frac{nh}{2\pi} = n\hbar$$

where h is Planck's constant and n is an integer.

2. Radiation of frequency $f = (E_i - E_f)/h$ occurs when the

electron jumps from its initial orbit of energy E_i to its final orbit of energy E_f.

These postulates lead to

$$r_n = \frac{n^2 a_0}{Z}$$

where $a_0 = \hbar^2/mke^2 = 0.529$ Å for the allowed electron orbits, and

$$E_n = -\frac{Z^2 E_0}{n^2}$$

where $E_0 = mk^2 e^4/2\hbar^2 = 13.6$ eV and $n = 1, 2, 3, \ldots$.

The Franck-Hertz experiment and Moseley's treatment of characteristic x-ray spectra provide strong corroboration of the Bohr picture.

The Wilson-Sommerfeld quantization rule for periodic systems

$$\oint P \, dq = nh$$

leads to

$$L = n\hbar \qquad \text{for circular motion}$$

$$E = nhf \qquad \text{for simple harmonic motion}$$

and

$$E = n^2 \left(\frac{h^2}{8mL^2} \right) \qquad \text{for a particle in a box}$$

Some useful combinations of physical constants are

$$\hbar c = \frac{hc}{2\pi} = 1973 \text{ eV-Å} = 197.3 \text{ eV-nm}$$

$$ke^2 = \frac{e^2}{4\pi\epsilon_0} = 14.4 \text{ eV-Å} = 1.44 \text{ eV-nm}$$

$$\alpha = \frac{ke^2}{\hbar c} = \frac{1}{137}$$

References

Many original papers, including those quoted in this and preceding chapters, can be found in *The World of the Atom* (Chapter 2, Reference 2) or in *Great Experiments in Physics* (Chapter 3, Reference 2).

1. A. Melissinos, *Experiments in Modern Physics*, New York: Academic Press, Inc., 1966. Many of the classic experiments that are now undergraduate laboratory experiments are described in detail in this text.

2. B. Cline, *The Questioners: Physicists and the Quantum Theory*, New York: Thomas Y. Crowell Company, 1965.

3. G. Gamow, *Thirty Years That Shook Physics: The Story of the Quantum Theory*, Garden City, N.Y.: Doubleday & Company, Inc., 1965.

References 2 and 3 are interesting accounts of the development of the quantum theory and the interactions of the people involved.

Exercises

Section 4-1, Empirical Spectral Formulas

There are no exercises for this section.

Section 4-2, Rutherford Scattering

1.   If a particle is deflected by 0.01° in each collision, about how many collisions would be necessary to produce an rms deflection of 10°? (Use the result of the one-dimensional random-walk problem.) Compare this result with the number of atomic layers in a gold foil of thickness 10^{-6} m, assuming that the thickness of each atom is 1 Å $= 10^{-10}$ m.

2.   Calculate the distance of closest approach r_d for a head-on collision between an α particle and a gold nucleus if the initial kinetic energy of the α particle is (*a*) 5 MeV, (*b*) 8 MeV.

3.   What energy α particle would be needed to just reach the surface of a gold nucleus in a head-on collision if the radius of the nucleus is 7 fm ($= 7 \times 10^{-15}$ m)?

4.   Do Exercise 2 for a head-on collision of an α particle and an Al nucleus ($Z = 13$).

5.   What energy α particle would be needed to just reach the surface of an Al nucleus if its radius is 4 fm?

6.   Show that Equation 4-4 for the impact parameter can be written

$$b = \frac{1.44\ Z}{E_\alpha}\cot\tfrac{1}{2}\theta$$

where E_α is the energy of the α particle in MeV and b is in fm.

✓7.   (*a*) Calculate the impact parameter b for scattering of a 5-MeV α particle by gold through an angle of 10°. (*b*) What fraction of the incident beam is scattered through angles greater than 10° by gold foil 10^{-6} m thick?

8.   Repeat Exercise 7 for scattering through 1° or more.

9.   (*a*) What is the ratio of the number of particles per unit area on the screen scattered at 10° to those at 1°? (*b*) What is the ratio at 30° to 1°?

Section 4-3, The Bohr Model of the Hydrogen Atom

10. Show that Equation 4-20 for the radius of the first Bohr orbit and Equation 4-24 for the magnitude of the lowest energy for the hydrogen atom can be written as

$$a_0 = \frac{\hbar c}{\alpha mc^2} = \frac{\lambda_c}{2\pi\alpha}$$

$$E_0 = \tfrac{1}{2}\alpha^2 mc^2$$

where $\lambda_c = h/mc$ is the Compton wavelength of the electron and $\alpha = ke^2/\hbar c$ is the fine-structure constant. Use these expressions to check the numerical values of the constants a_0 and E_0.

✓11. Calculate the three longest wavelengths in the Lyman series ($n_f = 1$) in nm and indicate their position on a horizontal linear scale. Indicate the series limit (shortest wavelength) on this scale. Are any of these lines in the visible spectrum?

12. Show that the Rydberg constant can be written $R = \tfrac{1}{2}\alpha^2 mc^2/hc$ and

use this to calculate R in m^{-1} (see Exercise 10).

13. Calculate the longest three wavelengths and the series limit for the Paschen ($n_f = 3$) and the Brackett ($n_f = 4$) series, and plot both series on the same linear scale.

14. On the average, a hydrogen atom will exist in an excited state for about 10^{-8} sec before making a transition to a lower-energy state. About how many revolutions does an electron in the $n = 2$ state make in 10^{-8} sec?

√15. It is possible for a muon to be captured by a proton to form a muonic atom. A muon is identical to an electron except for its mass, which is 105.7 MeV/c^2. (a) Calculate the radius of the first Bohr orbit of a muonic atom. (b) Calculate the magnitude of the lowest energy. (c) What is the shortest wavelength in the Lyman series for this atom?

16. In the lithium atom ($Z = 3$) two electrons are in the $n = 1$ orbit and the third is in the $n = 2$ orbit. (Only two are allowed in the $n = 1$ orbit because of the exclusion principle, which will be discussed in Chapter 7.) The interaction of the inner electrons with the outer one can be approximated by writing the energy of the outer electron as

$$E = -Z'^2 \frac{E_0}{n^2}$$

where $E_0 = 13.6$ eV, $n = 2$, and Z' is the effective nuclear charge, which is less than 3 because of the screening effect of the two inner electrons. Using the measured ionization energy of 5.39 eV, calculate Z'.

Section 4-4, X-Ray Spectra

√17. (a) Calculate the next two longest wavelengths in the K series (after the K_α line) of molybdenum. (b) What is the wavelength of the shortest wavelength in this series?

18. The wavelength of the K_α x-ray line for an element is measured to be 0.794 Å. What is the element?

19. The wavelength of the K_α line for an element is 3.368 Å. What is the element?

20. Calculate the three longest wavelengths in the L series for molybdenum.

21. The L_α line for a certain element has a wavelength of 0.3617 nm. What is the element?

Section 4-5, The Franck-Hertz Experiment
There are no exercises for this section.

Section 4-6, The Wilson-Sommerfeld Quantization Rule

22. Make an energy-level diagram for a charged particle oscillator in simple harmonic motion using the Wilson-Sommerfeld result for the energies $E_n = nhf$, where f is the frequency oscillation. Indicate the transitions on this diagram for which $n_f = n_i \pm 1$ and show that the frequencies of radiation (given by the Bohr condition, Equation 4-13) are the same as those expected classically.

√23. (a) Use the Wilson-Sommerfeld result to calculate the energy of the ground state ($n = 1$) and of the first excited state ($n = 2$) for an electron in a one-dimensional box of size $L = 1$ Å. (b) Find the wavelength of the radiation given off if the electron makes a transition from $n = 2$ to $n = 1$. How does this result compare in order of magnitude with the wavelengths emitted by the hydrogen atom?

24. Use the Wilson-Sommerfeld result to calculate the magnitude of the lowest energy ($n = 1$) for a proton in a one-dimensional box of size $L = 16$ fm (about the diameter of a gold nucleus). (Use $m_p = 938$ MeV/c^2.)

Section 4-7, Critique of the Bohr Theory and of "Old Quantum Mechanics"

There are no exercises for this section.

Problems

1.   (a) Calculate the ground-state energy of helium ($Z = 2$) assuming that the two electrons are in the $n = 1$ orbit and neglecting any interaction between them. (b) The farthest apart these electrons can be while in the same orbit is twice the radius of the orbit. Assuming their mutual repulsion keeps them this far apart, calculate the positive energy of interaction $+ ke^2/2r$. (c) The ionization energy is the energy needed to remove one electron. Estimate this for helium based on your results for parts (a) and (b). The measured value is about 24.6 eV.

2.   (a) Show that $R_\infty/R_H = 1 + m_e/m_p$, where m_e is the mass of the electron and m_p that of the proton. (b) Calculate m_e/m_p from the values of R_∞ and R_H given in the text and compare with the ratio obtained from the known values of m_e and m_p.

3.   (a) The current i due to a charge q moving in a circle with frequency f_{rev} is qf_{rev}. Find the current due to the electron in the first Bohr orbit. (b) The magnetic moment of a current loop is iA, where A is the area of the loop. Find the magnetic moment of the electron in the first Bohr orbit in units of A-m^2. This magnetic moment is called a *Bohr magneton*.

4.   Show that the speed of the electron in the nth Bohr orbit in hydrogen is given by $v/c = \alpha/n$, where $\alpha = ke^2/\hbar c$ is the fine-structure constant.

5.   Show that a small change in the reduced mass of the electron produces a small change in a spectral line given by $\Delta\lambda/\lambda = \Delta\mu/\mu$. Use this to calculate the difference $\Delta\lambda$ in the Balmer red line $\lambda = 6563$ Å between hydrogen and deuterium, which has a nucleus with twice the mass of hydrogen.

6.   A small shot of negligible radius hits a stationary smooth hard sphere of radius R, making an angle β with the normal to the sphere, as shown in Figure 4-20. It is reflected at an equal angle to the normal. The scattering angle is $\theta = 180° - 2\beta$, as shown. (a) Show by the geometry of the figure that the impact parameter b is related to θ by $b = R \cos \frac{1}{2}\theta$. ($b$) If the incoming intensity of the shot is I_0 particles/sec-area, how many are scattered through angles greater than θ? (c) Show that the cross section for scattering through angles greater than 0° is πR^2. (d) Discuss the implication of the fact that the Rutherford cross section for scattering through angles greater than 0° is infinite.

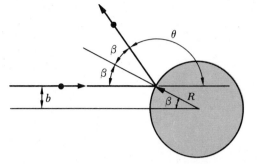

Figure 4-20
Small particle scattered by
a hard sphere of radius R.

7. Assume that a gold nucleus of radius 8 fm acts like a hard sphere for the scattering of uncharged particles such as neutrons. What fraction of uncharged point particles will be deflected at all by nuclei in a gold foil of thickness 10^{-6} m? (Use the result of Problem 6 that the total cross section for hard-sphere scattering is πR^2.)

8. A particle of mass m is attracted to the origin by a force proportional to the distance from the origin $\mathbf{F} = -K\mathbf{r}$. Assume that it moves in a circular orbit of radius r. (a) Use Newton's law to show that its kinetic energy $\frac{1}{2}mv^2$ equals its potential energy $\frac{1}{2}Kr^2$. (b) Use the Bohr quantum condition on the angular momentum to show that the allowed orbits are given by $r^2 = n\hbar/\sqrt{Km}$. (c) Find the allowed energies, and show that they can be written $E_n = nhf$, where f is the frequency of revolution.

9. A beam of 10-MeV protons are incident on a thin aluminum foil of thickness 10^{-6} m. Find the fraction of the particles that are scattered through angles greater than (a) 10°, (b) 90°.

10. Derive Equation 4-7 from Equation 4-4.

11. Calculate the lowest energy for an electron in a one-dimensional box of size $L = 16$ fm, using the Wilson-Sommerfeld result $p = nh/2L$ and the relativistic approximation $E \approx pc$.

12. In this problem you are to obtain the Bohr results for the energy levels in hydrogen without using the quantization condition of Equation 4-18. In order to relate Equation 4-14 to the Balmer-Ritz formula, assume that the radii of allowed orbits are given by $r_n = n^2 r_0$, where n is an integer and r_0 is a constant to be determined. (a) Show that the frequency of radiation for a transition from $n_i = n$ to $n_f = n - 1$ is given by $f \approx kZe^2/hr_0n^3$ for large n. (b) Show that the frequency of revolution is given by $f_{rev}^2 = kZe^2/4\pi^2 mr_0^3 n^6$. (c) Use the correspondence principle to determine r_0 and compare with Equation 4-20.

13. According to the variation of C_v with T for hydrogen (Figure 2-5), the H_2 molecule begins oscillating at about $T \sim 5000$ K. (a) Using this temperature, calculate the order of magnitude of the frequency of oscillation from $hf = kT$. (b) Assuming the energies of vibration to be quantized according to the Wilson-Sommerfeld results for a one-dimensional oscillator, calculate the wavelength of the radiation given off when the H_2 molecule makes a transition from one vibrational energy level to the next lower one.

14. For the energy levels of a nonrelativistic particle in a one-dimensional box given by Equation 4-41, show that the frequency of radiation emitted for the transition n to $n - 1$ equals the frequency of motion when $n \gg 1$, in accord with Bohr's correspondence principle.

CHAPTER 5   Electron Waves

Objectives

After studying this chapter you should:

---

1.   Know the de Broglie relations for the frequency and wavelength of electron waves.

2.   Be able to use the de Broglie relations and the standing-wave conditions for waves in a circle to derive the Bohr condition for the quantization of angular momentum in the hydrogen atom.

3.   Be able to describe the experimental determinations of electron wavelengths.

4.   Know the meaning of the terms wave function, wave equation, phase velocity, harmonic wave, amplitude, wavelength, period, frequency, intensity, group velocity, and standing-wave condition.

5.   Know the important features of the interference and diffraction of classical waves.

6.   Know the general properties of wave packets and in particular the relations between Δx and Δk, and between $\Delta \omega$ and Δt.

7.   Be able to discuss the uncertainty principle and some of its consequences.

9.   Be able to discuss particle-wave duality.

5-1   The de Broglie Relations

In 1924 a French student, Louis de Broglie, suggested in his dissertation that since light was known to have both wave and particle properties, perhaps matter—in particular, electrons—might also have both wave and particle characteristics. This suggestion was highly speculative; there was no evidence at that time for any wave aspects of electrons. For the frequency and wavelength of electron waves, de Broglie chose the equations

---

$$f = \frac{E}{h} \qquad\qquad 5\text{-}1$$

De Broglie relations

$$\lambda = \frac{h}{p} \qquad\qquad 5\text{-}2$$

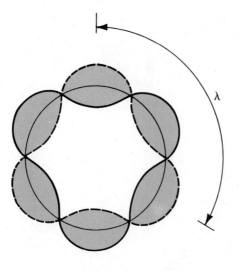

Figure 5-1
Standing waves around
the circumference of a
circle.

where p is the momentum and E the energy of the electron, by
analogy with the identical equations that hold for photons. He
pointed out that, on this assumption, the Bohr quantum condition for angular momentum was equivalent to a standing-wave
condition (Figure 5-1). We have

$$mvr = n\hbar = \frac{nh}{2\pi}$$

$$2\pi r = \frac{nh}{mv} = \frac{nh}{p}$$

or

$$n\lambda = 2\pi r = S \qquad\qquad\qquad 5\text{-}3$$

where S is the circumference of the circular Bohr orbit. The
idea of explaining discrete energy states in matter by standing
waves seemed quite promising. The Wilson-Sommerfeld quantum rule could be interpreted as a standing-wave requirement.
For example, the Wilson-Sommerfeld quantum condition for
the particle in a one-dimensional box of size L (Example 4-4,
in Section 4-6) is $p = nh/2L$. Using $p = h/\lambda$, this becomes
$h/\lambda = nh/2L$ or

$$n\frac{\lambda}{2} = L \qquad\qquad\qquad 5\text{-}4 \qquad\qquad \textit{Standing-wave condition}$$

This is the condition for standing waves on a string fixed at both
ends (Figure 5-2).

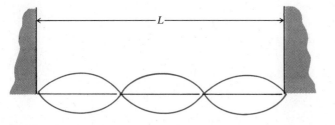

Figure 5-2
Standing waves in a one-
dimensional box.

Louis V. de Broglie, who first suggested that electrons might have wave properties.

The ideas of de Broglie were expanded and developed into a complete theory by Erwin Schrödinger later in 1924. In 1927, C. J. Davisson and L. H. Germer verified the de Broglie hypothesis directly by observing interference patterns with electron beams. We can see why the wave properties of matter were not readily observed if we recall that the wave properties of light were not noted until apertures or slits the same size as the wavelength of light could be obtained. When the wavelength of light is much smaller than any aperture, the diffraction and interference effects are not observable and the light obeys geometric or ray optics. Because of the smallness of Planck's constant, the wavelength given by Equation 5-2 is extremely small for any macroscopic object.

Example 5-1   What is the de Broglie wavelength for a very small but macroscopic object of mass 10^{-9} g moving with speed 3×10^{-8} m/sec?

$$\lambda = \frac{h}{mv} = \frac{6.63 \times 10^{-34} \text{ J-sec}}{(10^{-12} \text{ kg})(3 \times 10^{-8} \text{ m/sec})}$$
$$= 2.2 \times 10^{-14} \text{ m} = 2.2 \times 10^{-4} \text{ Å}$$

This is much smaller than any possible aperture.

The case is different for low energy electrons. Consider an electron that has been accelerated through V_0 volts. Its energy is then

$$E = \frac{p^2}{2m} = eV_0$$

and

$$\lambda = \frac{h}{p} = \frac{hc}{pc} = \frac{hc}{(2mc^2\,eV_0)^{1/2}}$$

Using $hc = 1.24 \times 10^4$ eV-Å and $mc^2 = 0.511 \times 10^6$ eV, we obtain

$$\lambda = \frac{12.26\ V^{1/2}}{V_0^{1/2}}\ \text{Å} \qquad \text{for } eV_0 \ll mc^2 \qquad\qquad 5\text{-}5$$

λ for nonrelativistic electrons

Example 5-2   What is the de Broglie wavelength of a 10-eV electron? Putting $V_0 = 10$, we have

$$\lambda = \frac{12.26}{\sqrt{10}}\ \text{Å} \approx 3.9\ \text{Å}$$

Though this wavelength is small, it is just the order of magnitude of the size of the atom and of the spacing of atoms in a crystal.

5-2   Measurements of Electron Wavelengths

The first measurements of the wavelengths of electrons were made in 1927 by Davisson and Germer, who were studying electron reflection from a nickel target at Bell Telephone Laboratories. After heating their target to remove an oxide coating that had accumulated during an accidental break in their vacuum system, they found that the scattered electron intensity as a function of the scattering angle showed maxima and minima. Their target had crystallized and they were observing electron diffraction. They then prepared a target consisting of a single crystal of nickel and extensively investigated the scattering of electrons from it. Figure 5-3 illustrates their experimental arrangement. Their data for 54-eV electrons, shown in Figure 5-4, indicate a strong maximum of scattering at $\phi = 50°$. Consider the scattering from a set of Bragg planes, as shown in Figure 5-5. The Bragg condition for constructive interference is $n\lambda = 2d \sin \theta$

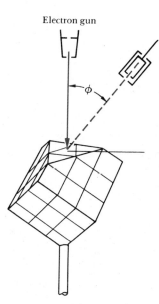

Electron gun

Figure 5-3
The Davisson-Germer experiment. Electrons scattered at angle ϕ from a nickel crystal are detected in an ionization chamber.

Figure 5-4
Polar plot of the scattered intensity versus angle for 54-eV electrons. (The intensity at each angle is indicated by the distance of the point from the origin.) There is a maximum intensity at $\phi = 50°$, as predicted for Bragg scattering of waves having wavelength $\lambda = h/p$. (*From* Nobel Prize Lectures: Physics, *Amsterdam and New York: Elsevier,* © *Nobel Foundation, 1964.*)

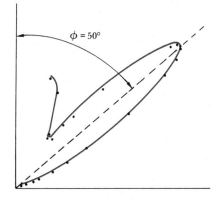

$\phi = 50°$

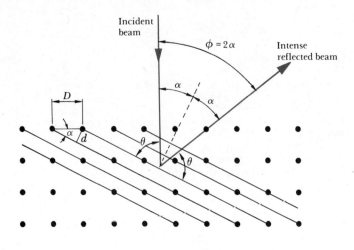

Figure 5-5
Scattering of electrons by a crystal. Electron waves are strongly scattered if the Bragg condition $n\lambda = 2d \sin \theta$ is met. This is equivalent to the condition $n\lambda = D \sin \phi$.

$= 2d \cos \alpha$. The spacing of the Bragg planes d is related to the spacing of the atoms D by $d = D \sin \alpha$; thus

$$n\lambda = 2D \sin \alpha \cos \alpha = D \sin 2\alpha$$

or

$$n\lambda = D \sin \phi \qquad\qquad 5\text{-}6$$

where $\phi = 2\alpha$ is the scattering angle.

The spacing D is known from x-ray diffraction to be 2.15 Å. The wavelength calculated from Equation 5-6 is, for $n = 1$,

$$\lambda = 2.15 \sin 50° = 1.65 \text{ Å}$$

This compares well with that calculated from the de Broglie relation

$$\lambda = \frac{12.26}{(54)^{1/2}} = 1.67 \text{ Å}$$

Figure 5-6 shows a plot of measured wavelengths versus $V_0^{-1/2}$. The wavelengths measured by diffraction are slightly lower than the theoretical predictions because the refraction of the electron waves at the crystal surface has been neglected. We have seen from the photoelectric effect that it takes work of the order of several eV to remove an electron from a metal. Electrons entering a metal gain kinetic energy; therefore, their de Broglie wavelength is slightly less inside the crystal. It is interesting to read Davisson's account of the connection between de Broglie's predictions and their experimental verification:[1]

> Perhaps no idea in physics has received so rapid or so intensive development as this one. De Broglie himself was in the van of this development, but the chief contributions were made by the older and more experienced Schroedinger.
>   In these early days—eleven or twelve years ago [i.e., 1925 or 1926]—attention was focused on electron waves in atoms.

[1] *Nobel Prize Lectures: Physics,* Amsterdam and New York: Elsevier, 1964.

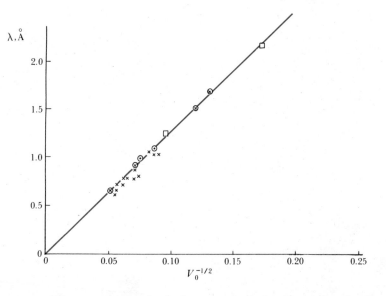

Figure 5-6
Test of the de Broglie
formula $\lambda = h/mv$. The
wavelength is computed
from a plot of the diffraction data plotted against
$V_0^{-1/2}$, where V_0 is the
accelerating voltage. The
straight line is
$12.26\,V_0^{-1/2}$ Å as predicted
from $\lambda = h(2mE)^{-1/2}$.
($\times$ From observations
with diffraction appa-
ratus; $\otimes$ same, particularly reliable; $\square$ same,
grazing beams. $\odot$ From
observations with reflec-
tion apparatus.) (*From
Nobel Prize Lectures:
Physics, Amsterdam and
New York: Elsevier,* ©
Nobel Foundation, 1964.)

The wave mechanics had sprung from the atom, so to speak,
and it was natural that the first applications should be to the
atom. No thought was given at this time, it appears, to electrons in free flight. It was implicit in the theory that beams of
electrons like beams of light would exhibit the properties of
waves, that scattered by an appropriate grating they would
exhibit diffraction, yet none of the chief theorists mentioned
this interesting corollary. The first to draw attention to it was
Elsasser, who pointed out in 1925 that a demonstration of diffraction would establish the physical existence of electron
waves. The setting of the stage for the discovery of electron
diffraction was now complete.

It would be pleasant to tell you that no sooner had Elsasser's
suggestion appeared than the experiments were begun in New
York which resulted in a demonstration of electron diffraction—pleasanter still to say that the work was begun the day
after copies of de Broglie's thesis reached America. The true
story contains less of perspicacity and more of chance. The
work actually began in 1919 with the accidental discovery that
the energy spectrum of secondary electron emission has, as its
upper limit, the energy of the primary electrons, even for pri-
maries accelerated through hundreds of volts; that there is, in
fact, an elastic scattering of electrons by metals.

Out of this grew an investigation of the distribution-in-
angle of these elastically scattered electrons. And then chance
again intervened; it was discovered, purely by accident, that
the intensity of elastic scattering varies with the orientations of
the scattering crystals. Out of this grew, quite naturally, an investigation of elastic scattering by a single crystal of predeter-
mined orientation. The initiation of this phase of the work occurred in 1925, the year following the publication of de
Broglie's thesis, the year preceding the first great developments in the wave mechanics. Thus the New York experiment
was not, at its inception, a test of the wave theory. Only in the
summer of 1926, after I had discussed the investigation in
England with Richardson, Born, Franck and others, did it take
on this character.

Bell Telephone Laboratories, Inc.

Clinton J. Davisson (left)
and Lester A. Germer at
Bell Laboratories, where
electron diffraction was
first observed.

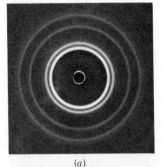

(a)

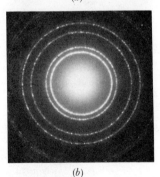

(b)

Figure 5-7
(a) Diffraction pattern
produced by x rays of
wavelength 0.71 Å and an
aluminum-foil target. (b)
Diffraction pattern pro-
duced by 600-eV elec-
trons (de Broglie wave-
length of about 0.5 Å)
and an aluminum-foil
target. The pattern has
been enlarged by 1.6
times to facilitate comparison with (a). (*Courtesy of
Film Studio, Education
Development Center.*)

The search for diffraction beams was begun in the autumn
of 1926, but not until early in the following year were any
found—first one and then twenty others in rapid succession.
Nineteen of these could be used to check the relationship
between wavelength and momentum, and in every case the
correctness of the de Broglie formula, $\lambda = h/p$, was verified to
within the limit of accuracy of the measurements.

Another demonstration of the wave nature of electrons was
provided in the same year by G. P. Thomson, who observed the
transmission of electrons through thin metallic foils (G. P.
Thomson, the son of J. J. Thomson, shared the Nobel Prize in
1937 with Davisson). The experimental arrangement was similar
to that used to obtain Laue patterns with x rays (see Figure
3-10). Because the metal foil consists of many tiny crystals randomly oriented, the diffraction pattern consists of concentric
rings. If a crystal is oriented at an angle θ with the incident
beam, where θ satisfies the Bragg condition, this crystal will
strongly scatter at an equal angle θ; thus there will be a scattered
beam making an angle 2θ with the incident beam. Figure 5-7a
and b shows the similarities in patterns produced by x rays and
electron waves. A similar pattern produced by neutrons is shown
in Figure 5-8.

In 1930 Stern and Estermann observed diffraction of hydrogen and helium atoms from a lithium fluoride crystal (Figure
5-9). Since then, diffraction of other atoms and of neutrons has
been observed (Figures 5-10 and 5-8). In all cases the measured
wavelengths agree with the de Broglie prediction.

Before we consider the implications of the wave properties of
electrons, we shall review some properties of classical waves.

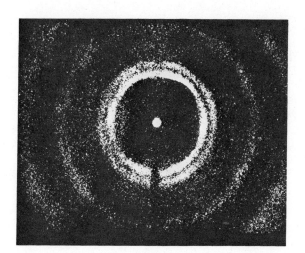

Figure 5-8
Diffraction pattern produced by 0.0568-eV neutrons (de Broglie wavelength of 1.20 Å) and a target of polycrystalline copper. Note the similarity in the patterns produced by x rays, electrons, and neutrons. (*Courtesy of C. G. Shull.*)

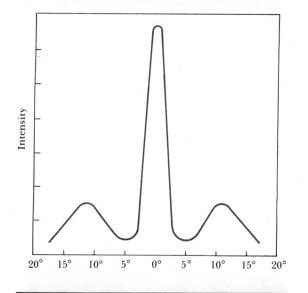

Figure 5-9
The diffraction pattern obtained with helium reflected from a lithium fluoride crystal. (*From H. Semat,* Introduction to Atomic and Nuclear Physics, *4th ed. Copyright 1939, 1946, 1954, 1962, 1967 by H. Semat. Reprinted by permission of Holt, Rinehart and Winston, Inc.*)

Figure 5-10
Neutron Laue pattern of NaCl. Compare this with the x-ray Laue pattern in Figure 3-10. (*Courtesy of E. O. Wollan and C. G. Shull.*)

5-3   Properties of Classical Waves

Despite the immense diversity of wave phenomena such as vibrating strings, sound waves, light waves, water waves, radio waves, etc., there are many features common to all wave phenomena, including electron waves and other "matter" waves. A thorough understanding of classical wave properties is therefore a great help in the understanding of electron waves. In fact, much of the difficulty experienced by students in learning quantum mechanics can be traced to a lack of familarity with classical wave phenomena. You are therefore urged to review what you learned about waves in elementary physics. In this section we will give a brief review or outline of some of the properties of waves that are important for the understanding of electron waves and quantum mechanics.[1]

Let us begin with the familiar example of waves on a very long string, which we assume to be along the x axis. Waves are usually produced on such a string by vibrating one end in some way so that the string has a displacement y, which depends on both position x and time t. The function $y(x,t)$ is called the *wave function*. The acceleration of a point on the string is $\partial^2 y/\partial t^2$, where we must use the partial-derivative notation because y depends on both x and t. A direct application of Newton's law, $\Sigma \mathbf{F} = m\mathbf{a}$, to a segment of the string shows that the wave function $y(x,t)$ obeys the *wave equation*[2]

Wave function

$$\frac{\partial^2 y}{\partial x^2} = \frac{1}{v^2}\frac{\partial^2 y}{\partial t^2} \qquad\qquad 5\text{-}7$$

Classical wave equation

For a perfectly flexible string the *phase velocity* v is given by $v = \sqrt{T/\mu}$, where T is the tension in the string and μ its mass per unit length. A similar application of Newton's laws, along with the gas laws, leads to a wave equation identical to Equation 5-7 for sound waves in a gas; in this case the wave function $y(x,t)$ can represent either the displacement of gas particles from their equilibrium position, the pressure, or the gas density. An equivalent wave equation for light and other electromagnetic waves can be obtained from Maxwell's equations of electrodynamics. Here the wave function $y(x,t)$ represents the electric or magnetic field and the phase velocity v is given by $v = c/n$, where c is the speed of light in vacuum and n the index of refraction of the medium. Any function $y(x,t)$ which depends on x and t only in the combination $x - vt$ or $x + vt$ is a solution of the wave equation (see Problem 4). For example, if $y = f(x)$ is the shape of the string at time $t = 0$, the function $f(x - vt)$ describes the situation in which this shape is propagated to the right with speed v, and $f(x + vt)$ describes propagation to the left. An important property of the wave equation (and of any linear equation) is that the sum of two solutions is also a solution. This is known as the *superposition principle*. The superposition of wave functions has important applications in the understanding of standing waves, inter-

Superposition principle

---

[1] A more complete elementary discussion of waves can be found in Reference 1.

[2] A derivation of this equation is given in Reference 1, pp. 529–531.

ference, diffraction, Bragg scattering, harmonic analysis and synthesis, and the propagation of wave pulses.

A particularly useful solution to the wave equation is the *harmonic wave:*

$$y(x,t) = y_0 \cos 2\pi \left(\frac{x}{\lambda} - \frac{t}{T}\right) = y_0 \cos \frac{2\pi}{\lambda}(x - vt) \qquad 5\text{-}8$$

Harmonic wave function

This describes a wave traveling in the positive x direction with *amplitude* y_0, *wavelength* λ, *period* T, *frequency* $f = 1/T$, and *phase velocity*

$$v = f\lambda \qquad 5\text{-}9$$

It is convenient to use the angular frequency ω and *wave number*[1] k, defined by

$$\omega = \frac{2\pi}{T} = 2\pi f \qquad 5\text{-}10$$

Angular frequency and wave number defined

$$k = \frac{2\pi}{\lambda} \qquad 5\text{-}11$$

The harmonic wave function is then written

$$y = y_0 \cos(kx - \omega t) \qquad 5\text{-}12a$$

Other forms of this type of solution are

$$y(x,t) = y_0 \sin(kx - \omega t) \qquad 5\text{-}12b$$

and

$$y(x,t) = y_0 e^{i(kx - \omega t)} \qquad 5\text{-}12c$$

Equation 5-12b differs from 5-12a only in the choice of origin of x or t. Equation 5-12c is a linear combination of Equations 5-12a and 5-12b since

$$e^{i\theta} = \cos\theta + i\sin\theta$$

Because $y(x,t)$ must be real, it is usually understood that either the real or the imaginary part of the right side of Equation 5-12c, is to be taken.[2]

Harmonic waves have the somewhat artificial property of extending throughout all space from $x = -\infty$ to $x = +\infty$ and throughout all time. They are quite useful for two reasons: (1) they approximate the waves of finite extent that occur in many real situations, for example, on a long string by moving one end with simple harmonic motion; and (2) any other type of wave, such as a wave pulse (which we shall consider in some detail later), can be represented by a superposition of harmonic waves of different frequencies and wavelengths.

[1] In spectroscopy, the quantity $\lambda^{-1} = k/2\pi$ is called the *wave number*. In the theory of waves, the term *wave number* is used for $k = 2\pi/\lambda$.

[2] Every complex number z can be written in the form $z = a + bi$, where the "real part" a, and the "imaginary part" b, are both real numbers and $i = \sqrt{-1}$. The real part of $e^{i\theta}$ is therefore $\cos\theta$ and the imaginary part is $\sin\theta$. Since both $\cos\theta$ and $\sin\theta$ satisfy the wave equation if $\theta = kx \pm \omega t$, the linear combination $e^{i\theta}$ also satisfies the wave equation.

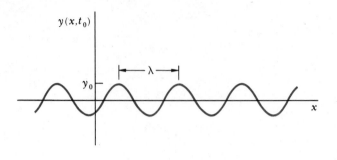

Figure 5-11
A harmonic wave function at some fixed time t_0.

If we look at a harmonic wave on a string at a certain instant $t = t_0$, the wave function $y(x,t_0)$ describes the shape of the string as it would look in a "snapshot" (Figure 5-11). The shape is a cosine wave (or sine wave, depending on the position of the origin). If we look at a particular point x_0, the wave function, $y(x_0,t)$, describes the motion of that point on the string. Since y is a sinusoidal function, the motion of any point x_0 is simple harmonic motion, of amplitude y_0 and frequency f.

An important property of all waves is that the energy density (energy per unit volume) in the wave, η, is proportional to the square of the amplitude: $\eta \propto y_0^2$. The *intensity* of a wave I, which is the energy transported per unit area per unit time (measured in watts per square meter), is related to the energy density and the wave speed. For a plane wave the relation is

Intensity

$$I = \eta v \qquad\qquad 5\text{-}13$$

The intensity is therefore also proportional to the square of the amplitude.

If a string fixed at both ends is set into oscillation, there are certain resonance frequencies for which standing waves result. The *standing-wave condition* (Equation 5-4) is that an integral number of half wavelengths fit into the length of the string (Figure 5-12). In terms of the frequency, this condition is

$$f_n = \frac{v}{\lambda_n} = n\,\frac{v}{2L} \qquad\qquad 5\text{-}14$$

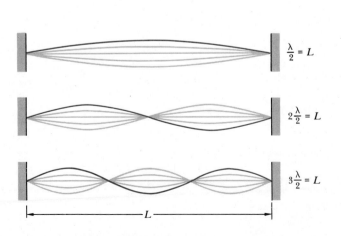

Figure 5-12
Standing waves on a string for $n = 1, 2,$ and 3. The wavelengths obey the standing-wave condition $n\lambda/2 = L$.

The standing waves can be considered as a superposition of two travelling waves of equal frequency and wavelength moving in opposite directions. Standing sound waves can be set up in air columns, as in a flute or organ pipe. Similarly, standing electromagnetic waves can be set up in a cavity, for example, in a microwave oscillator. The discrete or quantized frequencies that occur naturally in the classical theory of standing waves were a strong suggestion to de Broglie, Schrödinger, and others that the quantization of energy observed in many systems could be obtained from a wave theory applied to those systems.

An important example of the superposition of waves is the interference of two waves which differ in phase. Figure 5-13 shows the classic interference pattern due to waves from two sources which are coherent and in phase. At points on the screen for which the path difference from the sources is an integral number of wavelengths, the waves are in phase and the amplitude is twice that from either source separately. The intensity is therefore four times that from either source. At points for which the path difference is an integral number of wavelengths $\pm\frac{1}{2}\lambda$, the waves are 180° out of phase, the two waves cancel, and the resultant amplitude and intensity are zero.

We are often interested in determining when the waves from several sources will give complete destructive interference, i.e., will add to zero. It can be shown[1] that for N waves of equal amplitude with phase difference δ between each successive pair of waves, complete cancellation occurs when $\delta = 360°/N$ (or any integer times $360°/N$). The phase difference between the first and last waves is then $(N - 1)360°/N = 360° - 360°/N$. For a very large number of waves, we have the important result that complete destructive interference occurs when the first and last

Interference

Figure 5-13
Two-source interference pattern. If the sources are coherent and in phase, the waves from the sources interfere constructively at points for which the path difference $d \sin \theta$ is an integral number of wavelengths.

[1] See, for example, p. 589 of Reference 1.

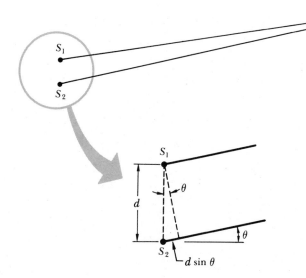

waves differ in phase by $360° = 2\pi$ rad. Let us illustrate this by considering 100 waves with a phase difference $\delta = 360°/100$ between each successive pair. The first wave and the fifty-first wave will then differ in phase by $180°$ and therefore cancel. Similarly, the second and fifty-second waves will also differ in phase by $180°$ and cancel, as will the third and fifty-third waves, etc. The resulting amplitude and the intensity due to the superposition of all of the waves will therefore be zero.

An important wave property that can be understood in terms of the superposition of many waves is *diffraction,* the spreading or bending of waves around corners. The single-slit diffraction pattern is illustrated in Figure 5-14. This pattern can be analyzed using Huygens' principle by replacing the slit opening with a large number of sources, and calculating the resulting interference pattern on the screen. At any angle θ, there will be a phase difference between waves from two sources a distance d apart due to the path difference $d \sin \theta$. If a is the width of the slit, the path difference for waves from the top and bottom of the slit is $a \sin \theta$. If this path difference is one wavelength, the first and last waves will differ in phase by $360°$ and the resulting amplitude due to the superposition of all the waves will be zero. The location of the first minimum in the single-slit diffraction pattern is thus given by

Diffraction

$$a \sin \theta = \lambda \qquad\qquad 5\text{-}15$$

We see from this result that the width of the diffraction pattern on the screen varies inversely with the width of the slit since the larger the width a, the smaller the angle θ to the first minimum.

Many important wave phenomena, such as Bragg scattering (already discussed), involve both diffraction and interference. Interference and diffraction are the primary wave properties that distinguish wave propagation from the propagation of particles.

Our final example of the superposition of classical waves, the motion of a wave packet or wave group, will be discussed in the next section. Although this subject falls naturally in our review of properties of classical waves, it is often not covered in elementary physics courses. Because of its importance for the understanding of electron waves we shall devote a complete section to this topic.

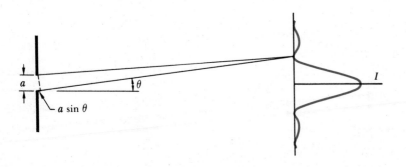

Figure 5-14
Single-slit Fraunhofer diffraction pattern. At the first minimum, the path difference of light from the top and bottom of the slit $a \sin \theta$ is one wavelength.

Questions

1. Can you distinguish between wave propagation and particle propagation in reflection or refraction?

2. At the second minimum in the single-slit diffraction pattern the waves from the top and bottom of the slit differ in phase by $2 \times 360°$. By imagining the slit as replaced by a large number of Huygens sources, explain how the waves from these sources cancel in pairs at the second minimum.

5-4   Wave Packets

A familiar wave phenomenon which cannot be described by a single harmonic wave is that of a pulse, such as a flip of one end of a long string (Figure 5-15), a sudden noise, or the brief opening of a shutter in front of a light source. The main characteristic of a pulse is that of localization in time and space. As we have mentioned, a single harmonic wave is not localized in either time or space. The description of a pulse, therefore, requires a group of waves of different frequencies and wavelengths. Such a group is called a *wave packet*. The range of wavelengths or frequencies of the harmonic waves needed to form a wave packet depends on the extent in space and duration in time of the pulse. In general, if the extent in space Δx is small, the range Δk of wave numbers must be large. Similarly, if the duration in time Δt is small, the range of frequencies $\Delta \omega$ must be large. It can be shown that for a general wave packet, Δx and Δk are related by

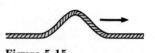

Figure 5-15
Wave pulse moving along a string. A pulse has a beginning and an end; i.e., it is localized, unlike a pure harmonic wave, which goes on forever in space and time.

$$\Delta k \, \Delta x \sim 1 \qquad\qquad 5\text{-}16$$

Similarly,

$$\Delta \omega \, \Delta t \sim 1 \qquad\qquad 5\text{-}17$$

We have written these as order-of-magnitude equations because the exact value of the products $\Delta x \, \Delta k$ and $\Delta t \, \Delta \omega$ depends on how these ranges are defined, as well as on the particular shape of the packets. Equation 5-17 is sometimes known as the response-time–bandwidth relation, expressing the result that a circuit component such as an amplifier must have a large bandwidth ($\Delta \omega$) if it is to be able to respond to signals of short duration.

The mathematics of representing arbitrarily shaped pulses by sums of sine or cosine functions involves Fourier series and Fourier integrals. We shall merely illustrate the phenomenon of wave packets by considering some simple and somewhat artificial examples and discussing the general properties qualitatively. Wave groups are particularly important because a wave description of an electron must include the important property of localization.

Consider a very simple group consisting of only two waves of equal amplitude and nearly equal frequencies and wavelengths. Such a group occurs in the phenomenon of beats and is described in most elementary textbooks. Let the wave numbers be k_1 and k_2 and the angular frequencies ω_1 and ω_2. The sum of the two waves is

$$
\begin{aligned}
y(x,t) &= y_0 \cos\left(k_1 x - \omega_1 t\right) + y_0 \cos\left(k_2 x - \omega_2 t\right) \\
&= 2y_0 \cos\left[\tfrac{1}{2}(k_1 - k_2)x - \tfrac{1}{2}(\omega_1 - \omega_2)t\right] \\
&\quad \times \cos\left[\tfrac{1}{2}(k_1 + k_2)x - \tfrac{1}{2}(\omega_1 + \omega_2)t\right] \\
&= \left[2y_0 \cos\left(\tfrac{1}{2}\,\Delta k x - \tfrac{1}{2}\,\Delta \omega t\right)\right] \cos\left(\overline{k}x - \overline{\omega}t\right) \qquad \text{5-18}
\end{aligned}
$$

where we have written Δk and $\Delta \omega$ for the differences in wave numbers and frequencies, and $\overline{k}$ and $\overline{\omega}$ for the mean values. Figure 5-16 shows a sketch of $y(x,t_0)$ versus x at some time t_0. The dashed curve is the envelope of the group of two waves, given by the term in brackets above. The individual waves move with the speed $\overline{\omega}/\overline{k}$, the phase velocity v. If we write the modulating term in brackets as $\cos\{\tfrac{1}{2}\Delta k[x - (\Delta\omega/\Delta k)\,t]\}$ we see that the envelope moves with speed $\Delta\omega/\Delta k$. The speed of the envelope is called the *group velocity* v_g. If x_2 and x_1 are two consecutive values of x for which the envelope is zero, we may take $\Delta x = x_2 - x_1$ as the spatial extent of the group. Then

$$
\tfrac{1}{2}\,\Delta k x_2 - \tfrac{1}{2}\,\Delta k x_1 = \pi
$$

or

$$
\Delta x\,\Delta k = 2\pi
$$

For a particular value of x, the function $y(x,t)$ versus t looks like Figure 5-16, with t replacing x. The extent in time Δt is thus related to $\Delta\omega$ by

$$
\Delta t\,\Delta\omega = 2\pi
$$

These results are in accord with Equations 5-16 and 5-17 although the ranges Δx and Δt that we have used for this particular group of two waves are somewhat artificial because the envelope does not remain small outside these ranges.

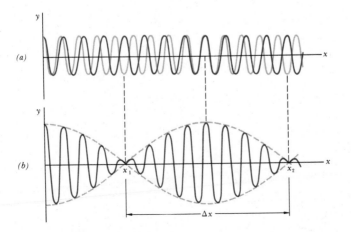

Figure 5-16
Two waves of slightly different wavelength and frequency produce beats. (a) Shows $y(x)$ at a given instant for each of the two waves. The waves are in phase at first, but because of the difference in wavelength, they become out of phase and then in phase again. The sum of these waves is shown in (b). The spatial extent of the group Δx is inversely proportional to the difference in wave numbers Δk, where k is related to the wavelength by $k = 2\pi/\lambda$. Identical figures are obtained if y is plotted versus time t at a fixed point x. In that case the extent in time Δt is inversely proportional to the frequency difference $\Delta\omega$.

Example 5-3   Consider a large number of harmonic waves of equal amplitude with wave numbers which range from k_1 to k_N in equal steps. At some time t_0 the waves are all in phase at point x_1. Where is the nearest point x_2 for which the waves completely cancel at the same time? According to our discussion in the previous section, the waves will cancel when the phase difference between the first and Nth wave is 2π, if N is large. If we move from point x_1 to x_2, the phase of the first wave changes by $k_1(x_2 - x_1)$. Similarly, the phase of the Nth wave changes by $k_N(x_2 - x_1)$. Since these waves are in phase at x_1, the phase difference at x_2 will be the difference in these phase changes, which is $(k_N - k_1)(x_2 - x_1)$. When this phase difference is 2π the waves will completely cancel. We see again from this example that the extent in space $x_2 - x_1$ of the wave group varies inversely with the range of wave numbers $k_N - k_1$.

We can construct a more general wave packet if we allow the amplitudes of the various harmonic waves to be different. Such a packet can be represented by an equation of the form

$$y(x, t) = \sum_i A_i \cos (k_i x - \omega_i t) \qquad 5\text{-}19$$

where A_i is the amplitude of the wave of wave number k_i and angular frequency ω_i. The calculation of the amplitudes A_i needed to construct a wave packet of some given shape $y(x, t_0)$ at some fixed time is a problem in Fourier series.

If we are restricted to a finite number of waves, it is not possible to obtain a wave packet that is small everywhere outside a well-defined range. The larger the number of waves, the larger the region in which destructive interference makes the envelope small, but eventually all the waves will again be in phase, the envelope will be large, and the pattern will repeat. To represent a pulse which is zero everywhere outside some range, such as that shown in Figure 5-15, we must construct a wave packet from a continuous distribution of waves. We can do this by replacing A_i in Equation 5-19 by $A(k)\, dk$ and changing the sum to an integral. The quantity $A(k)$ is called the distribution function for the wave number k. Either the shape of the wave packet at some fixed time $y(x)$ or the distribution of wave numbers $A(k)$ can be found from the other by methods of Fourier analysis.[1]

Figure 5-17 shows a Gaussian-shaped wave packet and the corresponding wave-number distribution function for a narrow packet and a wide packet. For this special case $A(k)$ is also a Gaussian function. The standard deviations of these Gaussian functions are related by

$$\sigma_x \sigma_k = \tfrac{1}{2} \qquad 5\text{-}20$$

It can be shown that the product of the standard deviations is greater than $\tfrac{1}{2}$ for a wave packet of any other shape.

[1] If you are familiar with Fourier analysis, you will recognize that $y(x)$ and $A(k)$ are essentially Fourier transforms of each other.

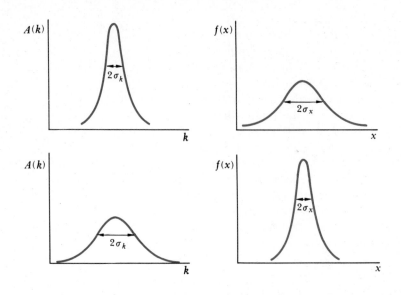

Figure 5-17
A Gaussian distribution of
wave numbers $A(k)$ leads
to a Gaussian wave packet.
The standard deviations
of these packets are re-
lated by $\sigma_k \sigma_x = \frac{1}{2}$.

For our simple group of only two waves, we found that the
envelope moved with the velocity $v_g = \Delta\omega/\Delta k$. For a general
wave packet the group velocity is given by

$$v_g = \frac{d\omega}{dk}$$ 5-21   *Group velocity*

where the derivative is evaluated at the central wave number.
The group velocity of a pulse can be related to the phase veloc-
ities of the individual harmonic waves making up the packet.
The phase velocity of a harmonic wave is

$$v = f\lambda = \left(\frac{\omega}{2\pi}\right)\left(\frac{2\pi}{k}\right) = \frac{\omega}{k}$$

so that

$$\omega = kv$$

Differentiating, we obtain

$$v_g = v + k\frac{dv}{dk}$$ 5-22

If the phase velocity is the same for all frequencies and wavelengths, $dv/dk = 0$ and the group velocity is the same as the
phase velocity. A medium for which the phase velocity is the
same for all frequencies is said to be *nondispersive*. Examples
are waves on a perfectly flexible string, sound waves in air, and
electromagnetic waves in vacuum. An important characteristic
of a nondispersive medium is that, since all the harmonic waves
making up a wave packet move with the same speed, the packet
maintains its shape as its moves. Conversely, if the phase velocity   *Dispersion*
is different for different frequencies, the shape of the pulse will
change as it travels. If the phase velocity does depend on frequency, the group velocity and phase velocity are not the same.
Such a medium is called a *dispersive* medium; examples are water
waves, waves on a wire that is not perfectly flexible, light waves

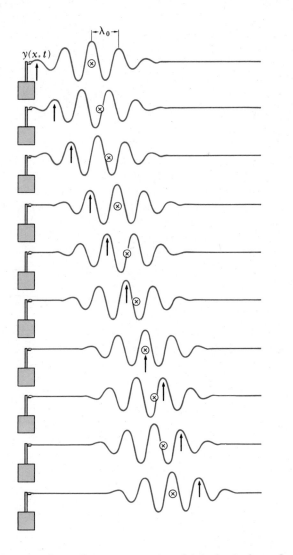

$y(x,t)$

λ_0

Figure 5-18
Wave packet for which
the group velocity is half
the phase velocity. Water
waves whose wavelengths
are a few centimeters, but
much less than the water
depth, have this property.
The arrow travels at the
phase velocity, following a
point of constant phase
for the dominant wave-
length. The cross at the
center of the group trav-
els at the group velocity.
*(Adapted from F. S. Craw-
ford, Jr., Berkeley Physics
Course, New York:
McGraw-Hill Book Com-
pany, 1965, vol. 3, p. 294.
Courtesy of Education Development Center, Inc.,
Newton, Mass.)*

in a medium such as glass or water in which the index of refraction has a slight dependence on frequency, and electron waves.
Figure 5-18 shows a wave packet for which the group velocity is
half the phase velocity.

Question

3. Which is more important for communication, the phase
velocity or the group velocity?

5-5   Electron Wave Packets

The quantity analogous to the displacement $y(x,t)$ for waves on a
string, to the pressure $P(x,t)$ for a sound wave, or to the electric
field $\mathscr{E}(x,t)$ for electromagnetic waves, is called the *wave function
for electron waves* and is usually designated $\Psi(x,t)$. Consider an
electron wave consisting of a single frequency and wavelength;
we could represent such a wave by $\Psi(x,t) = A \cos(kx - \omega t)$,
$\Psi(x,t) = A \sin(kx - \omega t)$, or $\Psi(x,t) = Ae^{i(kx - \omega t)}$.

The phase velocity is given by $v = f\lambda = (E/h)(h/p) = E/p$, where we have used the de Broglie relations for the energy and momentum. If we use the nonrelativistic expression $E = p^2/2m$, we see that the phase velocity is $v = E/p = p/2m$, which is half the velocity of an electron with momentum p. The phase velocity does not equal the velocity of the electron. Moreover, a wave of a single frequency and wavelength is spread out in space and is not localized. For an electron to have the property of being localized, $\Psi(x,t)$ must be a wave packet containing more than one wave number k and frequency ω. The position of the electron corresponds to the position of the maximum of the wave packet. Thus the wave packet should move with the same velocity as the electron, and the *group* velocity rather than the phase velocity should be equal to the velocity of the electron. We obtain the group velocity from Equation 5-21 by writing the angular frequency ω in terms of the wave number k. The de Broglie relations (Equations 5-1 and 5-2) can be written

$$E = hf = h\,\frac{\omega}{2\pi} = \hbar\omega \qquad\qquad\qquad 5\text{-}23$$

and   *De Broglie relations*

$$p = \frac{h}{\lambda} = \frac{hk}{2\pi} = \hbar k \qquad\qquad\qquad 5\text{-}24$$

The nonrelativistic expression for kinetic energy, $E = p^2/2m$, then becomes

$$\hbar\omega = \frac{\hbar^2 k^2}{2m}$$

Then

$$v_g = \frac{d\omega}{dk} = \frac{\hbar k}{m} = \frac{p}{m}$$

The wave packet moves with the particle velocity p/m. This was one of de Broglie's reasons for choosing Equations 5-1 and 5-2. (De Broglie used the relativistic expression relating energy and momentum, which also leads to the equality of the group velocity and particle velocity. See Problem 2.)

5-6   The Probabilistic Interpretation of the Wave Function

Let us consider in more detail the relation between the wave function $\Psi(x,t)$ and the location of the electron. We can get a hint as to this relation from the case of light. The wave equation that governs light is Equation 5-7, with $y = \mathscr{E}$, the electric field, as the wave function. The energy per unit volume in a light wave is proportional to $\mathscr{E}^2$, but the energy in a light wave is quantized in units of hf for each photon. We expect, therefore, that the number of photons in a unit volume is proportional to $\mathscr{E}^2$.

Consider the famous double-slit interference experiment (Figure 5-13). The pattern observed on the screen is determined by the interference of the waves from the slits. At a point on the screen where the wave from one slit is 180° out of phase with that from the other, the resultant electric field is zero; there is no light energy at this point; and the point is dark. If we reduce the intensity to a very low value, we can still observe the interference pattern if we replace the screen by a film and wait a sufficient length of time to expose the film.

The interaction of light with film is a quantum phenomenon. If we expose the film for only a very short time with a low-intensity source, we do not see merely a weaker version of the high-intensity pattern; we see, instead, "dots" on the film caused by the interactions of individual photons (Figure 5-19). At points where the waves from the slits interfere destructively there are no dots, and at points where the waves interfere constructively there are many dots. However, if the exposure is short and the source weak, random fluctuations from the average predictions of the wave theory are clearly evident. If the exposure is long enough so that many photons interact with the film, the fluctuations average out and the quantum nature of light is not noticed. The interference pattern depends only on the total number of photons interacting with the film and not on the rate. Even if the intensity is so low that only one photon at a time hits the film, the wave theory predicts the correct average pattern. For low intensities, we therefore interpret $\mathscr{E}^2$ to be proportional to the *probability* of detecting a photon in a unit volume. At points on the film or screen where $\mathscr{E}^2$ is zero, photons are never observed, whereas they are most likely to be observed at points where $\mathscr{E}^2$ is large.

Figure 5-19
Growth of two-slit interference pattern. The photo (*d*) is an actual two-slit electron interference pattern in which the film was exposed to millions of electrons. The pattern is identical to that usually obtained with photons. If the film were to be observed at various stages, such as after being struck by 28 electrons, then after about 1000 electrons, and again after about 10,000 electrons, the patterns of individually exposed grains would be similar to those shown in (*a*), (*b*), and (*c*), except that the exposed dots would be smaller than the dots drawn here. Note that there are no dots in the region of the interference minima. The probability of any point of the film being exposed is determined by wave theory, whether the film is exposed by electrons or photons. [*Parts (a), (b), and (c) from E. R. Huggins, Physics 1, copyright © by W. A. Benjamin, Inc., Menlo Park, California. Photo (d) courtesy of C. Jonsson.*]

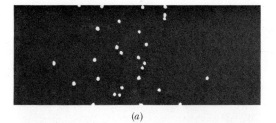

(a)

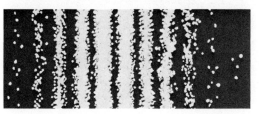

(c)

(b)

(d)

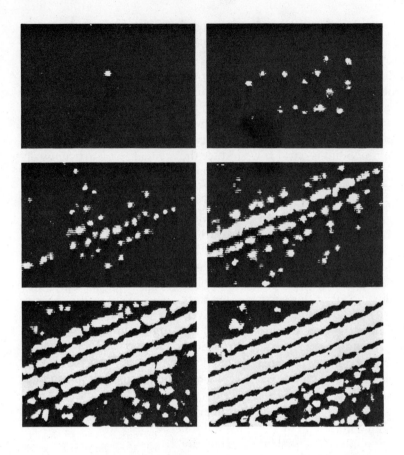

Actual electron interference pattern filmed from a TV monitor at increasing densities. [*From G. F. Missiroli and G. Pozzi,* American Journal of Physics, **44,** *no. 3, 306 (1976).*]

It is not necessary to use light waves to produce an interference pattern. Such patterns can be produced with electrons as well. In the wave theory of electrons, the motion of a *single* electron is described by a wave function Ψ. The quantity Ψ^2 is proportional to the probability of detecting an electron in a unit volume. In one dimension, $\Psi^2\, dx$ is the probability of an electron being in the interval dx. If we call this probability $P(x)\, dx$, where $P(x)$ is the probability distribution function, we have

$$P(x)\, dx = \Psi^2\, dx \qquad\qquad 5\text{-}25$$

Probability of electron being in dx

5-7   The Uncertainty Principle

Consider a wave packet $\Psi(x,t)$ representing an electron. The most probable position of the electron is the value of x for which $\Psi^2(x,t)$ is a maximum. Since $\Psi^2(x,t)$ is proportional to the probability that the electron is at x, and $\Psi^2(x,t)$ is nonzero for a range of values of x, there is an *uncertainty* in the value of the position of the electron. If we make a number of position measurements on identical electrons—electrons with the same wave function—we shall not always obtain the same result. In fact, the distribution function for the results of such measurements will be given by $\Psi^2(x,t)$. If the wave packet is very narrow, the uncertainty in position will be small. However, a narrow wave packet must contain a wide range of wave numbers k. Since the momentum is related to the wave number by $p = \hbar k$, a wide range of k values means a wide range of momentum values. We have seen that for all wave packets the ranges Δx and Δk are related by

$$\Delta x \, \Delta k \sim 1 \qquad\qquad 5\text{-}26$$

(Since the exact value for the product depends on the kind of packet and the exact specification of the meaning of Δx and Δk, we have taken the value 1 as a typical product.) Similarly, a packet that is localized in time Δt must contain a range of frequencies $\Delta\omega$, where these ranges are related by

$$\Delta\omega \, \Delta t \sim 1 \qquad\qquad 5\text{-}27$$

Equations 5-26 and 5-27 are inherent properties of waves. If we multiply these equations by $\hbar$ and use $p = \hbar k$ and $E = \hbar\omega$, we obtain

$$\Delta x \, \Delta p \sim \hbar \qquad\qquad 5\text{-}28$$

and

$$\Delta E \, \Delta t \sim \hbar \qquad\qquad 5\text{-}29$$

Equations 5-28 and 5-29 provide a statement of the *uncertainty principle* first enunciated in 1927 by Werner Heisenberg. Equation 5-28 expresses the fact that the distribution functions for position and momentum cannot both be made arbitrarily narrow; thus measurements of position and momentum will have uncertainties which are related by Equation 5-28. Of course, because of inaccurate measurements, the product of Δx and Δp can be, and usually is, much larger than $\hbar$. The lower limit is not due to any technical problem in the design of measuring equipment that might be solved at some later date; it is instead due to the wave and particle nature of both matter and light.

If we define precisely what we mean by the uncertainty in the measurements of position and momentum, we can give a precise statement of the uncertainty principle. We saw in Section 5-4 that, if σ_x is the standard deviation for measurements of position and σ_k is the standard deviation for measurements of the wave number k, the product $\sigma_x\sigma_k$ has its minimum value of $\frac{1}{2}$ when the distribution functions are Gaussian. If we define Δx and Δp to be the standard deviations, the minimum value of their product is $\frac{1}{2}\hbar$. Thus

$$\Delta x \, \Delta p \geqslant \tfrac{1}{2}\hbar \qquad\qquad 5\text{-}30$$

Uncertainty principle

Similarly

$$\Delta E \, \Delta t \geqslant \tfrac{1}{2}\hbar \qquad\qquad 5\text{-}31$$

Example 5-4   Let us see how a classical physicist might attempt to violate the uncertainty principle. A common way to measure the position of an object is to look at it with light. The momentum can be obtained by looking at it again, a short time later, and computing what velocity it must have had. Since our classical physicist knows that, because of diffraction effects, he cannot hope to make distance measurements smaller than the wavelength of the light he uses, he will use the shortest-wavelength x ray or ultraviolet light that he can obtain. (There is, in principle, no limit to how short a wavelength of electromagnetic radiation can be found.) He also knows the light carries momentum and energy, so that when it scatters off the electron, the motion of the electron will be disturbed, spoiling the momentum measurement. He therefore substantially reduces the intensity of the light. Though this seems like a good idea, it is not effective. Reducing the intensity merely decreases the number of photons; but at least one photon must be scattered to observe the electron. The momentum of the photon is $hf/c = h/\lambda$. The smaller he makes λ to measure the position, the more the photon will disturb the electron and spoil the momentum measurement.

The only way to increase the accuracy of the momentum measurement is to use longer wavelengths, which decreases the accuracy of the position measurement. Figure 5-20 illustrates the problem. (This illustration was first given by Niels Bohr.) The position of the electron is to be determined by viewing it through a microscope. We shall assume that only one photon is needed. We can take for the uncertainty in position, the minimum separation distance for which two objects can be resolved; this is[1]

$$\Delta x = \frac{\lambda}{2 \sin \theta}$$

where θ is the half angle subtended by the aperture, as shown in

[1] The resolving power of a microscope is discussed in some detail in F. A. Jenkins and H. E. White, *Fundamentals of Optics,* 4th ed., New York: McGraw-Hill Book Company, 1976, pp. 332–334. The expression for Δx used here is determined by Rayleigh's criterion that two points are just resolved if the central maximum of the diffraction pattern from one falls at the first minimum of the diffraction pattern of the other.

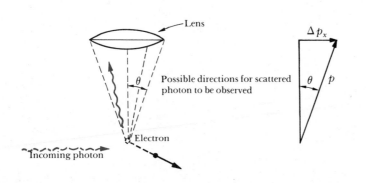

Figure 5-20
"Seeing an electron" with a microscope. Because of the size of the lens, the momentum of the scattered photon is uncertain by $\Delta p_x \approx p \sin \theta = h \sin \theta/\lambda$. Thus the recoil momentum of the electron is also uncertain by at least this amount. The position of the electron cannot be resolved better than $\Delta x \approx \lambda/\sin \theta$ because of diffraction. The product of the uncertainties $\Delta p_x \, \Delta x$ is therefore of the order of Planck's constant h.

Lens

Δp_x

θ

Possible directions for scattered photon to be observed

θ p

Electron

Incoming photon

the figure. Let us assume that the x component of momentum of the incoming photon is known precisely. The scattered photon can have any x component of momentum from 0 to $p_x = p \sin \theta$, where p is the total momentum of the scattered photon. By conservation of momentum, the uncertainty in the momentum of the electron after the scattering must be at least equal to that of the scattered photon (it would be equal, of course, if the electron's initial momentum were known precisely); thus we write

$$\Delta p_x \geqslant p \sin \theta = \frac{h}{\lambda} \sin \theta$$

and

$$\Delta x \, \Delta p_x \geqslant \frac{\lambda}{2 \sin \theta} \frac{h \sin \theta}{\lambda} = \tfrac{1}{2}h$$

This example illustrates the essential point of the uncertainty principle—that this product of uncertainties cannot be less than $\hbar$ in *principle,* that is, even in an ideal situation. If electrons rather than photons were used to locate the object, the analysis would not change, since the relation $\lambda = h/p$ is the same for both.

The example given above provides some insight into the difficulties in measurement imposed by the wave-particle duality of light and matter, but it can be misleading. In this example we assumed that the electron had a definite position and momentum, and in the process of measuring the position the momentum uncertainty was introduced. But, because of the wave nature of the electron, it must be represented by a wave packet; this implies that there is already an uncertainty in position and momentum corresponding to the spread in x of the packet and the spread in k of the wave numbers. In the very process of measuring the position of the electron, such as by irradiating it with photons, the wave packet of the electron is changed. If photons of very small wavelength are used, the position can be determined accurately. This means that the new wave packet describing the electron is now very narrow in x, but of course this also means that the distribution in k is correspondingly wide. Thus the uncertainty in momentum is now large.

Questions

4. What are Δx and Δk for a purely harmonic wave?

5. Does the uncertainty principle say that the momentum of an electron can never be precisely known?

5-8   Particle-Wave Duality

We have seen that electrons, which we usually think of as particles, exhibit the wave properties of diffraction and interference. In previous chapters we saw that light, which we ordinarily think of as a wave motion, also has particle properties in its interaction with matter, as in the photoelectric effect or the Compton effect. All phenomena—electrons, atoms, light, sound, etc.—have both particle and wave characteristics. It is sometimes said that an electron, for example, is both a wave and a particle. This is rather confusing since, in classical physics, the concepts of waves and particles are mutually exclusive. A *classical particle* behaves like a BB shot. It can be localized and scattered, it exchanges energy suddenly in a lump, and it obeys the laws of conservation of energy and momentum in collisions; but it does *not* exhibit interference and diffraction. A *classical wave* behaves like a water wave. It exhibits diffraction and interference patterns and has its energy spread out continuously in space and time. Nothing can be both a classical particle and a classical wave.

Until the twentieth century, it was thought that light was a classical wave and an electron was a classical particle. We now see that the concepts of classical waves and classical particles do not adequately describe either phenomenon. Each behaves like a classical wave when propagation is considered and like a classical particle when its energy exchange is considered. Let us elaborate on this statement.

Every phenomenon is describable by a wave function that is the solution of a wave equation. The wave function for light is the electric field $\mathcal{E}(x,t)$ (in one dimension), which is the solution of a wave equation like Equation 5-7. We have called the wave function for an electron $\Psi(x,t)$. We shall study the wave equation of which Ψ is the solution, called the *Schrödinger equation,* in the next chapter. The square of the wave function gives the probability (per unit volume) that the electron is in a given region. The wave function exhibits the classical wave properties of interference and diffraction. In order to determine where an electron is likely to be, we must find the wave function by methods similar to those of classical wave theory. When the electron (or light) interacts and exchanges energy and momentum, the wave function is changed by the interaction. The interaction can be described by classical particle theory, as is done in the Compton effect.

There are times when classical particle theory and classical wave theory give the same results. If the wavelength is much smaller than any object or aperture, particle theory can be used as well as wave theory to describe wave propagation, because diffraction and interference effects are too small to be observed. Common examples are geometrical optics, which is really a particle theory, and the motion of baseballs. If one is interested only in time averages of energy and momentum exchange, the classical wave theory works as well as the classical particle theory. For example, the wave theory of light correctly predicts that the total electron current in the photoelectric effect is proportional to the intensity of the light.

Example 5-5   Let us again consider the double-slit interference example. The slits are separated by a distance d and there is a screen or film far enough away from the slits that we can use small-angle approximations. We shall consider the case of a light beam, though the analysis is identical for an electron beam. The interference pattern has a maximum at $\theta = 0$, and the first minimum at θ given by $d \sin \theta = \lambda/2$. In the usual case θ is small, so we can approximate $\sin \theta$ by θ and write $\theta \approx \lambda/2d$ for the position of the first minimum.

As we have discussed before, the interference pattern does not depend on the intensity of the light, even if only one photon at a time hits the film. If we consider the light to be a beam of photons, we have the following difficulty. If a photon goes through slit A, how does it "know" if slit B is open or closed? If slit B is open, the photon never goes to the interference minimum (point P in Figure 5-21), but if slit B is closed, there is no interference minimum, and some photons go to point P.

Let us examine the statement "the photon goes through slit A." In this problem, in terms of classical wave theory no difficulty exists because a wave goes through both slits if they are open, so that a wave certainly knows if slit B is open or closed. Light cannot act like a classical particle and a classical wave at the same time. Let us see what happens if we try to observe the particle aspect of light as it goes through the slits. We shall try to *measure* which slit the photon goes through. In Figure 5-21 we have placed electrons as detectors in the region just beyond the slits. We can tell which slit the photon passed through by observing the recoil electron. To do this we must know the vertical position y of the electron within a distance $d/2$. If $\Delta y < d/2$, the

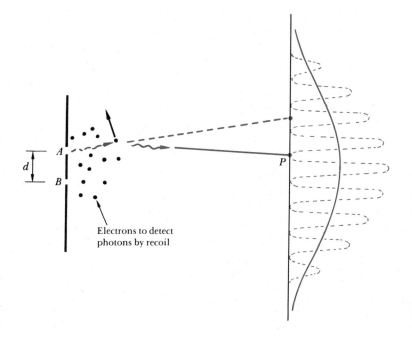

Electrons to detect
photons by recoil

Figure 5-21
The dashed curve is a typical two-slit interference pattern observed when there are no detectors present (electrons indicated by dots). No photons hit the screen at point P, an interference minimum. If detectors are placed near the slits to detect which slit a photon passes through, the interference pattern is destroyed by the scattering of the photons. The resulting pattern is the sum of two independent single-slit diffraction patterns, as shown by the solid curve.

uncertainty in the vertical momentum of the recoil electron must be greater than $2\hbar/d$; and by conservation of momentum, the vertical momentum of the scattered photon must be uncertain by $\Delta p_y \gtrsim 2\hbar/d$. If the photon was originally heading toward the interference maximum at $\theta = 0$ with momentum $p = h/\lambda$, it will be deflected through an angle that is uncertain by

$$\Delta\theta \approx \frac{\Delta p_y}{p} \gtrsim \frac{2\hbar d}{h/\lambda} = \frac{\lambda}{\pi d}$$

Comparing this with the angle of the interference minimum, we see that in the process of measuring which slit the photon goes through, the photon is scattered at least enough so that the interference pattern is washed out.

From this example we see that it is impossible to measure both the particle and wave aspects of light at the same time. When we place detectors near the slits to measure the particle properties of light, the wave properties of interference cannot be observed. With no detectors near the slit, the experiment is designed to measure the wave properties of light. We cannot, then, say that a photon passed through one slit or the other. This result is known as *Bohr's principle of complementarity*—the particle aspects and wave aspects complement each other. Both are needed, but both cannot be observed at the same time. Whether the wave aspect or the particle aspect is observed depends on the experimental arrangement.

Complementarity principle

5-9   Some Consequences of the Uncertainty Principle

In the next chapter we shall see that the Schrödinger wave equation provides a straightforward method of solving problems in atomic physics. However, the solution of the Schrödinger equation is often laborious and difficult. Much semiquantitative information about the behavior of atomic systems can be obtained from the uncertainty principle alone without a detailed solution of the problem. Some examples follow.

Example 5-6   Minimum Energy of a Particle in a Box   An important consequence of the uncertainty principle is that a particle confined to a finite space cannot have zero kinetic energy. Let us consider the case of a one-dimensional "box" of length L. If we know that the particle is in the box, Δx is not larger than L. This implies that Δp is at least $\hbar/L$. (Since we are interested in orders of magnitude, we shall ignore the $\frac{1}{2}$ in the minimum uncertainty product. In general, distributions are not Gaussian anyway, so $\Delta p\,\Delta x$ will be larger than $\frac{1}{2}\hbar$.)

Let us take the standard deviation as a measure of Δp,

$$(\Delta p)^2 = (p - \bar{p})_{av}^2 = (p^2 - 2p\bar{p} + \bar{p}^2)_{av} = \overline{p^2} - \bar{p}^2$$

If the box is symmetric, $\bar{p}$ will be zero since the particle moves to the left as often as to the right. Then

$$(\Delta p)^2 = \overline{p^2} \geqslant \left(\frac{\hbar}{L}\right)^2$$

and the average kinetic energy is

$$\overline{E} = \frac{\overline{p^2}}{2m} \geqslant \frac{\hbar^2}{2mL^2}$$

(Compare this with the lowest energy state for a particle in a one-dimensional box, obtained in Section 4-6 from the Wilson-Sommerfeld quantum condition.)

Let us calculate some numerical examples. Consider a small but macroscopic particle. Let $m = 10^{-6}$ g and $L = 10^{-6}$ m. Then the minimum kinetic energy is about

$$\overline{E} \approx \frac{\hbar^2}{2mL^2} = \frac{(1.055 \times 10^{-34} \text{ J-sec})^2}{2(10^{-9} \text{ kg})(10^{-6} \text{ m})^2} = 5.57 \times 10^{-48} \text{ J}$$

$$= 3.47 \times 10^{-29} \text{ eV}$$

The speed corresponding to this kinetic energy is

$$v = \sqrt{\frac{2E}{m}} = \sqrt{\frac{2(5.57 \times 10^{-48} \text{ J})}{10^{-9} \text{ kg}}} = 1.06 \times 10^{-19} \text{ m/sec}$$

We can see from this calculation that the minimum energy is certainly not observable for macroscopic systems even as small as 10^{-6} g. If we take m to be the mass of an electron and L to be 1 Å (the size of an atom), we find

$$E \approx \frac{(\hbar c)^2}{2mc^2 L^2} = \frac{(1973 \text{ eV-Å})^2}{2(0.511 \times 10^6 \text{ eV})(1\text{Å})^2} = 3.81 \text{ eV}$$

This is the correct order of magnitude for the kinetic energy of an electron in an atom.

Example 5-7   Minimum Energy of a Simple Harmonic Oscillator Consider a particle of mass m on a spring of force constant K. The potential energy is

$$V(x) = \tfrac{1}{2}Kx^2 = \tfrac{1}{2}m\omega^2 x^2$$

where $\omega = \sqrt{K/m}$ is the angular frequency. The total energy is constant and therefore equal to its average value:

$$E = \overline{E} = \frac{\overline{p^2}}{2m} + \tfrac{1}{2}m\omega^2 \overline{x^2}$$

As in the previous example, $\bar{p}$ and $\bar{x}$ are zero by symmetry. Then

$$(\Delta x)^2 = \overline{x^2}$$

and

$$(\Delta p)^2 = \overline{p^2}$$

Using the uncertainty relation $(\Delta p)^2 \geqslant \hbar^2/(\Delta x)^2$, the total energy can be written

$$E \geqslant \frac{\hbar^2}{2m(\Delta x)^2} + \tfrac{1}{2}m\omega^2(\Delta x)^2$$

When Δx is small, the first term is large; when Δx is large, the second term is large. The minimum value of E can easily be found (see Problem 11). The result is

$$E_{\min} = \hbar\omega$$

The quantum-mechanical solution of the simple harmonic oscillator problem shows that the wave function is actually a Gaussian function. The uncertainty product therefore actually has its minimum value $\Delta x\, \Delta p = \tfrac{1}{2}\hbar$. If this value were used rather than $\Delta x\, \Delta p = \hbar$, the calculation would yield for the minimum energy $\tfrac{1}{2}\hbar\omega$. It is easy to see that, for the low frequencies observed for macroscopic particles on springs, $E_{\min}$ is negligible; but for typical frequencies of $f = \omega/2\pi \sim 10^{13}/\text{sec}$ for molecular vibrations, the minimum energy is of the order of 0.1 eV.

Example 5-8   Size of the Hydrogen Atom   The energy of an electron of momentum p a distance r from a proton is

$$E = \frac{p^2}{2m} - \frac{ke^2}{r}$$

If we take for the order of magnitude of the position uncertainty $\Delta x = r$, we have $(\Delta p)^2 = \overline{p^2} \geqslant \hbar^2/r^2$. The energy is then

$$E = \frac{\hbar^2}{2mr^2} - \frac{ke^2}{r}$$

There is a radius r_m at which E is minimum. Setting $dE/dr = 0$ yields r_m and E_m:

$$r_m = \frac{\hbar^2}{ke^2m} = a_0 = 0.529 \ \text{Å}$$

and

$$E_m = -\frac{k^2e^4m}{2\hbar^2} = -13.6 \ \text{eV}$$

The fact that r_m came out to be exactly the radius of the first Bohr orbit is due to the judicious choice of $\Delta x = r$ rather than $2r$ or $r/2$, which are just as reasonable. It should be clear, however, that any reasonable choice for Δx gives the correct order of magnitude of the size of an atom.

Example 5-9   Widths of Spectral Lines   Equation 5-31 implies that the energy of a system cannot be measured exactly unless an infinite time is available for the measurement. If an atom is in an excited state, it does not remain in that state indefinitely but makes transitions to lower energy states until it reaches the ground state. The decay of an excited state is a statistical process.

We can take the mean time for decay τ, called the *lifetime*, to be a measure of the time available to determine the energy of the state. For atomic transitions, τ is of order of 10^{-8} sec. The uncertainty in the energy corresponding to this time is

$$\Delta E \geq \frac{\hbar}{\tau} = \frac{6.58 \times 10^{-16} \text{ eV-sec}}{10^{-8} \text{ sec}} \approx 10^{-7} \text{ eV}$$

This uncertainty in energy causes a spread in the wavelength of the light emitted, $\Delta\lambda$. For transitions to the ground state, which has a perfectly certain energy E_0 because of its infinite lifetime, the percentage spread in wavelength can be calculated from

$$E - E_0 = \frac{hc}{\lambda}$$

$$dE = -hc\,\frac{d\lambda}{\lambda^2}$$

$$|\Delta E| \approx hc\,\frac{|\Delta\lambda|}{\lambda^2}$$

thus

$$\frac{\Delta\lambda}{\lambda} \approx \frac{\Delta E}{E - E_0}$$

The energy width $\Gamma_0 = \hbar/\tau$ is called the *natural line width*. Other effects that cause broadening of spectral lines are the Doppler effect, the recoil effect, and atomic collisions (see Problems 13, 14, and 15). For optical spectra in the eV energy range, the Doppler width D is about 10^{-6} eV at room temperature, i.e., roughly 10 times the natural width, and the recoil width is negligible. For nuclear transitions in the MeV range, both the Doppler width and the recoil width are of the order of eV, much larger than the natural line width. We shall see in Chapter 11 that in some special cases of atoms in solids at low temperatures, the Doppler and recoil widths are essentially zero and the width of the spectral line is just the natural width. This effect, called the *Mössbauer effect* after its discoverer in 1958, is extremely important for it provides photons of well-defined energy which are useful in experiments demanding extreme precision. For example, the 14.4-keV photon from ^{57}Fe has a natural width of the order of 10^{-11} of its energy.

Summary

Electrons and other "particles" exhibit the usual wave properties of interference and diffraction. The frequency and wavelength of electron waves are related to the energy and momentum by the de Broglie relations, $E = hf = \hbar\omega$ and $p = h/\lambda = \hbar k$.

If the wavelength is small compared with all apertures and obstacles, the wave properties of diffraction and interference can be generally neglected for electrons as well as for light. All phenomena can be described by wave equations that relate the time and space behavior of the wave function appropriate to the phenomena. The square of the wave function at x and t is proportional to the probability of observing a particle in the region dx at x and t. The description of a wave localized in a region Δx requires a wave function, or wave packet, that contains a range of wave numbers Δk given by $\Delta k \approx 1/\Delta x$. Because k is related to the momentum by $p = \hbar k$, this implies that the product of the uncertainty in position and uncertainty in momentum cannot be less than $\hbar$. One consequence of this uncertainty principle is that a particle confined in a region of space cannot have zero kinetic energy.

References

1. P. Tipler, *Physics,* New York: Worth Publishers, Inc., 1976. Chapters 20 to 27 include a complete discussion of classical waves. Chapter 23 includes an elementary discussion of the superposition of waves of different frequency and wavelength to form wave packets.

2. E. Goldwasser, *Optics, Waves, Atoms, and Nuclei,* New York: W. A. Benjamin, Inc., 1965.

3. C. Andrews, *Optics of the Electromagnetic Spectrum,* Englewood Cliffs, N.J.: Prentice-Hall, Inc., 1960.

4. G. R. Fowles, *Introduction to Modern Optics,* New York: Holt, Rinehart and Winston, Inc., 1968.

The above references, as well as other elementary physics and optics textbooks, can be consulted for a review of the properties of classical waves.

5. L. de Broglie, *Matter and Light: The New Physics,* New York: Dover Publications, Inc., 1939. In this collection of studies is de Broglie's lecture on the occasion of receiving the Nobel Prize, in which he describes his reasoning leading to the prediction of the wave nature of matter.

6. Filmed lecture by Richard Feynman. "Probability and Uncertainty—The Quantum-Mechanical View of Nature," available from Educational Services, Inc., Film Library, Newton, Mass.

Exercises

Section 5-1, The de Broglie Relations

1. Calculate the de Broglie wavelength of an α particle which has kinetic energy (a) 1 MeV, (b) 5 MeV, (c) 8 MeV.

2. Find the de Broglie wavelength of a neutron of kinetic energy 0.02 eV (this is of the order of magnitude of kT at room temperature).

3. Find the kinetic energy of an electron if its de Broglie wavelength is (a) 1 Å, (b) 500 nm, (c) 1 cm.

4. Electrons in an electron microscope are accelerated from rest through a potential difference V_0 so that their de Broglie wavelength is 0.04 nm. What is V_0?

5. Find the kinetic energy of (a) α particles, (b) neutrons whose de Broglie wavelength is 1 Å.

6.   (a) What is the de Broglie wavelength of a 1-g mass moving at a speed of 1 m per year? (b) What should be the speed of such a mass if its de Broglie wavelength is to be 1 cm?

✓7.   If the kinetic energy of a particle is much greater than its rest energy the relativistic approximation $E \approx pc$ holds. Use this approximation to find the de Broglie wavelength of an electron of energy 100 MeV.

8.   (a) Calculate the de Broglie wavelength of an electron with kinetic energy 13.6 eV. (b) What is the ratio of this wavelength to the radius of the first Bohr orbit?

Section 5-2, Measurements of Electron Wavelengths

9.   What is the Bragg scattering angle ϕ for electrons scattered from a nickel crystal if their energy is (a) 75 eV, (b) 100 eV?

10. What is the Bragg scattering angle ϕ for the scattering of thermal neutrons of kinetic energy 0.02 eV from a nickel crystal?

✓11. (a) The scattering angle for 50-eV electrons from MgO is 55.6°. What is the crystal spacing D? (b) What would be the scattering angle for 100-eV electrons?

Section 5-3, Properties of Classical Waves

12. Show by direct substitution that the wave function $y(x,t) = y_0 \cos (kx - \omega t)$ satisfies Equation 5-7.

13. The phase velocity of sound waves in air is about 340 m/sec at ordinary temperatures and that of electromagnetic waves in vacuum is 3×10^8 m/sec. (a) Find the frequency range of visible electromagnetic waves of wavelength from 400 to 700 nm. (b) Find the range of wavelengths of audible sound waves of frequencies from 20 Hz to 20 kHz. (c) What is the wavelength of FM radio waves of frequency 100 MHz? (d) What is the frequency of microwaves of wavelength 3 cm?

14. By using the trigonometric identity for the sum of two sine waves the following can be shown: $y_0 \sin (kx - \omega t) + y_0 \sin (kw + \omega t) = 2y_0 \sin kx \cos \omega t$. (a) Show that the standing-wave function $y(x,t) = A \sin kx \cos \omega t$ satisfies the wave equation (Equation 5-7). (b) Find a condition on the wave number k such that the wave function is zero at $x = 0$ and $x = L$ for all time. Show that this implies the standing-wave condition of Equation 5-4.

✓15. A double-slit interference pattern is observed on a screen that is a very large distance R from two sources which are separated by a small distance d. (a) Show that the distance between successive interference maxima on the screen x is given approximately by $x = R\lambda/d$. (Hint: For small θ, $\sin \theta \approx \tan \theta \approx \theta$.) (b) Calculate x for visible light of wavelength 600 nm, a source separation of $d = 1$ cm, and screen distance $R = 2$ m. (c) From your results of (b) discuss why it is difficult to observe interference with waves of very small wavelength.

16. Two sources are coherent and in phase. A point P is a distance x_1 from one source and x_2 from another source. Show that the waves from the two sources differ in phase by the amount $2\pi(x_2 - x_1)/\lambda$.

17. A Fraunhofer diffraction pattern is observed with light of wavelength 600 nm incident on a slit of width a. (a) If the angle between the

central maximum of the pattern and the first minimum is to be 1°, what should be the width of the slit a? (b) What is the angle between the central maximum and the first minimum if the slit width is 1 mm?

Section 5-4, Wave Packets

18. What is the order of magnitude of the bandwidth $\Delta\omega$ of an amplifier if it is to amplify a pulse of width (a) 1 sec, (b) 1 μsec, (c) 1 nsec?

✓19. Two harmonic waves travel simultaneously along a long wire. Their wave functions are $y_1 = 0.002 \cos (8.0x - 400t)$ and $y_2 = 0.002 \cos (7.6x - 380t)$, where y and x are in meters and t in seconds. (a) Write the wave function for the resultant wave in the form of Equation 5-18. (b) What is the phase velocity of the resultant wave? (c) What is the group velocity? (d) Is the medium dispersive or nondispersive? (e) Calculate the range Δx between successive zeros of the group and relate it to Δk.

20. Repeat Exercise 19 for the two wave functions $y_1 = 0.003 \cos (8.0x - 400t)$ and $y_2 = 0.003 \cos (8.2x - 420t)$.

21. Information is transmitted along a cable in the form of short electric pulses at 100,000 pulses/sec. (a) What is the longest duration of the pulses such that they do not overlap? (b) What is the range of frequencies to which the receiving equipment must respond for this duration?

22. Two sound waves have frequencies 500 Hz and 505 Hz and both travel at 340 m/sec. (a) Compute the wave numbers k_1 and k_2 for these waves, and the approximate spatial spread $\Delta x = 1/(k_2 - k_1)$. (b) Compute $\Delta\omega$ and the approximate spread in time $\Delta t = 1/\Delta\omega$. (c) Compare your answer in (a) with that computed from $\Delta x = v\,\Delta t$.

Section 5-5, Electron Wave Packets, and Section 5-6, The Probabilistic Interpretation of the Wave Function

There are no exercises for these sections.

Section 5-7, The Uncertainty Principle

23. A mass of 1 μg has a speed of 1 cm/sec. If its speed is uncertain by 1 percent, what is the order of magnitude of the minimum uncertainty in its position?

✓24. A tuning fork of frequency f_0 vibrates for a time Δt and sends out a waveform which looks like that in Figure 5-22. This wave function is similar to a harmonic wave except that it is confined to a time Δt and space $\Delta x = v\,\Delta t$, where v is the phase velocity. Let N be the approximate number of cycles of vibration. We can measure the frequency by counting the cycles and dividing by Δt. (a) The number of cycles is un-

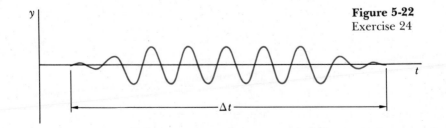

Figure 5-22
Exercise 24

certain by approximately ± 1 cycle. Explain why (see the figure). What uncertainty does this introduce in the determination of the frequency f? (b) Write an expression for the wave number k in terms of Δx and N. Show that the uncertainty in N of ± 1 leads to an uncertainty in k of $\Delta k = 2\pi/\Delta x$.

Section 5-8, Particle-Wave Duality, and Section 5-9, Some Consequences of the Uncertainty Principle

25. The energy of a certain nuclear state can be measured with an uncertainty of 1 eV. What is the minimum lifetime of this state?

26. Show that the relation $\Delta p_s \, \Delta s > \hbar$ can be written $\Delta L \, \Delta\phi > \hbar$ for a particle moving in a circle about the z axis, where p_s is the linear momentum tangential to the circle, s is the arc length, and L is the angular momentum. How well can the angular position of the electron be specified in the Bohr atom?

27. From the uncertainty principle, estimate the minimum kinetic energy of a proton confined to a region of (a) $\Delta x \approx 1$ Å, (b) $\Delta x \approx 5$ fm.

28. From the uncertainty principle, estimate the minimum kinetic energy of an electron confined to a region of space $\Delta x \approx 5$ fm. (Since the energy is much greater than the rest energy, use $E \approx pc$.)

Problems

1. Show that, in general, if the energy of a particle is much greater than its rest energy, its de Broglie wavelength is approximately the same as that of a photon of the same energy.

2. Using the relativistic expression $E^2 = p^2c^2 + m^2c^4$, (a) show that the phase velocity of an electron wave is greater than c. (b) Show that the group velocity of an electron wave equals the particle velocity of the electron.

3. Show that if y_1 and y_2 are solutions of Equation 5-7, the function $y_3 = C_1 y_1 + C_2 y_2$ is also a solution for any values of the constants C_1 and C_2.

4. Show that Equation 5-7 is satisfied by $y = f(\phi)$, where $\phi = x - vt$, for any function f.

Some understanding of wave groups made up of just a few waves can be obtained by adding the waves graphically. The tedious plotting of sine or cosine waves can be avoided without loss of understanding by using "triangular waves," as shown in Figure 5-23. In Problems 5 and 9, the notation Trn λ refers to such a triangular wave. Define $k = 2\pi/\lambda$, just as in sine waves.

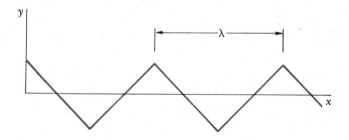

Figure 5-23
Triangular wave Trn λ
for Problems 5 and 9.

5.   (a) Obtain by graphical addition

$$f(x) = \text{Trn } \lambda_1 + \text{Trn } \lambda_2 + \text{Trn } \lambda_3$$

(start with all the waves at maximum at $x = 0$ and plot for positive x only) for $\lambda_1 = 2$ units, $\lambda_2 = 4$ units, $\lambda_3 = 6$ units, and sketch f^2 versus x. (b) What is the smallest positive value x_1 for which $f(x_1) = 0$? (c) Find the maximum and minimum values of $k = 2\pi/\lambda$. Using $\Delta k = k_{max} - k_{min}$ and $\Delta x = 2x_1$, find $\Delta x \, \Delta k$ for this group. (d) What is the first value of x after $x = 0$ for which all the waves are in phase? (e) Add a fourth wave, $\lambda_4 = 5$ units, to this group. Now what is the first value of x after $x = 0$ for which all four waves are in phase?

6.   Show that the group velocity v_g is related to the phase velocity v by $v_g = v - \lambda \, dv/d\lambda$.

7.   (a) Show that the uncertainty in position is related to that of wavelength by $\Delta x \, \Delta \lambda \geqslant \lambda^2$. (b) Use this result to compute Δx if $\Delta\lambda/\lambda = 1$ percent and $\lambda = 500$ nm. (c) Compute Δx if $\Delta\lambda/\lambda = 1$ percent and $\lambda = 1$ m (typical sound wave).

8.   In Equation 5-7 use the trial solution $y(x,t) = g(t)f(x)$, where f is a function of x only and g is a function of t only. Show that

$$\frac{g''(t)}{g(t)} = v^2 \, \frac{f''(x)}{f(x)}$$

If a function of t only equals a function of x only, neither function can depend on x or t. Set $g''(t)/g(t)$ equal to the constant $-\omega^2$ and solve for $g(t)$ and $f(x)$.

9.   (See Problem 5.) Obtain by graphical addition the group

$$f(x) = \sum_{i=1}^{5} A_i \, \text{Trn } \lambda_i$$

where $A_1 = A_5 = 1$, $A_2 = A_4 = 2$, $A_3 = 3$, and $\lambda_1 = 5$, $\lambda_2 = 6$, $\lambda_i = 4 + i$ up to $i = 5$. (a) Sketch f^2 versus x. (b) What is the smallest positive value x_1 for which $f(x_1) = 0$? (c) Sketch the distribution function A_i versus k_i. (d) What is the standard deviation σ_k? (e) Using $\Delta x = 2x_1$ and $\Delta k = \sigma_k$, find $\Delta x \, \Delta k$ for this group.

10.   If we have a continuous distribution of wave numbers $A(k)$, Equation 5-19 (with $t = 0$) can be written $y(x) = \int A(k) \cos kx \, dk$. Consider the uniform distribution $A(k) = C$ for $k_1 \leqslant k \leqslant k_2$ and $A(k) = 0$ for other k. (a) Show that if $A(k)$ is normalized so that $\int A(k) \, dk = 1$, the value of the constant C is given by $C = 1/(k_2 - k_1) = 1/\Delta k$. (b) Calculate $y(x)$ for this distribution and show that it can be written

$$y(x) = \left(\frac{\sin \frac{1}{2} \Delta kx}{\frac{1}{2}\Delta kx} \right) \cos \bar{k}x$$

where $\bar{k} = \frac{1}{2}(k_1 + k_2)$. (Hint: You will need to use a trigonometric identity for the sum of two sine functions.) (c) Sketch the envelope function in brackets for small Δk and for large Δk. At what value of x is $y(x)$ maximum? What is the smallest value of x for which $y(x) = 0$?

✓11.   Show that if $\Delta x \, \Delta p = \frac{1}{2}\hbar$, the minimum energy of a simple harmonic oscillator is $\frac{1}{2}\hbar\omega = \frac{1}{2}hf$. What is the minimum energy in joules for a mass of 10^{-2} kg oscillating on a spring of force constant $K = 1$ N/m?

12. A particle is on a table in a uniform gravitational field. The energy is $E = mgz + p^2/2m$, where $z = 0$ at the table. Classically the minimum energy is $E = 0$. Assume that the particle moves in a small range Δz above $z = 0$. Take for the average height $\bar{z} = \frac{1}{2}\Delta z$ and $\Delta p \geqslant \hbar/2\,\Delta z$. (a) Write $\bar{E}$ as a function of $\bar{z}$ and show that $\bar{E}$ has the minimum value

$$\bar{E}_{min} = \frac{3}{4}\left(\frac{mg^2\hbar^2}{2}\right)^{1/3} = \tfrac{3}{2}mg\bar{z}_{min}$$

at $\bar{z}_{min} = \frac{1}{2}(\hbar^2/2m^2g)^{1/3}$. (b) Find numerical values for E_{min} and z_{min} for a mass of 10^{-6} g and for the mass of a proton and that of an electron.

13. Using the first-order Doppler-shift formula $f' = f_0(1 + v/c)$, calculate the energy shift of a 1-eV photon emitted from an iron atom moving toward you with energy $\frac{3}{2}kT$ at $T = 300$ K. Compare this Doppler line broadening with the natural line width calculated in Example 5-9. Repeat the calculation for a 1-MeV photon from a nuclear transition.

14. Calculate the order of magnitude of the shift in energy of a (a) 1-eV photon and (b) 1-MeV photon resulting from the recoil of an iron nucleus. Do this by first calculating the momentum of the photon, and then by calculating $p^2/2m$ for the nucleus using that value of momentum. Compare with the natural line width calculated in Example 5-9.

15. If an atom collides with another while in the process of radiating, the frequency of the radiation may change. Assume the line width due to collisions is given by

$$\Gamma_c \approx \frac{\hbar}{\tau_c}$$

where τ_c is the mean time between collisions, which is related to the mean speed $\bar{v}$ and mean free path ℓ by $\tau_c = \ell/\bar{v}$. Using reasonable values for ℓ and $\bar{v}$ for oxygen gas, calculate Γ_c at standard conditions. (See Table 2-3 for ℓ determined from viscosity.) At what pressure is $\tau_c \approx 10^{-8}$ sec?

16. The angular frequency of water waves whose wavelength is greater than a few centimeters but much less than the depth of the water is related to the wave number by $\omega = \sqrt{gk}$. Find the phase velocity v and the group velocity v_g and show that $v_g = \frac{1}{2}v$.

CHAPTER 6 The Schrödinger Equation

Objectives

After studying this chapter you should:

---

1. Be able to write the time-dependent and time-independent Schrödinger equations.

2. Be able to solve the time-independent Schrödinger equation for the infinite-square-well problem and discuss how energy quantization arises.

3. Be able to sketch $\psi(x)$ and $\psi^2(x)$ for the infinite- and finite-square-well potentials.

4. Know the meaning of the terms quantum number, ground state, stationary state, expectation value, matrix element, selection rule, degeneracy, and exclusion principle.

5. Know how expectation values are calculated from the wave function.

6. Know the origin of selection rules for allowed transitions.

7. Be able to draw energy-level diagrams for the infinite-square-well and harmonic-oscillator potentials.

8. Be able to discuss qualitatively the reflection and transmission of waves at barrier and square-well potentials.

9. Know how quantum numbers arise in the solution of the Schrödinger equation in more than one dimension.

10. Be able to discuss the general features of the application of the Schrödinger equation to two particles in an infinite square well.

The success of the de Broglie relations in predicting the diffraction of electrons, and the realization that classical standing waves lead to a discrete set of frequencies, prompted a search for a wave theory of electrons analogous to the wave theory of light. In this electron-wave theory, classical mechanics should appear as the short-wavelength limit, just as geometric optics is the short-wavelength limit of the wave theory of light.

Pfaundler

(a)

(b)

(a) Erwin Schrödinger.
(b) Werner Heisenberg
and James Franck. [*Both
courtesy of the Niels Bohr
Library, American Institute
of Physics. Part (b) from
the Franck collection.*]

In 1925, Erwin Schrödinger published his now-famous wave equation which governs the propagation of electron waves. At about the same time, Werner Heisenberg published a seemingly different theory to explain atomic phenomena. In the Heisenberg theory, only measurable quantities appear. Dynamical quantities such as energy, position, and momentum are represented by matrices, the elements of which are the possible results of measurement. Though the Schrödinger and Heisenberg theories appear to be different, it was eventually shown that they were equivalent, in that each could be derived from the other. The resulting theory, now called *wave mechanics* or *quantum mechanics*, has been amazingly successful. Though its principles may seem strange to us whose experiences are limited to the macroscopic world, and though the mathematics required to solve even the simplest problem is quite involved, there seems to be no alternative to describe correctly the experimental results in atomic and nuclear physics. In this book we shall confine our study to the Schrödinger theory because it is easier to learn and is a little less abstract than the Heisenberg theory. We shall begin by restricting our discussion to problems in one dimension.

6-1   The Schrödinger Equation in One Dimension

The wave equation governing the motion of electrons (and other particles with mass), which is analogous to the classical wave equation (Equation 5-7), was found by Schrödinger in

1925 and is now known as the Schrödinger equation. Like the classical wave equation, the Schrödinger equation relates the time and space derivatives of the wave function. The reasoning followed by Schrödinger is somewhat difficult and not important for our purposes. In any case, we can't derive the Schrödinger equation just as we can't derive Newton's laws of motion. The validity of any fundamental equation lies in its agreement with experiment. Although it would be logical merely to postulate the Schrödinger equation, it is helpful to get some idea of what to expect by first considering the wave equation for photons, which is Equation 5-7 with speed $v = c$ and with $y(x,t)$ replaced by the wave function for light, namely, the electric field $\mathcal{E}(x,t)$.

$$\frac{\partial^2 \mathcal{E}}{\partial x^2} = \frac{1}{c^2}\frac{\partial^2 \mathcal{E}}{\partial t^2}$$ 6-1   *Classical wave equation*

As discussed in Chapter 5, a particularly important solution of this equation is the harmonic wave function $\mathcal{E}(x,t) = \mathcal{E}_0 \cos(kx - \omega t)$. Differentiating this function twice we obtain $\partial^2 \mathcal{E}/\partial t^2 = -\omega^2 \mathcal{E}_0 \cos(kx - \omega t) = -\omega^2 \mathcal{E}(x,t)$ and $\partial^2 \mathcal{E}/\partial x^2 = -k^2 \mathcal{E}(x,t)$. Substitution into Equation 6-1 then gives

$$-k^2 = -\frac{\omega^2}{c^2}$$

or

$$\omega = kc$$ 6-2

Using $\omega = E/\hbar$ and $p = \hbar k$, we have

$$E = pc$$ 6-3   *Energy-momentum relation for photon*

which is the relation between the energy and momentum of a photon.

Let us now use the de Broglie relations for a particle such as an electron to find the relation between ω and k for electrons which is analogous to Equation 6-2 for photons. We can then use this relation to work backwards and see how the wave equation for electrons must differ from Equation 6-1. The energy of a particle of mass m is

$$E = \frac{p^2}{2m} + V$$ 6-4

where V is the potential energy. Using the de Broglie relations we obtain

$$\hbar\omega = \frac{\hbar^2 k^2}{2m} + V$$ 6-5

This differs from Equation 6-2 for a photon because it contains the potential energy V and because the angular frequency ω does not vary linearly with k. Note that we get a factor of ω when we differentiate a harmonic wave function with respect to time and a factor of k when we differentiate with respect to position. We expect therefore that the wave equation that applies to electrons will relate the *first* time derivative to the *second* space derivative, and will also involve the potential energy of the electron.

We are now ready to postulate the Schrödinger equation. In one dimension, it has the form

$$-\frac{\hbar^2}{2m}\frac{\partial^2 \Psi(x,t)}{\partial x^2} + V(x)\Psi(x,t) = i\hbar\frac{\partial \Psi(x,t)}{\partial t} \qquad \text{6-6}$$

Time-dependent Schrödinger equation

We will now show that this equation is satisfied by a harmonic wave function in the special case in which there are no forces, so that the potential energy is constant, $V(x) = V_0$. We first note that a function of the form $\cos(kx - \omega t)$ does not satisfy this equation because differentiation with respect to time changes the cosine to a sine, but the second derivative with respect to x gives back a cosine. Similar reasoning rules out the form $\sin(kx - \omega t)$. However, the exponential form of the harmonic wave function (Equation 5-12c) does satisfy the equation. Let

$$\Psi(x,t) = Ae^{i(kx-\omega t)}$$
$$= A[\cos(kx - \omega t) + i\sin(kx - \omega t)] \qquad \text{6-7}$$

Harmonic wave function (complex form)

where A is a constant. Then

$$\frac{\partial \Psi}{\partial t} = -i\omega Ae^{i(kx-\omega t)} = -i\omega \Psi$$

and

$$\frac{\partial^2 \Psi}{\partial x^2} = (ik)^2 Ae^{i(kx-\omega t)} = -k^2\Psi$$

Substituting these derivatives into the Schrödinger equation with $V(x) = V_0$ gives

$$\frac{-\hbar^2}{2m}(-k^2\Psi) + V_0\Psi = i\hbar(-i\omega)\Psi$$

or

$$\frac{\hbar^2 k^2}{2m} + V_0 = \hbar\omega$$

which is Equation 6-5.

An important difference between the Schrödinger equation and the classical wave equation is the explicit appearance of the imaginary number $i = \sqrt{-1}$. The wave functions which satisfy the Schrödinger equation are not necessarily real, as we see from the case of the free-particle wave function of Equation 6-7. Evidently the wave function $\Psi(x,t)$ is not a measurable function like the classical wave functions $\mathscr{E}(x,t)$ or $y(x,t)$, since measurements always yield real numbers. The probability of finding the electron in dx is certainly measurable, however, just as is the probability that a flipped coin will turn up heads. The probability that an electron is in region dx can be measured by counting the fraction of times this occurs in a very large number of identical trials. Thus we must modify slightly the interpretation of the wave function discussed in Chapter 5 so that the probability of finding the electron in dx is real. We take for this probability

$$P(x,t) \, dx = |\Psi(x,t)|^2 \, dx = \Psi^*\Psi \, dx \qquad \text{6-8}$$

where Ψ^*, the complex conjugate of Ψ, is obtained from Ψ by replacing i by $-i$ wherever it appears.[1]

The probability of finding the electron in dx at x_1 or in dx at x_2 is the sum of the separate probabilities, $P(x_1) \, dx + P(x_2) \, dx$. Since the electron must certainly be somewhere, the sum of the probabilities over all possible values of x must equal 1. That is,

$$\int_{-\infty}^{+\infty} \Psi^*\Psi \, dx = 1 \qquad \text{6-9}$$

Normalization condition

Equation 6-9 is the *normalization condition,* familiar from the distribution functions studied in Chapter 2. This condition plays an important role in quantum mechanics, for it places a restriction on the possible solutions of the Schrödinger equation. If the integral in Equation 6-9 is to be finite, the wave function $\Psi(x,t)$ must approach zero as $x \to \pm\infty$. As we will see in Section 6-3, it is this restriction that leads to energy quantization.

Schrödinger's first application of his wave equation was to problems such as the hydrogen atom and the simple harmonic oscillator, in which he showed that energy quantization can be explained naturally in terms of standing waves. For such problems we need not consider the time dependence of the wave function. For any wave function which describes a particle in a state of definite energy, the general Schrödinger equation (Equation 6-6) can be greatly simplified by factoring out the time dependence and writing the wave function in the form[2]

$$\Psi(x,t) = \psi(x) \, e^{-i\omega t} \qquad \text{6-10}$$

where $\psi(x)$ is a function of x only. The harmonic wave function of Equation 6-7 is a special case of this form, since $Ae^{i(kx-\omega t)} = Ae^{ikx} e^{-i\omega t} = \psi(x)e^{-i\omega t}$ with $\psi(x) = e^{ikx}$. The right side of Equation 6-6 is then

$$i\hbar \, \frac{\partial \Psi(x,t)}{\partial t} = i\hbar(-i\omega)\psi(x)e^{-i\omega t} = + \hbar\omega\psi(x)e^{-i\omega t}$$

$$= E\psi(x)e^{-i\omega t}$$

Substituting $\psi(x) \, e^{-i\omega t}$ into Equation 6-6 and cancelling the

---

[1] Every complex number can be written in the form $z = a + bi$, where a and b are real numbers and $i = \sqrt{-1}$. The magnitude or absolute value of z is defined as $(a^2 + b^2)^{1/2}$. The complex conjugate of z is $z^* = a - bi$, so $z^*z = (a - bi)(a + bi) = a^2 + b^2 = |z|^2$.

[2] This result can be derived by the general method of solving partial differential equations known as "separation of variables." The derivation is outlined in Problem 2.

common factor $e^{-i\omega t}$, we obtain an equation for $\psi(x)$ called the time-independent Schrödinger equation:

$$-\frac{\hbar^2}{2m}\frac{d^2\psi(x)}{dx^2} + V(x)\psi(x) = E\psi(x) \qquad \text{6-11}$$

Time-independent Schrödinger equation

The time-independent Schrödinger equation in one dimension is an ordinary differential equation in one variable x and is therefore much easier to handle than the general form of Equation 6-6, which is a partial differential equation containing the two variables x and t. The normalization condition of Equation 6-9 can be expressed in terms of $\psi(x)$, since the time dependence of the absolute square of the wave function cancels. We have

$$\Psi^*(x,t)\Psi(x,t) = \psi^*(x)e^{+i\omega t}\psi(x)e^{-i\omega t} = \psi^*(x)\psi(x)$$

Equation 6-9 then becomes

$$\int_{-\infty}^{+\infty} \psi^*(x)\psi(x)\ dx = 1 \qquad \text{6-12}$$

The form of the wave function $\psi(x)$ which satisfies Equation 6-11 depends on the form of the potential-energy function $V(x)$. In the next few sections we shall study some simple but important problems in which $V(x)$ is specified. In some cases, the potential energy may be discontinuous, e.g., it may have one form in one region of space and another form in another. [This is a useful mathematical approximation to real situations in which $V(x)$ varies rapidly over a small region of space, such as at the surface boundary of a metal.] The procedure in such cases is to solve the Schrödinger equation separately in each region of space, and then require that the solutions join smoothly at the point of discontinuity. Since the probability of finding a particle cannot vary discontinuously from point to point, the wave function $\psi(x)$ must be continuous. Since the Schrödinger equation involves the second derivative $d^2\psi/dx^2 = \psi''(x)$, the first derivative ψ' (which is the slope) must also be continuous. That is, the graph of $\psi(x)$ versus x must be smooth. [In a special case in which the potential energy becomes infinite, this restriction is relaxed. Since no particle can have infinite potential energy, $\psi(x)$ must be zero in regions where $V(x)$ is infinite. Then, at the boundary of such a region, ψ' may be discontinuous.] A final restriction on the form of the wave function $\psi(x)$ is that in order to obey the normalization condition, $\psi(x)$ must approach zero as $x \to \pm\infty$. For future reference, we list the following conditions that the wave function $\psi(x)$ must meet to be acceptable:

1. $\psi(x)$ must satisfy the Schrödinger equation.

2. $\psi(x)$ must be continuous.

3. $\psi'(x)$ must be continuous [unless $V(x)$ is infinite].

4. $\psi(x) \to 0$ as $x \to \pm\infty$, so that $\psi(x)$ can be normalized.

Conditions for acceptable wave function

Questions

1. Like the classical wave equation, the Schrödinger equation is linear. Why is this important?

2. There is no factor $i = \sqrt{-1}$ in Equation 6-11. Does this mean that $\psi(x)$ must be real?

3. Why must the electric field $\mathscr{E}(x,t)$ be real? Is it possible to find a nonreal wave function which satisfies the classical wave equation?

6-2   The Infinite Square Well

One of the easiest problems to solve using the time-independent Schrödinger equation is that of the infinite square well, sometimes called the particle in a box, illustrated in Figure 6-1. For this problem the potential energy is of the form

$$V(x) = 0 \qquad 0 < x < L$$
$$V(x) = \infty \qquad x < 0 \quad \text{or} \quad x > L$$

6-13

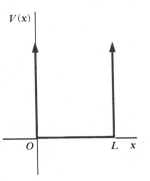

Figure 6-1
Infinite-square-well potential energy. For $0 < x < L$ the potential energy $V(x)$ is zero. Outside this region, $V(x)$ is infinite. The particle is confined to the region in the well $0 < x < L$.

Although such a potential is artificial, the problem is worth careful study for several reasons: (1) exact solutions to the Schrödinger equation can be obtained without the difficult mathematics which usually accompanies its solution for more realistic potential functions; (2) the problem is closely related to the vibrating-string problem familiar in classical physics; (3) many of the important features of all quantum-mechanical problems can be illustrated; and finally (4) this potential is a relatively good approximation to some real situations; e.g., the motion of a free electron inside a metal.

Since the potential energy is infinite outside the well, we require the wave function to be zero there; the particle must be inside the well. We then need only solve Equation 6-11 for the region inside the well $0 < x < L$, subject to the condition that since the wave function must be continuous, $\psi(x)$ must be zero at $x = 0$ and at $x = L$. Such a condition on the wave function at a boundary (here, the discontinuity of the potential-energy function) is called a *boundary condition*. We shall see that, mathematically, it is the boundary conditions that lead to the quantization of energy. A classical example is the case of a vibrating string fixed at both ends. In that case the wave function $y(x,t)$ is the displacement of the string. If the string is fixed at $x = 0$ and $x = L$ we have the same boundary condition on the vibrating-string wave function: namely, that $y(x,t)$ be zero at $x = 0$ and $x = L$. These boundary conditions lead to discrete allowed frequencies of vibration of the string. It was this quantization of frequencies (which always occurs for standing waves in classical physics), along with de Broglie's hypothesis, which motivated Schrödinger to look for a wave equation for electrons.

The standing-wave condition for waves on a string of length

L fixed at both ends is that *an integral number of half wavelengths fit into the length L:*

$$n \frac{\lambda}{2} = L \qquad n = 1, 2, 3, \ldots \qquad \text{6-14}$$

Standing-wave condition

We shall show below that the same condition follows from the solution of the Schrödinger equation for a particle in an infinite square well. Since the wavelength is related to the momentum of the particle by the de Broglie relation $p = h/\lambda$, and the total energy of the particle in the well is just the kinetic energy $p^2/2m$, this quantum condition on the wavelength implies that the energy is quantized and given by

$$E = \frac{p^2}{2m} = \frac{h^2}{2m\lambda^2} = \frac{h^2}{2m(2L/n)^2} = n^2 \frac{h^2}{8mL^2} \qquad \text{6-15}$$

Since the energy depends on the quantum number n, it is customary to label it E_n. In terms of $\hbar = h/2\pi$ the energy is given by

$$E_n = n^2 \frac{\pi^2 \hbar^2}{2mL^2} = n^2 E_1 \qquad n = 1, 2, 3, \ldots \qquad \text{6-16}$$

Allowed energies for infinite well

where E_1 is the lowest energy given by

$$E_1 = \frac{\pi^2 \hbar^2}{2mL^2} \qquad \text{6-17}$$

We now derive this result from the time-independent Schrödinger equation (Equation 6-11), which for $V(x) = 0$ is

$$-\frac{\hbar^2}{2m} \frac{d^2\psi(x)}{dx^2} = E\psi(x)$$

or

$$\psi''(x) = -\frac{2mE}{\hbar^2} \psi(x) = -k^2\psi(x) \qquad \text{6-18}$$

where the wave number k is defined by

$$k^2 = \frac{2mE}{\hbar^2} \qquad \text{6-19}$$

and we have written $\psi''(x)$ for the second derivative $d^2\psi(x)/dx^2$. Equation 6-18 has solutions of the form

$$\psi(x) = A \sin kx \qquad \text{6-20}a$$

and

$$\psi(x) = B \cos kx \qquad \text{6-20}b$$

where A and B are constants. The boundary condition $\psi(x) = 0$ at $x = 0$ rules out the cosine solution (Equation 6-20b) because $\cos 0 = 1$. The boundary condition $\psi(x) = 0$ at $x = L$ gives

$$\psi(L) = A \sin kL = 0 \qquad \text{6-21}$$

This condition is satisfied if kL is π, or any integer times π, i.e., if k is restricted to the values k_n given by

$$k_n = n\frac{\pi}{L} \qquad n = 1, 2, 3, \ldots \qquad\qquad 6\text{-}22$$

If we write the wave number k in terms of the wavelength $\lambda = 2\pi/k$, we see that Equation 6-22 is the same as Equation 6-14 for standing waves on a string. The quantized energy values are found from Equation 6-19, replacing k by k_n as given by Equation 6-22. We thus have

$$E_n = \frac{\hbar^2 k_n^2}{2m} = n^2\frac{\hbar^2\pi^2}{2mL^2} = n^2 E_1$$

which is the same as Equation 6-16.

Figure 6-2 shows the energy-level diagram for the infinite square-well potential. These energy levels are the same as those found in Chapter 4 from the Wilson-Sommerfeld quantization rule. Note that the lowest energy E_1 is not zero. This is consistent with the result from our discussion of the uncertainty principle in Chapter 5 that the kinetic energy of a particle confined in space must be greater than zero.

The constant A in the wave function of Equation 6-20a is determined by the normalization condition.

$$\int_{-\infty}^{+\infty} \psi_n^* \psi_n \, dx = \int_0^L A_n^2 \sin^2\left(\frac{n\pi x}{L}\right) dx = 1 \qquad 6\text{-}23$$

Integrating, we obtain $A_n = \sqrt{2/L}$ independent of n. The normalized-wave-function solutions for this problem are then

$$\psi_n(x) = \sqrt{\frac{2}{L}} \sin\frac{n\pi x}{L} \qquad n = 1, 2, 3, \ldots \qquad 6\text{-}24$$

Wave functions for infinite well

These wave functions are exactly the same as the standing-wave functions $y_n(x)$ for the vibrating-string problem. The wave func-

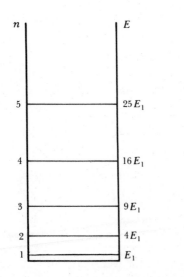

Figure 6-2
Energy-level diagram for the infinite-square-well potential. Classically, a particle can have any value of energy. Quantum-mechanically, only certain values of energy given by $E_n = n^2(\hbar^2\pi^2/2mL^2)$ yield well-behaved solutions of the Schrödinger equation.

(handwritten margin note:) mathematically, this is result of choosing n to be > 0 in eq. 6-22, as the boundry condition would be satisfied @ $n=0$ (?) ($n=0$ is used in Laplace's for fields)

(figure labels:) n ... E

5 ——— $25 E_1$

4 ——— $16 E_1$

3 ——— $9 E_1$

2 ——— $4 E_1$

1 ——— E_1

$E_n = n^2 E_1$
$E_1 = \dfrac{\hbar^2\pi^2}{2mL^2}$

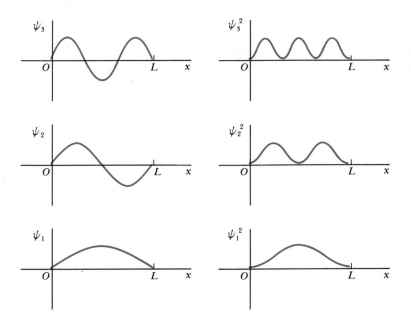

Figure 6-3
Wave functions $\psi_n(x)$ and probability densities $P_n(x) = \psi_n^2(x)$ for $n = 1, 2,$ and 3 for the infinite-square-well potential.

tions and the probability distribution functions $P_n(x)$ are sketched in Figure 6-3 for the lowest energy state $n = 1$, called the *ground state,* and for the first two excited states, $n = 2$ and $n = 3$. [Since these wave functions are real, $P_n(x) = \psi_n^* \psi_n = \psi_n^2.$]

The number n is called a *quantum number*. It specifies both the energy and the wave function. Given any value of n we can immediately write down the wave function and the energy of the system. The quantum number n occurs because of the boundary conditions $\psi(x) = 0$ at $x = 0$ and $x = L$. We shall see in Section 6-8 that for problems in three dimensions, three quantum numbers arise, one associated with boundary conditions on each coordinate.

Quantum numbers

We now compare our quantum-mechanical solution of this problem with the classical solution. In classical mechanics, if we know the potential-energy function $V(x)$, we can find the force from $F_x = -dV/dx$, and thereby obtain the acceleration $a_x = d^2x/dt^2$ from Newton's second law. We can then find the position x as a function of time t if we know the initial position and velocity. In this problem there is no force when the particle is between the walls of the well because $V = 0$ there. The particle therefore moves with constant speed in the well. Near the edge of the well the potential energy rises discontinuously to infinity—we may describe this as a very large force that acts for a very short distance and turns the particle around at the wall so that it moves away at its initial speed. Any speed, and therefore any energy, is permitted classically. The classical description breaks down because, according to the uncertainty principle, we can never precisely specify both the position and momentum (and therefore velocity) at the same time. We can therefore never specify the initial conditions precisely, and cannot assign a definite position and momentum to the particle. Of course, for a macroscopic particle moving in a macroscopic box, the energy is much larger than E_1 of Equation 6-17, and the minimum uncertainty of momentum, which is of the order of $\hbar/L$, is much less than the momentum and less than experimental uncertainties. Then the difference in energy between adjacent states will be a small fraction of the total energy, quantization will be unnoticed, and the classical description will be adequate.

Let us compare the classical prediction for the distribution of measurements of position with those from our quantum-mechanical solution. Classically, the probability of finding the particle in some region dx is proportional to the time spent in dx, which is dx/v, where v is the speed. Since the speed is constant, the classical distribution function is just a constant inside the well. The normalized classical distribution function is

$$P_c(x) = \frac{1}{L}$$

Classical probability distribution

In Figure 6-3 we see that for the lowest-energy states the quantum distribution function is very different from this. According to Bohr's correspondence principle, the quantum distributions should approach the classical distribution when n is large, that is, at large energies. For any state n, the quantum distribution has n peaks. The distribution for $n = 10$ is shown in Figure 6-4. For very large n, the peaks are close together, and if there are many peaks in a small distance Δx only the average value will be observed. But the average value of $\sin^2 k_n x$ over one or more cycles is $\frac{1}{2}$. Thus

$$[\psi_n{}^2(x)]_{av} = \left[\frac{2}{L}\sin^2 k_n x\right]_{av} = \frac{2}{L}\frac{1}{2} = \frac{1}{L}$$

which is the same as the classical distribution.

The complete wave function, including its time dependence, is found by multiplying the space part by $e^{-i\omega t} = e^{-i(E_n/\hbar)t}$ according to Equation 6-10. As mentioned previously, <u>a wave function corresponding to a single energy oscillates with angular frequency $\omega_n = E_n/\hbar$, but the probability distribution $|\Psi_n(x,t)|^2$ is independent of time. Such a state is therefore called a *stationary state.*</u> It is instructive to look at the complete wave function for one particular state.

$$\Psi_n(x,t) = \sqrt{\frac{2}{L}}\sin k_n x \; e^{-i\omega_n t}$$

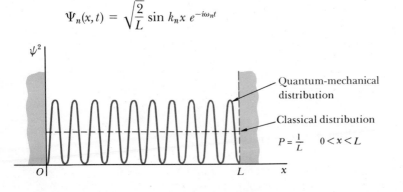

Figure 6-4
Probability distribution for $n = 10$ for the infinite-square-well potential. The dashed line is the classical probability density $P = 1/L$, which is equal to the quantum-mechanical distribution averaged over a region Δx containing several oscillations. A physical measurement with resolution Δx will yield the classical result if n is so large that $\psi^2(x)$ has many oscillations in Δx.

If we use the identity $\sin k_n x = (e^{ik_n x} - e^{-ik_n x})/2i$, we can write this wave function as

$$\Psi_n(x,t) = \frac{1}{2i} \sqrt{\frac{2}{L}} [e^{i(k_n x - \omega_n t)} - e^{-i(k_n x + \omega_n t)}]$$

Just as in the case of the standing-wave function for the vibrating string, we can consider this stationary-state wave function to be the superposition of a wave traveling to the right and a wave traveling to the left.

6-3   The Finite Square Well

The quantization of energy that we found for a particle in an infinite square well is a general result that follows from the solution of the Schrödinger equation for any particle confined in some region of space. We shall illustrate this by considering the qualitative behavior of the wave function for a slightly more general potential-energy function, the finite square well shown in Figure 6-5. The solutions of the Schrödinger equation for this type of potential energy are quite different, depending on whether the total energy E is greater or less than V_0. We shall defer discussion of the case $E > V_0$ to Section 6-6 except to remark that in that case the particle is not confined and any value of the energy is allowed; i.e., there is no energy quantization. Here we shall assume that $E < V_0$.

Inside the well, $V(x) = 0$ and the time-independent Schrödinger equation (Equation 6-11) is Equation 6-18, the same as for the infinite well:

$$\psi''(x) = -k^2 \psi(x) \qquad k^2 = \frac{2mE}{\hbar^2}$$

The solutions are sines and cosines (Equation 6-20) except that now we do not require $\psi(x)$ to be zero at the well boundaries, but rather we require that $\psi(x)$ and $\psi'(x)$ be continuous at these points. Outside the well Equation 6-11 becomes

$$\psi''(x) = \frac{2m}{\hbar^2} (V_0 - E)\psi(x) = \alpha^2 \psi(x) \qquad\qquad \text{6-25}$$

where

$$\alpha^2 = \frac{2m}{\hbar^2} (V_0 - E) > 0 \qquad\qquad \text{6-26}$$

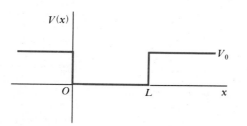

Figure 6-5
The finite-square-well potential.

The straightforward method of finding the wave functions and allowed energies for this problem is to solve Equation 6-25 for $\psi(x)$ outside the well and then require that $\psi(x)$ and $\psi'(x)$ be continuous at the boundaries. The solution of Equation 6-25 is not difficult [it is of the form $\psi(x) = Ce^{-\alpha x}$ for positive x], but applying the boundary conditions involves much tedious algebra which is not important for our purpose. The important feature of Equation 6-25 is that the second derivative ψ'', which is related to the curvature of the wave function, has the same sign as the wave function ψ. If ψ is positive, ψ'' is also positive and the wave function curves away from the axis, as shown in Figure 6-6(a). Similarly, if ψ is negative, ψ'' is negative and again, ψ curves away from the axis. This behavior is different from that inside the well, where ψ and ψ'' have opposite signs so that ψ always curves toward the axis like a sine or cosine function. Because of this behavior outside the well, for most values of the energy the wave function becomes infinite as $x \rightarrow \pm\infty$, i.e., $\psi(x)$ is not well behaved. Such functions, though satisfying the Schrödinger equation, are not proper wave functions because they cannot be normalized.

Figure 6-7 shows the wave function for the energy $E = p^2/2m = h^2/2m\lambda^2$ for $\lambda = 4L$. Figure 6-8 shows a well-behaved wave function corresponding to wavelength $\lambda = \lambda_1$, which is the ground-state wave function for the finite well, and the behavior of the wave functions for two nearby energies and wavelengths. The exact determination of the allowed energy levels in a finite square well can be obtained from a detailed solution of the problem. Figure 6-9 shows the wave functions and the probability distributions for the ground state and for the first two excited states. From this figure we see that the wavelengths inside the well are slightly longer than the corresponding wavelengths for the infinite well, so the corresponding energies are slightly less than those of the infinite well. Another feature of the finite-well problem is that there are only a finite number of allowed energies, depending on the size of V_0. For very small V_0 there is only one allowed energy level, i.e., only one bound state can exist.

Note that, in contrast to the classical case, there is some probability of finding the particle outside the box, in the regions $x > L$ or $x < 0$. In these regions, the total energy is less

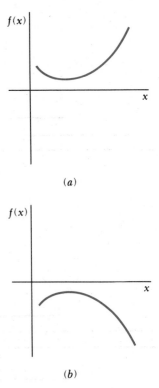

(a)

(b)

Figure 6-6
(a) Positive function with positive curvature; (b) negative function with negative curvature.

Figure 6-7
The function that satisfies the Schrödinger equation with $\lambda = 4L$ inside the well is not an acceptable wave function because it becomes infinite at large x. Although at $x = L$ the function is heading toward zero (slope is negative), the rate of increase of the slope ψ'' is so great that the slope becomes positive before the function becomes zero, and the function then increases. Since ψ'' has the same sign as ψ, the slope always increases and the function increases without bound. (See Reference 5.) (*This computer-generated plot courtesy of Paul Doherty, Oakland University.*)

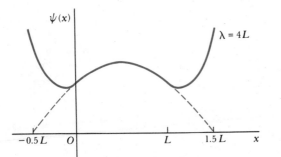

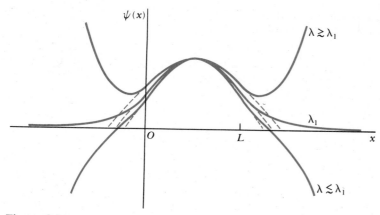

Figure 6-8
Functions satisfying the Schrödinger equation with wavelengths near the critical wavelength λ_1. If λ is slightly greater than λ_1, the function approaches infinity like that in Figure 6-7. At the critical wavelength λ_1, the function and its slope approach zero together. This is an acceptable wave function corresponding to the energy $E_1 = h^2/2m\lambda_1^2$. If λ is slightly less than λ_1, the function crosses the x axis while the slope is still negative. The slope becomes more negative because its rate of change ψ'' is now negative. This function approaches negative infinity at large x. (*This computer-generated plot courtesy of Paul Doherty, Oakland University.*)

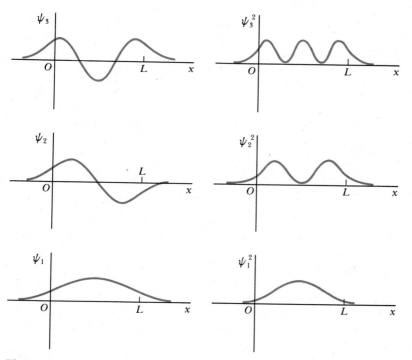

Figure 6-9
Wave functions $\psi_n(x)$ and probability distributions $\psi_n^2(x)$ for $n = 1, 2,$ and 3 for the finite square well. Compare these with Figure 6-3 for the infinite square well, where the wave functions are zero at $x = 0$ and $x = L$. The wavelengths are slightly longer than the corresponding ones for the infinite well, so the allowed energies are somewhat smaller.

than the potential energy so it would seem that the kinetic energy must be negative. Since negative kinetic energy has no meaning in classical physics, it is interesting to speculate about the meaning of this penetration of wave function beyond the well. Does quantum mechanics predict that we could measure a negative kinetic energy? If so this would be a serious defect in the theory. Fortunately, we are saved by the uncertainty principle. We can understand this qualitatively as follows (we shall consider the region $x > L$ only). Since the wave function decreases as $e^{-\alpha x}$, with α given by Equation 6-26, the probability density $\psi^2 = e^{-2\alpha x}$ becomes very small in a distance of the order of $\Delta x \approx \alpha^{-1}$. If we consider $\psi(x)$ to be negligible beyond $x = L + \alpha^{-1}$ we can say that finding the particle in the region $x > L$ is roughly equivalent to localizing it in a region $\Delta x \approx \alpha^{-1}$. Such a measurement introduces an uncertainty in momentum of the order of $\Delta p \approx \hbar/\Delta x = \hbar\alpha$ and a minimum kinetic energy of the order of $(\Delta p)^2/2m = \hbar^2\alpha^2/2m = V_0 - E$. This kinetic energy is just enough to prevent us from measuring a negative kinetic energy! The penetration of the wave function into a classically forbidden region does have important consequences in tunneling or barrier penetration, which we shall discuss in Section 6-7.

Much of our discussion of the finite-well problem applies to any problem in which $E < V(x)$ in some region and $E > V(x)$ outside that region. Consider the potential energy $V(x)$ shown in Figure 6-10. Inside the well, the Schrödinger equation is of the form

$$\psi''(x) = -k^2\psi(x)$$

where $k^2 = 2m[E - V(x)]/\hbar^2$ now depends on x. The solutions of this equation are no longer simple sine or cosine functions because the wave number $k = 2\pi/\lambda$ varies with x, but since ψ'' and ψ have opposite signs, ψ will always curve toward the axis and the solutions will oscillate. Outside the well, ψ will curve away from the axis so there will be only certain values of E for which solutions exist that approach zero as x approaches infinity.

6-4   Expectation Values and Operators

The solution of a classical mechanics problem is typically specified by giving the position of a particle or particles as a function of time. As we have discussed, the wave nature of matter prevents us from doing this for microscopic systems. Instead, we find the wave function $\Psi(x,t)$ and the probability distribution function $|\Psi(x,t)|^2$. The most that we can know is the probability of measuring a certain value of position x. The *expectation value* of x is defined as

$$\langle x \rangle = \int_{-\infty}^{+\infty} x\, \Psi^*(x,t)\, \Psi(x,t)\, dx \qquad \text{6-27}$$

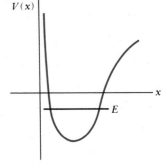

Figure 6-10
Arbitrary well-type potential with possible energy E. Inside the well $[E < V(x)]$ $\psi(x)$ and $\psi''(x)$ have opposite signs, and the wave function will oscillate. Outside the well, $\psi(x)$ and $\psi''(x)$ have the same sign and, except for certain values of E, the wave function will not be well behaved.

The expectation value is the same as the average value of x that we would expect to obtain from a measurement of the positions of a large number of particles with the same wave function $\Psi(x,t)$. As we have seen, for a particle in a state of definite energy the probability distribution is independent of time. The expectation value is then given by

$$\langle x \rangle = \int_{-\infty}^{+\infty} x \psi^*(x) \psi(x) \, dx \qquad \text{6-28}$$

Expectation value

From the infinite-square-well wave functions, we can see by symmetry (or by direct calculation) that $\langle x \rangle$ is $L/2$, the midpoint of the well.

The expectation value of any function $f(x)$ is given by

$$\langle f(x) \rangle = \int_{-\infty}^{+\infty} f(x) \psi^* \psi \, dx \qquad \text{6-29}$$

For example, $\langle x^2 \rangle$ can be calculated from the wave functions, above, for the infinite square well of width L. It is left as an exercise to show that for that case

$$\langle x^2 \rangle = \frac{L^2}{3} - \frac{L^2}{2n^2\pi^2} \qquad \text{6-30}$$

We should note that we don't necessarily expect to measure the expectation value. For example, for even n, the probability of measuring x in some range dx at the midpoint of the well $x = L/2$ is zero because the wave function $\sin(n\pi x/L)$ is zero there. We get $\langle x \rangle = L/2$ because the probability function $\psi^*\psi$ is symmetrical about that point.

Optional

Operators

If we knew the momentum p of a particle as a function of x, we could calculate the expectation value $\langle p \rangle$ from Equation 6-29. However, it is impossible in principle to find p as a function of x since, according to the uncertainty principle, both p and x cannot be determined at the same time. To find $\langle p \rangle$ we need to know the distribution function for momentum, which is equivalent to the distribution function $A(k)$ discussed in Section 5-4. As discussed there, if we know $\psi(x)$ we can find $A(k)$ and vice versa by Fourier analysis. Fortunately we need not do this each time. It can be shown from Fourier analysis that $\langle p \rangle$ can be found from

Expectation value of momentum

$$\langle p \rangle = \int_{-\infty}^{+\infty} \Psi^* \left(\frac{\hbar}{i} \frac{\partial}{\partial x} \right) \Psi \, dx \qquad \text{6-31}$$

Similarly $\langle p^2 \rangle$ can be found from

$$\langle p^2 \rangle = \int_{-\infty}^{+\infty} \Psi^* \left(\frac{\hbar}{i} \frac{\partial}{\partial x} \right) \left(\frac{\hbar}{i} \frac{\partial}{\partial x} \Psi \right) dx$$

Example 6-1   Find $\langle p \rangle$ and $\langle p^2 \rangle$ for the ground-state wave function of the infinite square well. We can neglect the time dependence. We then have

$$\langle p \rangle = \int_0^L \left(\sqrt{\frac{2}{L}} \sin \frac{\pi x}{L} \right)\left(\frac{\hbar}{i} \frac{\partial}{\partial x} \right)\left(\sqrt{\frac{2}{L}} \sin \frac{\pi x}{L} \right) dx$$

$$= \frac{\hbar}{i} \frac{2}{L} \frac{\pi}{L} \int_0^L \sin \frac{\pi x}{L} \cos \frac{\pi x}{L} \, dx = 0$$

The particle is equally as likely to be moving in the $-x$ as in the $+x$ direction, so its *average* momentum is zero.

Similarly, since

$$\frac{\hbar}{i} \frac{\partial}{\partial x} \left(\frac{\hbar}{i} \frac{\partial}{\partial x} \right)\psi = -\hbar^2 \frac{\partial^2 \psi}{\partial x^2} = -\hbar^2 \left(-\frac{\pi^2}{L^2} \sqrt{\frac{2}{L}} \sin \frac{\pi x}{L} \right)$$

$$= +\frac{\hbar^2 \pi^2}{L^2} \psi$$

we have

$$\langle p^2 \rangle = \frac{\hbar^2 \pi^2}{L^2} \int_0^L \psi^* \psi dx = \frac{\hbar^2 \pi^2}{L^2} \int_0^L \psi^* \psi dx = \frac{\hbar^2 \pi^2}{L^2}$$

Note that $\langle p^2 \rangle$ is simply $2mE$ since, for the infinite square well, $E = p^2/2m$. The quantity $(\hbar/i)\partial/\partial x$, which operates on the wave function in Equation 6-31, is called the *momentum operator* P_{op}:

Momentum operator

$$P_{op} = \frac{\hbar}{i} \frac{\partial}{\partial x} \qquad 6\text{-}32$$

The time-independent Schrödinger equation (Equation 6-11) can be written conveniently in terms of P_{op}:

$$\left(\frac{1}{2m} \right) P_{op}{}^2 \psi(x) + V(x)\psi(x) = E\psi(x) \qquad 6\text{-}33$$

where

$$P_{op}{}^2 \psi(x) = \frac{\hbar}{i} \frac{\partial}{\partial x} \left[\frac{\hbar}{i} \frac{\partial}{\partial x} \psi(x) \right] = -\hbar^2 \frac{\partial^2 \psi}{\partial x^2}$$

In classical mechanics, the total energy written in terms of the position and momentum variables is called the Hamiltonian function H. If we replace the momentum by the momentum operator P_{op} we obtain the Hamiltonian operator H_{op}. The Schrödinger equation can then be written

$$H_{op}\psi = E\psi \qquad 6\text{-}34$$

In one dimension, the Hamiltonian operator is simply

$$H_{op} = \frac{P_{op}{}^2}{2m} + V(x) \qquad 6\text{-}35$$

The advantage of writing the Schrödinger equation in this formal way is that it allows for easy generalization to more complicated problems such as those with several particles moving in three dimensions. We simply write the total energy of the system in terms of position and momentum and replace the momentum variables by the appropriate operators to obtain the Hamiltonian operator for the system.

Questions

4. For what kind of probability distribution would you expect to get the expectation value in a single measurement?

5. Is $\langle x^2 \rangle$ the same as $\langle x \rangle^2$?

Optional

6-5 Transitions between Energy States

We have seen that the Schrödinger equation leads to energy quantization for bound systems. The existence of these energy levels is determined experimentally by observation of the energy emitted or absorbed when the system makes a transition from one level to another. In this section we shall consider some aspects of these transitions in one dimension. The results will be readily applicable to more complicated situations.

In classical physics, a charged particle radiates when it is accelerated. If the charge oscillates, the frequency of the radiation emitted equals the frequency of oscillation. A stationary charge distribution does not radiate.

Consider a particle with charge q in a quantum state n described by the wave function

$$\Psi_n(x,t) = e^{-i(E_n/\hbar)t}\psi_n(x)$$

where E_n is the energy and $\psi_n(x)$ is a solution of the time-independent Schrödinger equation for some potential energy $V(x)$. The probability of finding the charge in dx is $\Psi_n^*\Psi_n \, dx$. If we make many measurements on identical systems (i.e., particles with the same wave function), the average amount of charge found in dx will be $q\Psi_n^*\Psi_n \, dx$. We therefore identify $q\Psi_n^*\Psi_n$ with the *charge density* ρ. As we have pointed out, the probability density is independent of time if the wave function contains a single energy, so the charge density for this state is also independent of time:

$$\rho_n = q\Psi_n^*(x,t)\Psi_n(x,t) = q\psi_n^*(x)\psi_n(x) = q\psi_n^*\psi_n\ddagger$$

We should therefore expect that this stationary charge distribution would not radiate. (This argument, in the case of the hydrogen atom, is the quantum-mechanical explanation of Bohr's postulate of nonradiating orbits.) However, we do observe that systems make transitions from one energy state to another with the emission or absorption of radiation. The cause of the transition is the interaction of the electromagnetic field with

‡ To simplify the notation in this section we shall sometimes omit the functional dependence and merely write ψ_n for $\psi_n(x)$ and Ψ_n for $\Psi_n(x,t)$.

the charged particle. A detailed treatment of this interaction is necessary in order to obtain rates of emission and absorption. Such a treatment is too difficult for us to consider; however, we can learn a great deal from a semiclassical treatment, which we shall now discuss.

Let us write the wave function for a particle which is making a transition from state n to state m as a mixture of the two states Ψ_n and Ψ_m:

$$\Psi_{nm}(x,t) = a\Psi_n(x,t) + b\Psi_m(x,t) \qquad 6\text{-}36$$

We need not be concerned with the numbers a and b. We wish only to show that if neither a nor b is zero, the probability density and charge density oscillate with the angular frequency ω_{nm}, given by the Bohr relation $hf = \hbar\omega_{nm} = E_n - E_m$, or

$$\omega_{nm} = \frac{E_n - E_m}{\hbar} \qquad 6\text{-}37$$

To simplify the notation, we shall assume that the time-independent functions $\psi_n(x)$ and $\psi_m(x)$ are real. The probability density for the wave function $\Psi_{nm}(x,t)$ is then

$$\Psi_{nm}^*\Psi_{nm} = a^2\Psi_n^*\Psi_n + b^2\Psi_m^*\Psi_m \\ + ab(\Psi_n^*\Psi_m + \Psi_m^*\Psi_n) \qquad 6\text{-}38$$

The first two terms are independent of time. The third term in Equation 6-38 contains the quantities

$$\Psi_n^*(x,t)\Psi_m(x,t) = e^{i(E_n/\hbar)t}\psi_n^* e^{-i(E_m/\hbar)t}\psi_m \\ = e^{i\omega_{nm}t}\psi_n\psi_m$$

and

$$\Psi_m^*(x,t)\Psi_n(x,t) = e^{i(E_m/\hbar)t}\psi_m^* e^{-i(E_n/\hbar)t}\psi_n \\ = e^{-i\omega_{nm}t}\psi_m\psi_n$$

where ω_{nm} is the Bohr angular frequency given by Equation 6-37. Adding these and using $e^{i\omega_{nm}t} + e^{-i\omega_{nm}t} = 2\cos\omega_{nm}t$, we see that the probability density is

$$|\Psi_{nm}(x,t)|^2 = a^2\psi_n^2 + b^2\psi_m^2 \\ + 2ab\cos(\omega_{nm}t)\psi_n\psi_m \qquad 6\text{-}39$$

Thus the wave function consisting of a mixture of two energy states does lead to a charge distribution which oscillates with the Bohr frequency. We can describe the radiation of a system somewhat loosely as follows. At some time, a system is in an excited state n described by Equation 6-36 with $a = 1$ and $b = 0$. Because of an interaction of the system with the electromagnetic field (which we have not included in the Schrödinger equation), a decreases and b is no longer zero. At this time the charge density oscillates with angular frequency ω_{nm}. The system does not, however, radiate energy continuously, as predicted by classical theory. Instead, the oscillating charge density implies a *probability* that a photon of energy $\hbar\omega_{nm} = E_n - E_m$ will be emitted, after which the system will be in the state m with $a = 0$, $b = 1$.

The emission of an individual photon is a statistical process.

The most elementary classical radiation system is an oscillating electric dipole. The dipole moment qx for a particle with wave function Ψ has the expectation value

$$q\langle x \rangle = q \int x \Psi^* \Psi \, dx$$

It can be seen from the previous discussion that, if the wave function corresponds to a stationary state containing a single energy, the expectation value of the dipole moment will be independent of time. However, if Ψ is the mixture given by Equation 6-36, $q\langle x \rangle$ will have time-dependent terms which oscillate with the Bohr frequency. Using Equation 6-39 the dipole moment can be written

$$q\langle x \rangle = 2qab \cos \omega_{nm} t \int x \psi_n \psi_m \, dx$$
$$+ \text{ stationary terms} \qquad 6\text{-}40$$

Matrix element

The integral in Equation 6-40 is called a *matrix element*. There are many cases for which this integral is zero. For example, if ψ_n and ψ_m are wave functions for the infinite square well, a direct calculation shows that the matrix element in Equation 6-40 is zero if n and m are both odd numbers or both even numbers. For these cases, electric dipole transitions are forbidden between these states. The absence of a transition between two states due to the fact that the matrix element is zero is usually

Selection rules

described by a *selection rule*. For example, a selection rule for the infinite square well is that the quantum number n must change by 1, 3, 5, . . . (and not by 2, 4, 6, . . .). We shall give other, more important examples of selection rules in the next section when we study the simple harmonic oscillator, and in Chapter 7 when we consider transitions between stationary states of atoms.

The transitions we have been discussing—those resulting from a perturbation of a charged system caused by interaction with the system's own electromagnetic field—are called spontaneous transitions. If a system (such as an atom) is in its ground state and exposed to external radiation of frequency ω_{nm} corresponding to the Bohr frequency for a transition to an excited state, it can make such a transition by absorbing a photon from the external radiation. If such a system is in an excited state and exposed to external radiation of frequency corresponding to the Bohr frequency for a transition to a lower energy state, the system can be stimulated to make such a transition and emit a photon of exactly the same energy as those in the external radiation. Such stimulated emission, which occurs in masers and lasers, is important because the emitted photons are coherent and in phase with those stimulating the transition. Lasers will be discussed in more detail in Chapter 8.

Question

6. How does $e|\psi|^2$ differ from a classical charge density ρ?

6-6   The Simple Harmonic Oscillator

One of the problems solved by Schrödinger in his first paper was that of the simple-harmonic-oscillator potential, given by

$$V(x) = \tfrac{1}{2}Kx^2 = \tfrac{1}{2}m\omega^2x^2$$

where K is the force constant and ω the angular frequency of vibration defined by $\omega = \sqrt{K/m} = 2\pi f$. The solution of the Schrödinger equation for this potential is particularly important as it can be applied to such problems as the vibrations of molecules in gases and solids. This potential-energy function is shown in Figure 6-11, with a possible total energy E indicated.

In classical mechanics, a particle in such a potential is in equilibrium at the origin $x = 0$, where $V(x)$ is minimum and the force $F_x = -dV/dx$ is zero. If disturbed, the particle will oscillate back and forth between $x = -A$ and $x = +A$, the points at which the kinetic energy is zero and the total energy is just equal to the potential energy. These points are called the classical turning points. The distance A is related to the total energy E by

$$E = \tfrac{1}{2}m\omega^2A^2 \qquad\qquad 6\text{-}41$$

Classically, the probability of finding the particle in dx is proportional to the time spent in dx, which is dx/v. The speed of the particle can be obtained from the conservation of energy:

$$\tfrac{1}{2}mv^2 + \tfrac{1}{2}m\omega^2x^2 = E$$

The classical probability is thus

$$P_c(x)\,dx \propto \frac{dx}{v} = \frac{dx}{\sqrt{(2/m)(E - \tfrac{1}{2}m\omega^2x^2)}} \qquad 6\text{-}42$$

Any value of the energy E is possible. The lowest energy is $E = 0$, in which case the particle is at rest at the origin.

The Schrödinger equation for this problem is

$$-\frac{\hbar^2}{2m}\frac{d^2\psi(x)}{dx^2} + \frac{1}{2}m\omega^2x^2\psi(x) = E\psi(x) \qquad 6\text{-}43$$

The mathematical techniques involved in solving this type of differential equation are standard in mathematical physics, but unfamiliar to most students at this level. We will, therefore, discuss the problem qualitatively. We first note that since the potential is symmetric about the origin $x = 0$, the probability distribution function $|\psi(x)|^2$ must also be symmetric about the origin, i.e., it must have the same value at $-x$ as at $+x$.

$$|\psi(-x)|^2 = |\psi(x)|^2$$

Then the wave function $\psi(x)$ must be either symmetric $\psi(-x) = +\psi(x)$, or antisymmetric $\psi(-x) = -\psi(x)$. We can therefore simplify our discussion by considering positive x only, and find the solutions for negative x by symmetry.

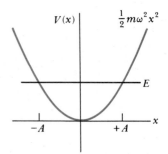

Figure 6-11
Potential-energy function for simple harmonic oscillator. Classically, the particle is confined between the "turning points" $-A$ and $+A$.

Consider some value of total energy E. For x less than the classical turning point A defined by Equation 6-41, the potential energy $V(x)$ is less than the total energy E, whereas for $x > A$, $V(x)$ is greater than E. Our discussion in Section 6-3 applies directly to this problem. For $x < A$, the Schrödinger equation can be written

$$\psi''(x) = -k^2\psi(x)$$

where

$$k^2 = \frac{2m}{\hbar^2} [E - V(x)]$$

and $\psi(x)$ curves towards the axis and oscillates. For $x > A$, the Schrödinger equation becomes

$$\psi''(x) = +\alpha^2\psi(x)$$

with

$$\alpha^2 = \frac{2m}{\hbar^2} [V(x) - E]$$

and $\psi(x)$ curves away from the axis. Only certain values of E will lead to solutions which are well behaved, i.e., which approach zero as x approaches infinity. The allowed values of E must be determined by solving the Schrödinger equation; in this case they are given by

$$E_n = (n + \tfrac{1}{2})\hbar\omega \qquad n = 0, 1, 2, \ldots \qquad \text{6-44}$$

Allowed energies for oscillator

These values differ from the values $E_n = n\hbar\omega$ assumed by Einstein to derive the temperature dependence of specific heats (Chapter 3) and derived from the Wilson-Sommerfeld quantization rule (Chapter 4) only in the presence of the term $\frac{1}{2}$. This term changes the ground-state energy from zero to $\frac{1}{2}\hbar\omega$ and shifts the other energies up by the amount $\frac{1}{2}\hbar\omega$, but the constant spacing of $\hbar\omega$ between any two successive levels is unchanged.

The wave functions of the simple harmonic oscillator in the ground state and in the first excited state ($n = 0$ and $n = 1$) are sketched in Figure 6-12. The ground-state wave function has the shape of a Gaussian curve, and the lowest energy $E_0 = \frac{1}{2}\hbar\omega$ is the minimum energy consistent with the uncertainty principle. The

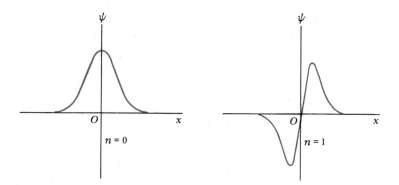

Figure 6-12
Wave functions for the ground state and first excited state of the simple-harmonic-oscillator potential.

allowed wave-function solutions to the Schrödinger equation can be written

$$\psi_n(x) = C_n e^{-m\omega x^2/2\hbar} f_n(x) \qquad\qquad 6\text{-}45$$

where the constants C_n are determined by normalization, and the functions f_n are polynomials[1] of order n. For even values of n the wave functions are symmetric about the origin; for odd values of n, they are antisymmetric. In Figure 6-13 the probability distributions $\psi_n^2(x)$ are sketched for $n = 0, 1, 2, 3,$ and 10 for comparison with the classical distribution.

A property of these wave functions that we shall state without proof is that

$$\int_{-\infty}^{+\infty} x\psi_n^*\psi_m\,dx = 0 \qquad \text{unless } n = m \pm 1 \qquad 6\text{-}46$$

This property leads to a selection rule for (electric dipole) radiation emitted or absorbed by a simple harmonic oscillator:

The quantum number of the final state must be 1 less than or 1 greater than that of the initial state.

[1] These polynomials, called Hermite polynomials, are known functions which are tabulated in most books on quantum mechanics.

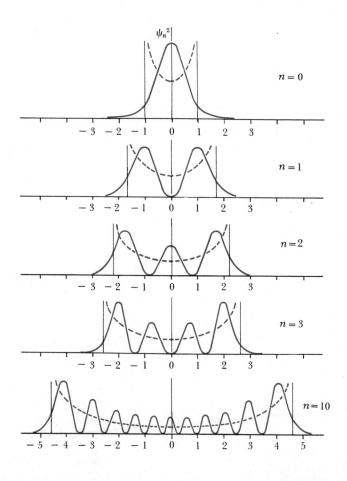

Figure 6-13
Probability density ψ_n^2 for the simple harmonic oscillator plotted against the dimensionless variable $u = (m\omega/\hbar)^{1/2}x$. The dashed curves are the classical probability densities for the same energy, and the vertical lines indicate the classical turning points $x = \pm A$. (*From Chalmers W. Sherwin*, Introduction to Quantum Mechanics. © *1959 by Holt, Rinehart and Winston, Inc. Reprinted by permission.*)

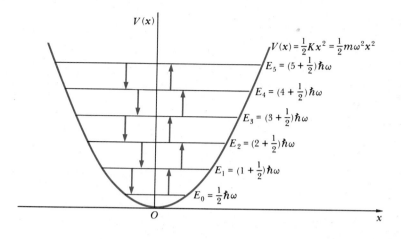

$$V(x) = \tfrac{1}{2}Kx^2 = \tfrac{1}{2}m\omega^2 x^2$$

$$E_5 = (5 + \tfrac{1}{2})\hbar\omega$$

$$E_4 = (4 + \tfrac{1}{2})\hbar\omega$$

$$E_3 = (3 + \tfrac{1}{2})\hbar\omega$$

$$E_2 = (2 + \tfrac{1}{2})\hbar\omega$$

$$E_1 = (1 + \tfrac{1}{2})\hbar\omega$$

$$E_0 = \tfrac{1}{2}\hbar\omega$$

Figure 6-14
Energy levels in the simple-harmonic-oscillator potential. Transitions obeying the selection rule $\Delta n = \pm 1$ are indicated by the arrows (those pointing up indicate absorption). Since the levels have equal spacing, the same energy $\hbar\omega$ is emitted or absorbed in all allowed transitions. For this special potential, the frequency of the emitted or absorbed photon equals the frequency of oscillation, as predicted by classical theory.

This selection rule is usually written

$$\Delta n = \pm 1 \qquad\qquad 6\text{-}47$$

Selection rule for simple harmonic oscillator

Since the difference in energy between two successive states is $\hbar\omega$, this is the energy of the photon emitted or absorbed in an electric dipole transition. The frequency of the photon is therefore equal to the classical frequency of the oscillator, as was assumed by Planck in his derivation of the blackbody radiation formula. Figure 6-14 shows an energy-level diagram for the simple harmonic oscillator, with the allowed energy transitions indicated by vertical arrows.

6-7   Reflection and Transmission of Waves

Up to this point, we have been concerned with bound-state problems in which the potential energy is larger than the total energy for large values of x. In this section, we shall consider some simple examples of unbound states for which E is greater than $V(x)$. For these problems, $\psi''(x)$ always has the opposite sign of the wave function so $\psi(x)$ everywhere curves toward the axis and does not become infinite at large values of x. Any value of E is allowed. The wave nature of the Schrödinger equation leads, even so, to some interesting consequences.

Step Potential

Consider a particle of energy E moving in a region in which the potential energy is the step function

$$V(x) = 0 \qquad \text{for } x < 0$$

$$V(x) = V_0 \qquad \text{for } x > 0$$

as shown in Figure 6-15. We are interested in what happens when a particle moving from left to right encounters the step.

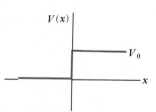

Figure 6-15
Step potential. A classical particle incident from the left, with total energy E greater than V_0, is always transmitted. The potential change at $x = 0$ merely provides an impulsive force which reduces the speed of the particle. A wave incident from the left is partially transmitted and partially reflected because the wavelength changes abruptly at $x = 0$.

The classical answer is simple. For $x < 0$, the particle moves with speed $v = \sqrt{2E/m}$. At $x = 0$, an impulsive force acts on it. If the original energy E is less than V_0, the particle will be turned around and will move to the left at its original speed; that is, it will be reflected by the step. If E is greater than V_0, the particle will continue moving to the right but with reduced speed, given by $v = \sqrt{2(E - V_0)/m}$. We might picture this classical problem as a ball rolling along a level surface and coming to a steep hill of height y_0, given by $mgy_0 = V_0$. If its original kinetic energy is less than V_0, the ball will roll part way up the hill and then back down and to the left along the level surface at its original speed. If E is greater than V_0 the ball with roll up the hill and proceed to the right at a smaller speed.

The quantum-mechanical result is similar for E less than V_0 but quite different for E greater than V_0. We consider a wave incident on such a barrier. Figure 6-16 shows the wave function for the case $E < V_0$. The wave function does not go to zero at $x = 0$ but decays exponentially, as does the wave function for the bound state in a finite-square-well problem. The wave penetrates slightly into the classically forbidden region $x > 0$ but eventually is completely reflected. (As discussed in Section 6-3, there is no prediction of *measurement* of negative kinetic energy in such a region, because to locate the particle in such a region introduces an uncertainty in the momentum corresponding to a minimum kinetic energy greater than $V_0 - E$.) This problem is somewhat similar to that of total internal reflection in optics.

For $E > V_0$ the quantum-mechanical result differs from the classical prediction. At $x = 0$, the wavelength changes abruptly from $\lambda_1 = h/p_1 = h/\sqrt{2mE}$ to $\lambda_2 = h/p_2 = h/\sqrt{2m(E - V_0)}$. We know from optics that when the wavelength changes suddenly (in a distance small compared with the wavelength), part of the wave is reflected and part transmitted. Since the motion of an electron is governed by a wave equation, the electron (or other "particle") will likewise be sometimes transmitted and sometimes reflected. The probabilities of reflection and transmission can be calculated by solving the Schrödinger equation in each region in space and comparing the amplitudes of the transmitted and reflected waves with that of the incident wave. This calculation and its result is similar to finding the fraction of light reflected from an air-glass surface. If R is the probability of re-

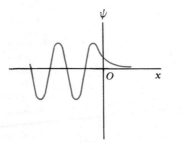

Figure 6-16
The wave function for total energy E less than V_0 penetrates slightly into the region $x > 0$. However, the probability of reflection for this case is 1 and no energy is transmitted.

flection, called the reflection coefficient, such a calculation gives

$$R = \frac{(k_1 - k_2)^2}{(k_1 + k_2)^2} \qquad \text{6-48}$$

Reflection coefficient

where k_1 and k_2 are the original and final wave numbers. This result is the same as that in optics for reflection at normal incidence. The transmission coefficient can be calculated from the reflection coefficient, since the probability of transmission plus that of reflection must equal 1:

$$T + R = 1 \qquad \text{6-49}$$

Transmission coefficient

Square-Well Potential

We now consider the square-well potential shown in Figure 6-17 for the case $E > 0$. (This problem is not significantly different if the well is replaced by a barrier of height V_0, as long as E is greater than the barrier.) The classical solution is again simple: a particle approaching the well from the left is accelerated and moves at a greater velocity inside the well. It is decelerated as it leaves the well and it continues to the right at its original speed. It is never reflected. In the quantum-mechanical or wave treatment, on the other hand, there are two abrupt changes in the wavelength, at $x = 0$ and $x = L$, and reflection occurs at each of these points.

A wave incident from the left is partially transmitted and partially reflected at $x = 0$. The transmitted wave proceeds to $x = L$, where again there is partial transmission and partial reflection. The reflected wave from $x = L$ returns to $x = 0$, where the process is repeated. Figure 6-18 shows the time development of a Gaussian wave packet incident from the left on a square-well potential. Eventually, two packets are formed, a transmitted packet moving to the right and a reflected packet moving back to the left. For the case shown, in which the mean energy of the packet is $\frac{1}{2}V_0$, the probability of transmission is greater than that of reflection, as indicated by the relative sizes of the packets.

An interesting phenomenon occurs if the well size L is just equal to half the wavelength of the particle in the well. The wave

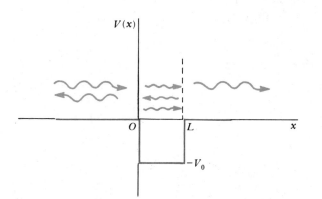

Figure 6-17
Square-well potential. A wave from the left is partially reflected and partially transmitted at each boundary. (Only the first few reflections are indicated on this figure.)

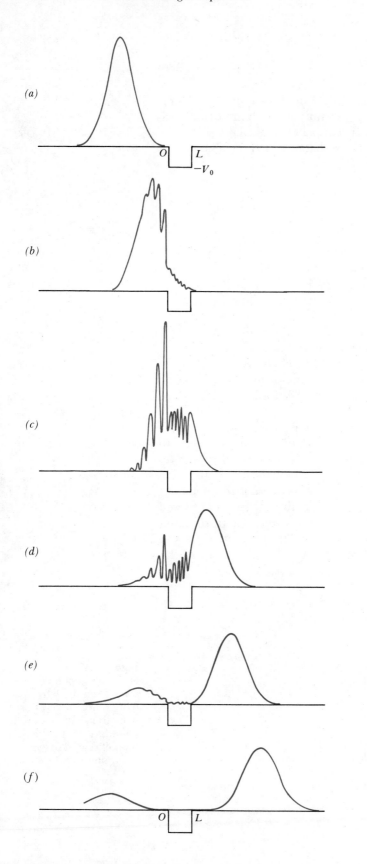

Figure 6-18
Behavior of a Gaussian wave packet incident from the left on a square-well potential. The oscillations evident in (b), (c), (d), and (e) are due to multiple reflections at the edges of the well. Eventually two packets are formed: a transmitted packet moving to the right, and a reflected one moving to the left. For this case, in which the mean energy was chosen to be $\frac{1}{2}V_0$, the probability of transmission is larger than that of reflection, as indicated by the relative size of the packets. (*From the film by A. Goldberg, H. Schey, and J. Schwartz,* Scattering in One Dimension, *Lawrence Radiation Laboratory, University of California. Work performed under the auspices of the U.S. Atomic Energy Commission.*)

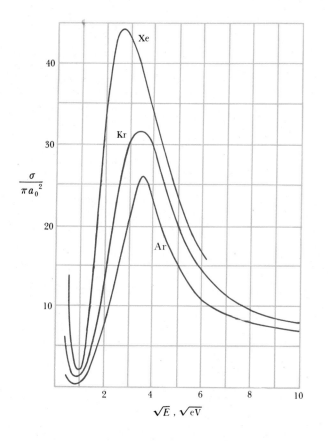

Figure 6-19
The cross section for scattering of electrons by inert gases. The minimum in the cross section at $E \approx 1$ eV corresponds to nearly perfect transmission. This effect, called the Ramsauer-Townsend effect, can be understood in terms of a one-dimensional model of scattering of a wave by a square-well potential. (*From H. Massey and E. Burhop,* Electronic and Ionic Impact Phenomena, *Oxford: Clarendon Press, 1952.*)

reflected at $x = L$ has traveled just one full wavelength when it returns to $x = 0$ and combines with a wave reflected at $x = 0$. Because there is a phase change of 180° in the reflection at $x = 0$ (analogous to the phase change when light is reflected from glass), the two reflected waves are out of phase and tend to cancel one another. (These waves are not, in general, of equal amplitude, so the cancellation is not complete.) For a well of given size, there will be a certain energy for which the reflection is nearly zero and the transmission perfect. Although atoms do not usually look like square-well potentials to incoming electron waves, the atoms of the inert gases have relatively sharp boundaries because of their closed-shell electron structure. Figure 6-19 shows the scattering cross section versus speed for electrons incident on argon, krypton, and xenon. At a speed corresponding to an energy of 1 eV, the cross section dips sharply, corresponding to nearly perfect transmission: this is called the *Ramsauer-Townsend effect*.

Barrier Penetration

We now consider one of the most interesting quantum-mechanical phenomena, barrier penetration or tunneling. For simplicity we first consider a particle of energy E incident on a

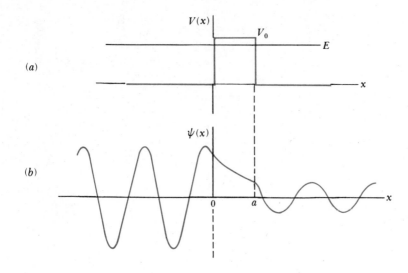

(a)

(b)

Figure 6-20
(a) Square-barrier potential. (b) Penetration of the barrier by a wave with energy less than the barrier energy. Part of the wave is transmitted by the barrier even though, classically, the particle cannot enter the region $0 < x < a$ in which the potential energy is greater than the total energy.

rectangular barrier of height V_0 and width a, as shown in Figure 6-20a for the case E less than V_0. Classically the particle would always be reflected. However, a wave incident from the left does not decrease immediately to zero at the barrier but instead will decay exponentially in the region of the barrier, as in the first case treated. Upon reaching the far wall of the barrier, the wave function must join smoothly to a sinusoidal wave function to the right of the barrier, as shown in Figure 6-20b. This implies that there will be some probability of the particle (which is represented by the wave function) being found on the far side of the barrier, although classically it should never be able to get through it. For the case in which the quantity $\alpha a = \sqrt{2ma^2(V_0 - E)/\hbar^2} \gg 1$, the transmission coefficient is proportional to $e^{-2a\alpha}$.

A variation on this problem is to consider two such barriers separated by some distance L, i.e., a square well with walls of finite height V_0 and finite thickness a. A particle originally in the well moves back and forth, striking the walls periodically. Each time it strikes a barrier, it has a small but finite probability of tunneling through it and escaping. Such *quantum-mechanical tunneling* is at the heart of a number of physical phenomena or devices such as the tunnel diode, the superconducting Josephson junction, and the phenomenon of α decay.

The penetration of a barrier is not a property unique to matter waves. In Figure 6-21 a light ray is incident on a glass-to-air surface. At an angle greater than the critical angle, total reflection occurs. Because of the wave nature of light, the electric field $\mathscr{E}$ is not zero at the surface, but decreases exponentially and becomes negligible within a few wavelengths of the surface. If another piece of glass is brought near the surface, the situation is analogous to that shown in Figure 6-20. Some of the light is transmitted across the barrier. (This can be demonstrated with a laser beam and two prisms.) Figure 6-22 shows barrier penetration by water waves in a ripple tank.

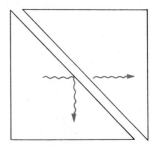

Figure 6-21
Optical barrier penetration. Because of the presence of the second prism, part of the wave penetrates the air barrier even though the angle of incidence in the first prism is greater than the critical angle. This effect can be demonstrated with two 45° prisms and a laser.

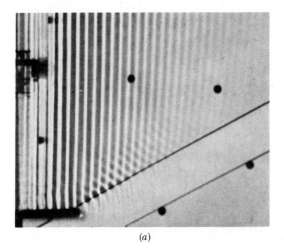

(a)

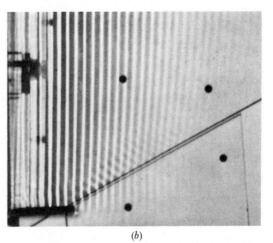

(b)

Figure 6-22
Barrier penetration by
waves in a ripple tank. In
(a) the waves are totally
reflected from a gap of
deeper water. When the
gap is very narrow, as in
(b), a transmitted wave
appears. (*Courtesy of Film
Studio, Education Development Center.*)

Figure 6-23
Model of potential-energy
function for an α particle
and a nucleus. The
strong attractive nuclear
force for r less than the
nuclear radius R can be
approximately described
by the potential well
shown. Outside the nu-
cleus the nuclear force is
negligible, and the potential is given by Coulomb's
law, $V(r) = +kZze^2/r$,
where Ze is the nuclear
charge and ze is the
charge of the α particle.
An α particle inside the
nucleus oscillates back
and forth, being reflected
at the barrier at R. Because of its wave proper-
ties, when the α particle
hits the barrier there is a
small chance that it will
penetrate and appear
outside the well at $r = r_1$.
The wave function is sim-
ilar to that shown in Figure 6-20b.

The theory of barrier penetration was used by George
Gamow in 1928 to explain the enormous variation in the mean
life for α decay of radioactive nuclei. In general, the smaller
the energy of the emitted α particle, the larger the mean life.
The energies of α particles from natural radioactive sources
range from about 4 to 7 MeV, whereas the mean lifetimes range
from about 10^{10} years to 10^{-6} sec. Gamow represented the radioactive nucleus by a potential well containing an α particle, as
shown in Figure 6-23. For r less than the nuclear radius R, the α
particle is attracted by the nuclear force. Without knowing much
about this force, Gamow represented it by a square well. Outside
the nucleus, the α particle is repelled by the Coulomb force. This
is represented by the Coulomb potential energy $+kZze^2/r$, where
$z = 2$ for the α particle and Ze is the remaining nuclear charge.
The energy E is the measured kinetic energy of the emitted α
particle since when it is far from the nucleus its potential energy is zero. We see from the figure that a small increase in E reduces the relative height of the barrier $V - E$ and also reduces
the thickness. Because the probability of transmission varies
exponentially with the relative height and barrier thickness, a
small increase in E leads to a large increase in the probability of
transmission, and in turn to a shorter lifetime. Gamow was able
to derive an expression for the mean lifetime as a function of energy E in excellent agreement with experimental results.

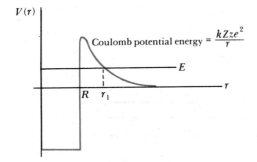

6-8   The Schrödinger Equation in Three Dimensions

So far we have been considering motion in just one dimension, whereas of course the real world is three-dimensional. While there are many cases in which the one-dimensional form brings out the essential physical features, there are some considerations introduced in three-dimensional problems which we want to examine. In rectangular coordinates, the time-independent Schrödinger equation is

$$-\frac{\hbar^2}{2m}\left(\frac{\partial^2\psi}{\partial x^2} + \frac{\partial^2\psi}{\partial y^2} + \frac{\partial^2\psi}{\partial z^2}\right) + V\psi = E\psi \qquad \text{6-50}$$

The wave function and the potential energy are generally functions of all three coordinates x, y, and z.

Let us consider the three-dimensional version of the particle in a box. We consider the potential-energy function $V = 0$ for $0 < x < L$, $0 < y < L$, and $0 < z < L$ and V infinite outside this cubical region. For this problem, the wave function must be zero at the walls of the box and will be a sine function inside the box. In fact, if we consider just one coordinate such as x, the solution will be the same as in the one-dimensional box. That is, the x dependence of the wave function will be of the form $\sin k_1 x$ with the restriction $k_1 L = n_1 \pi$, where n_1 is an integer. The complete wave function $\psi(x,y,z)$ can be written as a product of a function of x only, a function of y only, and a function of z only.

$$\psi(x,y,z) = \psi_1(x)\psi_2(y)\psi_3(z) \qquad \text{6-51}$$

where each of the functions ψ_n is a sine function as in one dimension. For example, if we try the solution

$$\psi(x,y,z) = A \sin k_1 x \sin k_2 y \sin k_3 z \qquad \text{6-52}$$

we find by inserting this function into Equation 6-50 that the energy is given by

$$E = \frac{\hbar^2}{2m}(k_1{}^2 + k_2{}^2 + k_3{}^2)$$

which is equivalent to $E = (p_x{}^2 + p_y{}^2 + p_z{}^2)/2m$ with $p_x = \hbar k_1$, etc. Using the restrictions on the wave numbers $k_i = n_i\pi/L$ from the boundary condition that the wave function be zero at the walls, we obtain for the total energy

$$E_{n_1 n_2 n_3} = \frac{\hbar^2 \pi^2}{2mL^2}(n_1{}^2 + n_2{}^2 + n_3{}^2) \qquad \text{6-53}$$

where n_1, n_2, and n_3 are integers.

As previously noted, the energy and wave function are characterized by three quantum numbers, each arising from a boundary condition for one of the coordinates. In this case the quantum numbers are independent of each other, but in more general problems the value of one quantum number may affect the possible values of the others. For example, we shall find that in problems such as the hydrogen atom that have spherical symmetry, the Schrödinger equation is most readily solved in spherical coordinates r, θ, and ϕ. The quantum numbers associated with the boundary conditions on these coordinates are interdependent.

The lowest energy state, the ground state for the cubical box, is given by Equation 6-53 with $n_1 = n_2 = n_3 = 1$. The first excited energy level can be obtained in three different ways: either $n_1 = 2$, $n_2 = n_3 = 1$ or $n_2 = 2$, $n_1 = n_3 = 1$, or $n_3 = 2$, $n_1 = n_2 = 1$. Each has a different wave function. For example, the wave function for $n_1 = 2$ and $n_2 = n_3 = 1$ is of the form

$$\psi_{2,1,1} = A \sin \frac{2\pi x}{L} \sin \frac{\pi y}{L} \sin \frac{\pi z}{L}$$

An energy level with which more than one wave function is associated is said to be *degenerate*. In this case there is threefold degeneracy. The degeneracy is related to the *symmetry* of the problem. If, for example, we considered a noncubical box $V = 0$ for $0 < x < L_1$, $0 < y < L_2$, and $0 < z < L_3$, the boundary conditions at the walls would lead to the quantum conditions $k_1 L_1 = n_1 \pi$, $k_2 L_2 = n_2 \pi$, and $k_3 L_3 = n_3 \pi$, and the total energy would be

$$E_{n_1 n_2 n_3} = \frac{\hbar^2 \pi^2}{2m} \left(\frac{n_1^2}{L_1^2} + \frac{n_2^2}{L_2^2} + \frac{n_3^2}{L_3^2} \right) \qquad 6\text{-}54$$

Figure 6-24 shows the energy levels for the ground state and first two excited states when $L_1 = L_2 = L_3$, for which the excited states are degenerate, and when L_1, L_2, and L_3 are slightly different, in which case the excited levels are slightly split apart and the degeneracy is removed.

6-9   The Schrödinger Equation for Two or More Particles

Our discussion of quantum mechanics so far has been limited to situations in which a single particle moves in some force field characterized by a potential energy function V. The most important physical problem of this type is the hydrogen atom, in which

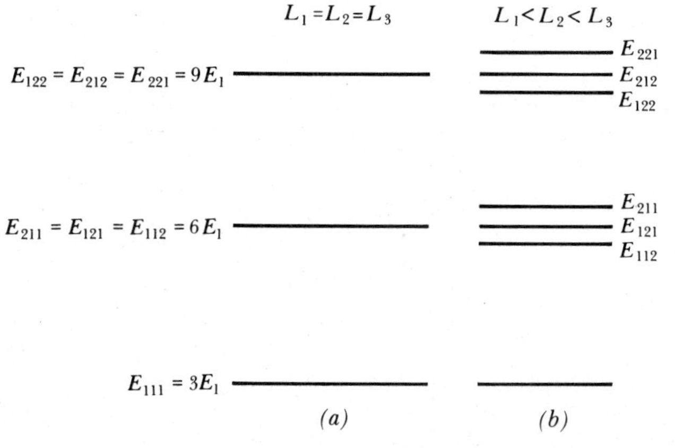

$$L_1 = L_2 = L_3 \qquad\qquad L_1 < L_2 < L_3$$

$E_{122} = E_{212} = E_{221} = 9E_1$ E_{221}, E_{212}, E_{122}

$E_{211} = E_{121} = E_{112} = 6E_1$ E_{211}, E_{121}, E_{112}

$E_{111} = 3E_1$

(a)   *(b)*

Figure 6-24
Energy-level diagram for *(a)* cubic-infinite-well potential and *(b)* noncubic infinite well. In the cubic well, the energy levels are degenerate, i.e., there are two or more wave functions having the same energy. The degeneracy is removed when the symmetry of the potential is removed, as in *(b)*.

a single electron moves in the Coulomb potential of the proton nucleus. This problem, which we shall consider in some detail in the next chapter, is actually a two-body problem, as the proton also moves in the Coulomb potential of the electron. However, as in classical mechanics, we can treat this as a one-body problem by considering the proton to be at rest and replacing the electron mass with the reduced mass. When we consider more complicated atoms we must face the problem of applying quantum mechanics to two or more electrons moving in an external field. Such problems are complicated by the interaction of the electrons with each other, and also by the fact that the electrons are identical.

The interaction of the electrons with each other is electromagnetic, and essentially the same as that expected classically for two charged particles. The Schrödinger equation for an atom with two or more electrons cannot be solved exactly, and approximate methods must be used. This is not very different from the situation in classical problems with three or more particles. The complications arising from the identity of electrons are purely quantum-mechanical and have no classical counterpart. They are due to the fact that it is impossible to keep track of which electron is which. Classically, identical particles can be identified by their positions, which can be determined with unlimited accuracy. This is impossible quantum-mechanically because of the uncertainty principle. Figure 6-25 offers a schematic illustration of the problem.

The indistinguishability of identical particles has important consequences related to the Pauli exclusion principle. We shall illustrate the origin of this important principle in this section by considering the very simple case of two noninteracting identical particles in a one-dimensional infinite square well. A more complete discussion of the exclusion principle requires a discussion of electron spin, which we shall defer until the next chapter.

The time-independent Schrödinger equation for two particles of mass m is

$$-\frac{\hbar^2}{2m}\frac{\partial^2\psi(x_1,x_2)}{\partial x_1{}^2} - \frac{\hbar^2}{2m}\frac{\partial^2\psi(x_1,x_2)}{\partial x_2{}^2}$$
$$+ V\psi(x_1,x_2) = E\psi(x_1,x_2) \qquad 6\text{-}55$$

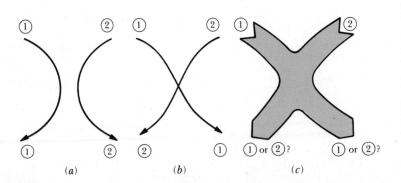

(a) (b) (c)

Figure 6-25
Two possible classical electron paths are shown in (a) and (b). The electrons can be distinguished classically. Because of the wave properties of the electrons, the paths are uncertain, as indicated by the shaded region in (c). It is impossible to distinguish which electron is which after they separate.

where x_1 and x_2 are the coordinates of the two particles. If the particles are interacting, the potential energy V contains terms with both x_1 and x_2, which cannot usually be separated. For example, if the particles are charged, their mutual electrostatic potential energy (in one dimension) is $+ke^2/|x_2 - x_1|$. If they do not interact, however, we can write V as $V_1(x_1) + V_2(x_2)$. For the case of an infinite-square-well potential, we need solve the Schrödinger equation only inside the well where $V = 0$ and require the wave function to be zero at the walls of the well. With $V = 0$ we note that Equation 6-55 has the same form as that for a single particle in a two-dimensional well (Equation 6-50 without z and with y replaced by x_2). Solutions to this equation can be written in the form.

$$\psi_{nm}(x_1, x_2) = \psi_n(x_1)\psi_m(x_2) \qquad\qquad 6\text{-}56$$

where ψ_n and ψ_m are the single-particle wave functions for a particle in an infinite well. For example, for $n = 1$ and $m = 2$ the wave function is

$$\psi_{1,2} = C \sin\frac{\pi x_1}{L} \sin\frac{2\pi x_2}{L} \qquad\qquad 6\text{-}57$$

Single-particle wave functions

The probability of finding particle 1 in dx_1 *and* particle 2 in dx_2 is $|\psi(x_1,x_2)|^2\, dx_1\, dx_2$, which is just the product of the separate probabilities $\psi(x_1)\, dx_1$ and $\psi(x_2)\, dx_2$. However, even though we have labeled the particles 1 and 2, if they are identical we cannot distinguish which is in dx_1 and which in dx_2. For identical particles, therefore, we must construct the wave function so that the probability density is the same if we interchange the labels:

$$|\psi(x_1,x_2)|^2 = |\psi(x_2,x_1)|^2 \qquad\qquad 6\text{-}58$$

Equation 6-58 holds if $\psi(x_1,x_2)$ is either symmetric or antisymmetric on exchange of particles. That is

$$\psi(x_2,x_1) = +\psi(x_1,x_2) \qquad \text{symmetric}$$

$$\psi(x_2,x_1) = -\psi(x_1,x_2) \qquad \text{antisymmetric}$$

We note that the general wave function of the form of Equation 6-56 and the example (Equation 6-57) are neither symmetric nor antisymmetric. If we interchange x_1 and x_2 we get a different wave function, implying that the particles can be distinguished. These forms are thus *not* consistent with the indistinguishability of identical particles. However, if ψ_{nm} and ψ_{mn} are added or subtracted we form symmetric or antisymmetric wave functions which are solutions of the Schrödinger equation.

$$\psi_A = C[\psi_n(x_1)\psi_m(x_2) + \psi_n(x_2)\psi_m(x_1)] \qquad \text{symmetric}$$

$$\psi_A = C[\psi_n(x_1)\psi_m(x_2) - \psi_n(x_2)\psi_m(x_1)] \qquad \text{antisymmetric}$$

There is an important difference between antisymmetric and symmetric combinations. If $n = m$, the antisymmetric wave function is identically zero for all x_1 and x_2 whereas the symmetric function is not. It is found that electrons (and other particles including protons and neutrons) can only have antisymmetric wave functions. Single-particle wave functions such as $\psi_n(x_1)$ and $\psi_m(x_1)$ for two such particles thus cannot have the same quantum numbers. This is an example of the *Pauli exclusion principle*, to be discussed more fully in Chapter 7. (For the case of electrons in an atom, four quantum numbers describe the state of each electron, one for each space coordinate and one associated with spin.) The Pauli exclusion principle states that no two electrons in an atom can have the same set of values for their four quantum numbers. Particles such as α particles, deuterons, photons, and mesons have symmetric wave functions and do not obey the exclusion principle.

Pauli exclusion principle

Summary

The time-dependent Schrödinger equation in one dimension is

$$-\frac{\hbar^2}{2m}\frac{\partial^2 \Psi(x,t)}{\partial x^2} + V(x)\Psi(x,t) = i\hbar\,\frac{\partial \Psi(x,t)}{\partial t}$$

The distribution function for position measurements is $|\Psi|^2$. The time-independent Schrödinger equation is

$$-\frac{\hbar^2}{2m}\psi''(x) + V(x)\psi(x) = E\psi(x)$$

Given the potential-energy function $V(x)$, the time-independent Schrödinger equation can be solved for the wave functions $\psi_n(x)$ and the allowed energies E_n. When $V(x)$ is greater than the total energy E outside some range of x, only certain discrete values of E lead to well-behaved wave functions and thus the energy is quantized. For the infinite-square-well potential of length L, the allowed energies are

$$E_n = n^2\,\frac{h^2}{8mL^2} = n^2\,\frac{\pi^2\hbar^2}{2mL^2} \qquad n = 1, 2, 3, \ldots$$

This is equivalent to quantizing the de Broglie wavelength

$$n\left(\frac{\lambda}{2}\right) = L$$

For the simple-harmonic-oscillator potential, the allowed energies are

$$E_n = (n + \tfrac{1}{2})\hbar\omega \qquad n = 0, 1, 2, \ldots$$

where ω is the classical angular frequency.

The expectation value of x is obtained from the wave function by

$$\langle x \rangle = \int x\psi^*\psi\,dx$$

A system can make a transition from state n_1 to state n_2 by the emission or absorption of electric dipole radiation only if the matrix element

$$\int x\psi_{n_1}^*\,\psi_{n_2}\,dx$$

is not zero. This requirement leads to selection rules. An example of a selection rule is $\Delta n = \pm 1$ for the simple harmonic oscillator. Transitions not obeying a selection rule either do not occur or involve some process other than electric dipole radiation, and they are usually much less probable than electric dipole transitions.

When the potential changes abruptly in a distance that is small compared with the de Broglie wavelength, a particle may be reflected even though $E > V(x)$. A particle may also penetrate a region in which $E < V(x)$. Reflection and barrier penetration of de Broglie waves are similar to those for other kinds of waves.

The solution of the Schrödinger equation in three dimensions leads to three quantum numbers, one associated with each dimension. If there is more than one wave function associated with a single value of the energy, the system is said to be degenerate. Degeneracy is associated with symmetry: an example is a three-dimensional square well with all three sides of equal size.

When the Schrödinger equation is applied to two identical particles, the wave function must be either symmetric or antisymmetric on exchange of the particles. Particles such as electrons, protons, and neutrons, which obey the Pauli exclusion principle, must have antisymmetric wave functions. Particles such as α particles or deuterons, which do not obey the Pauli exclusion principle, have symmetric wave functions.

References

1. R. Eisberg, *Fundamentals of Modern Physics,* New York: John Wiley & Sons, Inc., 1961.

2. C. Sherwin, *Introduction to Quantum Mechanics,* New York: Holt, Rinehart and Winston, Inc., 1960.

3. P. Matthews, *Introduction to Quantum Mechanics,* New York: McGraw-Hill Book Company, 1963.

4. V. Rojansky, *Introductory Quantum Mechanics,* Englewood Cliffs, N.J.: Prentice-Hall, Inc., 1938.

5. J. R. Merrill, *Using Computers in Physics,* Boston: Houghton Mifflin Company, 1976. Computers are very helpful in solving the Schrödinger equation for problems in which the mathematical solution in closed form is either very difficult or impossible. Computer programs similar to those used to generate Figures 6-7 and 6-8 can be found in this useful book.

Exercises

Section 6-1, The Schrödinger Equation in One Dimension

1. (a) Show that the wave function $\Psi(x,t) = A \sin (kx - \omega t)$ does *not* satisfy the time-dependent Schrödinger equation. (b) Show that $\Psi(x,t) = A \cos (kx - \omega t) + iA \sin (kx - \omega t)$ does satisfy this equation.

2. Show that $\mathcal{E}(x,t) = \mathcal{E}_0 e^{ikx} e^{-i\omega t}$ satisfies the classical wave equation (Equation 6-1).

Section 6-2, The Infinite Square Well

√3.   A particle is in the ground state of an infinite well of size L. Find the probability of finding the particle in the interval $\Delta x = 0.01L$ at (a) $x = \frac{1}{2}L$, (b) $x = \frac{3}{4}L$, (c) $x = L$. (Since Δx is very small, you need not do any integration.)

4.   Do Exercise 3 for a particle in the first excited state of an infinite-square-well potential.

5.   Do Exercise 3 for a particle in the second excited state ($n = 3$) of an infinite-square-well potential.

6.   Sketch the wave function $\psi(x)$ and the probability distribution $\psi^2(x)$ for the state $n = 4$ of the infinite-square-well potential.

7.   A mass of 10^{-6} g is moving with speed of about 10^{-1} cm/sec in a box of length 1 cm. Treating this as a one-dimensional infinite-square-well problem, calculate the approximate value of the quantum number n.

8. (a) A point where the wave function crosses the axis is called a node. How is the number of nodes in the wave function for the infinite square well related to the quantum number n? (Do not count the two nodes at the ends of the well.) (b) Use this relation to sketch the wave functions for $n = 5$ and $n = 6$.

9. (a) For the classical particle of Exercise 7 find $\Delta x\, \Delta p$, assuming that $\Delta x/L = 0.01$ percent and $\Delta p/p = 0.1$ percent. (b) What is $\Delta x\, \Delta p/\hbar$?

Section 6-3, The Finite Square Well

10.   Sketch the wave function $\psi(x)$ and the probability distribution for $n = 4$ for the finite-square-well potential.

✔ 11.   Sketch the wave function and the probability distribution for $n = 5$.

12.   Using arguments about the curvature of a function similar to those in Section 6-3, sketch the solution $f(t)$ to the equation $f''(t) = -\omega^2 f(t)$.

Section 6-4, Expectation Values and Operators

√13.   Find $\langle x^2 \rangle$ for the ground state of the infinite square well.

14.   Show directly from the time-independent Schrödinger equation that $\langle p^2 \rangle = \langle 2m[E - V(x)] \rangle$ in general and that $\langle p^2 \rangle = \langle 2mE \rangle$ for the infinite square well. Use this result to compute $\langle p^2 \rangle$ for the ground state of the infinite square well.

15.   Find $\sigma_x = \sqrt{\langle x^2 \rangle - \langle x \rangle^2}$, $\sigma_p = \sqrt{\langle p^2 \rangle - \langle p \rangle^2}$, and $\sigma_x \sigma_p$ for the ground-state wave function of an infinite square well. (Use the fact that $\langle p \rangle = 0$ by symmetry and $\langle p^2 \rangle = \langle 2mE \rangle$ from Exercise 14.)

16.   From the graphs of the probability distributions $\psi_1{}^2(x)$ and $\psi_2{}^2(x)$ for the infinite square well compare (a) $\langle x \rangle$ and (b) $\langle x^2 \rangle$ for these two states.

Section 6-5, Transitions between Energy States

There are no exercises for this section.

Section 6-6, The Simple Harmonic Oscillator

✓ 17. For the ground state $n = 0$ the polynomial f_n is given by $f_0 = 1$. Find (a) the normalization constant C_0, (b) $\langle x^2 \rangle$, (c) $\langle V(x) \rangle$ for this state. (*Hint:* Use Table 2-2 to compute the needed integrals.)

18. For the first excited state, $f_1 = x$. Find (a) the normalization constant C_1, (b) $\langle x \rangle$, (c) $\langle x^2 \rangle$, (d) $\langle V(x) \rangle$ for this state (see Exercise 17).

Section 6-7, Reflection and Transmission of Waves

19. The reflection coefficient for light reflected from an air-glass surface at normal incidence is

$$R = \frac{(n_1 - n_2)^2}{(n_1 + n_2)^2}$$

where n_1 and n_2 are the indexes of refraction of air and glass. Demonstrate that this is the same as Equation 6-48 by showing that for light waves $k = \omega n / c$.

Section 6-8, The Schrödinger Equation in Three Dimensions

20. (a) Show that the energy relation $E = \hbar^2 (k_1^2 + k_2^2 + k_3^2)/2m$ results from a direct substitution of the wave function of Equation 6-52 into the Schrödinger equation (Equation 6-50 with $V = 0$). (b) Apply the boundary conditions on the wave function of Equation 6-52 for the case of the noncubic well to derive Equation 6-54.

✓ 21. Find the energies E_{311}, E_{222}, and E_{321} and construct an energy-level diagram for the three-dimensional infinite cubic well which includes the third, fourth, and fifth excited states. Which of the states on your diagram are degenerate?

Section 6-9, The Schrödinger Equation for Two or More Particles

22. Show that the wave function of Equation 6-57 satisfies the Schrödinger equation 6-55 with $V = 0$ and find the energy of this state.

Problems

✓ 1. (a) Show that $\langle x^2 \rangle = L^2/3$ for the classical probability distribution function in the infinite-square-well potential $P = 1/L$. (b) For the wave functions

$$\psi_n(x) = \sqrt{\frac{2}{L}} \sin \frac{n\pi x}{L} \qquad n = 1, 2, 3, \ldots$$

corresponding to an infinite square well of size L, show that

$$\langle x^2 \rangle = \frac{L^2}{3} - \frac{L^2}{2n^2\pi^2}$$

2. In this problem you will obtain the time-independent Schrödinger equation from the time-dependent equation by the methods of separation of variables. (a) Substitute the trial function $\Psi(x,t) = \psi(x)f(t)$ into Equation 6-6, and divide each term by $\psi(x)f(t)$ to obtain the equation

$$i\hbar \frac{f'(t)}{f(t)} = -\frac{\hbar^2}{2m} \frac{\psi''(x)}{\psi(x)} + V(x)$$

(b) Since the left side of the equation in (a) does not vary with x, the right side cannot vary with x. Similarly, neither side can vary with t; thus they both must equal some constant C. Show that this implies that $f(t)$ is given by $f(t) = e^{-iCt/\hbar}$. Use the de Broglie relation to argue that C must be the total energy E. (c) Use the result of (b) to obtain Equation 6-11.

√3.   A particle of mass m on a table at $z = 0$ can be described by the potential energy

$$V = mgz \quad \text{for } z > 0$$

$$V = \infty \quad \text{for } z < 0$$

For some positive value of total energy E, indicate the classically allowed region on a sketch of $V(z)$ versus z. Sketch also the kinetic energy versus z. The Schrödinger equation for this problem is quite difficult to solve. Using arguments similar to those in Section 6-3 about the curvature of the wave function as given by the Schrödinger equation, sketch your "guesses" for the shape of the wave function for the ground state and the first two excited states.

4.   Show that Equations 6-48 and 6-49 imply that the transmission coefficient for particles of energy E incident on a step barrier $V_0 < E$ is given by

$$T = \frac{4k_1 k_2}{(k_1 + k_2)^2} = \frac{4r}{(1 + r)^2} \quad \text{where } r = \frac{k_2}{k_1}$$

5.   (a) Show that for the case of a particle of energy E incident on a step barrier $V_0 < E$, the wave numbers k_1 and k_2 are related by

$$\frac{k_2}{k_1} = r = \sqrt{1 - \left(\frac{V_0}{E}\right)}$$

Use this and the results of Problem 4 to calculate the transmission coefficient $T = 4r/(1 + r)^2$ and the reflection coefficient $R = (1 - r)^2/(1 + r)^2$ for the case (b) $E = 1.2V_0$, (c) $E = 2V_0$, (d) $E = 10V_0$.

6.   Find $\langle p \rangle$ and $\langle p^2 \rangle$ for the ground state in the simple-harmonic-oscillator potential and use your results, along with those of Exercise 17, to show that $\sigma_x \sigma_p = \frac{1}{2}\hbar$ for this state. (See Equation 6-31.)

7.   Plot a few points and then sketch the probability density $\psi^2(x)$ versus x for the ground state of the simple-harmonic-oscillator problem ($f_0 = 1$). Estimate the probability that the particle is outside the classically allowed region by counting squares on your graph.

CHAPTER 7   Atomic Physics

Objectives

After studying this chapter you should:

1. Know the origin of the quantum numbers m, l, and n; know the possible values for these numbers; and know their relation to the quantization of angular momentum and energy.

2. Be able to sketch the wave function and probability distribution functions for the ground state of hydrogen.

3. Be able to discuss the similarities and differences between the Bohr model and Schrödinger-equation treatment of the hydrogen atom.

4. Know the connection between magnetic moment and angular momentum, and be able to describe the Stern-Gerlach experiment.

5. Know the rules for the combination of two angular-momentum vectors, and be able to discuss qualitatively the spin-orbit effect.

6. Be able to discuss the effect of electron spin on the periodic table, and understand the origin of the electron configurations for the ground states of atoms.

7. Understand the general features of the energy-level diagram of a one-electron atom such as sodium, and of a two-electron atom such as mercury.

8. Be able to discuss the Zeeman effect and compare the normal and anomalous effects both in theory and experiment.

In this chapter we shall apply quantum theory to atomic systems. For all atoms except hydrogen, the Schrödinger equation cannot be solved exactly. Despite this, it is in the realm of atomic physics that the Schrödinger equation has had its greatest success because the electromagnetic interaction of the electrons with each other and with the atomic nucleus is well understood. With powerful approximation methods and high-speed computers, many features of complex atoms such as their energy levels and the wavelengths and intensities of their spectra can be calculated, often to whatever accuracy is desired. The Schrödinger equation for the hydrogen atom was first solved in Schrödinger's first paper in 1924. This problem is of considerable importance not only because the Schrödinger equation can be solved exactly in this case, but also because the solutions obtained form the basis for the approximate solutions for other atoms. We shall therefore discuss this problem in some detail. Because of the mathematical difficulties that usually arise in solving the Schrödinger equation, we shall often present results without proof and discuss important features of these results qualitatively. Whenever possible, we shall give simple arguments to make important results plausible.

7-1   The Schrödinger Equation in Spherical Coordinates

We can treat the hydrogen atom as a single particle, an electron moving with kinetic energy $p^2/2m_e$ and potential energy $V(r)$ due to the electrostatic attraction of the proton:

$$V(r) = -\frac{Zke^2}{r} \qquad 7\text{-}1$$

As in the Bohr theory, we include the factor Z, which is 1 for hydrogen, so we can apply our results to other similar systems, such as ionized helium He^+, where $Z = 2$. We also note that we can account for the motion of the nucleus by replacing the electron mass m_e by the reduced mass $\mu = m_e/(1 + m_e/M_N)$, where M_N is the mass of the nucleus. The Schrödinger equation for a particle of mass μ moving in three dimensions is Equation 6-50, with m replaced by μ:

$$-\frac{\hbar^2}{2\mu}\left(\frac{\partial^2\psi}{\partial x^2} + \frac{\partial^2\psi}{\partial y^2} + \frac{\partial^2\psi}{\partial z^2}\right) + V\psi = E\psi \qquad 7\text{-}2$$

Since the potential energy $V(r)$ depends only on the radial distance $r = (x^2 + y^2 + z^2)^{1/2}$, the problem is most conveniently treated in spherical coordinates r, θ, and ϕ. These are related to x, y, z by

$$z = r\cos\theta$$
$$x = r\sin\theta\cos\phi \qquad 7\text{-}3$$
$$y = r\sin\theta\sin\phi$$

Rectangular and spherical coordinates

These relations are shown in Figure 7-1. The transformation of the three-dimensional Schrödinger equation into spherical coordinates is straightforward but involves much tedious calculation which we shall omit. The result is

$$-\frac{\hbar^2}{2\mu}\frac{1}{r^2}\frac{\partial}{\partial r}\left(r^2\frac{\partial\psi}{\partial r}\right)$$
$$-\frac{\hbar^2}{2\mu r^2}\left[\frac{1}{\sin\theta}\frac{\partial}{\partial\theta}\left(\sin\theta\frac{\partial\psi}{\partial\theta}\right) + \frac{1}{\sin^2\theta}\frac{\partial^2\psi}{\partial\phi^2}\right]$$
$$+ V\psi = E\psi \qquad 7\text{-}4$$

Schrödinger equation in spherical coordinates

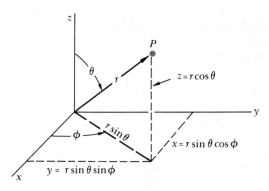

Figure 7-1
Geometric relation
between spherical and
rectangular coordinates.

Despite the formidable appearance of this equation, it was not difficult for Schrödinger to solve because it is similar to other partial differential equations which arise in classical physics, and such equations had been thoroughly studied. We shall outline the solution of this equation, giving just enough details to make some of the results plausible. More details are given in Appendix C.

7-2   Quantization of Angular Momentum and Energy in the Hydrogen Atom

In this section we shall discuss how quantization of angular momentum and energy arise in the solution of the Schrödinger equation for the hydrogen atom and, in particular, the origin and interpretation of the quantum numbers n, l, and m. The mathematical form of the hydrogen-atom wave functions will be discussed in Section 7-3.

The first step in the solution of a partial differential equation such as Equation 7-4 is to separate the variables by writing the wave function $\psi(r,\theta,\phi)$ as a product of functions of each single variable. We write

$$\psi(r,\theta,\phi) = R(r)f(\theta)g(\phi) \qquad\qquad 7\text{-}5$$

Separation of variables

where R depends only on the radial coordinate r, f depends only on θ, and g depends only on ϕ. When this form of $\psi(r,\theta,\phi)$ is substituted into Equation 7-4, the partial differential equations can be transformed into three ordinary differential equations, one for $R(r)$, one for $f(\theta)$, and one for $g(\phi)$.

The potential energy $V(r)$ appears only in the equation for $R(r)$, called the radial equation. The particular form of $V(r)$ as given in Equation 7-1 therefore has no effect on the solution of the equations for $f(\theta)$ and $g(\phi)$. These solutions are applicable to any central field problem, i.e., any problem in which the potential energy depends only on r. As we have mentioned, the requirement that the wave function be well-behaved so that it is continuous and can be normalized introduces three quantum numbers, each associated with one of the three variables. The quantum number associated with ϕ is called m and is related to the z component of the angular momentum by

$$L_z = m\hbar \qquad\qquad 7\text{-}6$$

The requirement that $g(\phi)$ must have the same value at $\phi = 0$ and $\phi = 360°$, which refer to the same point in space, leads to the restriction that m be a positive or negative integer or zero. The boundary condition on ϕ thus leads to a quantum condition on the z component of angular momentum. (The z axis is somewhat special because it is the axis used to define the spherical coordinate θ. In order to observe the quantization of the z component of angular momentum in a physical system such as an atom, there must be some way of defining the z direction in space such as an external magnetic field.)

The quantum number associated with θ is called l and is related to the magnitude L of the angular momentum by

$$L = \sqrt{l(l + 1)}\hbar \qquad\qquad 7\text{-}7$$

The requirement that $f(\theta)$ not be infinite at $\theta = 0$ or $\theta = 180°$ leads to the requirement that l be a positive integer or zero, with the added restriction that $l \geq m$. The boundary condition on θ therefore leads to the quantization of the magnitude of the angular momentum given by Equation 7-7.

The solution of the angular parts of the Schrödinger equation therefore leads to the quantization of the magnitude of the angular momentum L and of the z component L_z, as given by Equations 7-6 and 7-7. The importance of angular momentum in the quantum-mechanical solution of central-force problems should not be too surprising. In the analogous problem in classical mechanics, when V depends on r only, the force is directed toward (or away from) the origin and the angular momentum $\mathbf{L}$ is a constant of the motion. Furthermore, in the simple Bohr model of the hydrogen atom, it is the postulate of quantization of angular momentum that leads to discrete energy values which agree with the experimentally observed spectra. The correct results from the Schrödinger equation, however, differ from the postulate of the simple Bohr model. Although the z component of the angular momentum is an integer times $\hbar$, the magnitude is not. Instead, it is restricted to the values given by $\sqrt{l(l + 1)}\hbar$, where l is an integer.

Figure 7-2 shows a vector-model diagram illustrating the possible orientations of the angular-momentum vector. We have the somewhat peculiar result that the angular-momentum vector never points in the z direction, for the maximum z component $m\hbar$ is always less than the magnitude $\sqrt{l(l + 1)}\hbar$. This result is related to an uncertainty principle for angular momentum (which we shall not derive) that implies that no two components of angular momentum can be precisely known, except in the case of zero angular momentum. It is worth noting that for a given value of l there are $2l + 1$ possible values of m, ranging from $-l$ to $+l$ in integral steps.

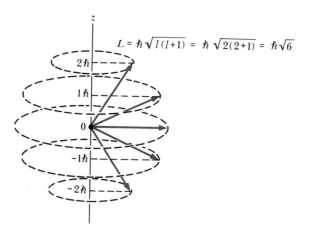

$$L = \hbar \sqrt{l(l+1)} = \hbar \sqrt{2(2+1)} = \hbar\sqrt{6}$$

Figure 7-2
Vector model illustrating the possible values of the z component of angular momentum for the case $l = 2$.

Example 7-1   If a system has angular momentum characterized by the quantum number $l = 2$, what are the possible values of L_z, what is the magnitude L, and what is the smallest possible angle between **L** and the z axis?

The possible values of L_z are $m\hbar$ where the ($2l + 1 = 5$) values of m are $m = -2, -1, 0, +1$, and $+2$. The magnitude L is $L = \sqrt{l(l + 1)}\,\hbar = \sqrt{6}\,\hbar$.

From Figure 7-2 the angle between **L** and the z axis is given by

$$\cos\theta = \frac{L_z}{L} = \frac{m\hbar}{\sqrt{l(l + 1)}\,\hbar} = \frac{m}{\sqrt{l(l + 1)}}$$

The smallest angle occurs when $m = \pm l$, which for $l = 2$ gives $\cos\theta = 2/\sqrt{6} = .816$, or $\theta = 35.3°$.

The results discussed so far apply to any system which is spherically symmetric, so that the potential energy depends on r only. The solution of the radial equation for $R(r)$, on the other hand, depends on the detailed form of $V(r)$. The quantum number associated with the coordinate r is called the *principal quantum number n.* This quantum number is related to the energy in the hydrogen atom. Figure 7-3 shows a sketch of the potential-energy function of Equation 7-1. If the total energy is positive, the electron is not bound to the atom. We are interested here only in bound-state solutions, for which E is negative. For this case, the potential-energy function is greater than E for large r, as shown in the figure. As we have discussed previously, for bound systems only certain values of the energy E lead to well-behaved solutions. These values, found by solving the radial equation, are

$$E_n = -\frac{Z^2 E_0}{n^2} \qquad\qquad 7\text{-}8$$

where $E_0 = \frac{1}{2}(ke^2/\hbar c)^2 \mu c^2 \approx 13.6$ eV and the principal quantum number n can take on the values $n = 1, 2, 3, \ldots$, with the further restriction that n must be greater than l. These energy values are identical with those found from the Bohr model.

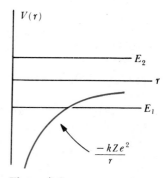

Figure 7-3
Potential energy of an electron in a hydrogen atom. If the total energy is greater than zero, as E_2, the electron is not bound and the energy is not quantized. If the total energy is less than zero, as E_1, the electron is bound. Then, as in one-dimensional problems, only certain discrete values of the total energy lead to well-behaved wave functions.

The restrictions on the quantum numbers n, l, and m associated with the variables r, θ, and ϕ are summarized in Equations 7-9.

$n = 1, 2, 3, \ldots$

$l = 0, 1, 2, \ldots, (n - 1)$ 7-9

$m = -l, -l + 1, \ldots, 0, 1, 2, \ldots, +l$

The fact that the energy of the hydrogen atom depends only on the principal quantum number n and not on l is a peculiarity of the inverse-square force. It is related to the classical-mechanics result that the energy in an elliptical orbit in an inverse-square force field depends only on the major axis of the orbit and not on the eccentricity. The largest value of angular momentum ($l = n - 1$) corresponds most nearly to a circular orbit, whereas a small value of l corresponds to a highly eccentric orbit. (Zero angular momentum corresponds to oscillation through the force center.) For central forces that do not obey an inverse-square law, the energy does depend on the angular momentum (both classically and quantum-mechanically), so it depends on both n and l.

The m quantum number is related to the z component of angular momentum. Since there is no preferred direction for the z axis for any central force, the energy cannot depend on m. We shall see later that if we place an atom in an external magnetic field, there is a preferred direction in space and the energy then does depend on the value of m. (This effect, called the Zeeman effect, will be discussed in Section 7-9.)

Figure 7-4 shows an energy-level diagram for hydrogen. This diagram is similar to Figure 4-15 except that the states with the same n but different l are shown separately. These states (called terms) are referred to by giving the value of n, along with a code letter: S standing for $l = 0$, P for $l = 1$, D for $l = 2$, and F for $l = 3$. These code letters are remnants of the spectroscopist's descriptions of various series of spectral lines as Sharp, Principal, Diffuse, and Fundamental. (For values greater than 3 the letters follow alphabetically; thus G for $l = 4$, etc.) The allowed electric dipole transitions obey the selection rules[1]

$$\Delta m = 0 \text{ or } \pm 1$$

                                    7-10   *Selection rules for Δm*

$$\Delta l = \pm 1$$

                                                     and Δl

The fact that the quantum number l of the atom must change by ± 1 when the atom emits a photon is related to conservation of angular momentum and the fact that the photon itself has an intrinsic angular momentum which has a maximum component along any axis of $1\hbar$.

---

[1] These selection rules can be derived from the wave functions for these states (discussed in the next section) by computing the matrix elements between various states, as discussed in Section 6-5.

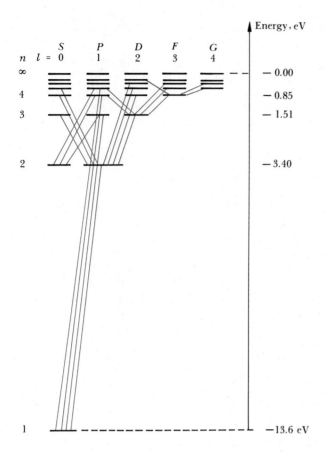

Figure 7-4
Energy-level diagram for hydrogen, showing transition obeying the selection rule $\Delta l = \pm 1$. States with the same n value but different l value have the same energy, E_0/n^2, where $E_0 = 13.6$ eV, as in the Bohr theory.

Optional

The connection between the quantum numbers l and m and angular momentum can be made more plausible if we look at the solution of the Schrödinger equation in somewhat more detail. The three differential equations for $R(r)$, $f(\theta)$, and $g(\phi)$ are obtained in Appendix C by substituting Equation 7-5 into Equation 7-4. The simplest of these is that for $g(\phi)$, which is

$$\frac{d^2g}{d\phi^2} = -m^2 g(\phi) \qquad 7\text{-}11$$

where m is a constant that arises in the separation of the variables. The solutions of this equation are of the form $e^{im\phi}$. (If we allow m to be negative as well as positive, this form includes the solution $e^{-im\phi}$.) This function has the same value at $\phi = 0$ and $\phi = 2\pi$ only if m is 0 or an integer. The function $e^{im\phi}$ is similar to the one-dimensional wave function e^{ikx} for a particle of linear momentum $\hbar k$ moving in the x direction. We can bring out this similarity if we consider a particle moving around the z axis in a circle of radius ρ with linear momentum $\hbar k_s$. Its angular momentum is then $L_z = \rho \hbar k_s$. Writing $\phi = s/\rho$, where s is the distance along the arc of the circle, we have

$$e^{im\phi} = e^{i(m/\rho)s} = e^{ik_s s} \qquad 7\text{-}12$$

where $m/\rho = k_s$. Then $m = \rho k_s$ and $m\hbar = \rho\hbar k_s$ is the z component of the angular momentum.

The differential equation for $f(\theta)$ is much more complicated than Equation 7-11, but it is related to a standard equation of mathematical physics called Legendre's equation which has complicated but well-known solutions called Legendre functions (see Appendix C). These functions are characterized by the number l. The requirement that these functions be finite at $\theta = 0$ and $\theta = 180°$ restricts l to integral values satisfying $l \geqslant m$.

The differential equation for $R(r)$ is

Radial equation

$$-\frac{\hbar^2}{2\mu}\frac{1}{r^2}\frac{d}{dr}\left(r^2\frac{dR}{dr}\right) + \left[\frac{l(l+1)\hbar^2}{2\mu r^2} + V(r)\right]R = ER \qquad 7\text{-}13$$

It is instructive to compare this equation with the classical equation for the total energy $E = p^2/2\mu + V(r)$. Let us write the momentum $\mathbf{p}$ in terms of its component p_r along the radius r and its component $p_\perp$ perpendicular to $\mathbf{r}$ (Figure 7-5). The angular momentum of the particle has the magnitude $L = rp_\perp$. In terms of L, we have then

$$p^2 = p_r{}^2 + p_\perp{}^2 = p_r{}^2 + \frac{L^2}{r^2}$$

and the total energy is

$$E = \frac{p_r{}^2}{2\mu} + \frac{L^2}{2\mu r^2} + V(r) \qquad 7\text{-}14$$

Comparing Equations 7-13 and 7-14, we should not be surprised to learn that $l(l+1)\hbar^2 = L^2$ is the square of the angular momentum.

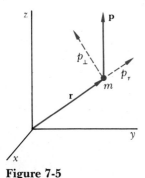

Figure 7-5
Resolution of momentum of classical particle into radial and perpendicular components. The angular momentum of the particle is $rp_\perp$.

Questions

1. Why wasn't quantization of angular momentum noticed in classical physics?

2. What are the similarities and differences between the quantization of angular momentum in the Schrödinger theory and in the Bohr model?

3. Why doesn't the energy of the hydrogen atom depend on l? Why doesn't it depend on m?

7-3   The Hydrogen-Atom Wave Functions

The wave functions satisfying the Schrödinger equation for the hydrogen atom are rather complicated functions of r, θ, and ϕ. In this section we shall write some of these functions and display some of their more important features graphically.

The ϕ dependence of the wave function is simply $e^{im\phi}$. The dependence on θ is a somewhat complicated function of θ called an associated Legendre function. It depends on both l and $|m|$ and is written $f_{l|m|}(\theta)$. [For the special case $m = 0$, $f_{l0}(\theta)$ is a polynomial in $\cos \theta$ called a Legendre polynomial.] Some of the Legendre functions for small values of l and m are tabulated in Appendix C. The radial dependence is of the form

$$R_{nl} = e^{-Zr/na_0} \left(\frac{2Zr}{na_0}\right)^l \mathscr{L}_{nl}\left(\frac{2Zr}{na_0}\right) \qquad 7\text{-}15$$

where a_0 is the first Bohr radius and $\mathscr{L}_{nl}$ is a polynomial called a Laguerre polynomial. The complete wave function for the hydrogen atom is therefore written

$$\psi_{nlm}(r,\theta,\phi) = C_{nlm}R_{nl}(r)f_{l|m|}(\theta)e^{im\phi} \qquad 7\text{-}16$$

where R_{nl} is given by Equation 7-15 and C_{nlm} is a constant that is determined by normalization.

We see from the form of these complicated expressions that the wave function depends on the quantum numbers n, l, and m that arise because of the boundary conditions on $R(r)$, $f(\theta)$, and $g(\phi)$. [Those conditions are: $R(r) \to 0$ as $r \to \infty$, and $R(r)$ does not diverge as $r \to 0$; $f(\theta)$ is not infinite at $\theta = 0°$ or $180°$; and $g(\phi)$ is continuous so that $g(\theta + 2\pi) = g(\phi)$.] The energy, however, depends only on the value of n. From Equation 7-9 we see that for any value of n there are n possible values of l ($l = 0, 1, 2, \ldots, n - 1$); and for each value of l there are $2l + 1$ possible values of m ($m = -l, -l + 1, \ldots, + l$). Except for the lowest energy level (for which $n = 1$, and therefore l and m can only be zero) there are generally many different wave functions corresponding to the same energy. As discussed in the previous section, the origins of this degeneracy are the $1/r$ dependence of the potential energy and the fact that there is no preferred direction in space.

The lowest-energy state, called the ground state, has $n = 1$. Then l and m must both be zero. The Laguerre polynomial $\mathscr{L}_{10}$ in Equation 7-15 is 1, and the wave function is

$$\psi_{100} = C_{100}e^{-Zr/a_0} \qquad 7\text{-}17$$

Ground-state wave function

The constant C_{100} is determined by normalization:

$$\int \psi^*\psi \, d\tau = \int_0^\infty \int_0^\pi \int_0^{2\pi} \psi^*\psi r^2 \sin \theta \, d\phi \, d\theta \, dr = 1$$

using for the volume element in spherical coordinates

$$d\tau = (r \sin \theta \, d\phi)(r \, d\theta)(dr)$$

Because $\psi^*\psi$ is spherically symmetric the integration over angles gives 4π. Carrying out the integration over r gives

$$C_{100} = \frac{1}{\sqrt{\pi}}\left(\frac{Z}{a_0}\right)^{3/2}$$

The probability of finding the electron in volume $d\tau$ is $\psi^*\psi \, d\tau$.

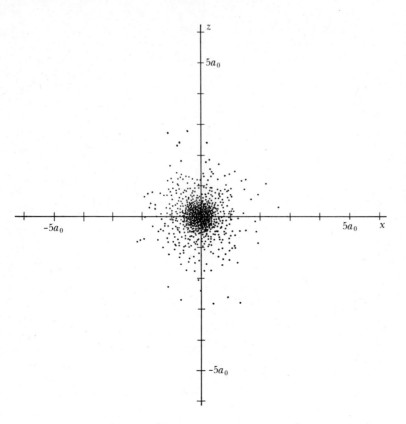

Figure 7-6
Probability density $\psi^*\psi$ for the ground state in hydrogen. The quantity $e\psi^*\psi$ can be thought of as the electron charge density in the atom. The density is spherically symmetric, is greatest at the origin, and decreases exponentially with r. *(This computer-generated plot courtesy of Paul Doherty, Oakland University.)*

The probability density $\psi^*\psi$ is illustrated in Figure 7-6. Note the similarity of this plot and the velocity distribution function shown in Figure 2-13. The probability density is maximum at the origin. It is often more interesting to determine the probability of finding the electron between r and $r + dr$. This probability, $P(r)\, dr$, is just the probability density $\psi^*\psi$ times the volume of the spherical shell of thickness dr:

$$P(r)\, dr = \psi^*\psi 4\pi r^2\, dr = 4\pi r^2 C_{100}^2 e^{-2Zr/a_0}\, dr \qquad 7\text{-}18$$

(Note the similarity of this calculation to the determination of the speed distribution from the velocity distribution in Chapter 2.) Figure 7-7 shows a sketch of $P(r)$ versus r/a_0. It is left as an exercise (see Exercise 6) to show that $P(r)$ has its maximum value at $r_m = a_0/Z$. In contrast to the Bohr model, in which the electron stays in a well-defined orbit at $r = a_0$, we see that it is *possible* for the electron to be found at any distance from the nucleus. However, the most probable distance is a_0, and the chance of finding the electron at a much different distance is small. It is useful to think of the electron as a charged cloud of charge density $\rho = e\psi^*\psi$. (We must remember, though, that the electron is always observed as one charge.) The angular momentum in the ground state is zero, contrary to the Bohr-model assumption of $1\hbar$.

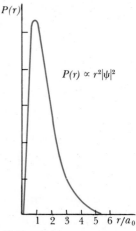

Figure 7-7
Radial probability density $P(r)$ versus r/a_0 for the ground state of the hydrogen atom. $P(r)$ is proportional to $r^2\psi^*\psi$. The most probable distance r is the Bohr radius a_0.

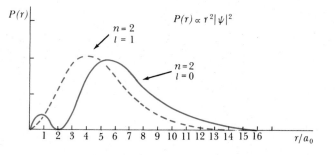

Figure 7-8
Radial probability density $P(r)$ versus r/a_0 for the $n = 2$ states in hydrogen. $P(r)$ for $l = 1$ has a maximum at the Bohr value 2^2a_0. For $l = 0$ there is a maximum near this value and a smaller submaximum near the origin.

In the first excited state $n = 2$, and l can be either 0 or 1. For $l = 0$, $m = 0$ and again we have a spherically symmetric wave function, given by

$$\psi_{200} = C_{200} \left(2 - \frac{Zr}{a_0} \right) e^{-Zr/2a_0} \qquad\qquad 7\text{-}19$$

For $l = 1$, m can be $+1$, 0, or -1. The corresponding wave functions are [see Appendix C for $f_{l|m|}(\theta)$]

$$\psi_{210} = C_{210} \frac{Zr}{a_0} e^{-Zr/2a_0} \cos\theta \qquad\qquad 7\text{-}20$$

$$\psi_{21\pm1} = C_{211} \frac{Zr}{a_0} e^{-Zr/2a_0} \sin\theta \, e^{\pm i\phi} \qquad\qquad 7\text{-}21$$

Figure 7-8 shows $P(r)$ for these wave functions. The distribution for $n = 2$, $l = 1$ is maximum at the second Bohr radius,

$$r_m = 2^2a_0$$

while for $n = 2$ and $l = 0$, $P(r)$ has two maxima, the larger of which is near this radius.

Radial probability distributions can be obtained in the same way for the other excited states of hydrogen. The main radial dependence of $P(r)$ is contained in the factor e^{-Zr/na_0}, except near the origin. A detailed examination of the Laguerre polynomials shows that $\psi \rightarrow r^l$ as $r \rightarrow 0$. Thus, for a given n, ψ_{nlm} is greatest near the origin when l is small.

An important feature of these wave functions is that for $l = 0$, the probability densities are spherically symmetric, whereas for $l \neq 0$ they depend on the angle θ. The probability density plots of Figure 7-9 illustrate this result for the first excited state $n = 2$. These angular distributions of the electron charge density depend only on the value of l and not on the radial part of the wave function. Similar charge distributions for the valence electrons in more complicated atoms play an important role in the chemistry of molecular bonding.

Question

4. At what value of r is $\psi^*\psi$ maximum for the ground state of hydrogen? Why is $P(r)$ maximum at a different value of r?

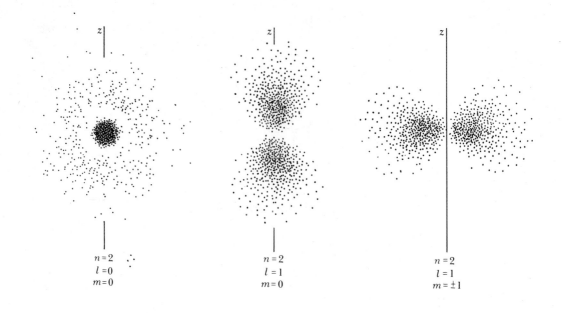

$n = 2$
$l = 0$
$m = 0$

$n = 2$
$l = 1$
$m = 0$

$n = 2$
$l = 1$
$m = \pm 1$

7-4   Electron Spin

As was mentioned in Chapter 4, when a spectral line of hydrogen or other atoms is viewed with high resolution it shows a *fine structure;* that is, it is seen to be split into two or more closely-spaced lines. Although Sommerfeld's relativistic calculation based on the Bohr model agrees with the experimental measurements of this fine structure for hydrogen, it predicts fewer lines than are seen for other atoms. In order to explain fine structure and to clear up a major difficulty with the quantum-mechanical explanation of the periodic table (Section 7-6), W. Pauli in 1925 suggested that in addition to the quantum numbers n, l, and m the electron has a fourth quantum number, which can take on just two values.

As we have seen, quantum numbers arise from boundary conditions on some coordinate. Pauli originally expected that the fourth quantum number would be associated with the time coordinate in a relativistic theory, but this idea was not pursued. In the same year, S. Goudsmit and G. Uhlenbeck, graduate students at Leiden, suggested that this fourth quantum number was the z component, m_s, of an intrinsic angular momentum of the electron called *spin*. If this intrinsic spin angular momentum is described by a quantum number s like the orbital angular-momentum quantum number l, we expect $2s + 1$ possible values of the z component just as there are $2l + 1$ possible z components of the orbital angular momentum. If m_s is to have only two values, s must be $\frac{1}{2}$. In addition to explaining fine structure and the periodic table, this proposal of electron spin explained an interesting experiment by O. Stern and W. Gerlach in 1922. Before we describe this experiment, we must review the connection

Figure 7-9
Probability densities $\psi^*\psi$ for the $n = 2$ states in hydrogen. The probability is spherically symmetric for $l = 0$. It is proportional to $\cos^2 \theta$ for $l = 1$, $m = 0$, and to $\sin^2 \theta$ for $l = 1$, $m = \pm 1$. The probability densities have rotational symmetry about the z axis. The shapes of these distributions are typical for all atoms in S states ($l = 0$) and P states ($l = 1$) and play an important role in molecular bonding. (*This computer-generated plot courtesy of Paul Doherty, Oakland University.*)

Spin

between the angular momentum and the magnetic moment of a charged system.

If a system of charged particles is rotating, it has a *magnetic moment* proportional to its angular momentum. This result is sometimes known as the *Larmor theorem*. Consider a particle of mass M and charge q moving in a circle of radius r with speed v and frequency $f = v/2\pi r$. The angular momentum of the particle is $L = Mvr$. The magnetic moment of a current loop is the product of the current and the area of the loop. For a circulating charge, the current is the charge times the frequency,

$$i = qf = \frac{qv}{2\pi r}$$

and the magnetic moment μ is

$$\mu = iA = q\left(\frac{v}{2\pi r}\right)(\pi r^2) = \frac{1}{2}qvr = \frac{1}{2}q\left(\frac{L}{M}\right) \qquad 7\text{-}22$$

From Figure 7-10 we see that, if q is positive, the magnetic moment is in the same direction as the angular momentum. Then Equation 7-22 can be written as a vector equation:

$$\boldsymbol{\mu} = \frac{q}{2M}\mathbf{L} \qquad 7\text{-}23$$

Equation 7-23, which we have derived for a single particle moving in a circle, also holds for a system of particles in any type of motion if the charge-to-mass ratio q/M is the same for each particle in the system.

Applying this result to the hydrogen atom, we have for the magnitude and z component of the magnetic moment

$$\mu = \frac{e}{2m_e}L = \frac{e\hbar}{2m_e}\sqrt{l(l+1)} = \sqrt{l(l+1)}\,\mu_B \qquad 7\text{-}24$$

and

$$\mu_z = -\frac{e\hbar}{2m_e}m = -m\mu_B \qquad 7\text{-}25$$

where m_e is the mass of the electron,[1] $m\hbar$ is the z component of the angular momentum, and μ_B is a natural unit of magnetic moment called a *Bohr magneton*, which has the value

$$\mu_B = \frac{e\hbar}{2m_e} = 9.27 \times 10^{-24}\text{ joule/tesla}$$

$$= 5.79 \times 10^{-9}\text{ eV/gauss} \qquad 7\text{-}26 \qquad Bohr\ magneton$$

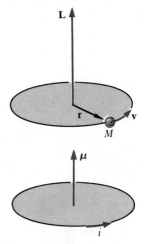

Figure 7-10
A particle moving in a circle has angular momentum **L**. If the particle has a positive charge, the magnetic moment due to the current is parallel to **L**.

---

[1] Since the same symbol μ is used for both the reduced mass and the magnetic moment, some care is needed to keep these unrelated concepts clear. The symbol m is sometimes used to designate the magnetic moment, but there is confusion enough with that symbol for the z component of angular momentum and with m_e as the electron mass.

There is a minus sign in Equation 7-25 because the electron has a negative charge $-e$. The magnetic moment and the angular momentum vectors are therefore oppositely directed. We see that quantization of angular momentum implies quantization of magnetic moments.

The behavior of a system with a magnetic moment in a magnetic field can be visualized by considering a small bar magnet (Figure 7-11). When placed in an external magnetic field **B** there is a torque $\boldsymbol{\tau} = \boldsymbol{\mu} \times \mathbf{B}$ that tends to align the magnet with the field **B**. If the magnet is spinning about its axis, the effect of the torque is to make the spin axis precess about the direction of the external field.

To change the orientation of the magnet relative to the applied field direction, work must be done on it. If it moves through angle $d\theta$, the work required is

$$dW = \tau \, d\theta = \mu B \sin \theta \, d\theta = d(-\mu B \cos \theta)$$
$$= d(-\boldsymbol{\mu} \cdot \mathbf{B})$$

The potential energy of the system can thus be written

$$U = -\boldsymbol{\mu} \cdot \mathbf{B} \qquad \text{7-27}$$

If **B** is in the z direction, the potential energy is

$$U = -\mu_z B \qquad \text{7-28}$$

If the magnetic field is not homogeneous, the force on one pole will be greater or less than that on the other, depending on the orientation; and so there will be a net force on the magnet. This effect was used by Stern and Gerlach in 1922 to measure the possible orientations in space of the magnetic moments of silver atoms. The experiment was repeated in 1927 by Phipps and Taylor using hydrogen atoms.

The experimental setup is shown in Figure 7-12. Atoms from an oven are collimated and sent through a magnet whose poles are shaped so that the magnetic field B_z increases slightly with z. The atoms then strike a collector plate. Figure 7-13 illustrates the effect of such a field on several bar magnets of different orientations. In addition to the torque, which merely causes the magnetic moment to precess about the field direction, there is a net force in the positive or negative z direction. This force deflects the magnet up or down by an amount that depends on the gradient of the field component B_z and on the z component of the magnetic moment μ_z. Classically, one would expect a continuum of deflections corresponding to the continuum of possible orientations of the magnetic moments. However, since the magnetic moment is quantized, quantum mechanics predicts that μ_z can have only the $2l + 1$ values corresponding to the $2l + 1$ possible values of m. We therefore expect $2l + 1$ deflections (counting 0 as a deflection). For example, for $l = 0$ there should be one line on the collector plate corresponding to no deflection, and for $l = 1$ there should be three lines corresponding to the three values $m = -1$, $m = 0$, and $m = +1$.

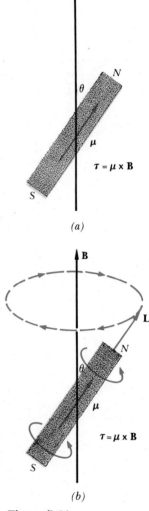

Figure 7-11
Bar-magnet model of magnetic moment. (*a*) In an external magnetic field, the moment experiences a torque which tends to align it with the field. If the magnet is spinning (*b*), the torque causes the system to precess around the external field.

Stern-Gerlach experiment

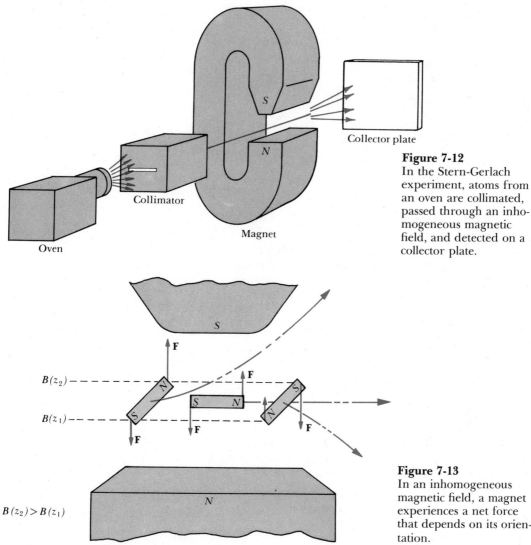

Figure 7-12
In the Stern-Gerlach experiment, atoms from an oven are collimated, passed through an inhomogeneous magnetic field, and detected on a collector plate.

Figure 7-13
In an inhomogeneous magnetic field, a magnet experiences a net force that depends on its orientation.

When the experiment was done with either silver or hydrogen atoms, there were two lines. Since the ground state of hydrogen has $l = 0$ we should expect only one line were it not for the electron spin. If the electron has spin angular momentum of magnitude $\sqrt{s(s + 1)}\,\hbar$, where $s = \frac{1}{2}$, the z component can be either $+\frac{1}{2}\hbar$ or $-\frac{1}{2}\hbar$. Since the orbital angular momentum is zero, the total angular momentum of the atom is simply the spin.[1]

---

[1] The nucleus of an atom also has angular momentum and therefore a magnetic moment; but the mass of the nucleus is about 2000 times that of the electron for hydrogen, and greater still for other atoms. From Equation 7-23 we expect the magnetic moment of the nucleus to be on the order of $\frac{1}{2000}$ of a Bohr magneton since M is now m_p rather than m_e. This small effect does not show up in the Stern-Gerlach experiment.

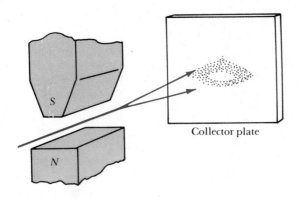

Figure 7-14
Results of the Stern-
Gerlach experiment. The
atomic beam is split into
two lines, indicating that
the magnetic moments of
the atoms are quantized
in space to two orienta-
tions. The shape of the
upper line is due to the
greater inhomogeneity of
the magnetic field near
the upper-pole face.

Collector plate

The quantization of the magnetic moment of the electron to two orientations in space is called *space quantization*. Figure 7-14 shows a sketch of the pattern observed by Stern and Gerlach. From quantitative measurement of the deflection, the magnitude of the magnetic moment due to the spin angular momentum can be determined. The result is *not* $\frac{1}{2}$ Bohr magneton, as predicted by Equation 7-25 with $m = m_s = \frac{1}{2}$, but twice this value. (This type of experiment is not an accurate way to measure magnetic moments, although the measurement of angular momentum is accurate because it involves only counting the number of lines.) This result, and the fact that s is not an integer like the orbital quantum number l, makes it clear that the classical model of the electron as a spinning ball is not to be taken literally. Like the Bohr model of the atom, the classical picture is useful in describing results of quantum-mechanical calculations, and it often gives useful guidelines as to what to expect from an experiment. It is customary to write the relation between the z component of the angular momentum, J_z, and the z component of the magnetic moment, μ_z, as

$$\mu_z = -g\left(\frac{e\hbar}{2m_e}\right)\frac{J_z}{\hbar} \qquad 7\text{-}29$$

where g is called the *gyromagnetic ratio*, which has the values $g_l = 1$ for orbital angular momentum and $g_s = 2$ for spin. More precise measurements indicate that $g_s = 2.00232$. The phenomenon of spin and the value of g_s are predicted by the Dirac relativistic wave equation.

Our description of the hydrogen-atom wave functions in the previous section is not complete because we did not include the spin of the electron. The hydrogen-atom wave functions are also characterized by the spin quantum number m_s, which can be $+\frac{1}{2}$ or $-\frac{1}{2}$. (We need not include the quantum number s because it always has the value $s = \frac{1}{2}$.) A general wave function is then written $\psi_{nlm_lm_s}$ where we have included the subscript l on m_l to distinguish it from m_s. There are now two wave functions for the ground state of the hydrogen atom, $\psi_{100+1/2}$ and $\psi_{100-1/2}$, corresponding to an atom with its electron spin "parallel" or "antiparallel" to the z axis (as defined by an external magnetic field).

In general, the ground state of a hydrogen atom is a linear combination of these wave functions:

$$\psi = C_1 \psi_{100+1/2} + C_2 \psi_{100-1/2}$$

The probability of measuring $m_s = +\frac{1}{2}$ (for example, by observing to which spot the atom goes in the Stern-Gerlach experiment) is $|C_1|^2$. Unless the atoms have been preselected in some way (such as by passing them through a previous inhomogeneous magnetic field), $|C_1|^2$ and $|C_2|^2$ will each be $\frac{1}{2}$, so that measuring the spin "up" ($m_s = +\frac{1}{2}$) and measuring the spin "down" ($m_s = -\frac{1}{2}$) are equally likely.

Questions

5. Does a system have to have a net charge to have a magnetic moment?

6. Consider the two beams of hydrogen atoms emerging from the magnetic field in the Stern-Gerlach experiment. How does the wave function for an electron in one beam differ from that of an electron in the other beam? How does it differ from the wave function for an electron in the incoming beam before passing through the magnetic field?

7-5 Addition of Angular Momenta and the Spin-Orbit Effect

In general an electron in an atom has both orbital angular momentum characterized by the quantum number l and spin angular momentum characterized by the quantum number s. Analogous classical systems which have two kinds of angular momentum are the earth, which is spinning about its axis of rotation in addition to revolving about the sun, or a precessing gyroscope, which has angular momentum of precession in addition to its spin. Classically the total angular momentum

$$\mathbf{J} = \mathbf{L} + \mathbf{S} \qquad\qquad 7\text{-}30$$

is an important quantity because the resultant torque on a system equals the rate of change of the total angular momentum, and in the case of central forces, the total angular momentum is conserved. For a classical system, the magnitude of the total angular momentum J can have any value between $L + S$ and $L - S$. We have already seen that in quantum mechanics, angular momentum is more complicated; both $\mathbf{L}$ and $\mathbf{S}$ are quantized and their directions are restricted. The quantum-mechanical rules for combining orbital and spin angular momenta or any two angular momenta (such as for two particles) are somewhat difficult to derive, but they are not difficult to understand. For the case of orbital and spin angular momenta, the total angular momentum $\mathbf{J}$ has the magnitude

$\sqrt{j(j + 1)}\ \hbar$, where the quantum number j can be either

$$j = l + s$$
or
$$j = l - s \quad (l \neq 0)$$
7-31

(If $l = 0$, the total angular momentum is simply the spin, and $j = s$.) Figure 7-15 is a vector model illustrating the two possible combinations $j = l + \frac{1}{2} = \frac{3}{2}$ and $j = l - \frac{1}{2} = \frac{1}{2}$ for the case of an electron with $l = 1$. The lengths of the vectors are proportional to $\sqrt{l(l + 1)}$, $\sqrt{s(s + 1)}$, and $\sqrt{j(j + 1)}$. The spin and orbital angular-momentum vectors are said to be "parallel" when $j = l + s$ and "antiparallel" when $j = l - s$.

Equation 7-31 is a special case of a more general rule for combining two angular momenta which is useful when dealing with more than one particle. For example, there are two electrons in the helium atom, each with spin, orbital, and total angular momentum. The general rule is:

If $\mathbf{J}_1$ is one angular momentum (orbital, spin, or a combination) and $\mathbf{J}_2$ is another, the resulting total angular momentum $\mathbf{J} = \mathbf{J}_1 + \mathbf{J}_2$ has the value $\sqrt{j(j + 1)}\ \hbar$ for its magnitude, where j can be any of the values

$$j_1 + j_2, j_1 + j_2 - 1, \ldots, |j_1 - j_2|$$
7-32

Example 7-2 Two electrons each have zero orbital angular momentum. What are the possible quantum numbers for the total angular momentum of the two-electron system? In this case $j_1 = j_2 = \frac{1}{2}$. Equation 7-32 then gives two possible results, $j = 1$ and $j = 0$. These results are commonly called parallel spins and antiparallel spins.

In spectroscopic notation, the total angular-momentum quantum number of an atomic state is written as a subscript after a code letter describing its orbital angular momentum. For example, the ground state of hydrogen is written $1S_{1/2}$ where the one indicates the value of n. The $n = 2$ states can have either $l = 0$ or $l = 1$. These states are thus denoted by $2S_{1/2}$, $2P_{3/2}$, and $2P_{1/2}$.

Atomic states with the same n and l values but different j values have slightly different energies because of the interaction of the spin of the electron with its orbital motion. This effect is called the *spin-orbit effect*. The resulting splitting of the spectral line such as the one that results from the transition $2P \rightarrow 1S$ in hydrogen is called *fine-structure splitting*. We can understand the spin-orbit effect qualitatively from a simple Bohr-model picture, as shown in Figure 7-16. In this picture, the electron moves in a circular orbit with speed v around a fixed proton. In the figure, the orbital angular momentum $\mathbf{L}$ is up. In the frame of reference of the electron, the proton moves in a circle around it, thus constituting a circular loop current which produces a magnetic field $\mathbf{B}$ at the position of the electron. The direction of $\mathbf{B}$ is also

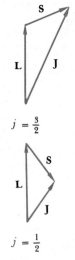

Figure 7-15
Vector model illustrating the addition of orbital and spin angular momenta for the case $l = 1$ and $s = \frac{1}{2}$. There are two possible values of the quantum number for the total angular momentum: $j = l + s = \frac{3}{2}$ and $j = l - s = \frac{1}{2}$.

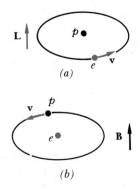

Figure 7-16
(*a*) An electron moving about a proton with angular momentum $\mathbf{L}$ up. The magnetic field seen by the electron due to the apparent relative motion of the proton (*b*) is also up. When the electron spin is parallel to $\mathbf{L}$, the magnetic moment is antiparallel to $\mathbf{L}$ and to $\mathbf{B}$ so the spin-orbit energy is greatest.

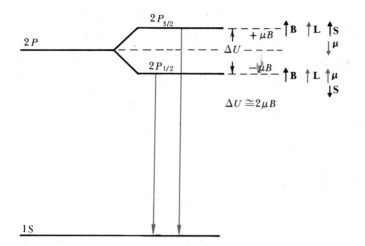

Figure 7-17
Fine-structure energy-level diagram. Because of the spin-orbit force, the $2P$ level is split into two energy levels, with the $j = \frac{3}{2}$ level having slightly greater energy than the $j = \frac{1}{2}$ level. The spectral line due to the transition $2P \to 1S$ is thereby split into two lines of slightly different wavelengths.

up, parallel to **L**. The energy of a magnetic moment in a magnetic field depends on its orientation. Since the torque tends to align the moment parallel to **B** (Figure 7-11), the energy is lowest when the magnetic moment is parallel to the field and highest when it is antiparallel. Since the magnetic moment of the electron is directed opposite to its spin (due to the negative charge of the electron), the energy is lowest when the spin is antiparallel to **B** and thus to **L**. The energy of the $2P_{1/2}$ state in hydrogen, in which **l** and **s** are "antiparallel," is therefore slightly lower than the $2P_{3/2}$ state, in which **l** and **s** are parallel (see Figure 7-17).

Optional

We shall now estimate the magnitude of the spin-orbit energy splitting from the Bohr model. We calculate the magnetic field at the center of a circle resulting from a proton moving with speed v in a circle of radius r. The magnetic field due to a current element $i\,dl$ at a distance r along the line perpendicular to the current element is given by the Biot-Savart law as

$$B = k_m \frac{i\,dl}{r^2}$$

where $k_m = 10^{-7}$ T-m/A is the magnetic constant in SI units. For a moving point charge we can replace the current element $i\,dl$ by qv. Then

$$B = k_m \frac{qv}{r^2} \qquad\qquad 7\text{-}33$$

For the $2P$ state in hydrogen, $r = 2^2 a_0 = 2.12 \times 10^{-10}$ m and the speed (which is the same as the speed of the electron in the rest frame of the proton) can be found from the kinetic energy. The total energy is $(-13.6/2^2)$ eV $= -3.4$ eV, which consists of -6.8 eV potential energy and $+3.4$ eV kinetic energy. Using $\frac{1}{2}m_e v^2 = 3.4$ eV gives $v = 1.1 \times 10^6$ m/sec. Putting these numbers into Equation 7-33 gives

$$B = \frac{(10^{-7})(1.6 \times 10^{-19})(1.1 \times 10^6)}{(2.12 \times 10^{-10})^2} = 0.4 \text{ T}$$

$$= 4 \times 10^3 \text{ G}$$

The potential energy of a magnetic moment $\boldsymbol{\mu}$ in a magnetic field is given by Equation 7-27:

$$U = -\boldsymbol{\mu} \cdot \mathbf{B} = -\mu_z B$$

where we have taken the z direction as that of $\mathbf{B}$. The difference in energy between parallel and antiparallel orientations is therefore of the order of $2\mu_z B$. Since the magnetic moment of the electron is a Bohr magneton μ_B, this energy difference is of the order of

$$\Delta U \approx 2\mu_B B = 2(5.79 \times 10^{-9} \text{ eV/G})(4 \times 10^3 \text{ G})$$
$$= 4.6 \times 10^{-5} \text{ eV}$$

This calculation is rather crude, but the result agrees with the measured splitting of about 4.5×10^{-5} eV for the $2P_{1/2}$ and $2P_{3/2}$ levels in hydrogen. For other atoms, the fine-structure splitting is larger than this. For example, for sodium it is about 2×10^{-3} eV, as will be discussed in Section 7-7.

It is left as an exercise to show that the result we calculated numerically from the Bohr model can be expressed as

$$\Delta U = 2\mu_B B = \frac{2Z^2\alpha^2}{n^3} |E_n| \qquad 7\text{-}34$$

where E_n is the energy of the nth orbit and $\alpha = ke^2/\hbar c \approx \frac{1}{137}$ is the fine-structure constant.

7-6   Ground States of Atoms; The Periodic Table

We now consider qualitatively the wave functions and energy levels for atoms more complicated than hydrogen. As we have mentioned, the Schrödinger equation for atoms other than hydrogen cannot be solved exactly because of the interaction of the electrons with each other, so approximate methods must be used. We shall discuss the energies and wave functions for the ground states of atoms in this section, and consider the excited states and spectra for some of the less complicated cases in the next two sections. To a good approximation we can describe the wave function for a complex atom in terms of single-particle wave functions obtained by neglecting the interaction energy of the electrons. These wave functions are similar to those of the hydrogen atom and are characterized by the quantum numbers n, l, m_l, and m_s. The energy of an electron is determined mainly by the quantum numbers n (which is related to the radial part of the wave function) and l (which characterizes the orbital angular momentum). The specification of n and l for each electron in an atom is called the *electron configuration*. The specification of l is customarily done by giving a letter rather than the numerical value. The code is

$$s\ p\ d\ f\ g\ h$$

l value: 0   1   2   3   4   5

This is the same code as that used to label the states of the hydrogen atom.[1] The n values are referred to as shells, using another letter code: $n = 1$ is called the K shell, $n = 2$ the L shell, and so on.

An important principle that governs the electron configuration of atoms is the *Pauli exclusion principle,* discussed briefly in Section 6-9.

No two electrons in an atom can be in the same quantum state; i.e., they cannot have the same set of values for the quantum numbers n, l, m_l, and m_s.

Pauli exclusion principle

Helium ($Z = 2$)

The energy of the two electrons in the helium atom consists of the kinetic energy of each electron, a potential energy of the form $-kZe^2/r_i$ for each electron corresponding to its attraction to the nucleus, and a potential energy of interaction V_{int} corresponding to the mutual repulsion of the two electrons. If $\mathbf{r}_1$ and $\mathbf{r}_2$ are the position vectors for the two electrons, V_{int} is given by

$$V_{int} = + \frac{ke^2}{|\mathbf{r}_2 - \mathbf{r}_1|} \qquad 7\text{-}35$$

Because this interaction term mixes the variables of the two electrons, its presence in the Schrödinger equation prevents the separation of the equation into separate equations for each electron. If we neglect the interaction term, however, the Schrödinger equation can be separated and solved exactly. We then obtain separate equations for each electron, with each equation identical to that for the hydrogen atom except that $Z = 2$. The allowed energies are then given by

$$E = - \frac{Z^2 E_0}{n_1{}^2} - \frac{Z^2 E_0}{n_2{}^2} \qquad \text{where } E_0 = 13.6 \text{ eV} \qquad 7\text{-}36$$

George Gamow and Wolfgang Pauli on a Swiss lake in 1930. (*Courtesy of George Gamow.*)

The lowest energy, $E_1 = -2(2)^2 E_0 \approx -108.8$ eV, occurs for $n_1 = n_2 = 1$. For this case, $l_1 = l_2 = 0$. The total wave function, neglecting the spin of the electrons, is of the form

$$\psi_{100}(r_1,\theta_1,\phi_1)\psi_{100}(r_2,\theta_2,\phi_2) \qquad 7\text{-}37$$

The quantum numbers n, l, and m_l can be the same for the two electrons only if the fourth quantum number m_s is different, i.e., if one has $m_s = +\frac{1}{2}$ and the other $m_s = -\frac{1}{2}$. The resultant spin of the two electrons must therefore be zero.

We can obtain a first-order correction to the ground-state energy by using the approximate wave function of Equation 7-37 to calculate the average value of the interaction energy V_{int}, which is simply the expectation value $\langle V_{int} \rangle$. The result of this calculation is

$$\langle V_{int} \rangle = +34 \text{ eV}$$

---

[1] Capital letters are used to specify atomic states; lowercase letters are used for individual electron states.

With this correction, the ground-state energy is

$$E \approx -108.8 + 34 = -74.8 \text{ eV}$$

This approximation method, in which we neglect the interaction of the electrons to find an approximate wave function and then use this wave function to calculate the interaction energy, is called *first-order perturbation theory*. The approximation can be continued to higher orders: for example, the next step is to use the new ground-state energy to find a correction to the ground-state wave function. This approximation method is similar to that used in classical mechanics to calculate the orbits of the planets about the sun. In the first approximation, the interaction of the planets is neglected and the elliptical orbits are found for each planet. Then using this result for the position of each planet, the perturbing effects of the nearby planets can be calculated.

The experimental value of the energy needed to remove both electrons from the helium atom is about 79 eV. The discrepancy between this result and the value 74.8 eV is due to the inaccuracy of the approximation used to calculate $\langle V_{\text{int}} \rangle$, as indicated by the rather large value of the correction (about 30 percent). (It should be pointed out that there are better methods of calculating the interaction energy for helium that give much closer agreement with experiment.) The helium *ion* He^+, formed by removing one electron, is identical to the hydrogen atom except that $Z = 2$; so the ground-state energy is

$$-Z^2(13.6) = -54.4 \text{ eV}$$

The energy needed to remove the first electron from the helium atom is 24.6 eV. The corresponding potential, 24.6 V, is called the *first ionization potential* of the atom.

The configuration of the ground state of the helium atom is written $1s^2$. The 1 signifies $n = 1$, the s signifies $l = 0$, and the 2 signifies that there are two electrons in this state. Since l can only be zero for $n = 1$, the two electrons fill the K shell ($n = 1$).

Lithium (Z = 3)

Lithium has three electrons. Two are in the K shell ($n = 1$), but the third cannot have $n = 1$ because of the exclusion principle. The next-lowest energy state for this electron has $n = 2$. The possible l values are $l = 1$ or $l = 0$.

In the hydrogen atom, these l values have the same energy because of the degeneracy associated with the inverse-square nature of the force. This is not true in lithium and other atoms because the charge "seen" by the outer electron is not a point charge. The positive charge of the nucleus $+Ze$ can be considered to be approximately a point charge, but the negative charge of the K-shell electrons $-2e$ is spread out in space over a volume of radius of the order of a_0/Z. We can in fact take for the charge density of each inner electron $\rho = -e|\psi|^2$, where ψ is a hydrogenlike $1s$ wave function (neglecting the interaction of the two electrons in the K shell). The probability distribution for the outer electron in the $2s$ or $2p$ states is similar to that shown in Figure 7-8. We see that the probability distributions in both cases have a large maximum well outside the inner K-shell electrons, but that the $2s$ distribution also has a small bump near the origin. We could describe this by saying that the electron in the $2p$ state is nearly always outside the shielding of the two electrons in the K shell so that it sees an effective central charge of $Z_{eff} \approx 1$; whereas in the $2s$ state the electron penetrates this "shielding" more often, and therefore sees a slightly larger effective positive central charge. The energy of the outer electron is therefore lower in the $2s$ state than in the $2p$ state, and the lowest-energy configuration of the lithium atom is $1s^2 2s$.

The total angular momentum of the electrons in this atom is $\frac{1}{2}$ due to the spin of the outer electron, since each of the electrons has zero orbital angular momentum, and the inner K-shell electrons are paired to give zero spin. The first ionization potential for lithium is only 5.39 V. We can use this result to calculate the effective positive charge seen by the $2s$ electron. For $Z = Z_{eff}$ and $n = 2$ we have

$$E = -\frac{Z^2 E_0}{n^2} = -\frac{Z_{eff}^2 \,(13.6 \text{ eV})}{2^2} = -5.39 \text{ eV}$$

which gives $Z_{eff} \approx 1.3$. It is generally true that the smaller the value of l, the greater the penetration of the wave function into the inner shielding cloud of electrons: The result is that for given n, the energy of the electron increases with increasing l.

Beryllium (Z = 4)

The fourth electron has the least energy in the $2s$ state. The exclusion principle requires that its spin be antiparallel to the other electron in this state, so that the angular momentum of the four electrons in this atom is 0. The electron configuration of beryllium is $1s^2 2s^2$. The first ionization potential is 9.32 V. This is greater than that for lithium because of the greater value of Z.

Boron to Neon (Z = 5 to Z = 10)

Since the $2s$ subshell is filled, the fifth electron must go into the $2p$ subshell, that is, $n = 2$ and $l = 1$. Since there are three possible values of m_l (+1, 0, and −1) and two values of m_s for each, there can be six electrons in this subshell. The electron configuration for boron is $1s^2 2s^2 2p$. Although it might be expected that boron would have a greater ionization potential than beryllium because of the greater Z, the $2p$ wave function penetrates the shielding of the core electrons to a lesser extent and the ionization potential of boron is actually about 8.3 V, slightly less than that of beryllium. The electron configuration of the elements carbon ($Z = 6$) to neon ($Z = 10$) differs from boron only by the number of electrons in the $2p$ subshell. The ionization potential increases slightly with Z for these elements, reaching the value of 21.6 V for the last element in the group, neon. Neon has the maximum number of electrons allowed in the $n = 2$ shell. The electron configuration of neon is $1s^2 2s^2 2p^6$. Because of its very high ionization potential, neon, like helium, is chemically inert. The element just before this, fluorine, has a "hole" in this shell, that is, it has room for one more electron. It readily combines with elements such as lithium, which has one outer electron that is donated to the fluorine atom to make a F$^-$ ion and a Li$^+$ ion, which bond together. This is an example of ionic bonding, to be discussed in the next chapter.

Sodium to Argon ($Z = 11$ to $Z = 18$)

The eleventh electron must go into the $n = 3$ shell. Since this electron is weakly bound in the Na atom, Na combines readily with atoms such as F. The ionization potential for sodium is only 5.14 V. Because of the lowering of the energy due to penetration of the electronic shield formed by the other 10 electrons—similar to that discussed for Li—the $3s$ state is lower than the $3p$ or $3d$ states. (With $n = 3$, l can have the values 0, 1, or 2.) This energy difference between subshells of the same n value becomes greater as the number of electrons increases. The configuration of Na is thus $1s^2 2s^2 2p^6 3s^1$. As we move to higher Z elements, the $3s$ subshell and then the $3p$ subshell begin to fill up. These two subshells can accommodate $2 + 6 = 8$ electrons. The configuration of argon ($Z = 18$) is $1s^2 2s^2 2p^6 3s^2 3p^6$. One might expect that the nineteenth electron would go into the third subshell; but the shielding or penetration effect is now so strong that the energy is lower in the $4s$ shell than in the $3d$ shell. There is another large energy difference between the eighteenth and nineteenth electrons, and argon, with its full $3p$ subshell is stable and inert.

Atoms with $Z > 18$

The nineteenth electron in potassium ($Z = 19$) and the twentieth electron in calcium ($Z = 20$) go into the $4s$ rather than the $3d$ subshell. The electron configurations of the next 10 elements, scandium ($Z = 21$) through zinc ($Z = 30$), differ only in the number of electrons in the $3d$ shell except for chromium ($Z = 24$) and copper ($Z = 29$), each of which has only one $4s$ electron. These elements are called *transition elements*. Since their chemical properties are mainly due to their $4s$ electrons, they are quite similar chemically.

Figure 7-18 shows a plot of the first ionization potential of an atom versus Z up to $Z = 60$. The sudden decrease in ionization potential after the Z numbers 2, 10, 18, 36, and 54 mark the closing of a shell or subshell.

Table 7-1 gives the electron configuration of all the elements.

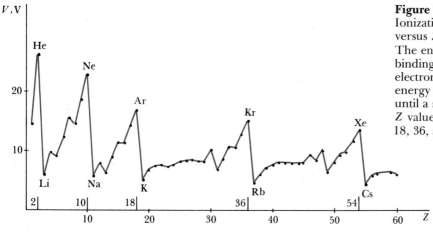

Figure 7-18
Ionization potential V versus Z up to $Z = 60$. The energy eV is the binding energy of the last electron in the atom. This energy increases with Z until a shell is closed at Z values of 2, 10, 18, 36, and 54.

Questions

7. Why is the energy of the $3s$ state considerably lower than that of the $3p$ state for sodium, whereas in hydrogen these states have essentially the same energy?

8. Discuss the evidence from the periodic table of the need for a fourth quantum number. How would the properties of He differ if there were only three quantum numbers, n, l, and m?

7-7 Excited States and Spectra of Alkali Atoms

In order to understand atomic spectra we need to understand the excited states of atoms. The situation for an atom with many electrons is, in general, much more complicated than that of hydrogen. An excited state of the atom may involve a change in the state of any one of the electrons, or even two or more electrons. Even in the case of the excitation of only one electron, the change in the state of this electron changes the energies of the others. Fortunately, there are many cases in which this effect is negligible, and the energy levels can be calculated accurately from a relative simple model of one electron plus a stable core. This model works particularly well for the alkali metals: Li, Na, K, Rb, and Cs. These elements are in the first column of the periodic table (Appendix E). The optical spectra of these elements are similar to that of hydrogen.

Another simplification is possible because of the wide difference in energies between the excitation of a core electron and the excitation of an outer electron. Consider the case of sodium, which has a neon core (except with $Z = 11$ rather than 10) and an outer $3s$ electron. If this electron did not penetrate the core, it would see an effective nuclear charge of $Z_{eff} = 1$. The ionization energy would be the same as the energy of the $n = 3$ electron in hydrogen, about 1.5 eV. Penetration into the core increases Z_{eff} and so lowers the energy of the outer electron,

Table 7-1

Electron configurations of the atoms in their ground states. For some of the rare-earth elements ($Z = 57$ to 71) and the heavy elements ($Z > 89$) the configurations are not firmly established.

			K	L		M			N				O				P			Q
		n:	1	2		3			4				5				6			7
Z	Element	*l:*	*s*	*s*	*p*	*s*	*p*	*d*	*s*	*p*	*d*	*f*	*s*	*p*	*d*	*f*	*s*	*p*	*d*	*s*
1	H hydrogen		1																	
2	He helium		2																	
3	Li lithium		2	1																
4	Be beryllium		2	2																
5	B boron		2	2	1															
6	C carbon		2	2	2															
7	N nitrogen		2	2	3															
8	O oxygen		2	2	4															
9	F fluorine		2	2	5															
10	Ne neon		2	2	6															
11	Na sodium		2	2	6	1														
12	Mg magnesium		2	2	6	2														
13	Al aluminum		2	2	6	2	1													
14	Si silicon		2	2	6	2	2													
15	P phosphorus		2	2	6	2	3													
16	S sulfur		2	2	6	2	4													
17	Cl chlorine		2	2	6	2	5													
18	Ar argon		2	2	6	2	6													
19	K potassium		2	2	6	2	6	.	1											
20	Ca calcium		2	2	6	2	6	.	2											
21	Sc scandium		2	2	6	2	6	1	2											
22	Ti titanium		2	2	6	2	6	2	2											
23	V vanadium		2	2	6	2	6	3	2											
24	Cr chromium		2	2	6	2	6	5	1											
25	Mn manganese		2	2	6	2	6	5	2											
26	Fe iron		2	2	6	2	6	6	2											
27	Co cobalt		2	2	6	2	6	7	2											
28	Ni nickel		2	2	6	2	6	8	2											
29	Cu copper		2	2	6	2	6	10	1											
30	Zn zinc		2	2	6	2	6	10	2											
31	Ga gallium		2	2	6	2	6	10	2	1										
32	Ge germanium		2	2	6	2	6	10	2	2										
33	As arsenic		2	2	6	2	6	10	2	3										
34	Se selenium		2	2	6	2	6	10	2	4										
35	Br bromine		2	2	6	2	6	10	2	5										
36	Kr krypton		2	2	6	2	6	10	2	6										

Table 7-1 (Continued)

Z	Element		K ⟨n:1, l:s⟩	L ⟨n:2⟩ s	p	M ⟨n:3⟩ s	p	d	N ⟨n:4⟩ s	p	d	f	O ⟨n:5⟩ s	p	d	f	P ⟨n:6⟩ s	p	d	Q ⟨n:7⟩ s
37	Rb	rubidium	2	2	6	2	6	10	2	6	.	.	1							
38	Sr	strontium	2	2	6	2	6	10	2	6	.	.	2							
39	Y	yttrium	2	2	6	2	6	10	2	6	1	.	2							
40	Zr	zirconium	2	2	6	2	6	10	2	6	2	.	2							
41	Nb	niobium	2	2	6	2	6	10	2	6	4	.	1							
42	Mo	molybdenum	2	2	6	2	6	10	2	6	5	.	1							
43	Tc	technetium	2	2	6	2	6	10	2	6	6	.	1							
44	Ru	ruthenium	2	2	6	2	6	10	2	6	7	.	1							
45	Rh	rhodium	2	2	6	2	6	10	2	6	8	.	1							
46	Pd	palladium	2	2	6	2	6	10	2	6	10	.	.							
47	Ag	silver	2	2	6	2	6	10	2	6	10	.	1							
48	Cd	cadmium	2	2	6	2	6	10	2	6	10	.	2							
49	In	indium	2	2	6	2	6	10	2	6	10	.	2	1						
50	Sn	tin	2	2	6	2	6	10	2	6	10	.	2	2						
51	Sb	antimony	2	2	6	2	6	10	2	6	10	.	2	3						
52	Te	tellurium	2	2	6	2	6	10	2	6	10	.	2	4						
53	I	iodine	2	2	6	2	6	10	2	6	10	.	2	5						
54	Xe	xenon	2	2	6	2	6	10	2	6	10	.	2	6						
55	Cs	cesium	2	2	6	2	6	10	2	6	10	.	2	6	.	.	1		
56	Ba	barium	2	2	6	2	6	10	2	6	10	.	2	6	.	.	2		
57	La	lanthanum	2	2	6	2	6	10	2	6	10	.	2	6	1	.	2		
58	Ce	cerium	2	2	6	2	6	10	2	6	10	1	2	6	1	.	2		
59	Pr	praseodymium	2	2	6	2	6	10	2	6	10	3	2	6	.	.	2		
60	Nd	neodymium	2	2	6	2	6	10	2	6	10	4	2	6	.	.	2		
61	Pm	promethium	2	2	6	2	6	10	2	6	10	5	2	6	.	.	2		
62	Sm	samarium	2	2	6	2	6	10	2	6	10	6	2	6	.	.	2		
63	Eu	europium	2	2	6	2	6	10	2	6	10	7	2	6	.	.	2		
64	Gd	gadolinium	2	2	6	2	6	10	2	6	10	7	2	6	1	.	2		
65	Tb	terbium	2	2	6	2	6	10	2	6	10	9	2	6	.	.	2		
66	Dy	dysprosium	2	2	6	2	6	10	2	6	10	10	2	6	.	.	2		
67	Ho	holmium	2	2	6	2	6	10	2	6	10	11	2	6	.	.	2		
68	Er	erbium	2	2	6	2	6	10	2	6	10	12	2	6	.	.	2		
69	Tm	thulium	2	2	6	2	6	10	2	6	10	13	2	6	.	.	2		
70	Yb	ytterbium	2	2	6	2	6	10	2	6	10	14	2	6	.	.	2		
71	Lu	lutetium	2	2	6	2	6	10	2	6	10	14	2	6	1	.	2		
72	Hf	hafnium	2	2	6	2	6	10	2	6	10	14	2	6	2	.	2		
73	Ta	tantalum	2	2	6	2	6	10	2	6	10	14	2	6	3	.	2		
74	W	tungsten (wolfram)	2	2	6	2	6	10	2	6	10	14	2	6	4	.	2		
75	Re	rhenium	2	2	6	2	6	10	2	6	10	14	2	6	5	.	2		

(Continued)

Table 7-1 (Continued)

		K	L		M			N				O				P			Q
		n: 1	2		3			4				5				6			7
Z	Element	l: s	s	p	s	p	d	s	p	d	f	s	p	d	f	s	p	d	s
76	Os osmium	2	2	6	2	6	10	2	6	10	14	2	6	6	.	2			
77	Ir iridium	2	2	6	2	6	10	2	6	10	14	2	6	7	.	2			
78	Pt platinum	2	2	6	2	6	10	2	6	10	14	2	6	9	.	1			
79	Au gold	2	2	6	2	6	10	2	6	10	14	2	6	10	.	1			
80	Hg mercury	2	2	6	2	6	10	2	6	10	14	2	6	10	.	2			
81	Tl thallium	2	2	6	2	6	10	2	6	10	14	2	6	10	.	2	1		
82	Pb lead	2	2	6	2	6	10	2	6	10	14	2	6	10	.	2	2		
83	Bi bismuth	2	2	6	2	6	10	2	6	10	14	2	6	10	.	2	3		
84	Po polonium	2	2	6	2	6	10	2	6	10	14	2	6	10	.	2	4		
85	At astatine	2	2	6	2	6	10	2	6	10	14	2	6	10	.	2	5		
86	Rn radon	2	2	6	2	6	10	2	6	10	14	2	6	10	.	2	6		
87	Fr francium	2	2	6	2	6	10	2	6	10	14	2	6	10	.	2	6	.	1
88	Ra radium	2	2	6	2	6	10	2	6	10	14	2	6	10	.	2	6	.	2
89	Ac actinium	2	2	6	2	6	10	2	6	10	14	2	6	10	.	2	6	1	2
90	Th thorium	2	2	6	2	6	10	2	6	10	14	2	6	10	.	2	6	2	2
91	Pa protactinium	2	2	6	2	6	10	2	6	10	14	2	6	10	1	2	6	2	2
92	U uranium	2	2	6	2	6	10	2	6	10	14	2	6	10	3	2	6	1	2
93	Np neptunium	2	2	6	2	6	10	2	6	10	14	2	6	10	4	2	6	1	2
94	Pu plutonium	2	2	6	2	6	10	2	6	10	14	2	6	10	6	2	6	.	2
95	Am americium	2	2	6	2	6	10	2	6	10	14	2	6	10	7	2	6	.	2
96	Cm curium	2	2	6	2	6	10	2	6	10	14	2	6	10	7	2	6	1	2
97	Bk berkelium	2	2	6	2	6	10	2	6	10	14	2	6	10	8	2	6	1	2
98	Cf californium	2	2	6	2	6	10	2	6	10	14	2	6	10	10	2	6	.	2
99	Es einsteinium	2	2	6	2	6	10	2	6	10	14	2	6	10	11	2	6	.	2
100	Fm fermium	2	2	6	2	6	10	2	6	10	14	2	6	10	12	2	6	.	2
101	Md mendelevium	2	2	6	2	6	10	2	6	10	14	2	6	10	13	2	6	.	2
102	No nobelium	2	2	6	2	6	10	2	6	10	14	2	6	10	14	2	6	.	2
103	Lw lawrencium	2	2	6	2	6	10	2	6	10	14	2	6	10	14	2	6	1	2

thereby increasing the ionization energy. The measured ionization energy of sodium is about 5 eV. The energy needed to remove one of the outermost core electrons, a $2p$ electron, is about 31 eV, whereas that needed to remove one of the $1s$ electrons is about 1041 eV. An electron in the inner core cannot be excited to any of the filled $n = 2$ states because of the exclusion principle. Thus the minimum excitation of an $n = 1$ electron is to the $n = 3$ shell, which requires an energy only slightly less than that needed to remove this electron completely from the atom. Since the energies of photons in the visible range (about 400 to 800 nm) vary only from about 1.5 to 3 eV, the *optical* spectrum of sodium must be due to transitions involving only the outer electron. Transitions involving the core electrons produce line spectra in the x-ray region of the electromagnetic spectrum.

Figure 7-19 shows an energy-level diagram for the optical transitions in sodium. Since the spin angular momentum of the neon core adds up to zero, the spin of each state of sodium is $\frac{1}{2}$. Because of the spin-orbit effect, the states with $j = l - \frac{1}{2}$ have a slightly lower energy than those with $j = l + \frac{1}{2}$. Each state is therefore a doublet (except for the S states). The doublet splitting is very small and is not evident on the energy scale of this diagram. The states are labeled by the usual spectroscopic notation, with the superscript 2 before the letter indicating that the state is a doublet. Thus $^2P_{3/2}$, read as "doublet P three halves," denotes a state in which $l = 1$ and $j = \frac{3}{2}$. (The S states are customarily labeled as if they were doublets even though they are not. The number indicating the n value of the electron is often omitted.) In the first excited state, the outer electron is excited from the $3s$ level to the $3p$ level, which is about 2.1 eV above the ground state. The spin-orbit energy difference between the $P_{3/2}$ and $P_{1/2}$ states due to the spin-orbit effect is about 0.002 eV.

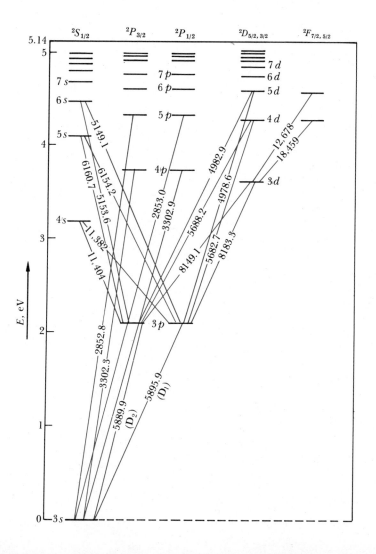

Figure 7-19
Energy-level diagram for sodium with some of the transitions indicated. The energy of the ground state has been chosen as zero for the scale on the left. (*From H. G. Kuhn,* Atomic Spectra, *New York: Academic Press, Inc., 1962.*)

Transitions from these states to the ground state give the familiar sodium yellow doublet

$$3p(^2P_{1/2}) \longrightarrow 3s(^2S_{1/2}) \qquad \lambda = 5896 \text{ Å}$$

$$3p(^2P_{3/2}) \longrightarrow 3s(^2S_{1/2}) \qquad \lambda = 5890 \text{ Å}$$

It is important to distinguish between doublet energy states and doublet spectral lines. All transitions beginning or ending on an S state give double lines because they involve one doublet state and one singlet state (the selection rule $\Delta l = \pm 1$ rules out transitions between two S states). There are four possible energy differences between two doublet states. One of these is ruled out by a selection rule on j, which is[1]

$$\Delta j = \pm 1 \text{ or } 0 \qquad \text{(but not } j = 0 \text{ to } j = 0\text{)} \qquad \text{7-38}$$

Transitions between doublet states therefore result in triplet spectral lines.

The energy levels and spectra of other alkali atoms are similar to those for sodium. Figure 7-20 shows the energy-level diagram for the potassium atom ($Z = 19$), which consists of an argon core plus one outer electron.

[1] We can think of this rule in terms of the conservation of angular momentum. The intrinsic-spin angular momentum of a photon has the quantum number $s = 1$. For electric dipole radiation, the photon spin is its total angular momentum relative to the center of mass of the atom. If the initial angular-momentum quantum number of the atom is j_1 and the final is j_2, the rules for combining angular momenta imply that $j_2 = j_1 + 1$, j_1, or $j_1 - 1$, if $j_1 \neq 0$. If $j_1 = 0$, j_2 must be 1.

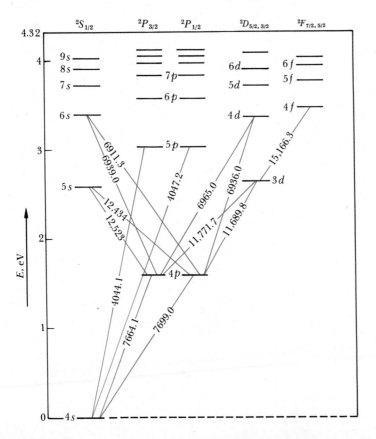

Figure 7-20
Energy-level diagram for potassium with some transitions indicated. (*From H. G. Kuhn,* Atomic Spectra, *New York: Academic Press, Inc., 1962.*)

Optional

7-8   Excited States and Spectra of Two-Electron Atoms

The energy levels and optical spectra are much more complicated for atoms with more than one electron in the outer shell. We shall discuss qualitatively in this section the energy levels for helium and the alkaline earths, atoms in the second column of the periodic table. These atoms all consist of a core of electrons plus two electrons in an outer s shell. Most of the observed spectra can be understood in terms of energy levels corresponding to the raising of one of these electrons to a shell or subshell of higher energy. These are called *normal levels*. Energy levels involving excitation of both outer electrons are called *anomalous* and will not be discussed here.

The model used to calculate the energy levels for these atoms consists of two identical electrons moving in a potential due to the nucleus and the core electrons. Consider magnesium ($Z = 12$) as a specific example. The ground-state electron configuration is $(1s^2 2s^2 2p^6)3s^2$. In the ground state both outer electrons have the same space quantum numbers so the resultant spin must be zero. When one of the electrons is excited to a higher energy state such as $3p$, the spatial quantum numbers are no longer the same so there is no restriction on the spin quantum numbers. The resultant spin S for two particles with spin $s = \frac{1}{2}$ can be either $S = 0$ (antiparallel spins) or $S = 1$ (parallel spins). If $S = 0$, the total angular momentum of the atom is due entirely to the orbital angular momentum of the excited electron, so $j = l$. The $S = 0$ states are called *singlet* states. If $S = 1$, there are three possible values for the total angular momentum: $j = l + 1, j = l,$ or $j = l - 1$ (except if $l = 0$, in which case $j = 1$ is the only possibility). Because of the spin-orbit effect, these three states have slightly different energies, i.e., there is fine-structure splitting. The states with $S = 1$ are therefore called *triplet* states.

Figure 7-21 is an energy-level diagram for magnesium with observed transitions indicated. On the scale of this diagram, the fine-structure splitting of the triplet states is not evident. Note that all but one of the transitions follow the selection rule $\Delta S = 0$. That is, the triplet and singlet states don't mix. The one transition indicated (from the triplet state $3s3p$ to the ground state) that does not obey this selection rule is called an *intercombination line*.

If we examine Figure 7-21 closely we note that the singlet energy levels are higher than the triplet levels with the same electron configuration. For example, consider the states, which have one electron in the $3p$ state. If it were not for the electrostatic interaction of the two electrons, the singlet state written $^1P_1$ ($j = 1$ since $S = 0$ and $l = 1$) and the triplet states written ^3P_j (with $j = 2, 1,$ or 0 for $l = 1$ and $S = 1$) would have the same energy, except for the small fine-structure splitting. Evidently the electrostatic interaction energy of the two electrons is considerably greater in the singlet states than in the triplet states.

Figure 7-21
Energy-level diagram for the two-electron atom magnesium. On this scale the fine-structure separation of the triplet levels is not evident. Note that the energy of each singlet level is greater than that of the corresponding triplet levels. This is because the average separation of the outer electrons is greater in the triplet states than in the singlet states, as indicated in Figure 7-22. (*From H. G. Kuhn,* Atomic Spectra, *New York: Academic Press, Inc., 1962.*)

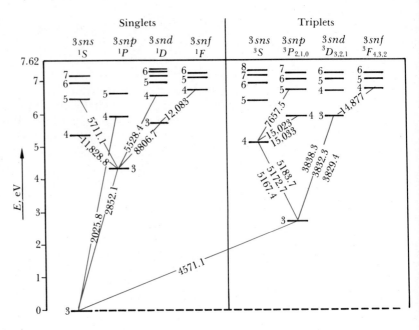

The cause of this energy difference is a rather subtle quantum-mechanical effect that has to do with the symmetry requirements on the total wave function for two identical particles. In Section 6-9 we wrote the wave function for two particles in one dimension, with one in state n and the other in state m, as

$$\psi(x_1,x_2) = C\left[\psi_n(x_1)\psi_m(x_2) \pm \psi_n(x_2)\psi_m(x_1)\right] \qquad 7\text{-}39$$

where the plus sign gives a function that is symmetric on exchange of the particles and the minus sign one that is antisymmetric. We stated that electrons have antisymmetric wave functions. We must now include spin in the wave function. The total wave function for two particles can be written as a product of an ordinary space part and a part that describes the spin. The spin part of the wave function turns out to be symmetric for the $S = 1$ triplet state and antisymmetric for the $S = 0$ singlet state. The *space part* of the wave function must therefore be antisymmetric in the triplet state and symmetric in the singlet state, so that the total wave function is antisymmetric. We note from Equation 7-39 that if $x_1 = x_2$ the antisymmetric space wave function is identically zero. This is an example of a general result illustrated in Figure 7-22 that, in an antisymmetric space state, the particles tend to be farther apart than in a symmetric space state. Since the interaction energy due to the electrostatic repulsion is positive and varies inversely as the separation distance, the energy is greater when the electrons are close together in the space-symmetric singlet state $S = 0$ than it is when the electrons are relatively far apart in the space antisymmetric triplet state. The energy difference is of the order of 1 eV, which is much greater than the fine-structure splitting.[1]

---

[1] This is true for nearly all two-electron atoms, such as He, Be, Mg, and Ca, except for the triplet P states in the very heavy atom mercury, where fine-structure splitting is of about the same order of magnitude as the single-triplet splitting.

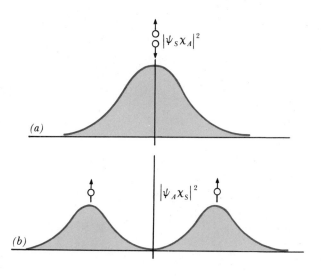

(a)

(b)

Figure 7-22
Probability distribution versus separation for two electrons. (a) In the singlet state, the space part ψ_S of the wave function is symmetric and the spin part χ_A is antisymmetric. (b) In the triplet state, the space part is antisymmetric and the spin part is symmetric. Because the average separation in the triplet state is greater, the energy of the system is lower in this state.

7-9   The Zeeman Effect

As we mentioned in Chapter 3, the splitting of spectral lines when an atom is placed in an external magnetic field was looked for by Faraday and first observed by Zeeman, for whom the effect is now named. Even without any knowledge of the internal structure of an atom, we can calculate classically the magnitude of the frequency change due to the action of the magnetic field on an oscillating charge. The result, which we shall derive below, is $\Delta\omega = \pm\frac{1}{2}(e/m)B$ or $\Delta\omega = 0$, where e is the charge, m the mass of the charge, and B the external field strength.

The observation of spectral lines split into three components by a magnetic field—one of higher frequency, one of lower frequency, and one unchanged—provided one of the earliest estimates of the charge-to-mass ratio of the electron. Though this classical calculation is in agreement with experiment in many cases, it is more usual to find that a spectral line is split into more than three components, with none of the frequency shifts given by the classical calculation. The splitting of a spectral line into three components in a magnetic field is called the *normal Zeeman effect,* whereas the more common splitting into more than three components is called the *anomalous Zeeman effect.* Such terminology is a bit confusing today, for, with the concept of electron spin, both effects are well understood.

We shall first discuss the classical calculation of the frequency shift in the normal Zeeman effect and then derive the energy splitting of the levels in the atom using quantum mechanics. The anomalous Zeeman effect is somewhat more difficult to treat and will be discussed only qualitatively.

Consider a point charge q oscillating along a line that we take to be the x axis. According to classical physics, the charge will radiate at the frequency of oscillation. The motion of the charge can be thought of as the superposition of two circular motions in a plane containing the x axis, one clockwise and one counterclockwise. Both circular motions are in phase and have radius R

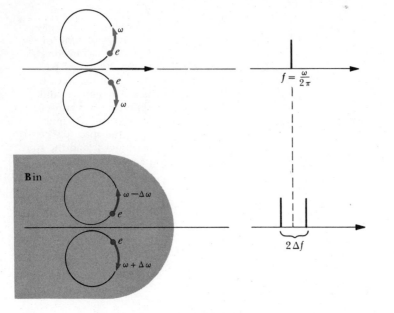

equal to one-half the amplitude of oscillation as in Figure 7-23. For the circular motion, $v = R\omega$ and

$$\frac{mv^2}{R} = mR\omega^2 = F \qquad\qquad 7\text{-}40$$

where F is the force. A magnetic field B perpendicular to the plane of the circle gives an additional force, $qvB = qR\omega B$, which causes ω in one circle to increase and in the other to decrease. Since the changes are small for even the largest B field produced in the laboratory, we can assume that the radius does not change and we can approximate as follows:

$$\Delta F = \Delta(mR\omega^2) \approx 2mR\omega\,\Delta\omega$$

Using $\Delta F = qvB = qR\omega B$, we have

$$qR\omega B = 2mR\omega\,\Delta\omega$$

or

$$\Delta\omega = \frac{1}{2}\frac{q}{m}B \qquad\qquad 7\text{-}41$$

Figure 7-23
Simple harmonic motion of a charge is resolved into two circular motions. The frequency of the emitted light is equal to the frequency of the motion according to classical physics. In the presence of an external magnetic field, one of the circular frequencies is slightly increased and the other is slightly decreased, leading to two new radiated frequencies. Since oscillation parallel to the plane of **B** is unaffected, the original frequency is also seen. The photographs show the spectral lines. (*From H. E. White,* Introduction to Atomic Spectra, *New York: McGraw-Hill Book Company, 1934. Used by permission of the publisher.*)

If the magnetic field is parallel to the plane of the orbits, i.e., parallel to the original line of oscillation, there is no shift, and the original frequency of oscillation persists. The comparison of these predictions with the measured wavelength differences allows a determination of the ratio q/m for the oscillating particle.

In quantum mechanics, a shift in the frequency and wavelength of a spectral line implies a shift in the energy level of one or both of the states involved in the transition. The normal Zeeman effect occurs for spectral lines resulting from a transition between *singlet* states. For these states, the spin is zero and the total angular momentum **J** is equal to the orbital angular momentum **L**. When placed in an external magnetic field the

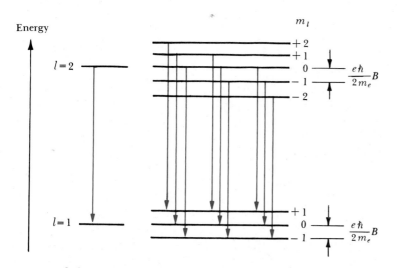

Energy

$l = 2$

m_l

$+2$
$+1$
0 —— $\dfrac{e\hbar}{2m_e}B$
-1
-2

$l = 1$

$+1$
0 —— $\dfrac{e\hbar}{2m_e}B$
-1

Figure 7-24
Energy-level splitting in the normal Zeeman effect for singlet levels $l = 2$ and $l = 1$. Each level is split into $2l + 1$ terms. The nine transitions consistent with the selection rule $\Delta m = 0, \pm 1$, give only three different energies because the energy difference between adjacent terms is $e\hbar B/2m_e$ independent of l.

energy of the atom changes because of the energy of its magnetic moment in the field, which is given by

$$\Delta E = -\boldsymbol{\mu} \cdot \mathbf{B} = -\mu_z B \qquad\qquad 7\text{-}42$$

where the z direction is defined by the direction of $\mathbf{B}$. Using Equation 7-25 for μ_z, we have $\mu_z = -m_l\mu_B = -m_l(e\hbar/2m_e)$, and

$$\Delta E = +m_l \frac{e\hbar}{2m_e} B \qquad\qquad 7\text{-}43$$

Since there are $2l + 1$ values of m_l, each energy level splits into $2l + 1$ levels. Figure 7-24 shows the splitting of the levels for the case of a transition between a state with $l = 2$ and one with $l = 1$. The selection rule $\Delta m_l = \pm 1$ or 0 restricts the number of possible lines to the nine shown.

Because of the uniform splitting of the levels, there are only three different transition energies: $E_0 + e\hbar B/2m_e$, E_0, and $E_0 - e\hbar B/2m_e$, corresponding to the transitions with $\Delta m_l = +1$, $\Delta m_l = 0$, and $\Delta m_l = -1$. We can see that there will be only these energies for any initial and final values of l. The change in the angular frequency of the emitted spectral line is the energy change divided by $\hbar$. The frequency changes are therefore $\pm eB/2m_e$ or 0, in agreement with the classical calculation.

Optional

The anomalous Zeeman effect occurs when the spin of either the initial or the final states, or both, is nonzero. The calculation of the energy-level splitting is quite complicated because the magnetic moment due to spin is 1 rather than $\frac{1}{2}$ Bohr magneton, and accordingly the total magnetic moment is not parallel to the total angular momentum. Consider an atom with orbital angular momentum $\mathbf{L}$ and spin $\mathbf{S}$. Its total angular momentum is

$$\mathbf{J} = \mathbf{L} + \mathbf{S}$$

whereas the total magnetic moment is

$$\boldsymbol{\mu} = -g_l\mu_B \frac{\mathbf{L}}{\hbar} - g_s\mu_B \frac{\mathbf{S}}{\hbar}$$

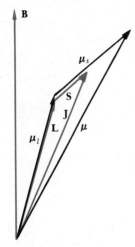

Figure 7-25
Vector diagram for the total magnetic moment when S is not zero. The moment is not parallel to the total angular momentum $\mathbf{J}$, because μ_s/S is twice μ_l/L. (The directions of $\boldsymbol{\mu}_l$, $\boldsymbol{\mu}_s$, and $\boldsymbol{\mu}$ have been reversed in this drawing for greater clarity.)

Since $g_l = 1$ and $g_s = 2$, we have

$$\boldsymbol{\mu} = -\frac{\mu_B}{\hbar}(\mathbf{L} + 2\mathbf{S}) \qquad 7\text{-}44$$

Figure 7-25 shows a vector-model diagram of the addition of $\mathbf{L} + \mathbf{S}$ to give $\mathbf{J}$. The magnetic moments are indicated by the black vectors. Such a vector model can be used to calculate the splitting of the levels, but as the calculation is rather involved we shall discuss only the results.[1]

Each energy level is split into $2j + 1$ levels, corresponding to the possible values of m_j. For the usual laboratory magnetic fields, which are weak compared with the internal magnetic field associated with the spin-orbit effect, the level splitting is small compared with the fine-structure splitting. Unlike the case of the singlet levels in the normal effect, the Zeeman splitting of these levels depends on j, l, and s, and in general there are more than three different transition energies. The level splitting can be written

$$\Delta E = g m_j \left(\frac{e\hbar B}{2m_e}\right) \qquad 7\text{-}45$$

where g, called the Landé g factor, is given by

$$g = 1 + \frac{j(j + 1) + s(s + 1) - l(l + 1)}{2j(j + 1)} \qquad 7\text{-}46$$

Note that for $s = 0$, $j = l$ and $g = 1$ and Equation 7-45 gives the splitting in the normal Zeeman effect. Figure 7-26 shows the splitting of sodium doublet levels $^2P_{1/2}$, $^2P_{3/2}$, and $^2S_{1/2}$. The se-

[1] This calculation can be found in H. White, Reference 3.

Figure 7-26
Energy-level splitting in a magnetic field for the $^2P_{3/2}$, $^2P_{1/2}$, and $^2S_{1/2}$ energy levels for sodium, showing the anomalous Zeeman effect. The splitting of the levels depends on L, S, and J, leading to more than the three lines seen in the normal effect. (*Photo from H. E. White,* Introduction to Atomic Spectra, *New York: McGraw-Hill Book Company, 1934. Used by permission of the publisher.*)

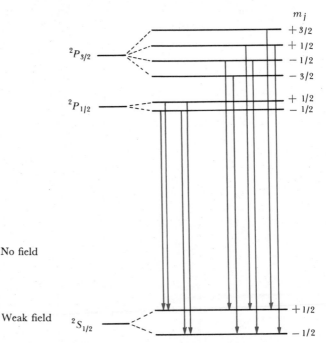

lection rule $\Delta m_j = \pm 1$ or 0 gives four lines for the transition $^2P_{1/2} \rightarrow {}^2S_{1/2}$ and six lines for the transition $^2P_{3/2} \rightarrow {}^2S_{1/2}$, as indicated. The energies of these lines can be calculated in terms of $e\hbar B/2m_e$ from Equations 7-45 and 7-46.

If the external magnetic field is sufficiently large, the Zeeman splitting is greater than the fine-structure splitting. If B is large enough so that we can neglect the fine-structure splitting, the Zeeman splitting is given by

Figure 7-27
Paschen-Back effect. When the external magnetic field is so strong that the Zeeman splitting is greater than the spin-orbit splitting, the level splitting is uniform for all atoms and only three spectral lines are seen, as in the normal Zeeman effect.

$$\Delta E = (m_l + 2m_s) \left(\frac{e\hbar B}{2m_e} \right)$$

The splitting is then similar to the normal Zeeman effect and only three lines are observed. This behavior in large magnetic fields is called the Paschen-Back effect after its discoverers, F. Paschen and E. Back. Figure 7-27 shows the transition of the splitting of the levels from the anomalous Zeeman effect to the Paschen-Back effect.

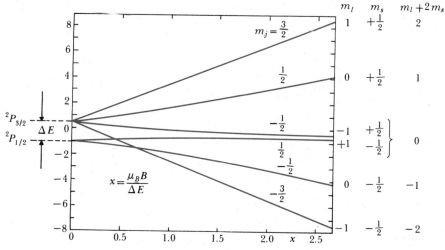

Summary

The Schrödinger equation in three dimensions is solved for the hydrogen atom by separating it into three ordinary differential equations, one for each coordinate r, θ, and ϕ. The quantum numbers n, l, and m arise from the application of the boundary conditions on the coordinates r, θ, and ϕ to the solutions of these equations. The quantum numbers l and m are related to the angular momentum of the atom; the magnitude of the angular momentum can take on only the values $\sqrt{l(l + 1)}\,\hbar$, where $l = 0$, $1, 2, \ldots$, and the z component to the values $m_l\hbar$, where m_l can take on any of the $2l + 1$ values from $-l$ to $+l$ in integer steps. The quantum number n arising from the solution of the radial equation is related to the total energy, which is restricted to the values $E_n = -Z^2E_0/n^2$, where $E_0 = \frac{1}{2}(ke^2/\hbar c)^2 \mu c^2 \approx$ 13.6 eV and $n = 1, 2, \ldots$. These energy values are the same as those obtained from the Bohr model. The values of l in the hydrogen atom are restricted to $l < n$.

The wave function for the ground state of the hydrogen atom is maximum at the origin. The radial probability density $r^2|\psi|^2$ for the ground state is maximum at the first Bohr radius $r = a_0$. For other states in the hydrogen atom the radial probability density is large in the region of the Bohr radii $r_n = n^2 a_0$. For a given n, the wave functions vary as r^l for small r and so are largest near the origin for $l = 0$.

In addition to its orbital angular momentum, an electron has an intrinsic spin angular momentum described by the quantum numbers $s = \frac{1}{2}$ and $m_s = \pm\frac{1}{2}$. The magnetic moment associated with its orbital angular momentum is $\boldsymbol{\mu}_l = -\mu_B(\mathbf{L}/\hbar)$ and that associated with its spin is $\boldsymbol{\mu}_s = -2\mu_B(\mathbf{S}/\hbar)$, where $\mu_B = e\hbar/2m_e$ is called a Bohr magneton. The orbital and spin angular-momentum vectors can combine to give a total angular momentum characterized by the quantum number j, which can have either of two values, $j = l + s$ or $j = l - s$. There is a slight difference in energy associated with these two possible j values because of the interaction of the spin and orbital angular momenta called the spin-orbit effect. This leads to the doublet structure of the energy levels in one-electron atoms and to the triplet structure in two-electron atoms, called fine structure.

The periodic table of the elements is built up by starting with hydrogen and adding one electron to the preceding atom (while increasing the nuclear charge by one) in the quantum state of lowest energy consistent with the exclusion principle. Because of the greater penetration of the outer-electron wave function into the inner-electron shielding of the nuclear charge, outer electrons with low angular momentum have lower energies than do those in the same n shell with larger angular momentum.

The energy levels of alkali atoms, with one electron outside a closed-shell core, are similar to those of the hydrogen atom except that the l degeneracy is removed because of the penetration of the outer electron into the core electrons. The energy levels of atoms with two electrons outside a closed-shell core can be separated into singlet states (spin 0) and triplet states (spin 1). Because the average separation of the electrons is greater when they are in the triplet state, the average interaction energy is lower, so the triplet-state energy levels have a lower energy than do the corresponding singlet-state levels. Normal excited states are due to the excitation of only one of the outer electrons.

The energy levels of an atom in a state of total angular momentum j are split into $2j + 1$ components in a magnetic field because of the interaction of the field and the magnetic moment. In the normal Zeeman effect, which occurs when the spin of the original state is zero, the splitting is independent of j and l, and the selection rules lead to only three different spectral lines. In the anomalous Zeeman effect, which occurs when the spin of the one or the other state is not zero, the splitting depends in a complicated way on j and l and more than three spectral lines are seen.

References

1. H. G. Kuhn, *Atomic Spectra,* New York: Academic Press, Inc., 1962.

2. L. Pauling and S. Goudsmit, *The Structure of Line Spectra,* New York: McGraw-Hill Book Company, 1930.

3. H. White, *Introduction to Atomic Spectra,* New York: McGraw-Hill Book Company, 1934.

4. G. Herzberg, *Atomic Spectra and Atomic Structure,* New York: Dover Publications, Inc., 1944.

5. L. Pauling and E. B. Wilson, *Introduction to Quantum Mechanics,* New York: McGraw-Hill Book Company, 1935.

Exercises

Section 7-1, The Schrödinger Equation in Spherical Coordinates

There are no exercises for this section.

Section 7-2, Quantization of Angular Momentum and Energy in the Hydrogen Atom

1. For $l = 1$ find the magnitude of L in units of $\hbar$ and find the possible angles between $\mathbf{L}$ and the z axis.

2. Draw a vector-model diagram illustrating the possible orientations of the angular-momentum vector $\mathbf{L}$ for (a) $l = 1$, (b) $l = 2$, (c) $l = 4$.

3. For $l = 2$, (a) what is the minimum value of $L_x^2 + L_y^2$? (b) What is the maximum value of $L_x^2 + L_y^2$? (c) What is $L_x^2 + L_y^2$ for $l = 2$ and $m = 1$? Can either L_x or L_y be determined from this?

4. In classical physics the angular momentum of a system can have any direction and its three components can be precisely known, e.g., it can point along the z axis. Discuss the application of the correspondence principle in connection with the possible directions of $\mathbf{L}$. In particular, show that the minimum angle between $\mathbf{L}$ and the z axis approaches zero as l approaches infinity.

5. The moment of inertia of a record is about 10^{-3} kg-m². (a) Find the angular momentum $L = I\omega$ when it rotates at $\omega/2\pi = 33\frac{1}{3}$ rev/min and find the approximate value of l. (b) Find the least nonzero value of $\omega/2\pi$ in revolutions per minute that the record can have.

Section 7-3, The Hydrogen-Atom Wave Functions

6. The radial probability distribution function for hydrogen in its ground state can be written $P(r) = Cr^2e^{-2Zr/a_0}$, where C is a constant. Show that $P(r)$ has its maximum value at $r = a_0/Z$.

7. (a) If spin is not included, how many different wave functions are there corresponding to the first excited energy level $n = 2$ for hydrogen? (b) List these functions by giving the quantum numbers for each state.

8. (a) Show that the radial probability distribution for the $n = 2$, $l = 1$ energy levels can be written $P(r) = Ar^4e^{Zr/a_0}$, where A depends on θ but not on r. (b) Show that $P(r)$ is maximum at $r = 4a_0/Z$.

9.   Show that $\psi^*\psi$ does not depend on ϕ for any state with n, l, and m specified.

Section 7-4, Electron Spin

10. If a classical system does not have a constant charge-to-mass ratio throughout the system, the magnetic moment can be written

$$\mu = g\,\frac{Q}{2M}\,L$$

where Q is the total charge, M is the total mass, and $g \neq 1$. (a) Show that $g = 2$ for a solid cylinder ($I = \frac{1}{2}MR^2$) that spins about its axis and has a uniform charge on its cylindrical surface. (b) Show that $g = 2\frac{1}{2}$ for a solid sphere ($I = \frac{2}{5}MR^2$) that has a ring of charge on the surface at the equator, as shown in Figure 7-28.

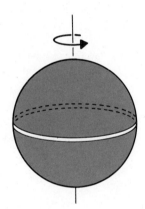

Figure 7-28
Solid sphere with charge Q uniformly distributed on ring for Exercise 10.

11. A convenient unit for the magnetic moment of nuclei is the nuclear magneton $e\hbar/2m_p$, where m_p is the mass of the proton. Calculate the magnitude of the nuclear magneton in (a) joules/tesla and (b) eV/gauss.

12. The angular momentum of the yttrium atom in the ground state is characterized by the quantum number $j = 1\frac{1}{2}$. How many lines would you expect to see if you could do a Stern-Gerlach-type experiment with yttrium atoms?

Section 7-5, Addition of Angular Momenta and the Spin-Orbit Effect

13. The total angular momentum of a hydrogen atom in a certain excited state has the quantum number $j = \frac{1}{2}$. What can you say about the orbital angular-momentum quantum number l?

14. The total angular momentum of a hydrogen atom in a certain excited state has the quantum number $j = 1\frac{1}{2}$. What can you say about the orbital angular-momentum quantum number l?

15. A hydrogen atom is in the $3D$ state ($n = 3$, $l = 2$). (a) What are the possible values of j? (b) What are the possible values of the magnitude of the total angular momentum including spin? (c) What are the possible z components of the total angular momentum?

16. A deuteron is a nucleus with one proton and one neutron, each having spin $\frac{1}{2}$. (a) What are the possible values of the total spin quantum number of the deuteron ($l = 0$)? (b) In the ground state the deuteron has $l = 0$ and $s = 1$. What is the magnitude of the angular momentum of the deuteron? (c) Draw a vector diagram illustrating the spins of the proton, neutron, and deuteron and find the angle between the spins of the neutron and proton. (d) What is the angle between the spin of the deuteron and the z axis when $m_s = 1$?

17. Consider a system of two electrons, each with $l = 1$ and $s = \frac{1}{2}$. (a) Neglecting spin, what are the possible values of the quantum number for the total orbital angular momentum $\mathbf{L} = \mathbf{L_1} + \mathbf{L_2}$? (b) What are the possible values of the quantum number S for the total spin $\mathbf{S} = \mathbf{S_1} + \mathbf{S_2}$? (c) Using the results of parts (a) and (b), find the possible quantum numbers j for the combination $\mathbf{J} = \mathbf{L} + \mathbf{S}$. (d) What are the possible quantum numbers j_1 and j_2 for the total angular momentum of each particle? (e) Use the results of part (d) to calculate the possible values of j from the combination of j_1 and j_2. Are these the same as in part (c)?

18. (*a*) A beam of free electrons passes through a uniform magnetic field of strength $B = 5000$ G. What is the difference in energy between those electrons whose spins are parallel and those whose spins are anti-parallel to **B**? Which have higher energy? (*b*) If these electrons are bombarded with photons of energy equal to this energy difference, "spin flip" transitions can be induced. Find the wavelength of the photons needed for such transitions. This phenomenon is called *electron spin resonance*.

19. The prominent yellow doublet lines in the spectrum of sodium result from transitions from the $3P_{3/2}$ and $3P_{1/2}$ states to the ground state. The wavelengths of these two lines are 589.6 nm and 589.0 nm. (*a*) Calculate the energies in eV of the photons corresponding to these wavelengths. (*b*) The difference in energy of these photons equals the difference in energy ΔE of the $3P_{3/2}$ and $3P_{1/2}$ states. This energy difference is due to the spin-orbit effect. Calculate ΔE. (*c*) If the $3p$ electron in sodium sees an internal magnetic field B, the spin-orbit energy splitting will be of the order of $\Delta E = 2\mu_B B$, where μ_B is the Bohr magneton. Estimate B from the energy difference ΔE found in part (*b*).

20. See Exercise 19. (*a*) Calculate the energy difference between the $4P_{3/2}$ and $4P_{1/2}$ states in potassium from the measured wavelengths of 766.41 nm and 769.90 nm for transitions to the ground state. (*b*) Calculate the order of magnitude of the internal magnetic field B seen by the $4p$ electron in potassium from $\Delta E = 2\mu_B B$.

Section 7-6, Ground States of Atoms; The Periodic Table

21. Use the Bohr model for helium and the result that the average interaction energy of the two electrons is about 30 eV to calculate the average separation of the electrons.

22. In Figure 7-18 there are small dips in the ionization potential curve at $Z = 31$ (gallium) and $Z = 49$ (indium) that are not labeled in the figure. Explain these dips, using the electron configuration of these atoms given in Table 7-1.

23. Which of the following atoms would you expect to have its ground state split by the spin-orbit interaction: Li, B, Na, Al, K, Ag, Cu, Ga? (*Hint:* Use Table 7-1 to see which elements have $l = 0$ in their ground state and which do not.)

24. The properties of iron and cobalt, which are adjacent in the periodic table ($Z = 26$ and $Z = 27$), are similar, whereas the properties of neon ($Z = 10$) and sodium ($Z = 11$), which also are adjacent, are very different. Explain why.

25. If the $3s$ electron in sodium did not penetrate the inner core its energy would be -13.6 eV$/3^2 = -1.51$ eV. Because it does penetrate it sees a higher effective Z and its energy is lower. Use the measured ionization potential of 5.14 V to calculate Z_{eff} for the $3s$ electron in sodium.

Section 7-7, Excited States and Spectra of Alkali Atoms and Section 7-8, Excited States and Spectra of Two-Electron Atoms

26. Which of the following elements should have an energy-level diagram similar to that of sodium and which should be similar to mercury: Li, He, Ca, Ti, Rb, Ag, Cd, Mg, Cs, Ba, Fr, Ra?

27. Since the P states and the D states of sodium are both doublets, there are four possible energies for transitions between these states. Indicate which three transitions are allowed and which one is not allowed by the selection rule of Equation 7-38.

28. The relative penetration of the inner-core electrons by the outer electron in sodium can be described by calculation of Z_{eff} from $E = -[Z_{\text{eff}}^2 \, (13.6 \text{ eV})]/n^2$ and comparing with $E = -13.6 \text{ eV}/n^2$ for no penetration (see Exercise 25). (a) Find the energies of the outer electron in the $3s$, $3p$, and $3d$ states from Figure 7-19. (*Hint:* This can be done simply but crudely by merely subtracting 5.14 eV from the energy scale in the figure. A more accurate method is to use -5.14 eV for the ground state as given and find the energy of the $3p$ and $3d$ states from the photon energies of the indicated transitions.) (b) Find Z_{eff} for the $3p$ and $3d$ states. (c) Is the approximation $-13.6 \text{ eV}/n^2$ good for any of these states?

29. Find (a) the energies and (b) Z_{eff} for the outer electron in the $4s$, $4p$, and $4d$, states of potassium. (See Exercise 28 and Figure 7-20.)

30. The quantum numbers n, l, and j for the outer electron in potassium have the values 4, 0, and $\frac{1}{2}$, respectively in the ground state; 4, 1, and $\frac{1}{2}$ in the first excited state; and 4, 1, and $\frac{3}{2}$ in the second excited state. Make a table giving the n, l, and j values for the 12 lowest energy states in potassium (see Figure 7-20).

31. Which of the following transitions in sodium do not occur as electric dipole transitions? (Give the selection rule that is violated.) $4S_{1/2} \to 3S_{1/2}$; $4S_{1/2} \to 3P_{3/2}$; $4P_{3/2} \to 3S_{1/2}$; $4D_{5/2} \to 3P_{1/2}$; $4D_{3/2} \to 3P_{1/2}$; $4D_{3/2} \to 3S_{1/2}$.

Section 7-9, The Zeeman Effect

32. Show that the change in wavelength $\Delta\lambda$ of a transition due to a small change in energy is

$$\Delta\lambda \approx -\frac{\lambda^2}{hc} \Delta E$$

(*Hint:* Differentiate $E = hc/\lambda$.)

33. Which of these elements—H, He, Li, Mg, and Na—might exhibit the normal Zeeman effect?

34. (a) Find the normal Zeeman energy shift $\Delta E = e\hbar B/2m_e$ for a magnetic field of strength $B = 5000$ G. (b) Use the result of Exercise 32 to calculate the wavelength changes for the singlet transition in mercury of wavelength $\lambda = 5790.7$ Å. (c) If the smallest wavelength change that can be measured in a spectrometer is 0.1 Å, what is the strength of the magnetic field needed to observe the Zeemann effect in this transition?

Problems

1. If a rigid body has moment of inertia I and angular velocity ω its kinetic energy is

$$E = \tfrac{1}{2}I\omega^2 = \frac{(I\omega)^2}{2I} = \frac{L^2}{2I}$$

where L is the angular momentum. The solution of the Schrödinger equation for this problem leads to quantized energy values given by

$$E_l = \frac{l(l+1)\hbar^2}{2I}$$

(a) Make an energy-level diagram of these energies, and indicate the transitions that obey the selection rule $\Delta l = \pm 1$. (b) Show that the allowed transition energies are E_1, $2E_1$, $3E_1$, $4E_1$, etc., where $E_1 = \hbar^2/I$. (c) The moment of inertia of the H_2 molecule is $I = \frac{1}{2}m_p r^2$, where m_p is the mass of the proton and $r \approx 0.74$ Å is the distance between the protons. Find the energy of the first excited rotational state $l = 1$ for H_2, assuming it is a rigid rotor. (d) What is the wavelength of the radiation emitted in the transition $l = 1$ to $l = 0$ for the H_2 molecule?

2.   (a) Show that Equation 7-13 with $l = 0$ can be written

$$R'' + \frac{2}{r}R' + \frac{2\mu k Z e^2}{\hbar^2}\frac{1}{r}R + \frac{2\mu E}{\hbar^2}R = 0$$

where μ is the reduced mass and $R' = dR/dr$, etc. (b) Substitute the trial solution $R = Ae^{-\lambda r}$, where A and λ are constants, and show that it satisfies the equation if

$$\lambda = \frac{\mu k Z e^2}{\hbar^2} = \frac{Z}{a_0} \quad \text{and} \quad E = \frac{-\lambda^2\hbar^2}{2\mu} = -Z^2 E_0$$

where a_0 is the Bohr radius and $E_0 = \mu k^2 e^4/2\hbar^2 = 13.6$ eV.

3.   Show that the expectation value of r for the ground state of the hydrogen atom is $\langle r \rangle = \frac{3}{2}a_0/Z$.

4.   Show that the expectation value of the potential energy $V(r) = -kZe^2/r$ for the ground state of the hydrogen atom is given by $\langle V(r) \rangle = -2Z^2 E_0$.

5.   The radius of a proton is about $R_0 = 10^{-15}$ m. The probability that the hydrogen-atom electron is inside the proton is

$$P = \int_0^{R_0} P(r)\, dr$$

where $P(r)$ is the radial probability density (Equation 7-18). Calculate the probability that, in the ground state, the electron is inside the proton. (Hint: Show that the approximation $e^{-r/a_0} \approx 1$ is valid for this calculation.)

6.   Show by direct substitution that the first-excited-state wave function of the hydrogen atom for $l = 0$ satisfies the radial part of the Schrödinger equation (see Problem 2).

7.   Because of the spin and magnetic moment of the proton, there is a very small splitting of the ground state of the hydrogen atom called *hyperfine splitting*. The splitting can be thought of as caused by the interaction of the electron magnetic moment with the magnetic field due to the magnetic moment of the proton, or vice versa. The magnetic moment of the proton is parallel to its spin and is about $2.8\mu_N$, where $\mu_N = e\hbar/2m_p$ is called the *nuclear magneton*. (a) The magnetic field at a distance r from a magnetic moment varies with angle, but is of the order of $B \sim 2k_m \mu/r^3$, where $k_m = 10^{-7}$ in SI units. Find B at $r = a_0$ if $\mu = 2.8\mu_N$. (b) Calculate the order of magnitude of the hyperfine splitting energy $\Delta E \approx 2\mu_B B$, where μ_B is one Bohr magneton and B is

your result from part (a). (c) Calculate the order of magnitude of the wavelength of radiation emitted if a hydrogen atom makes a "spin flip" transition between the hyperfine levels of the ground state. [Your result is greater than the actual wavelength of this transition, 21.11 cm, because $\langle r^{-3} \rangle$ is appreciably smaller than a_0^{-3}, making the energy ΔE found in part (b) greater. The detection of this radiation from hydrogen atoms in interstellar space is an important part of radio astronomy.]

8.   If the angular momentum of the nucleus is **I** and that of the atomic electrons is **J**, the total angular momentum of the atom is $\mathbf{F} = \mathbf{I} + \mathbf{J}$, and the total angular-momentum quantum number f ranges from $I + J$ to $|I - J|$. Show that the number of possible f values is $2I + 1$ if $I < J$ or $2J + 1$ if $J < I$. (If you can't find a general proof, show it for enough special cases to convince yourself of its validity.) (Because of the very small interaction of the nuclear magnetic moment with that of the electrons, a hyperfine splitting of the spectral lines is observed. When $I < J$, the value of I can be determined by counting the number of lines.)

9.   (a) Calculate the Landé g factor (Equation 7-46) for the $^2P_{1/2}$ and $^2S_{1/2}$ levels in a one-electron atom and show that there are four different energies for the transition between these levels in a magnetic field. (b) Calculate the Landé g factor for the $^2P_{3/2}$ level and show that there are six different energies for the transition $^2P_{3/2} \rightarrow {}^2S_{1/2}$ in a magnetic field.

10. In the anomalous Zeeman effect, the external magnetic field is much weaker than the internal field seen by the electron as a result of its orbital motion. In the vector model (Figure 7-25) the vectors **L** and **S** precess rapidly around **J** because of the internal field and **J** precesses slowly around the external field. The energy splitting is found by first calculating the component of the magnetic moment μ_J in the direction of **J** and then finding the component of **J** in the direction of **B**. (a) Show that $\mu_J = \dfrac{\boldsymbol{\mu} \cdot \mathbf{J}}{J}$ can be written

$$\mu_J = -\frac{\mu_B}{\hbar J}(L^2 + 2S^2 + 3\mathbf{S} \cdot \mathbf{L})$$

(b) From $J^2 = (\mathbf{L} + \mathbf{S}) \cdot (\mathbf{L} + \mathbf{S})$ show that $\mathbf{S} \cdot \mathbf{L} = \frac{1}{2}(J^2 - L^2 - S^2)$. (c) Substitute your result in part (b) into that of part (a) to obtain

$$\mu_J = -\frac{\mu_B}{2\hbar J}(3J^2 + S^2 - L^2)$$

(d) Multiply your result by J_z/J to obtain

$$\mu_z = -\mu_B \left(1 + \frac{J^2 + S^2 - L^2}{2J^2}\right)\frac{J_z}{\hbar}$$

(e) Replace J^2 by $j(j + 1)$, S^2 by $s(s + 1)$, L^2 by $l(l + 1)$, and $J_z/\hbar$ by m_j to obtain Equations 7-45 and 7-46.

PART 2 Applications

In Part 1 we saw how the ideas and methods of quantum mechanics developed and how their application to atomic physics yields an understanding of atomic structure and spectra. In Part 2 we shall extend the applications of quantum theory to a wide variety of systems of particular interest to engineers, chemists, and physicists.

Some of these applications are theoretical, such as the study of molecular bonding and spectra (Chapter 8), the structure of solids and their thermal and electrical properties (Chapter 9), the quantum mechanical modifications of the statistical distribution laws (Chapter 10), superconductors and superfluids (Chapters 9 and 10), nuclear structure, radioactivity, and nuclear reactions (Chapter 11), and elementary particles (Chapter 12). Other, more practical applications include the study of lasers (Chapter 8), semiconductors, semiconductor junctions, and transistors (Chapter 9), and carbon dating, nuclear fission, nuclear fusion and reactors (Chapter 11). Many of these applications have revolutionized contemporary society.

These chapters are independent of one another and can be studied in any order.

CHAPTER 8 Molecular Structure and Spectra

Objectives

After studying this chapter you should:

1. Be able to discuss various kinds of molecular bonding and, in particular, to compare ionic and covalent bonding.

2. Know the meaning of the terms atomic orbital, molecular orbital, and hybridization.

3. Know the general shapes of the H_2O and NH_3 molecules and understand the origin of these shapes.

4. Know the general features of the energy-level diagram for a diatomic molecule and be able to discuss the vibration-rotation spectrum.

5. Know why only certain lines in the emission spectrum of a molecule are seen in the absorption spectrum.

6. Know the meaning of the terms population inversion, optical pumping, and metastable state.

7. Be able to describe the operation of a three-level or four-level laser.

In this chapter we shall study some of the properties of molecules—systems of two or more atoms. Properly, a molecule is the smallest constituent of a substance which retains its chemical properties. The study of the properties of molecules forms the basis for theoretical chemistry. The application of quantum mechanics to molecular physics has been spectacularly successful in explaining the structure of molecules, the complexity of their spectra, and answering such puzzling questions as why two H atoms join together to form a molecule but three H atoms do not. As in atomic physics, the detailed quantum-mechanical calculations are quite difficult, so much of our discussion will be qualitative.

There are essentially two extreme views we can take of a molecule. Consider, for example, H_2. We can think of it either as two H atoms somehow joined together, or as a quantum-mechanical system of two protons and two electrons. The latter picture is more fruitful in this case because neither of the electrons in the H_2 molecule can be thought of as belonging to either proton. Instead, the wave function for each electron is spread out in space about the whole molecule. For more complicated molecules, however, an intermediate picture is useful. Consider the N_2 molecule as an example. We need not consider the complicated problem of two nuclei and 14 electrons. The electron configuration of an N atom in the ground state is $1s^2 2s^2 2p^3$. Of the three electrons in the $2p$ state, two have their spins paired (that is, with spins antiparallel so that their resultant is zero) and one does not. Only the electron with the unpaired spin is free to take part in the bonding of the N_2 molecule. We therefore can consider this molecule as two N^+ ions and two electrons that belong to the molecule as a whole. The molecular wave functions for these bonding electrons are called *molecular orbitals*. In many cases these molecular wave functions can be constructed from linear combinations of the atomic wave functions with which we are familiar.

8-1   Molecular Bonding

The two principal types of bonds that join two or more atoms together to form a molecule are called *ionic* and *covalent* bonds. Other types of bonds that are important in the bonding of liquids and solids are *van der Waals* bonds, *metallic* bonds, and *hydrogen* bonds. In many cases the bonding is a mixture of these mechanisms. We shall discuss all of these qualitatively in this section.

The Ionic Bond

The easiest type of bond to understand is the ionic bond, found in most salts. Consider NaCl as an example. The sodium atom has one $3s$ electron outside a neon core, $1s^2 2s^2 2p^6$. The ionization energy for Na is low, as it is for all the alkali metals; for Na, only 5.1 eV is required to remove the outer electron from the atom. The removal of one electron from Na leaves a positive ion with a spherically symmetric, closed-shell core. Chlorine, on the other hand, is only one electron short of having a closed argon core. The energy released by the acquisition of one electron is called the *electron affinity*, which in the case of Cl is 3.8 eV. The acquisition of one electron by chlorine leaves a negative ion with a spherically symmetric, closed-shell electron core. Thus the formation of a Na^+ ion and a Cl^- ion by the donation of one electron of Na to Cl requires just $5.1 - 3.8 = 1.3$ eV. The electrostatic potential energy of the two ions a distance r apart is $-ke^2/r$. When the separation of the ions is less than about 11 Å, the negative potential energy of attraction is of greater magnitude than the energy needed to create the ions.

Since the electrostatic attraction increases as the ions get closer, it would seem that equilibrium could not exist. For very small separation of ions, however, there is a strong repulsion

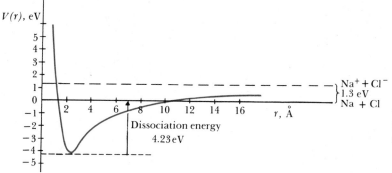

Figure 8-1
Potential energy for Na^+ and Cl^- ions as a function of separation distance. The energy at infinite separation is chosen to be 1.3 eV, corresponding to the energy needed to form the ions from the neutral Na and Cl atoms.

due to the exclusion principle. This "exclusion-principle repulsion" is responsible for the repulsion of the atoms in all molecules (except H_2), no matter what the bonding mechanism is. When the ions are very far apart, the wave function for a core electron of one ion does not overlap that of the other ion. We can distinguish the electrons by the ion to which they belong, and the electrons of one ion can have the same quantum numbers as in the other ion. However, when the ions are close, the core-electron wave functions begin to overlap, and some of the electrons must go into higher-energy quantum states because of the exclusion principle. This is not a sudden process; the energy states of the electrons are gradually changed as the ions are brought together. A sketch of the potential energy of the Na^+ and Cl^- ions versus separation is shown in Figure 8-1. The energy is lowest at an equilibrium separation of about 2.4 Å. At smaller separations, the energy rises steeply as a result of the exclusion principle. The energy required to separate the ions and form Na and Cl *atoms*, called the *dissociation energy*, is about 4.23 eV.

The separation distance of 2.4 Å is for gaseous diatomic NaCl (which can be obtained by evaporation of solid NaCl). Normally, NaCl exists in a cubic crystal structure, with Na^+ and Cl^- at alternate corners of a cube. The separation of the ions in a crystal is somewhat larger—about 2.8 Å. Because of the presence of neighboring ions of opposite charge, the Coulomb energy per ion pair is lower when the ions are in a crystal. This energy is usually expressed as $\alpha\, ke^2/r_0$, where r_0 is the separation distance and α, called the *Madelung constant*, depends on the crystal structure. For NaCl, α is about 1.75.

The Covalent Bond

A completely different mechanism is responsible for the bonding of such molecules as H_2, N_2, and CO. If we calculate the energy needed to form the ions H^+ and H^- by the transfer of an electron from one atom to the other and then add this energy to the electrostatic energy, we find that there is no separation distance for which the total energy is negative. The bond thus cannot be ionic. The attraction of two hydrogen atoms is instead an entirely quantum-mechanical effect. The decrease in

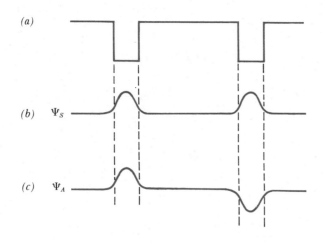

Figure 8-2
(*a*) Two square wells far apart. The electron wave function can be either (*b*) symmetric or (*c*) antisymmetric. The probability distributions and energies are the same for the two wave functions when the wells are far apart.

energy when two hydrogen atoms approach each other is due to the sharing of the two electrons by both atoms and is intimately connected with the symmetry properties of the electron wave functions. We can gain some insight into this phenomenon by studying a simple one-dimensional quantum-mechanics problem—that of two finite square wells.

Consider first a single electron that is equally likely to be in either well. Since the wells are identical, symmetry requires that Ψ^2 be symmetric about the midpoint of the wells. Then Ψ must be either symmetric or antisymmetric. These two possibilities for the ground state are shown in Figure 8-2. The energies for these two cases are the same when the wells are far apart. Figure 8-3 shows the symmetric and antisymmetric wave functions when the wells are very close together. Now the parts of the wave function describing the electron in one well or the other overlap, and the symmetric and antisymmetric resultant wave functions are quite different. In the limiting case of no separation, the symmetric wave function Ψ_S approaches the ground-state wave function for a particle in a well of size $2L$ and the antisymmetric wave function Ψ_A approaches that for the first excited state in such a well. There are two important results from this discussion:

1. The originally equal energies for Ψ_A and Ψ_S are split into two different energies as the wells become close.

2. The wave function for the symmetric state is large in the region between the wells, whereas that for the antisymmetric state is small.

Now consider adding a second electron to the two wells. The total wave function for the two electrons must be antisymmetric on exchange. Note that exchanging the electrons in the wells is the same as exchanging the wells, i.e., for a two-particle system, exchange symmetry is the same as space symmetry. The two electrons can therefore be in the space-symmetric state if the spins are antiparallel ($S = 0$) or in the space-antisymmetric state if their spins are parallel ($S = 1$).

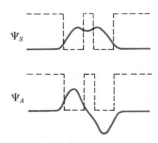

Figure 8-3
Symmetric and antisymmetric space wave functions for two square wells close together. The probability distributions and energies are not the same for the two wave functions in this case. The symmetric space wave function is larger between the wells than the antisymmetric space wave function.

Exchange symmetry and space symmetry

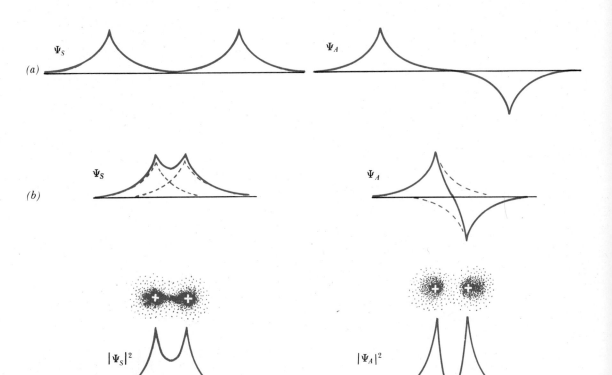

Figure 8-4
One-dimensional symmetric and antisymmetric electron space wave functions for (a) two protons far apart and (b) two protons close together. (c) Probability distributions for wave functions in (b).

We now turn to the problem of two hydrogen atoms. In the ground state, the hydrogen-atom wave function is proportional to e^{-r/a_0}. For a one-dimensional model, we shall write this as $e^{-|x|/a_0}$. Figure 8-4 shows the symmetric and antisymmetric combinations for two values of the distance between the protons. The results are similar to the square-well case: Ψ_S is large in the region between the protons. This concentration of charge between the protons for Ψ_S holds the protons together. It is interesting to compare this situation with our discussion of the He atom in Chapter 7. In the excited state of He, the energy of the space-symmetric state is *higher* because the electrons are close together and they repel each other. In H_2 the positive energy of repulsion for the electrons is also greater in the symmetric state because they are close; however, the negative energy of attraction of each electron for the two protons is more important. The potential energy versus separation for two H atoms can be calculated, approximately, by calculating the interaction energy of the two electrons and two protons, using the hydrogen-atom wave functions to compute the charge density $e|\Psi_S|^2$ or $e|\Psi_A|^2$. The results are sketched in Figure 8-5, where we see that the potential energy for the symmetric state is lower, as expected, and is of similar shape to that for ionic bonding. As the separation approaches zero, both curves approach $+\infty$. In the hydrogen molecule, this repulsion is due to the repulsion of the protons; for other molecules, it is due to the overlap of the core electrons

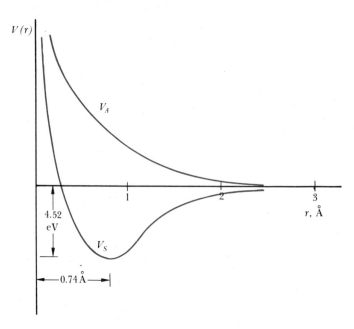

Figure 8-5
Potential energy versus
separation for two hy-
drogen atoms. V_S is for
the symmetric space wave
function, and V_A is for
the antisymmetric space
wave function.

and to the exclusion principle. The equilibrium separation for H_2 is $r_0 = 0.74$ Å, and the binding energy is 4.52 eV.

We can now see why three H atoms do not bond to form H_3. If a third H atom is brought near an H_2 molecule, the third electron cannot be in a $1s$ state and have its spin antiparallel to both the other electrons. If it is in an antisymmetric state with respect to exchange with one of the electrons, the repulsion of this atom is greater than the attraction of the other. As the three atoms are pushed together, the third electron is, in effect, forced into a higher quantum state by the exclusion principle. The bond between two H atoms is called a *saturated bond* because there is no room for another electron. The two electrons being shared essentially fill the $1s$ states of both atoms.

It is found that two identical atoms can bond with a one-electron bond. For example, two protons can share one electron to form the hydrogen-molecule ion H_2^+. Because the two protons are identical, the wave function of the electron must be symmetric or antisymmetric in space. The symmetric wave function is concentrated between the protons and leads to bonding. This bonding is weaker than the two-electron bond. The equilibrium separation of the protons in H_2^+ is 1.06 Å, and the binding energy is 2.65 eV. If the atoms are not identical, one-electron bonds are usually not formed. Since the wave function of the electron need not be symmetric in space, the electron tends to stay near the atom for which the potential energy is lower, and therefore the electron is not shared.

It should be clear now why He atoms do not bond together to form He_2. There are no valence electrons that can be shared. The electrons in the closed shell are forced into higher energy states when two atoms are brought together. At low temperatures or high pressures, He atoms do bond together; but the

Linus Pauling, who developed the theory of molecular bonding. (*Courtesy of the Linus Pauling Institute of Science and Medicine.*)

bonds are very weak and are due to van der Waals forces, which we shall discuss next. The bonding is so weak that at atmospheric pressure He boils at 4 K, and it does not form a solid at any temperature unless the pressure is greater than about 20 atm.

The van der Waals Bond

Any two separated molecules will be attracted towards one another by electrostatic forces. Similarly, atoms which do not otherwise form ionic or covalent bonds will be attracted to one another by the same sort of weak electrostatic bonds. The practical result of this is that at temperatures so low that the disruptive effects of thermal agitation are negligible, all substances will condense into a liquid and then a solid form. (Helium is the only element that does not solidify at any temperature at atmospheric pressure.) The relatively weak electrostatic forces responsible for this sort of intermolecular attraction are known as *van der Waals forces,* and they arise because of the electrostatic attraction of electric dipoles. It is not hard to see that molecules with permanent electric dipole moments—polar molecules such as NaCl or H_2O—will attract other polar molecules, as shown in Figure 8-6. Also, a nonpolar molecule will be polarized by the field of a polar molecule and thus have an induced dipole moment and be attracted to the polar molecule.

It is somewhat harder to see why two nonpolar molecules attract each other. Though the average dipole moment $\bar{p}$ of a nonpolar molecule is zero, the average-square dipole moment $\overline{p^2}$ is not, because of fluctuations of the charge. The instantaneous dipole moment of a nonpolar molecule is, in general, not zero. When two nonpolar molecules are nearby, the fluctuations in the instantaneous dipole moments tend to be correlated so as to produce attraction, as illustrated in Figure 8-7. This attractive force between nonpolar molecules is called a van der Waals force.

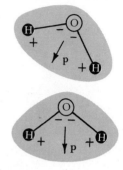

Figure 8-6
Bonding of H_2O molecules because of the attraction of the electric dipoles. The dipole moment of each molecule is indicated by **p.**

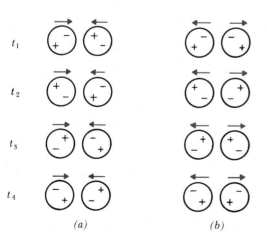

(a)   (b)

Figure 8-7
Van der Waals attraction of molecules with zero average dipole moments. (*a*) Possible orientations of instantaneous dipole moments at different times, leading to attraction. (*b*) Possible orientations leading to repulsion. The electric field of the instantaneous dipole moment of one molecule tends to polarize the other molecule; thus the orientations leading to attraction (*a*) are much more likely than those leading to repulsion (*b*).

The Hydrogen Bond

Another bonding mechanism of great importance is the hydrogen bond, which commonly provides much of the source of the binding that holds different groups of molecules together, and is responsible for the cross-linking that allows giant biological molecules and polymers to hold their fixed shapes. The well-known helical structure of DNA is due to hydrogen bonds linking across turns of the helix (Figure 8-8). The hydrogen bond is formed by the sharing of a proton between two electronegative atoms, frequently two oxygen atoms. This sharing is facilitated by the small mass of the proton, the absence of inner-core electrons in the case of hydrogen, and the presence of a valence-electron state.

The Metallic Bond

The nature of the bonding of atoms in a metal is different from the bonding of atoms in a molecule. In a metal, two atoms do not bond together by exchanging or sharing an electron to form a molecule. Instead, each valence electron is shared by many atoms. The bonding is thus distributed throughout the metal rather than being between two atoms. A metal can be thought of as a lattice of positive ions held together by a "gas" of essentially free electrons that roam throughout the solid. The number of free electrons varies from metal to metal but is of the order of one per atom.

Figure 8-8
The DNA molecule.
(*Courtesy of M. H. F. Wilkins, King's College, London.*)

Optional

8-2   Polyatomic Molecules

In this section we shall discuss briefly the structure of some polyatomic molecules. Such molecules may be small or large, ranging from such relatively simple molecules as water, with molecular weight 18, through such giants as proteins, with molecular weights of up to a hundred thousand or perhaps even a million. As with diatomic molecules, the structure of such molecules can be understood by applying basic quantum mechanics to the bonding of the individual atoms. The bonding mechanism for most polyatomic molecules is the covalent bond or the hydrogen bond. We shall discuss only some of the simplest polyatomic molecules, H_2O, NH_3, and CH_4, to illustrate both the simplicity and the complexity of the application of quantum mechanics to molecular bonding.

The basic requirement for the sharing of electrons in a covalent bond is that the wave functions of the valence electrons in the individual atoms overlap as much as possible. As our first example we consider the water molecule. The ground-state configuration of the oxygen atom is $1s^2 2s^2 2p^4$. The $1s$ and $2s$ electrons are in closed-shell states and do not contribute to the bonding. The $2p$ shell has room for six electrons, two in each of the three space states corresponding to $l = 1$. In an isolated atom, we described these space states by the hydrogenlike wave functions corresponding to $l = 1$, $m = 0, \pm 1$. These wave functions have angular dependence, proportional to $\cos \theta$ for $m = 0$ and $\sin \theta e^{\pm i\phi}$ for $m = \pm 1$. Since the energy is the same for these three space states, we could equally well use any linear combinations of these wave functions. When an atom participates in molecular bonding, certain linear combinations of these atomic wave functions, called *atomic orbitals*, are important. These combinations are called the p_x, p_y, and p_z atomic orbitals. The p_z orbital is just the atomic wave function with $m = 0$. The p_x and p_y orbitals are mixtures of the $m = \pm 1$ atomic wave functions. The angular dependence of these orbitals is

Atomic orbitals

$$p_z \propto \cos \theta$$

$$p_x \propto \sin \theta \left(\frac{e^{+i\phi} + e^{-i\phi}}{2}\right) = \sin \theta \cos \phi \qquad 8\text{-}1$$

$$p_y \propto \sin \theta \left(\frac{e^{+i\phi} - e^{-i\phi}}{2i}\right) = \sin \theta \sin \phi$$

The electron charge distribution is maximum along the z, x, or y axis for these orbitals, as shown in Figure 8-9. When oxygen is in an H_2O molecule, maximum overlap of the electron wave functions occurs when two of the $2p$ electrons are in one of the orbitals (e.g., p_z) with spins paired, one is in the p_x orbital, and the other is in the p_y orbital. Each of the unpaired electrons in the p_x and p_y orbitals forms a bond with a hydrogen-atom electron. If it were not for the repulsion of the two H atoms, we would therefore expect the O—H bonds to be at right angles to one another. The effect of this repulsion can be calculated and the result is in agreement with the measured angle of 104.5° (Figure 8-10).

As we have previously mentioned, the electrons that participate in the bonding are not confined to their original atoms, but instead belong to both atoms. A complete description of the molecular charge distribution would therefore require the use

Figure 8-9
The p_x, p_y, and p_z atomic orbitals. (*This computer-generated plot courtesy of Paul Doherty, Oakland University.*)

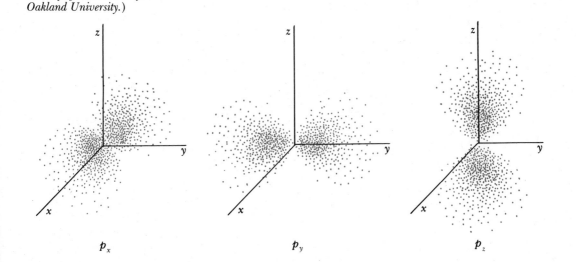

p_x p_y p_z

Figure 8-10
Electron charge distribu-
tion in the H_2O molecule.

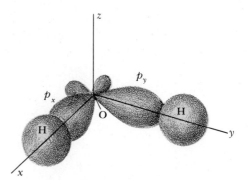

of molecular orbitals rather than atomic orbitals. However, the resultant molecular orbital has much the same character as the original atomic orbitals. For example, the molecular orbital occupied by the electron originally in the p_x atomic orbital of oxygen and one of the hydrogen-atom electrons is concentrated along the x axis, just as the original p_x orbital of oxygen is.

For our next example we consider the ammonia molecule, NH_3. The ground-state configuration of nitrogen is $1s^2 2s^2 2p^3$. Maximum overlap of bonding-electron wave functions is achieved when the three electrons in the p state of nitrogen are unpaired in the p_x, p_y, and p_z orbitals. Each electron bonds with a hydrogen-atom electron to form a three-dimensional structure. Again, because of the repulsion of the hydrogen atoms, the angles between the bonds are somewhat larger than 90°: in this case the angle is 107.3°. The structure is a pyramid, with the nitrogen atom at the vertex and the three hydrogen atoms at the base.

As our final example, we consider the more complicated but very important case of the bonding of carbon atoms to hydrogen and other atoms. The ground-state configuration of carbon is $1s^2 2s^2 2p^2$. From our previous discussion we might expect carbon to be divalent, with the two $2p$ electrons forming bonds at approximately 90° like the oxygen atom. The most important experimental feature of carbon chemistry, however, is that tetravalent carbon compounds such as CH_4 are overwhelmingly favored.

The observed valence of 4 for carbon comes about in an interesting way. One of the first excited states of carbon occurs when a $2s$ electron is excited to a $2p$ state, giving a configuration of $1s^2 2s 2p^3$ for this state. In this excited state we can have four unpaired electrons, one each in the $2s$, $2p_x$, $2p_y$, and $2p_z$ atomic orbitals. If the four bonding electrons were described by these atomic orbitals, we would expect there to be three similar bonds at angles of about 90° and one somewhat different bond corresponding to the $2s$ orbital, which has no directional character. Again, this is not what is observed. The four wave functions corresponding to these four atomic orbitals have approximately but not exactly the same energy. However, it is possible to find four linear combinations of these orbitals that do have the same energy and that have the necessary directional characteristics to describe the observed bonding of carbon. These *hybrid* wave functions are of the form

$$\psi_1 = \tfrac{1}{2}(s + p_x + p_y + p_z)$$

$$\psi_2 = \tfrac{1}{2}(s + p_x - p_y - p_z)$$

$$\psi_3 = \tfrac{1}{2}(s - p_x - p_y + p_z)$$

$$\psi_4 = \tfrac{1}{2}(s - p_x + p_y - p_z)$$

8-2

These are called the sp^3 hybrid orbitals.

Hybridization

This mixing of atomic orbitals, called *hybridization*, is probably the single most important new feature involved in the physics of complex molecular bonds. Figure 8-11 shows the tetrahedral structure of the methane CH_4 molecule, and Figure 8-12 shows the ethane molecule (CH_3—CH_3), which is similar to CH_4 except that one of the C-H bonds is replaced with a C-C bond.

Carbon orbitals can also hybridize in the sp^2 configuration, in which the s, p_x, and p_y orbitals combine to form three hybrid orbitals in the xy plane with 120° bonds, and the p_z orbital remains unmixed. Examples of this configuration include graphite, where the xy plane bonds provide the strongly layered structure characteristic of the material, and organic molecules such as ethylene (H_2C=CH_2) and benzene (C_6H_6).

Figure 8-11
Electron charge distribution in the CH_4 molecule.

Figure 8-12
Electron charge distribution in the CH_3—CH_3 molecule.

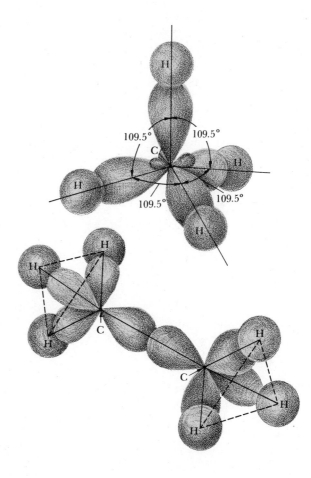

We do not have the space to discuss further the many interesting features of the molecular bonding in carbon compounds. The great family of carbon compounds, known as *organic* compounds because of their close connection with the chemistry of living things, arises not from different underlying principles but rather from carbon's ready hybridization to give bonds that may be incorporated into molecules of great diversity. Despite the complexity of polyatomic molecules, their features can be understood in the framework of the principles of quantum mechanics discussed in the previous chapters.

8-3   Energy Levels and Spectra of Diatomic Molecules

As might be expected, the energy levels of molecular systems are even more complex than those of atoms. For simplicity, we shall consider only diatomic molecules. The energy of a molecule can be conveniently separated into three parts: energy due to excitation of its electrons, energy of vibration of the molecule, and energy of rotation of the molecule. Fortunately, the magnitudes of these energies are sufficiently different that they can be treated separately. The energies of electronic excitations of a molecule are of the order of magnitude of 1 eV, the same as for the excitation of atoms. The energies of vibration and rotation are about 100 to 1000 times smaller than this.

Classically, the kinetic energy of rotation is

$$E = \frac{1}{2} I\omega^2 = \frac{(I\omega)^2}{2I} = \frac{L^2}{2I} \qquad\qquad 8\text{-}3$$

where I is the moment of inertia, ω the angular velocity of rotation, and L the angular momentum. The solution of the Schrödinger equation for the rotation of a rigid body leads to the quantization of the angular momentum, with values given by

$$L^2 = l(l + 1)\hbar^2 \qquad\qquad 8\text{-}4$$

just as in the case of the hydrogen atom. The energy levels are therefore given by

$$E = \frac{l(l + 1)\hbar^2}{2I} \qquad\qquad 8\text{-}5$$

Energy of rigid rotor

The moment of inertia about an axis through the center of mass of a diatomic molecule is (see Figure 8-13)

$$I = m_1 r_1^2 + m_2 r_2^2$$

Using the relations $m_1 r_1 = m_2 r_2$ for the distances to the center of mass and $r_0 = r_1 + r_2$ for the separation, the moment of inertia can be written (see Problem 2)

$$I = \mu r_0^2$$

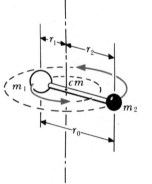

Figure 8-13
Diatomic molecule rotating about an axis through the center of mass.

where μ is the reduced mass $\mu = m_1 m_2/(m_1 + m_2)$. If the masses are equal, as in H_2 and O_2, $\mu = \frac{1}{2}m$ and

$$I = \tfrac{1}{2}mr_0^2$$

Let us calculate the order of magnitude of the rotational energy levels for H_2. We have

$$E = \frac{l(l + 1)\hbar^2}{mr_0^2} = \frac{l(l + 1)(\hbar c)^2}{mc^2 r_0^2}$$

Taking for the rest energy of the proton $mc^2 = 938$ MeV $\sim 10^9$ eV and $r_0 \sim 1$ Å, we obtain, using $\hbar c = 1973$ eV-Å,

$$E \approx \frac{l(l + 1)(1973 \text{ eV-Å})^2}{10^9 \text{ eV } (1 \text{ Å})^2} \sim l(l + 1)(4 \times 10^{-3} \text{ eV})$$

Other diatomic molecules have roughly the same separation but larger masses. We see that the rotational-energy levels are smaller than those due to electronic excitation by a factor of at least 1000, and transitions between pure rotational-energy levels yield photons in the infrared region.

The vibrational energies are a little harder to estimate. If we approximate the potential-energy curve of Figure 8-1 by a parabola near the equilibrium point, we can use the results of our study of the simple harmonic oscillator in Chapter 6. The energy levels are given by

$$E_n = (n + \tfrac{1}{2})\hbar\omega$$

where ω is the classical angular frequency of vibration. We could estimate ω by fitting the one-dimensional harmonic potential to the curve of Figure 8-1, but for simplicity we can get a rough idea of the order of magnitude of the vibrational energies by observing that the energy of an atom of mass m in a square well of length r_0 is (Figure 8-14)

$$E_n = n^2 \frac{h^2}{8mr_0^2} = n^2 \frac{4\pi^2\hbar^2}{8mr_0^2} = n^2 \frac{\pi^2}{2} \frac{\hbar^2}{mr_0^2}$$

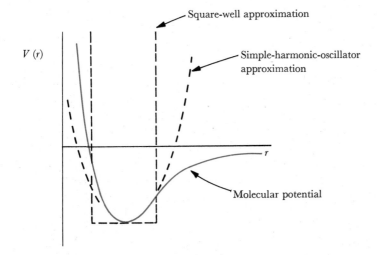

Square-well approximation

$V(r)$

Simple-harmonic-oscillator approximation

r

Molecular potential

Figure 8-14
Molecular potential. The simple-harmonic-oscillator approximation, used to calculate the energy levels, and a square-well approximation, used to estimate the order of magnitude of the energy levels, are each indicated by dashed curves.

Except for the factor $\pi^2/2 \approx 5$ (and n^2), this expression is the same as that for rotation; thus we can expect the vibrational energies to be closer in magnitude to the rotational energies than to the electronic energies. Actually, the vibrational energies are typically greater than the rotational energies by a factor of 10 or more.

Figure 8-15 is a schematic sketch of some electronic, vibrational, and rotational energy levels of a molecule. The levels are labeled by the quantum numbers n for vibration and l for rotation. The lower vibrational levels are evenly spaced, with $\Delta E_n \approx \hbar\omega$. For higher vibrational levels, the approximation that the potential energy is a simple quadratic is not as good. The actual potential spreads somewhat more rapidly, as can be seen from Figure 8-1, and the spacing of the vibrational levels becomes closer for large quantum numbers n. The energy differences between the rotational levels $l + 1$ and l is

$$E_{l+1} - E_l = [(l + 1)(l + 2) - l(l + 1)] \frac{\hbar^2}{2I}$$

$$= 2(l + 1) \frac{\hbar^2}{2I} \qquad\qquad 8\text{-}6$$

The level spacing therefore increases with l.

Because the energies of vibration or rotation of a molecule are so much smaller than the energies of excitation of an atomic electron, molecular vibration and rotation show up in optical transitions as a fine splitting of the lines. When this structure is not resolved, the spectrum appears as bands. Figure 8-16 shows a typical band spectrum.

Much of molecular spectroscopy is done by infrared absorption techniques in which only the vibrational and rotational energy levels are excited. We shall therefore concentrate on what is called the *vibration-rotation spectrum*. In Section 6-6 we saw that, for the simple harmonic oscillator, only transitions for which

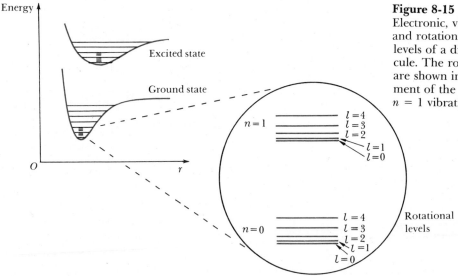

Figure 8-15
Electronic, vibrational, and rotational energy levels of a diatomic molecule. The rotational levels are shown in an enlargement of the $n = 0$ and $n = 1$ vibrational levels.

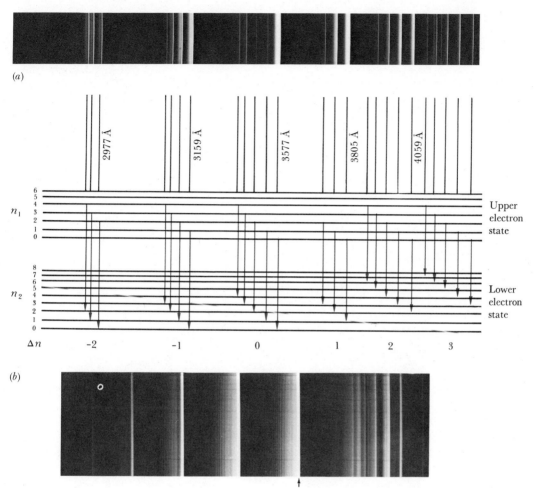

(a)

2977 Å   3159 Å   3577 Å   3805 Å   4059 Å

n_1 6 5 4 3 2 1 0   Upper electron state

n_2 8 7 6 5 4 3 2 1 0   Lower electron state

Δn -2   -1   0   1   2   3

(b)

3805 Å

Figure 8-16
Part of the emission spectrum of N_2. (a) These components of the band are due to transitions between the vibrational levels of two electronic states, as indicated in the diagram. (b) An enlargement of part of (a) shows that the apparent lines in (a) are in fact band heads with structure caused by rotational levels. (*Courtesy of J. A. Marquisee.*)

$\Delta n = \pm 1$ were allowed for electric dipole radiation. This selection rule is obeyed for transitions involving the lowest vibrational levels, but not for the higher levels because of the deviation of the molecular potential from that of a simple harmonic oscillator. For ordinary temperatures ($T \approx 300$ K), the vibrational energies are sufficiently large compared with kT that most of the molecules are in the lowest vibrational state. The transition $n = 0$ to $n = 1$ is predominant in absorption. The rotational energies, however, are sufficiently less than kT that the molecules are distributed among several rotational states. Electric dipole transitions between rotational states obey the selection rule

$$\Delta l = \pm 1$$

If the original rotational-vibrational energy is

$$E_l = E_e + \tfrac{1}{2}\hbar\omega + l(l + 1)E_{0r} \qquad 8\text{-}7$$

where E_e is the electronic energy, $n = 0$, and $E_{0r} = \hbar^2/2I$, the final energy for the transition l to $l + 1$ is

$$E_{l+1} = E_e + \tfrac{3}{2}\hbar\omega + (l + 1)(l + 2)E_{0r} \qquad 8\text{-}8$$

Energy

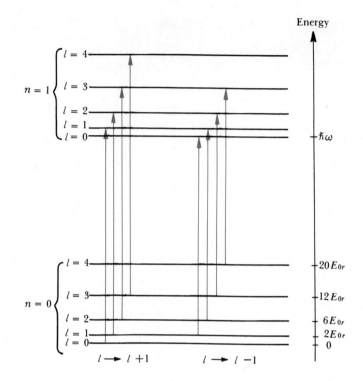

$n = 1$ $\begin{cases} l = 4 \\ l = 3 \\ l = 2 \\ l = 1 \\ l = 0 \end{cases}$

$\hbar\omega$

$n = 0$ $\begin{cases} l = 4 \\ l = 3 \\ l = 2 \\ l = 1 \\ l = 0 \end{cases}$

$20\,E_{0r}$

$12\,E_{0r}$

$6\,E_{0r}$
$2\,E_{0r}$
0

$l \rightarrow l + 1$ $l \rightarrow l - 1$

Figure 8-17
Absorptive transitions between the lowest rotational bands $n = 0$ and $n = 1$ in a diatomic molecule. These transitions obey the selection rule $\Delta l = \pm 1$ and fall into two bands. The energies of the $l \rightarrow l - 1$ band are $\hbar\omega - 2E_{0r}$, $\hbar\omega - 4E_{0r}$, $\hbar\omega - 6E_{0r}$, . . . , whereas the energies of the $l \rightarrow l + 1$ band are $\hbar\omega + 2E_{0r}$, $\hbar\omega + 4E_{0r}$, $\hbar\omega + 6E_{0r}$, . . . , where $E_{0r} = \hbar^2/2I$.

For the transition l to $l - 1$, the final energy is

$$E_{l-1} = E_e + \tfrac{3}{2}\hbar\omega + (l - 1)lE_{0r} \qquad\qquad 8\text{-}9$$

The energy differences are thus

$$\Delta E_{l \mapsto l+1} = \hbar\omega + 2(l + 1)E_{0r} \qquad\qquad 8\text{-}10$$

where $l = 0, 1, 2, \ldots$, and

$$\Delta E_{l \mapsto l-1} = \hbar\omega - 2lE_{0r} \qquad\qquad 8\text{-}11$$

where $l = 1, 2, \ldots$. (l begins at $l = 1$ in Equation 8-11, because from $l = 0$, only the transition $l \rightarrow l + 1$ is possible.) Figure 8-17 illustrates these transitions. As can be seen from Equations 8-10 and 8-11, the absorption spectrum contains frequencies equally spaced by $2E_{0r}/\hbar$ except that there is a gap of $4E_{0r}/\hbar$ at the vibrational frequency ω. A measurement of the position of the gap determines ω, while a measurement of the spacing of the peaks determines E_{0r}, which is inversely proportional to the moment of inertia. Figure 8-18 shows the absorption spectrum for HCl.

Symmetric molecules such as H_2 or O_2 have no electric dipole moment. The vibration or rotation of these molecules does not involve a changing dipole moment and there is no vibrational-rotational electric dipole absorption or radiation for these molecules.

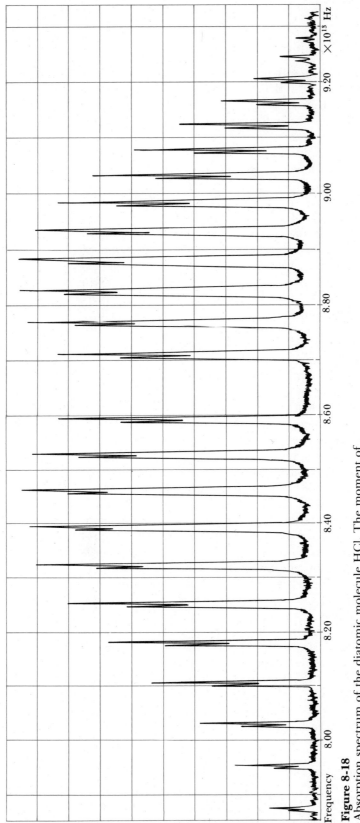

Figure 8-18
Absorption spectrum of the diatomic molecule HCl. The moment of inertia can be determined from the energy spacing of $2E_{0r} = 2(\hbar^2/2I)$ between the large peaks in each band. The bands are separated by a gap of $4E_{0r}$; the frequency at the center of this gap is the frequency of vibration. The double peak structure results from the two isotopes of chlorine, ^{35}Cl (abundance 75.5 percent) and ^{37}Cl (abundance 24.5 percent). *(Courtesy of T. Faulkner and T. Nestrick at Oakland University.)*

8-4   Absorption, Scattering, and Stimulated Emission

Information about the energy levels of an atom or molecule is usually obtained from the radiation emitted when the atom or molecule makes a transition from an excited state to a state of lower energy. As mentioned in the previous section, we can also obtain information about such energy levels from the absorption spectrum. When atoms and molecules are irradiated with a continuous spectrum of radiation, the transmitted radiation shows dark lines corresponding to absorption of light at discrete wavelengths. Absorption spectra of atoms were the first line spectra observed. Fraunhofer in 1817 labeled the most prominent absorption lines in the spectrum of sunlight; it is for this reason that the two intense yellow lines in the spectrum of sodium are called the *Fraunhofer D lines*. Since at normal temperatures atoms and molecules are in their ground states or in low-lying excited states, the absorption spectra are usually simpler than the emission spectra. For example, only those lines corresponding to the Lyman emission series are seen in the absorption spectrum of atomic hydrogen because nearly all the atoms are originally in their ground states.

Photon interactions

In addition to absorption, several other interesting phenomena occur when a photon is incident on an atom or molecule. These are illustrated in Figure 8-19. In Figure 8-19*a* the photon is absorbed and the system makes a transition to an excited state. Later, the system makes a transition to a lower state or back to the ground state with the emission of a photon. This two-step process is called *resonance absorption*. The emitted photon is not correlated with the incident photon. Figure 8-19*b* illustrates *elastic* or *Rayleigh scattering,* which is a one-step process in which the incident and emitted or scattered photon are correlated. The incident and scattered photons are also correlated in the *inelastic* scattering process illustrated in Figure 8-19*c*. Such scattering of light from molecules was first observed by the Indian physicist C. V. Raman and is known as *Raman scattering.* The scattered photon may have less energy than the incident photon, as illustrated, or it may have greater energy if the excited molecule is initially at an excited vibrational or rotational energy level. Figure 8-19*d* illustrates the photoelectric effect, in which the absorption of the photon ionizes the atom or molecule, and Figure 8-19*e* illustrates Compton scattering. These effects were discussed in Chapter 3.

Figure 8-19*f* illustrates stimulated emission, which we shall discuss more fully here because of its important applications. This process occurs if the atom is initially in an excited state, and if the energy of the incident photon is just $E_2 - E_1$, where E_2 is the excited energy of the atom and E_1 is the energy of a lower state or the ground state. In this case, the oscillating electromagnetic field of the incident frequency stimulates the excited atom, which emits a photon in the same direction as the incident photon and in phase with it. The relative probabilities of stimu-

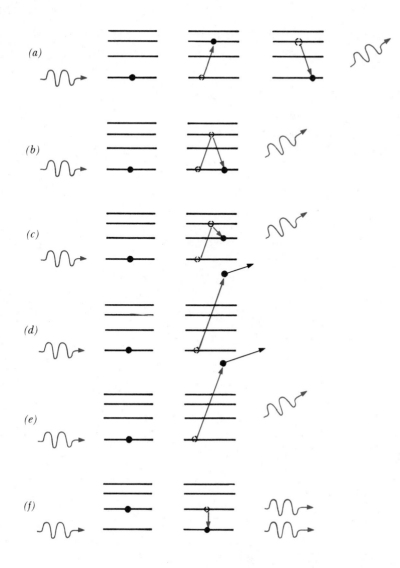

Figure 8-19
Description of photon interactions with an atom. In (*a*) the photon is absorbed and the atom, in an excited state, later emits a photon as it decays to a state of lower energy. This is a two-step process and the emitted photon is uncorrelated with the incident photon. The Rayleigh scattering process (*b*) and Raman scattering (*c*) differ from (*a*) in that they are single-step processes and there is a correlation between the incident and emitted photons. Parts (*d*) and (*e*) illustrate the photoelectric effect and Compton scattering discussed in Chapter 3. In (*f*) the atom, in an excited state, is stimulated to make a transition to a lower state by an incident photon of just the right energy. The emitted and incident photons have the same energy and are coherent.

lated emission and absorption were first worked out by Einstein. He showed that the probability of stimulating a transition from E_2 to E_1 is the same as that of absorption from state E_1 to E_2.

Stimulated emission is important because the resulting light is *coherent,* i.e., the phase of the light emitted from one atom is related to that from each other atom. Because of this phase relation, interference of light from different atoms can be observed. In the more usual case of spontaneous emission, the phase of the light from one atom is unrelated to that from another atom and the light is called *incoherent.* Important applications of stimulated emission are the *maser* (*m*icrowave *a*mplification by *s*timulated *e*mission of *r*adiation) and the *laser* (*l*ight *a*mplification by *s*timulated *e*mission of *r*adiation). We shall describe only the basic idea of these important devices. (For more detailed discussions, see References 4 and 5.)

The laser and maser do not differ in theory but in the range of frequency of the electromagnetic spectrum in which they operate. The frequency is determined by the energy levels involved. Consider a system of atoms with the fraction f_1 in the ground state and f_2 in the first excited state, and incident radiation of frequency $f = (E_3 - E_1)/h$. The three processes of interest are absorption transitions from E_1 to E_3, stimulated emission from E_2 to E_1, and spontaneous emission from E_2 to E_1. If the second state has a long mean lifetime or if the incident radiation is very intense, we can neglect spontaneous transitions. Since the probabilities of absorption and stimulated emission are equal, the number of events of each kind will depend on the fractions f_1 and f_2. If there are more atoms in the excited state, emissions with two coherent photons coming out will occur more often than will absorption with the disappearance of a photon; in this case amplification of the incident light will occur. However, the population of the upper state will be more reduced by the emissions than increased by the lesser number of absorptions, so that eventually both states will be equally populated and equilibrium will result. For energy differences of the order of 1 eV, corresponding to optical frequencies, the fraction in the upper state will be very small at ordinary temperatures, because kT is much less than 1 eV. The ratio of the fraction in the upper state to that in the lower state is

$$\frac{f_2}{f_1} = \frac{g_2}{g_1} e^{-(E_2 - E_1)/kT}$$

which is always less than g_2/g_1, approaching g_2/g_1 as T approaches infinity.

Amplification by stimulated emission depends on the possibility of obtaining more electrons in the higher state than in the lower state. This situation is called *population inversion*. The most common way of achieving population inversion is by a method called *optical pumping*, which is essentially the intense irradiation of the maser or laser with auxiliary radiation in order to excite many atoms into a higher energy state. Figure 8-20 shows the important energy levels for a three-level laser (such as the ruby laser). The energy level labeled (2) has a long lifetime; such a state is called a *metastable state*. Intense incident radiation of frequency $f = (E_3 - E_1)/h$ (this is green light for the ruby laser)

Population inversion

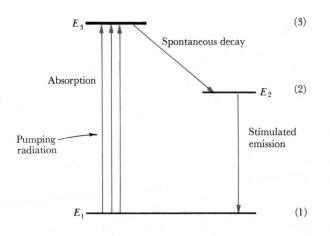

Figure 8-20
Energy levels of a three-level laser. In order to obtain amplification, the population of level (2) must be greater than that of the ground state. An intense auxiliary radiation is necessary to excite atoms from the ground state to the band of levels (3), which decay into level (2).

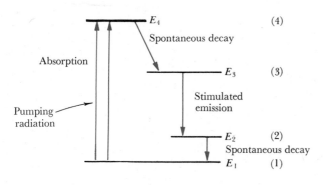

Figure 8-21
Energy levels of a four-
level laser. Since level (2)
is above the ground state,
it is sparsely populated if
the temperature is low
enough. Then population
inversion between levels
(2) and (3) is easily ob-
tained by exciting just a
few atoms from the
ground state to the band
of levels (4) which decay
into level (3).

causes many atoms initially in the ground state to make transitions to level (3), from which they decay quickly to the metastable level (2) or back to the ground state (1). The level labeled
(3) is actually a band of levels of nearly equal energy in a solid, so
the incident radiation need not be monochromatic. If the incident radiation is intense enough, more atoms will be transferred
to state (2) than remain in the ground state, with the result that
the population of these two states is inverted. Incident photons
of energy $E_2 - E_1$ from another source cause stimulated emission from the more populated level (2) to the ground state (1);
thus amplification takes place. In the ruby laser, the ends of the
crystal are polished, with one end totally reflecting and the other
partially reflecting, so that some of the beam is transmitted. If
the ends are parallel, an intense beam of coherent light emerges.

Figure 8-21 illustrates a four-level laser. In this case, transitions are stimulated between levels (3) and (2). Level (2) is sufficiently above the ground state so that there are few atoms in this
state at low temperatures. Atoms in the ground state are excited
to the level (or band of levels) (4) by absorption of the auxiliary
incident radiation of frequency $f = (E_4 - E_1)/h$. Spontaneous
transitions then populate level (3). Since (2) is not the ground
state, population inversion between levels (2) and (3) is
quickly accomplished. Incident radiation of frequency $f = (E_3 - E_2)/h$ stimulates transitions to state (2), and amplification results. In both types of laser, the population of the upper
state is maintained by "pumping" with the auxiliary radiation.

Questions

1. How does Rayleigh scattering differ from resonance absorption?

2. How does the photoelectric effect differ from all the other
processes illustrated in Figure 8-19?

3. Why is stimulated emission usually not observed?

4. What are the advantages of a four-level laser over a three-
level laser?

Summary

Atoms and molecules join together by means of one or more of several bonding mechanisms: ionic, covalent, van der Waals, hydrogen, or metallic. Ionic and covalent bonds are the strongest of these.

The shapes of such polyatomic molecules as H_2O and NH_3 can be understood from the spatial distribution of the atomic- or molecular-orbital wave functions. The tetravalent nature of the carbon atom is a result of the hybridization of the $2s$ and $2p$ atomic orbitals.

The energy states of diatomic molecules consist of rotational bands superimposed on the more widely spaced vibrational levels, which in turn are superimposed on the much more widely spaced energy levels due to excitation of the atomic electrons. Infrared absorption spectra involve only the rotational bands of the ground-state and the first-excited-state vibrational levels in the ground-state electronic level, because the molecules are all initially in the lowest vibrational state $n = 0$, and Δn must be ± 1. The absorption spectrum for diatomic molecules consists of equally spaced maxima separated by $\Delta E = 2(\hbar^2/2I)$, with a gap of twice this separation at the energy of vibration $\hbar\omega$.

A photon incident on an atom can be absorbed or scattered elastically or inelastically. If the photon energy is greater than the ionization energy of the atom, Compton scattering or the photoelectric effect can occur. If the atom is initially in an excited state, an incident photon of the proper energy can stimulate emission of another photon of the same energy. The incident and emitted photons are in phase and travel parallel to each other. Masers and lasers are important applications of stimulated emission. Amplification by stimulated emission depends on the possibility of obtaining population inversion, in which there are more atoms in an excited state than in the ground state or another excited state of lower energy. Population inversion is usually obtained by optical pumping.

References

1. M. Alonso and E. Finn, *Fundamental University Physics, Vol. III: Quantum and Statistical Physics*, Reading, Mass.: Addison-Wesley Publishing Company, Inc., 1968.

2. L. Pauling, *The Chemical Bond*, Ithaca, N.Y.: Cornell University Press, 1967.

3. W. Finkelnburg, *Structure of Matter*, New York: Academic Press, Inc., 1964.

4. B. Lengyel, *Introduction to Laser Physics*, New York: John Wiley & Sons, Inc., 1966.

5. The following *Scientific American* articles are available as reprints, published by W. H. Freeman and Company, San Francisco: J. Gordon, "The Masers," December 1958 (reprint 215); A.

Schawlow, "Optical Masers," June 1961 (reprint 274); "Advances in Optical Masers," July 1963 (reprint 294); E. Leith and J. Upatnieks, "Photography by Laser," June 1965 (reprint 300).

Exercises

Section 8-1, Molecular Bonding

1. The equilibrium separation of the K^+ and Cl^- ions in KCl is about 2.79 Å. (a) Calculate the potential energy of attraction of the ions, assuming them to be point charges. (b) The ionization energy of potassium is 4.34 eV, and the electron affinity of Cl is 3.82 eV. Find the dissociation energy, neglecting the energy of repulsion. (c) The measured dissociation energy is 4.43 eV. What is the energy due to repulsion of the ions?

√2. What type of bonding mechanism would you expect for (a) the HCl molecule, (b) the O_2 molecule, (c) Cu atoms in a solid?

3. The dissociation energy is sometimes expressed in kcal/mole. (a) Find the relation between the units eV/molecule and kcal/mole. (b) Find the dissociation energy of molecular NaCl in kcal/mole.

4. The equilibrium distance between the K^+ and F^- ions in KF is 2.55 Å. (a) Calculate the potential energy of attraction of the ions at this distance, assuming them to be point charges. (b) The ionization energy of potassium is 4.34 eV and the electron affinity of fluorine is 4.07 eV. Find the dissociation energy, neglecting the energy of repulsion.

Section 8-2, Polyatomic Molecules

5. From the electron configurations given in Table 7-1, what other elements might you expect to exhibit the same type of hybridization as carbon?

Section 8-3, Energy Levels and Spectra of Diatomic Molecules

6. Indicate the mean value of r for two vibration levels in the potential-energy curve for a diatomic molecule and show that because of the asymmetry in the curve, r_{av} increases with increasing vibration energy, and therefore solids expand when heated.

7. For the O_2 molecule, the separation of the atoms is 1.21 Å. Calculate the characteristic rotational energy $E_{0r} = \hbar^2/2I$ in eV.

8. The characteristic energy of rotation of the N_2 molecule is $E_{0r} = \hbar^2/2I = 2.48 \times 10^{-4}$ eV. From this find the separation distance of the N atoms in N_2.

9. The characteristic rotational energy $E_{0r} = \hbar^2/2I$ for KCl is 1.43×10^{-5} eV. (a) Find the reduced mass for the KCl molecule. (b) Find the separation distance of the K^+ and Cl^- ions.

10. (a) Explain why the moment of inertia of a diatomic molecule increases with increasing angular momentum for the same vibrational level. (b) Explain why the moment of inertia of a diatomic molecule increases with increasing vibration energy.

Section 8-4, Absorption, Scattering, and Stimulated Emission

√11. What is the minimum fraction of the atoms that must be excited in a ruby laser to obtain amplification? Is this also true for a four-level laser?

12. Why does cooling increase the efficiency of a four-level laser but not a three-level laser?

Problems

1. (a) Calculate the electrostatic potential energy of Na^+ and Cl^- ions at their equilibrium separation distance of 2.4 Å, assuming the ions to be point charges. (b) What is the energy of repulsion at this separation? (c) Assume that the energy of repulsion varies as C/r^n. From Figure 8-1, this energy equals ke^2/r at about $r = 1.4$ Å. Use this and your answer to part (b) to calculate n and C. (Though this calculation is not very accurate, the energy of repulsion does vary much more rapidly with r than does the energy of attraction.)

2. Derive the following expression for the moment of inertia of a diatomic molecule: $I = \mu r_0^2$, where r_0 is the distance between the masses, and μ is the reduced mass

$$\mu = \frac{m_1 m_2}{m_1 + m_2}$$

3. The central frequency for the absorption band of HCl, shown in Figure 8-18, is $f_0 = 8.66 \times 10^{13}$ Hz and the absorption peaks are separated by $\Delta f \approx 6 \times 10^{11}$ Hz. (a) What is the magnitude of the zero-point vibration energy for HCl? (b) What is the moment of inertia of HCl? (c) Calculate the reduced mass of HCl, and from part (b), find the equilibrium separation of the atoms.

4. What is the ratio of the number of H_2 molecules in states $l = 2$ and $l = 1$ to the number in the $l = 0$ rotational state at $T = 300$ K? The statistical weights are $g = 2l + 1$.

√5. The separation of the ^{127}I and ^{39}K ions in the KI molecule is about 2.79 Å. (a) What is the reduced mass of this molecule? (b) What is the characteristic energy of rotation $\hbar^2/2I$ in eV? (c) What do you expect for the separation frequency Δf between two adjacent lines in the absorption spectrum for this molecule?

6. Do Problem 4 for HCl, which has a characteristic rotational energy of $E_{0r} = \hbar^2/2I = 1.23 \times 10^{-3}$ eV.

7. (a) Calculate the reduced mass μ for the $H^{35}Cl$ and $H^{37}Cl$ molecules and the fractional difference $\Delta\mu/\mu$. (b) Show that the mixture of isotopes in HCl leads to a fractional difference in the frequency of a transition from one rotational state to another given by $\Delta f/f = \Delta\mu/\mu$.

8. The state E_2 (see Figure 8-21) of the Pr^{3+} ion used in a four-level laser is about 4.2×10^{-2} eV above the ground state. What is the ratio of the number of ions in this level to the number in the ground state at $T = 300$ K (with no excitation radiation)? What is this ratio at $T = 77$ K? (Assume equal statistical weights.)

CHAPTER 9   Some Properties of Solids

Objectives

After studying this chapter you should:

1. Know what the Madelung constant is and why it differs from 1.

2. Be able to discuss the successes and failures of the classical free-electron theory of metals.

3. Know what the Fermi energy is and be able to derive the expression for it in one dimension at $T = 0$.

4. Be able to discuss the modifications of the classical free-electron theory introduced because of the exclusion principle and the wave nature of electron scattering.

5. Know the origin of the band structure of solids and be able to discuss the differences in band structure of conductors, insulators, and intrinsic semiconductors.

6. Know why the temperature coefficient of electrical conduction is positive for some materials and negative for others.

7. Be able to discuss the effect of adding impurities to intrinsic semiconductors.

8. Know the general features of the current-voltage curve for a typical semiconductor pn junction.

9. Be able to describe the characteristics and operation of a transistor.

10. Be able to describe the BCS theory of superconductivity.

The many and varied properties of solids have fascinated and intrigued mankind for centuries. The technological development of metals and alloys has shaped the course of civilizations, while the symmetry and beauty of naturally occurring large single crystals have captured man's imagination. Yet the origins of the physical properties of solids were not understood even in rudimentary form until the development of quantum mechanics. The application of quantum mechanics to solids has provided the basis for much of the technological progress of modern times. We shall study briefly some aspects of the structure of solids in Section 9-1 and then concentrate on their electrical and thermal properties.

9-1   Structure of Solids

In our everyday world we see matter in three phases: gases, liquids, and solids. In a gas the average distance between two atoms or molecules is large compared with the size of an atom or molecule. The molecules have little influence on one another, except during their frequent but brief collisions. In a liquid or solid the atoms or molecules are close together and exert forces on one another comparable to the forces which bind atoms into molecules. (There is a fourth phase of matter, plasma, which occurs only at very high temperatures, inside stars and in the laboratory. A plasma is a gas of ions and electrons. The properties of a plasma are very different from those of an ordinary gas because of the long-range electrical and magnetic effects arising from the charges of the particles.) In a liquid, the molecules form temporary short-range bonds which are continually broken and reformed as a result of the thermal kinetic energy of the molecules. The strength of the bonds depends on the type of molecule. For example, the bonds between helium atoms are very weak van der Waals bonds, and He does not liquefy at atmospheric pressure until the very low temperature of 4.2 K is reached.

If a liquid is slowly cooled, the kinetic energy of its molecules is reduced and the molecules will arrange themselves in a regular crystalline array, producing the maximum number of bonds and leading to a minimum potential energy. However, if the liquid is cooled rapidly so that its internal energy is removed before the molecules have a chance to arrange themselves, a solid is often formed that is not crystalline but resembles a "snapshot" of a liquid. Such a solid is called *amorphous*; it displays short-range order but not the long-range order (over many atomic diameters) characteristic of a crystal. Glass is a typical amorphous solid. It is characteristic of the long-range ordering of a crystal that it has a well-defined melting point whereas an amorphous solid merely softens as its temperature is increased. Many materials may solidify in either an amorphous or a crystalline state, depending on how they are prepared. Others exist only in one form or the other. Most common solids are polycrystalline—i.e., they are collections of single crystals that exhibit long-range order over many atomic diameters. The size of such single crystals is typically a fraction of a millimeter; however, large single crystals occur naturally and can be produced artificially (Figure 9-1). We shall discuss only simple crystalline solids in this chapter.

The most important property of a single crystal is its symmetry and regularity of structure: it can be thought of as a single unit structure repeated throughout the solid. The smallest unit of a crystal is called the *unit cell*. The structure of the unit cell depends on the type of bonding between the atoms, ions, or molecules in the crystal. If more than one kind of atom is present, the structure will also depend on their relative size. The bonding mechanisms are those discussed in Section 8-1: ionic, covalent,

Amorphous solids and crystals

Sawyer Research Products, Inc.

Figure 9-1
Single quartz crystal.

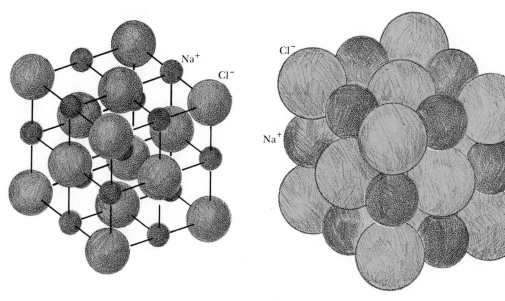

Figure 9-2
Structure of face-
centered-cubic NaCl
crystal.

metallic, hydrogen, and van der Waals. Figure 9-2 shows the
structure of the ionic crystal NaCl. The Na^+ and Cl^- ions are
spherically symmetric, with the Cl^- ion approximately twice as
large as the Na^+ ion. The minimum potential energy of this
crystal occurs when an ion of either kind has six nearest
neighbors of the other kind. This structure is called face-
centered cubic (fcc). Note that the Na^+ and Cl^- ions are *not*
paired into NaCl molecules in solid NaCl.

The net attractive part of the potential energy of an ion in a
crystal can be written

$$V_{att} = -\alpha \frac{ke^2}{r} \qquad \text{9-1}$$

where r is the separation distance between neighboring Na and
Cl ions (which is 2.81 Å), and α, called the *Madelung* constant,
depends on the geometry of the crystal. If only the six nearest
neighbors were important, α would be 6. However, each ion sees
not only 6 neighbors of the opposite charge at a distance r, but
also 12 ions of the same charge at a distance $\sqrt{2}r$, 8 ions of opposite charge at distance $\sqrt{3}r$, etc. The Madelung constant is
thus

Madelung constant

$$\alpha = +6 - \frac{12}{\sqrt{2}} + \frac{8}{\sqrt{3}} - \cdots \qquad \text{9-2}$$

The calculation of this sum is difficult because the convergence
is very slow. The result for face-centered-cubic structures is
$\alpha = 1.7476$.

When Na^+ and Cl^- ions are very close together they repel
each other because of the overlap of their electrons and the
Pauli exclusion principle. This repulsion mechanism is the same
as that for molecules discussed in Chapter 8. A simple empirical

expression for the repulsive part of the potential energy that works fairly well is

$$V_{rep} = +\frac{A}{r^n}$$

The total potential energy of an ion is then

$$V = -\alpha \frac{ke^2}{r} + \frac{A}{r^n} \qquad \text{9-3}$$

The equilibrium separation $r = r_0$ is that at which the force $F = -dV/dr$ is zero. Differentiating and setting $dV/dr = 0$ we obtain $A = \alpha ke^2 r_0^{n-1}/n$. The total potential energy can therefore be written

$$V = -\alpha \frac{ke^2}{r_0} \left[\frac{r_0}{r} - \frac{1}{n} \left(\frac{r_0}{r} \right)^n \right] \qquad \text{9-4}$$

Potential energy of an atom in a crystal

At $r = r_0$ we have

$$V(r_0) = -\alpha \frac{ke^2}{r_0} \left(1 - \frac{1}{n} \right) \qquad \text{9-5}$$

The value of n can be found approximately from the *dissociation energy* of the crystal, which is the energy needed to break up the crystal into atoms in the case of NaCl, it is approximately 184 kcal/mole. The result is $n \approx 9$.

Example 9-1   Calculate the equilibrium spacing r_0 for NaCl from the measured density $\rho = 2.16$ g/cm³. We consider each ion to occupy a cubic volume of r_0^3. The mass of Avogadro's number of Na⁺ ions plus Avogadro's number of Cl⁻ ions is the sum of the atomic weights of Na and Cl, which is 58.4 g. These ions occupy a volume of $2N_A r_0^3$. The density is thus

$$\rho = \frac{58.4 \text{ g}}{2(6.02 \times 10^{23}) r_0^3} = 2.16 \text{ g/cm}^3$$

Then

$$r_0^3 = \frac{58.4}{2(6.02 \times 10^{23})(2.16)} \text{ cm}^3 = 2.24 \times 10^{-23} \text{ cm}^3$$

$$r_0 = 2.82 \times 10^{-8} \text{ cm} = 2.82 \text{ Å}$$

Example 9-2   Calculate the dissociation energy of NaCl using $n = 9$ in Equation 9-5. Equation 9-5 gives the potential energy for each ion pair. There are N_A ion pairs in a mole of NaCl. Using $r_0 = 2.82$ Å and $n = 9$, and $ke^2 = 14.4$ eV-Å, we obtain

$$V = -(1.7476) \frac{14.4 \text{ eV-Å}}{2.82 \text{ Å}} \left(1 - \frac{1}{9} \right)$$

$$= -7.93 \text{ eV/ion pair}$$

The dissociation energy is then $D = -V = 7.93$ eV/ion pair. A useful conversion factor is 1 eV/molecule = 23.052 kcal/mole.

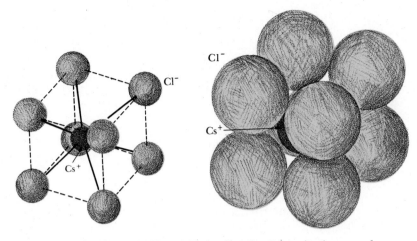

Figure 9-3
Structure of body-centered-cubic CsCl crystal.

Then $D = (7.93)(23.052) = 183$ kcal/mole. Most ionic crystals, such as LiF, NaI, KF, KCl, KI, AgCl, etc., have face-centered-cubic structure. Other solids that have this structure are Ag, Al, Au, Ca, Cu, Ni, Pb, etc.

Figure 9-3 shows the structure of CsCl, which is called body-centered cubic (bcc). In this structure, each ion has eight nearest neighbor ions of the opposite charge. The Madelung constant for ionic crystals with body-centered-cubic structure is 1.7627. Monatomic solids with this structure include Ba, Cs, Fe, K, Li, Mo, and Na.

Figure 9-4 illustrates another important crystal structure called hexagonal close-packed (hcp). This is the structure obtained by stacking identical spheres such as bowling balls. In one layer, each ball touches six others; hence the name hexagonal. In the next layer, each ball fits into the triangular depressions of the first layer. In the third layer, each ball fits into a triangular depression of the second layer such that it lies directly over a ball in the first layer. (There is another method of stacking the third layer, so that a ball does not lie directly over one in the first layer, but this structure can be shown to be equivalent to face-centered-cubic structure.) Elements with hexagonal close-packed crystal structure include Be, Cd, Ce, Mg, Os, Zn, Zr, and others.

In some solids with covalent bonding the crystal structure is determined by the directional nature of the bonds. Figure 9-5 illustrates the diamond structure of carbon (which is also the structure of Ge and Si), in which each atom is bonded to four others as a result of the sp^3 hybridization discussed in Chapter 8. The diamond structure can be considered to be two interpenetrating face-centered-cubic structures.

Figure 9-4
Hexagonal close-packed crystal structure.

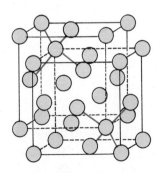

Figure 9-5
Diamond crystal structure showing how this structure can be considered to be a combination of two interpenetrating face-centered-cubic structures.

Questions

1. Why is r_0 different for solid NaCl than for the diatomic molecule?

2. Why doesn't NaCl have hexagonal close-packed structure?

9-2   Classical Free-Electron Theory of Metals

Because metals conduct electricity so readily, there must be charges in metals that are relatively free to move. The idea that metals contain electrons free to move about through a lattice of relatively fixed positive ions was proposed by Drude and Thomson around 1900, just three years after Thomson's discovery of the electron, and developed by Lorentz about 1909. This model successfully predicts Ohm's law of conduction and relates electrical conduction and heat conduction to the motion of free electrons in conductors. However, the model gives the wrong temperature dependence for electrical conductivity, and it predicts that the heat capacity of metals should be greater than that of insulators by $\frac{3}{2}R$ per mole, which is not observed. Despite these failures, the classical free-electron theory is a good starting point for a more sophisticated treatment of metals based on quantum mechanics. The main defects in the classical theory are the use of the classical Maxwell-Boltzmann distribution function for electrons in a metal and the treatment of the scattering of electrons by the lattice as a classical particle scattering.

In the Drude model, a metal is pictured as a regular three-dimensional array of atoms or ions with a large number of electrons free to move about the whole metal. In copper, for example, there is approximately one free electron per copper atom. The density of free electrons can be measured using the Hall effect (see Section 9-6). In the absence of an electric field, the free electrons move about the metal much like gas molecules in a container. Unlike an ordinary gas, however, thermal equilibrium is maintained by collisions of electrons with the lattice ions rather than by collision of electrons with each other. The mean speed of electrons can be calculated from the equipartition theorem. The result is the same as that for ideal-gas molecules, with the electron mass m_e instead of the molecular mass in Equation 2-31. For example, at temperature $T = 300$ K the mean speed is

$$\bar{v} = \sqrt{\frac{8kT}{\pi m_e}} = \sqrt{\frac{8(1.38 \times 10^{-23} \text{ J/K})(300 \text{ K})}{\pi (9.11 \times 10^{-31} \text{ kg})}}$$

$$= 1.08 \times 10^5 \text{ m/sec} \qquad\qquad 9\text{-}6$$

Electrical Conduction

In the presence of an electric field, there is a small *drift velocity* superimposed on the large thermal velocity. This drift velocity **u** is in the direction opposite to that of the field, since the electron charge is negative. This drift of negative charge opposite to the electric field constitutes an electric current along the field direction. It is well known that the current in a conductor is proportional to the applied voltage over a wide range of voltages. This relation, known as Ohm's law, can be written

$$I = \frac{V}{R}$$

where R is the *resistance*. The resistance of a wire is proportional to the length and inversely proportional to the cross-sectional area

$$R = \rho \frac{L}{A}$$

The constant of proportionality ρ, called the *resistivity*, is a property of the conductor. We can thus write Ohm's law

$$I = \frac{A}{\rho} \frac{V}{L}$$

For a uniform electric field $\mathscr{E}$, the voltage is just $V = \mathscr{E}L$. In terms of the current density $j = I/A$, Ohm's law is thus

$$j = \frac{1}{\rho} \mathscr{E} = \sigma \mathscr{E} \qquad\qquad 9\text{-}7 \qquad \textit{Ohm's law}$$

where σ is called the *conductivity*. The objective of the classical theory of conductivity is to find an expression for ρ or σ in terms of the properties of metals.

If there are n electrons per unit volume, moving with an average speed u (the drift velocity) parallel to the wire, the current density is

$$j = neu \qquad\qquad 9\text{-}8$$

We can estimate the drift velocity for a typical case.

Example 9-3   What is the drift velocity for a current of 1 A in a No. 14 copper wire (diameter of 0.064 in = 0.163 cm)? If there is one free electron per atom, n is the same as the number density of copper atoms:[1]

$$n = \frac{(6.02 \times 10^{23} \text{ atoms/mole})(8.96 \text{ g/cm}^3)}{63.5 \text{ g/mole}}$$
$$= 8.49 \times 10^{22} \text{ atoms/cm}^3$$

Then

$$u = \frac{j}{ne} = \frac{I}{Ane}$$
$$= \frac{1 \text{ C/sec}}{(\pi/4)(0.163 \text{ cm})^2(8.49 \times 10^{22} \text{ cm}^{-3})(1.60 \times 10^{-19} \text{ C})}$$
$$= 3.53 \times 10^{-3} \text{ cm/sec}$$

We see that typical drift velocities are very small.

---

[1] The measured value is 8.47×10^{22} electrons/cm^3, as given in Table 9-2, page 323.

At first glance it is surprising that any material obeys Ohm's law since, in the presence of an electric field, a free electron experiences a force of magnitude $e\mathscr{E}$. If this were the only force, the velocity of the electron would increase with acceleration $e\mathscr{E}/m_e$. Instead, Ohm's law implies that there is a steady-state situation in which the drift velocity u is proportional to the electric field $\mathscr{E}$, since j is proportional to $\mathscr{E}$ and to u. Of course, other forces act on the electrons when they make collisions with the lattice ions of the metal. In the classical model it is assumed that after an electron makes such a collision its velocity is completely unrelated to that before the collision. The justification for this assumption is that the drift velocity is very small compared with the random thermal velocity. We therefore picture the electron as being accelerated by the electric field for a short time between collisions with the ions and thereby acquiring a small drift velocity. After each collision that gives it a random velocity, the electron is again accelerated by the field. The motion of such an electron is illustrated in Figure 9-6, in which the drift velocity is greatly exaggerated. In Figure 9-7b a small drift velocity **u** has been added to the random electron velocities of Figure 9-7a. We can relate the drift velocity to the electric field by ignoring the random thermal velocities of the electrons and assuming the electron starts from rest after each collision.

The average time since the electron's last collision is called the *relaxation time*, τ; this is also the average time between collisions.[1] Since the acceleration is $e\mathscr{E}/m_e$, the speed of this electron will then be

$$u = \frac{e\mathscr{E}}{m_e}\tau \qquad\qquad 9\text{-}9$$

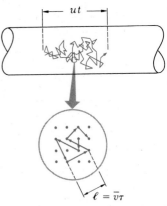

Figure 9-6
Path of electron in a wire. Superimposed on the random thermal motion is a slow drift at speed u in the direction of the electric force $e\mathscr{E}$. The mean free path ℓ, the mean time between collisions τ, and the mean speed $\bar{v}$ are related by $\ell = \bar{v}\tau$.

[1] It is tempting but incorrect to think that for an electron picked at random, the average time since its last collision is $\tfrac{1}{2}\tau$ where τ is the average time between collisions. If you find this statistical result confusing you may take comfort in the fact that Drude used the incorrect result $\tfrac{1}{2}\tau$ in his original work.

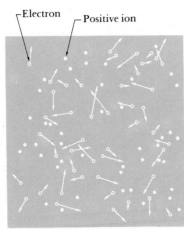

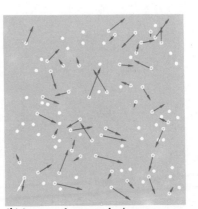

(a) Average electron velocity = 0   (b) Average electron velocity = ⟶

Figure 9-7
(a) Stationary positive ions and electrons with a random distribution of velocities. (b) A small drift velocity to the right has been added to each electron velocity, as indicated for the electron in the lower left corner. [*Adapted from E. M. Purcell,* Electricity and Magnetism, *Berkeley Physics Course, New York: McGraw-Hill Book Company, 1965, vol. II, p. 123. Courtesy of Education Development Center, Inc., Newton, Mass.*]

Using this result in Equation 9-8 we obtain

$$j = \frac{ne^2\tau}{m_e} \mathscr{E} \qquad\qquad 9\text{-}10$$

We thus have Ohm's law with the conductivity given by

$$\sigma = \frac{ne^2\tau}{m_e} \qquad\qquad 9\text{-}11$$

If $\bar{v}$ and ℓ are the mean speed and mean free path of the electrons, the relaxation time is

$$\tau = \frac{\ell}{\bar{v}} \qquad\qquad 9\text{-}12$$

Then

$$\sigma = \frac{ne^2\ell}{m_e\bar{v}} \qquad\qquad 9\text{-}13a$$

Classical predictions for σ and ρ

and the resistivity is

$$\rho = \frac{1}{\sigma} = \frac{m_e\bar{v}}{ne^2\ell} \qquad\qquad 9\text{-}13b$$

We can estimate the mean free path from Equation 2-39 (page 76), replacing d by r, the radius of the copper atom, and assuming the size of the electron to be negligible. Using $r \approx 10^{-10}$ m $= 10^{-8}$ cm we obtain

$$\ell = \frac{1}{n\pi r^2} \approx \frac{1}{(8.49 \times 10^{22} \text{ cm}^{-3})\pi(10^{-8} \text{ cm})^2}$$

$$\approx 3.7 \times 10^{-8} \text{ cm} = 3.7 \text{ Å}$$

Using $\bar{v} \approx 10^5$ m/sec from Equation 9-6 we obtain for the relaxation time:

$$\tau = \frac{\ell}{\bar{v}} \approx \frac{3.7 \times 10^{-8} \text{ cm}}{10^7 \text{ cm/sec}} = 3.7 \times 10^{-15} \text{ sec}$$

Using these results to compute the conductivity for copper at $T = 300$ K, we obtain $\sigma \approx 9 \times 10^6/\Omega$-m. This is about six times smaller than the experimental value. In addition to this discrepancy, the temperature dependence is incorrect because the measured resistance is proportional to the absolute temperature, whereas from Equation 9-13b,

$$\rho \propto \bar{v} \propto \sqrt{T}$$

The numerical discrepancy is even worse at lower temperatures. We see that, though this classical model does predict Ohm's law, both the numerical value and the temperature dependence of σ or ρ are incorrect.

Heat Conduction

Good conductors of electricity are also good conductors of heat. The classical theory assumes that this is because the electron gas is mainly responsible for heat conduction in metals. The coefficient of heat conduction or thermal conductivity K of a gas (Equation 2-49) is

$$K = \frac{1}{3} \frac{n\bar{v}\ell C_v}{N_A} \qquad 9\text{-}14$$

The molar heat capacity for an electron gas obeying the Maxwell-Boltzmann distribution is

$$C_v = \tfrac{3}{2}R = \tfrac{3}{2}N_A k$$

Using this, the thermal conductivity can be written

$$K = \tfrac{1}{2}n\bar{v}\ell k \qquad 9\text{-}15$$

Thus K and σ are related by

$$\frac{K}{\sigma} = \frac{\tfrac{1}{2}n\bar{v}\ell k}{ne^2\ell/m_e\bar{v}} = \frac{m_e\bar{v}^2 k}{2e^2} = \frac{4k^2 T}{\pi e^2} \qquad 9\text{-}16$$

where we have used Equation 9-6 for $\bar{v}$. The classical theory therefore predicts that the ratio of thermal to electrical conductivity is proportional to the absolute temperature, and that the proportionality constant is the same for any metal. This is known as the *Wiedemann-Franz law*. The ratio $K/\sigma T$ is called the *Lorentz number:*

Wiedemann-Franz law

$$L = \frac{K}{\sigma T} = \frac{4k^2}{\pi e^2} \approx 1.0 \times 10^{-8}\ \text{W-}\Omega/\text{K}^2 \qquad 9\text{-}17$$

Lorentz number

Table 9-1 shows that $K/\sigma T$ is indeed nearly the same for all metals and is independent of temperature, though the numerical value is somewhat higher than predicted.[1] Because of the simplicity of the model, we can only hope for an order-of-magnitude agreement. The important test of the model is that, though K and σ vary greatly with temperature and from metal to metal, the ratio $K/\sigma T$ does not.

[1] Drude's mistake of using $u = \tfrac{1}{2}e\mathscr{E}\tau/m$ led him to a prediction of the Lorentz number twice as large as this and, in fact, in much better agreement with experiment.

Table 9-1
Lorentz number $L = K/\sigma T$ in units of 10^{-8} W-Ω/K^2, for several metals at 0°C and 100°C

Metal	0°C	100°C	Metal	0°C	100°C
Ag	2.31	2.37	Pb	2.47	2.56
Au	2.35	2.40	Pt	2.51	2.60
Cd	2.42	2.43	Sn	2.52	2.49
Cu	2.23	2.33	W	3.04	3.20
Ir	2.49	2.49	Zn	2.31	2.33
Mo	2.61	2.79			

From C. Kittel, Reference 3.

Heat Capacity

If the electron gas has a Maxwellian distribution, it should have a mean kinetic energy $\frac{3}{2}kT$, and we would expect the molar heat capacity of a metal to be $\frac{3}{2}R$ greater than that of an insulator; that is,

$$C_v = (3R)_{\text{lattice vibrations}} + (\tfrac{3}{2}R)_{\text{electron gas}} = \tfrac{9}{2}R$$

This is not observed. The molar heat capacity of metals is very nearly $3R$. At higher temperatures it is slightly greater, but the increase is nowhere near the value of $\frac{3}{2}R$ predicted by the classical theory. The increase is, in fact, proportional to temperature, and at $T = 300$ K it is only about $0.02R$.

Question

3. In the classical free-electron model the electron loses energy (on the average) in a collision, since it loses the drift velocity it has picked up since the last collision. Where does this energy appear?

9-3 The Fermi Electron Gas

One of the difficulties of the classical free-electron theory of metals is connected with the assumption that the Maxwell-Boltzmann distribution applies to the electron gas. In this section we shall examine the effects of quantum mechanics on the energy distribution of the electrons. Because of the exclusion principle, the energy distribution of electrons in a metal is not even approximately Maxwellian. We shall first consider the energy distribution at $T = 0$. This can be calculated rather easily, and is a good approximation to the distribution at other temperatures. Even for temperatures as high as several thousand degrees, the energy distribution of an electron gas does not differ very much from that at $T = 0$.

The Fermi Energy at $T = 0$

Classically, at $T = 0$, all the electrons would have zero kinetic energy. As the conductor is heated, the lattice ions acquire an average kinetic energy of $\frac{3}{2}kT$, which is imparted to the electron gas by interactions of the lattice with the electrons (the interactions we have called *collisions*). The electrons are expected to have a mean kinetic energy of $\frac{3}{2}kT$ in equilibrium; those with larger energies lose energy, on the average, in a collision with a lattice ion, and those with less energy gain on the average.

Since the electrons are confined to the space occupied by the metal, it is clear from the uncertainty principle that even at $T = 0$ an electron cannot have zero kinetic energy. Furthermore, the exclusion principle prevents more than two electrons (with opposite spins) from being in the lowest energy level. At $T = 0$, we expect the electrons to have the lowest energies consistent with the exclusion principle. It is instructive to consider a one-dimensional model first.

Consider N electrons in a one-dimensional infinite square well of size L. The lowest energy state is

$$E_1 = \frac{h^2}{8m_e L^2}$$

and the energy levels are given by

$$E_n = n^2 E_1 = n^2 \frac{h^2}{8m_e L^2} \qquad \text{9-18}$$

We can put two electrons in the level $n = 1$, two in the level $n = 2$, etc. The N electrons will thus fill up $N/2$ levels (from $n = 1$ to $n = N/2$). The energy of the last-filled level (or half-filled if N is odd) is called the *Fermi energy* at $T = 0$. We can calculate this energy for N electrons by setting $n = N/2$ in Equation 9-18:

$$E_F = E_{N/2} = \left(\frac{N}{2}\right)^2 E_1 = \frac{h^2}{32m_e}\left(\frac{N}{L}\right)^2 \qquad \text{9-19}$$

Fermi energy in one dimension

We see that the Fermi energy is a function of the number of electrons per unit length, which is the *number density* in one dimension. The number density of electrons in copper is $8.49 \times 10^{22}/\text{cm}^3$, assuming one free electron per atom. In one dimension, this corresponds to

$$\frac{N}{L} = (8.49 \times 10^{22}/\text{cm}^3)^{1/3} = 4.40 \times 10^7/\text{cm} = 0.440/\text{Å}$$

The Fermi energy is then

$$\begin{aligned} E_F &= \frac{(hc)^2}{32m_e c^2}\left(\frac{N}{L}\right)^2 = \frac{(12{,}400 \text{ eV-Å})^2(0.44/\text{Å})^2}{(32)(5.11 \times 10^5 \text{ eV})} \\ &\approx 1.82 \text{ eV} \end{aligned}$$

The average energy is the total energy divided by the number of particles:

$$\overline{E} = \frac{1}{N}\sum_{n=1}^{N/2} 2n^2 E_1$$

Since $N/2 \gg 1$, we can approximate the sum by an integral:

$$\sum_1^{N/2} n^2 \approx \int_0^{N/2} n^2 \, dn = \frac{1}{3}\left(\frac{N}{2}\right)^3$$

Thus

$$\overline{E} = \frac{2E_1}{N}\frac{1}{3}\left(\frac{N}{2}\right)^3 = \frac{1}{3}\left(\frac{N}{2}\right)^2 E_1 = \tfrac{1}{3}E_F \qquad \text{9-20}$$

Our one-dimensional calculation thus gives an average energy of about 0.6 eV at $T = 0$. The temperature at which the average energy would be 0.6 eV for a one-dimensional Maxwell distribution is about 14,000 K, obtained from $\tfrac{1}{2}kT = 0.6$ eV.

Since the energy states are so close together, we can assume that they are continuous (Figure 9-8). Let $n(E) \, dE$ be the num-

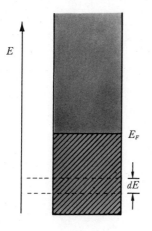

Figure 9-8
Energy levels in a one-dimensional square well. The Fermi energy at $T = 0$ is that of the highest occupied level. The levels are so closely spaced they can be assumed continuous. The density of states $g(E)$ is the number of states between E and $E + dE$ divided by dE.

ber of particles with energy between E and $E + dE$. We can write this distribution function as

$$n(E) \, dE = g(E) \, dE \, F \qquad \qquad 9\text{-}21$$

where $g(E) \, dE$ is the number of states in dE, and the *Fermi factor* F is the probability that a state will be occupied. At $T = 0$,

$$
\begin{aligned}
F &= 1 \qquad \text{for } E < E_F \\
F &= 0 \qquad \text{for } E > E_F
\end{aligned}
\qquad \qquad 9\text{-}22
$$

Fermi factor at $T = 0$

The density of states $g(E)$ in one dimension is given by

$$g(E) = 2 \frac{dn}{dE}$$

where $E = n^2 E_1$ and the 2 is for the two spin states per space state. Then $dE = 2nE_1 \, dn = 2E^{1/2}E_1^{1/2} \, dn$, and

$$g(E) = E_1^{-1/2} E^{-1/2} \qquad \qquad 9\text{-}23$$

The energy distribution function $n(E)$ is then

$$n(E) = E_1^{-1/2} E^{-1/2} F$$

In three dimensions it is a little more difficult to count the number of states. We shall state the three-dimensional results analogous to Equations 9-19, 9-20, and 9-23 and derive them at the end of this section. The Fermi energy is given by

$$E_F = \frac{h^2}{8m_e}\left(\frac{3N}{\pi V}\right)^{2/3} \qquad \qquad 9\text{-}24$$

where V is the volume of the metal. As in one dimension, the Fermi energy depends on the number density N/V. Table 9-2 lists the free-electron number densities N/V for several elements.

Fermi energy

Table 9-2
Free-electron number densities for selected elements

Element	N/V (units of $10^{22}/\text{cm}^3$)
Al	18.1
Ag	5.86
Au	5.90
Cu	8.47
Fe	17.0
K	1.40
Li	5.70
Mg	8.61
Mn	16.5
Na	2.65
Sn	14.8
Zn	13.2

Example 9-4   Calculate the Fermi energy at $T = 0$ for copper. From Table 9-2 we have $N/V = 8.47 \times 10^{22}/\text{cm}^3 = 8.47 \times 10^{-2}/\text{Å}^3$. Then using Equation 9-24 we have

$$E_F = \frac{(hc)^2}{8 m_e c^2} \left(\frac{3}{\pi}\right)^{2/3} \left(\frac{N}{V}\right)^{2/3}$$

$$= \frac{(12{,}400 \text{ eV-Å})^2}{8(5.11 \times 10^5 \text{ eV})} \left(\frac{3}{\pi}\right)^{2/3} (8.47 \times 10^{-2}/\text{Å}^3)^{2/3}$$

$$= 7.03 \text{ eV}$$

Again, we note that E_F is much larger than kT at ordinary temperatures.

We can again write the energy distribution $n(E)$ as

$$n(E) = g(E)F \qquad\qquad 9\text{-}25$$

Distribution function

where the Fermi factor F is the same as in Equation 9-22, and the density of states $g(E)$ in three dimensions is

$$g(E) = \frac{\pi V}{2} \left(\frac{8 m_e}{h^2}\right)^{3/2} E^{1/2} = \frac{3N}{2} E_F^{-3/2} E^{1/2} \qquad\qquad 9\text{-}26$$

Density of states in three dimensions

The average energy at $T = 0$ is

$$\bar{E} = \tfrac{3}{5} E_F \qquad\qquad 9\text{-}27$$

Average energy

At higher temperatures, some electrons will gain energy and occupy higher-energy states. However, electrons cannot move to a higher- or lower-energy state unless it is unoccupied. Since the kinetic energy of the lattice ions is of the order of kT, electrons cannot gain much more energy than kT in collisions with the lattice; therefore only those electrons with energies within about kT of the Fermi energy can gain energy as the temperature is increased. At $T = 300$ K, kT is only 0.026 eV, so the exclusion principle prevents all but a very few electrons near the top of the energy distribution from gaining energy by random collisions. Figure 9-9 shows sketches of the density of states (*a*), the Fermi factor (*b*), and the product of these which gives the energy distribution.

It is convenient to define the *Fermi temperature* T_F by

$$kT_F = E_F \qquad\qquad 9\text{-}28$$

Fermi temperature

For temperatures much lower than the Fermi temperature, the average energy of the lattice ions will be much less than the Fermi energy; thus the electron energy distribution will not differ greatly from that at $T = 0$. The Fermi temperature corresponding to $E_F = 7.0$ eV for copper is about 81,000 K.

The complete quantum-mechanical distribution function for electrons at any temperature is called the Fermi-Dirac distri-

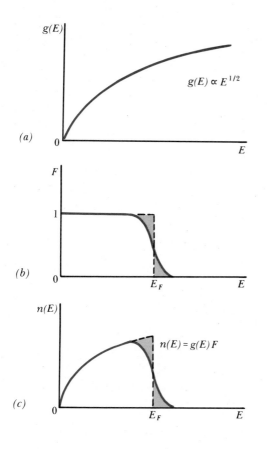

(a)

(b)

(c)

$g(E) \propto E^{1/2}$

$n(E) = g(E) F$

Figure 9-9
The Fermi energy-distribution function $n(E)$ is the product of (a) the density of states $g(E)$ and (b) the Fermi factor F. The dashed curves show the Fermi factor and energy distribution at $T = 0$. At higher temperatures, some electrons with energies near the Fermi energy are excited, as indicated by the shaded regions in (b) and (c).

bution. The energy distribution function $n(E)$ is given by Equation 9-25, where the Fermi factor F now differs slightly from that at $T = 0$. Figure 9-9b shows the Fermi factor for some temperature T much less than the Fermi temperature. Since there is now no energy below which all states are full and above which all states are empty, we must alter our definition of the Fermi energy. At any temperature, the Fermi energy E_F is defined as that energy at which the probability of a state being occupied is $\frac{1}{2}$, i.e. the energy for which the Fermi factor F has the value $\frac{1}{2}$. For all but extremely high temperatures, the difference between the Fermi energy at temperature T, E_F, and that at absolute zero, E_{F0}, is essentially negligible. From Figure 9-9b we see that $F = 1$ for energies much less than E_F, indicating that all these states are occupied. The Fermi factor at arbitrary T differs from that at $T = 0$ only for those energies within about kT of the Fermi energy. In this energy region, some of the energy states below E_F are unoccupied and some above E_F occupied, indicating that some of the electrons near the top of the energy distribution have been thermally excited to higher-energy states.

At temperatures much larger than the Fermi temperature (e.g., much larger than 81,000 K for copper) the Fermi factor can be shown to approach $e^{-E/kT}$ and the Fermi-Dirac distribution approaches the Maxwell-Boltzmann distribution. This result is not very important for the understanding of the behavior of \conductors since there are no conductors which remain as solids or even liquids at such extreme temperatures.

On the other hand, for electrons with energies much greater than the Fermi energy, the Fermi factor approaches $e^{-(E-E_F)/kT} = Ae^{-E/kT}$, where $A = e^{E_F/kT}$. That is, the high-energy tail of the Fermi-Dirac distribution decreases as $e^{-E/kT}$, as in the Maxwell-Boltzmann distribution. In this energy region, there are many unoccupied energy states and few electrons, so the Pauli exclusion principle is not important and the distribution approaches the classical distribution. This result is important as it applies to the conduction electrons in semiconductors, as will be discussed in Section 9-7.

Optional

Derivation of the Fermi Energy at $T = 0$ in Three Dimensions

For simplicity we assume the metal to be a cube of side L and volume $V = L^3$, and we approximate the potential energy of an electron in the metal by an infinite three-dimensional square well. The energy levels for such a potential are given by Equation 6-53 (see Section 6-8).

$$E = \frac{h^2}{8m_e L^2}(n_1{}^2 + n_2{}^2 + n_3{}^2)$$
$$= (n_1{}^2 + n_2{}^2 + n_3{}^2)E_1 \qquad \text{9-29}$$

where n_1, n_2, and n_3 are integers, and $E_1 = h^2/8m_e L^2$ as before. Each set of values for n_1, n_2, and n_3 corresponds to one energy level and two quantum states because of the spin of the electron. We now wish to count the number of states below a certain energy E. Consider the space formed by the axes n_1, n_2, and n_3 (see Figure 9-10). We are interested only in the octant for which $n_1 > 0$, $n_2 > 0$, and $n_3 > 0$. Each set of n_1, n_2, and n_3 corresponds to a point on a cubical lattice. The number of points in a volume in this space just equals the volume, since each cube has a volume of 1. From Equation 9-29 we see that the total energy is

$$E = R^2 \frac{h^2}{8m_e L^2} = R^2 E_1$$

where $R = (n_1{}^2 + n_2{}^2 + n_1{}^2)^{1/2}$. The number of states within radius R is

$$N = 2\frac{1}{8}\frac{4}{3}\pi R^3 = \frac{1}{3}\pi\left(\frac{E}{E_1}\right)^{3/2} \qquad \text{9-30}$$

The factor 2 is due to the two spin states per space state, and the factor $\frac{1}{8}$ is because only one octant of the sphere has positive values of n_1, n_2, and n_3. If we have N particles, they will fill the states up to the Fermi energy given by Equation 9-30. We thus have in three dimensions at $T = 0$,

$$E_F = \left(\frac{3N}{\pi}\right)^{2/3} E_1 = \frac{h^2}{8m_e}\left(\frac{3N}{\pi L^3}\right)^{2/3}$$

Figure 9-10
Three-dimensional N-space for counting states. Associated with each state, n_1, n_2, and n_3, is a cube of unit volume. The number of states within $R = (n_1^2 + n_2^2 + n_3^2)^{1/2}$ equals the volume of one octant of a sphere of radius R in N-space.

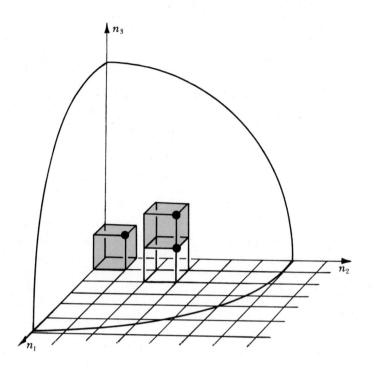

which is Equation 9-24 with $V = L^3$. The density of states is found by differentiating Equation 9-30:

$$dN = \frac{\pi}{2} E_1^{-3/2} E^{1/2} \, dE = \frac{\pi}{2} \left(\frac{8m_e L^2}{h^2}\right)^{3/2} E^{1/2} \, dE$$

Thus

$$g(E) = \frac{dN}{dE} = \frac{\pi}{2} \left(\frac{8m_e L^2}{h^2}\right)^{3/2} E^{1/2} = \frac{\pi V}{2} \left(\frac{8m_e}{h^2}\right)^{3/2} E^{1/2}$$

which is Equation 9-26. The average energy at $T = 0$ is

$$\overline{E} = \frac{1}{N} \int E \, n(E) \, dE = \frac{1}{N} \int_0^{E_F} \left(\frac{3N}{2}\right) E_F^{-3/2} E^{3/2} \, dE$$

$$= \tfrac{3}{5} E_F$$

9-4 Quantum Theory of Conduction

With two relatively simple but important modifications of the classical free-electron theory we can understand the electrical conductivity, heat conductivity, and heat capacity of metals. First we must replace the classical Maxwell-Boltzmann distribution with the Fermi distribution of energies in the electron gas, as discussed in the previous section. In addition, we shall discuss qualitatively the effect of the wave properties of the electrons on the scattering by the lattice ions.

Heat Capacity

Let us estimate the contribution of the electron gas to the molar heat capacity. At $T = 0$, the average energy of the electrons is $\frac{3}{5}E_F$, so the total energy is $U = \frac{3}{5}NE_F$. At a temperature T, only those electrons near the Fermi level can be excited by random collisions with the lattice ions, which have an average energy of the order of kT. The fraction of the electrons that are excited is of the order kT/E_F, and their energy is increased from that at $T = 0$ by an amount of the order of kT. We can thus write for the energy of the N electrons at temperature T,

$$U = \frac{3}{5}NE_F + \alpha N \frac{kT}{E_F} kT \qquad\qquad 9\text{-}31$$

where α is some constant, which we expect to be of the order of 1 if our reasoning is correct. The calculation of α requires the use of the complete Fermi electron distribution at an arbitrary temperature T and is quite difficult. Such a calculation, first carried out by Sommerfeld, shows that this equation is correct with $\alpha = \pi^2/4$. Using this result, the contribution of the electrons to the molar heat capacity is

$$C_v = \frac{dU}{dT} = 2\alpha Nk \frac{kT}{E_F} = \frac{\pi^2}{2} R \frac{T}{T_F} \qquad\qquad 9\text{-}32$$

where $Nk = R$ for 1 mole and $T_F = E_F/k$ is the Fermi temperature. We see that because of the large value of T_F, the contribution of the electron gas is a small fraction of R at ordinary temperatures. Using $T_F = 81{,}000$ K for copper, the molar heat capacity of the electron gas at $T = 300$ K is

$$C_v = \frac{\pi^2}{2} \left(\frac{300}{81{,}000} \right) R \approx 0.02R$$

which is in agreement with experiment.

Electrical Conduction

We might expect that most of the electrons would not participate in the conduction of electricity because of the exclusion principle, but this is not the case because the electric field accelerates all the electrons together. Figure 9-11 shows the

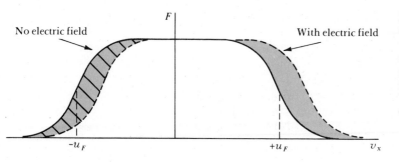

No electric field F With electric field

$-u_F$ $+u_F$ v_x

Figure 9-11
Occupation probability F versus velocity in one dimension, with no electric field and with an electric field in the $+x$ direction. The difference is greatly exaggerated.

Fermi factor versus velocity for some temperature that is small compared with T_F. The factor is approximately 1 for $-u_F < v_x < +u_F$, where u_F is the speed corresponding to the Fermi energy

$$u_F = \left(\frac{2E_F}{m}\right)^{1/2} \qquad\qquad 9\text{-}33 \qquad \textit{Fermi speed}$$

For copper, $u_F = 1.6 \times 10^8$ cm/sec. The dashed curve shows the factor after the electric field has acted for a time τ. The net effect is equivalent to shifting only the electrons near the Fermi level. We can use the classical equation (9-13) for the conductivity if we use the Fermi speed u_F in place of $\bar{v}$:

$$\sigma = \frac{1}{\rho} = \frac{ne^2\ell}{mu_F} \qquad\qquad 9\text{-}34$$

We now have two problems. First, since u_F is independent of temperature (to a very good approximation), the above expressions for σ and ρ are independent of temperature, unless the mean free path depends on it. The second problem concerns the magnitudes. We saw in Section 9-2 that the expression for σ was too small by about a factor of 6, using $\bar{v}$ calculated from the Maxwell-Boltzmann distribution. Since u_F is about 16 times this value of $\bar{v}$, the magnitude of σ predicted from Equation 9-34 will be even smaller by another factor of 16.

The resolution of both of these problems lies in the value of the mean free path. If we use u_F in Equation 9-34 and the experimental value $\sigma \approx 6 \times 10^7/\Omega\text{-m}$, we obtain for the mean free path $\ell \approx 370$ Å, about 100 times the value of 3.7 Å calculated from

$$\ell = \frac{1}{n\pi r^2}$$

where $r \approx 1$ Å is the radius of the copper atom.

We shouldn't be too surprised that the mean free path of electrons in the copper lattice is not given correctly by classical kinetic theory. The wave nature of the electron must be taken into account. The wave phenomenon analogous to the collision of an electron with a lattice ion is the scattering of an electron wave by the ion. If the wavelength is long compared with the crystal spacing, Bragg scattering cannot occur. A detailed calculation of the scattering of electron waves by a *perfectly* ordered crystal shows that there is *no scattering*, and the mean free path is infinite. The scattering of electron waves arises from imperfections in the crystal lattice. The most common imperfections are due to impurities or to thermal vibrations.

Let us estimate the mean free path of an electron, assuming that the lattice ions are *points* which are vibrating because of their thermal energy. We shall take for the scattering cross section πr^2, where $\overline{r^2} = \overline{x^2} + \overline{y^2}$ is the mean-square displacement of the point atom in a plane perpendicular to the direction of the electron. We can calculate $\overline{r^2}$ from the equipartition theorem.

We have

$$\tfrac{1}{2}K\overline{r^2} = \tfrac{1}{2}M\omega^2\overline{r^2} = kT \qquad\qquad 9\text{-}35$$

where K is the force constant, M the mass of the ion, and $\omega = (K/M)^{1/2}$ is the angular frequency of vibration. The mean free path is then

$$\ell = \frac{1}{n\pi\overline{r^2}} = \frac{M\omega^2}{2\pi nk}\frac{1}{T} \qquad\qquad 9\text{-}36$$

We thus see that this argument gives the correct temperature dependence for σ and ρ; that is, $\rho \propto T$ rather than $\rho \propto \sqrt{T}$, as was obtained from the classical calculation.

We can calculate the magnitude of $\overline{r^2}$, and therefore ℓ, using the Einstein model of a solid, which is fairly accurate except at very low temperatures. In the Einstein model (see Section 3-7) all the atoms vibrate with the same frequency. The Einstein temperature is defined by

$$kT_E = hf = \hbar\omega$$

Using this for ω, we have

$$\overline{r^2} = \frac{2kT}{M\omega^2} = \frac{2T\hbar^2}{MkT_E{}^2} = \frac{2(\hbar c)^2}{Mc^2 kT_E}\frac{T}{T_E} \qquad\qquad 9\text{-}37$$

The Einstein temperature for copper is about 200 K, corresponding to an energy of $kT_E = 0.0172$ eV. Using this and $Mc^2 = 63.5 \times 931$ MeV for the mass of a copper ion, the value of $\overline{r^2}$ at $T = 300$ K is

$$\overline{r^2} = \frac{2(1973 \text{ eV-Å})^2}{(63.5 \times 931 \times 10^6 \text{ eV})(0.0172 \text{ eV})}\frac{300 \text{ K}}{200 \text{ K}}$$
$$= 1.14 \times 10^{-2} \text{ Å}^2$$

Since this is about 100 times smaller than the area presented by a copper ion of radius 1 Å, the mean free path is about 100 times larger than that calculated from the classical model, in agreement with that calculated from the measured value of the conductivity. We see, therefore, that the free-electron model of metals gives a good account of electrical conduction if the classical mean speed $\overline{v}$ is replaced by the Fermi speed u_F and if collisions are interpreted in terms of the scattering of electron waves for which only deviations from a perfectly ordered lattice are important.

The presence of impurities in a metal also causes deviations from perfect regularity in the crystal. These are approximately independent of temperature. The resistivity of a metal containing impurities can be written $\rho = \rho_t + \rho_I$, where ρ_t is due to the thermal motion of the lattice and ρ_I is due to impurities. Figure 9-12 shows a typical resistance-versus-temperature curve for a metal with impurities. As the temperature approaches zero, ρ_t approaches zero and the resistivity approaches the constant ρ_I, which is due to impurities.

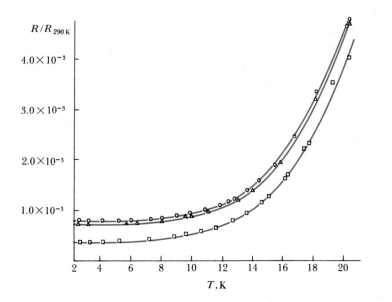

Figure 9-12
Relative resistance versus temperature for three samples of sodium. The three curves have the same temperature dependence but have different magnitudes because of differing amounts of impurities in the samples. [*D. Mac-Donald and K. Mendels-sohn,* Proceedings of the Royal Society, **A202,** *103 (1950).*]

Heat Conduction

If we replace $\bar{v}$ by u_F, Equation 9-14 for the thermal conductivity of a gas becomes

$$K = \frac{1}{3}\frac{nu_F \ell C_v}{N_A}$$

If we substitute from Equation 9-32 $C_v = \frac{1}{2}\pi^2 RT/T_F = \frac{1}{2}\pi^2 RkT/E_F$ this becomes

$$K = \frac{1}{3}\frac{nu_F\ell}{N_A}\frac{\pi^2}{2}\frac{RkT}{E_F}$$

Writing $R = N_A k$ and $E_F = \frac{1}{2}mu_F^2$ gives for the thermal conductivity

$$K = \frac{n\ell\pi^2 k^2 T}{3mu_F} \qquad\qquad 9\text{-}38$$

From Equations 9-34 and 9-38 we obtain for the Lorentz number

$$L = \frac{K}{\sigma T} = \frac{\pi^2 k^2}{3e^2} = 2.45 \times 10^{-8}\ \text{W-}\Omega/\text{K}^2 \qquad 9\text{-}39 \qquad \textit{Lorentz number}$$

Thus the Wiedemann-Franz law is also predicted by the quantum calculation, and the value of the Lorentz number $K/\sigma T = 2.45 \times 10^{-8}$ W-Ω/K² is in good agreement with the experimental values listed in Table 9-1.

Question

4. When the temperature is lowered from 300 K to 4 K, the resistivity of pure copper drops by a much greater factor than that of brass. Why?

9-5   Band Theory of Solids

We have seen that, if the electron gas is treated as a Fermi gas and the electron-lattice collisions treated as the scattering of electron waves, the free-electron model gives a good account of the thermal and electrical properties of conductors. This simple model, however, gives no indication of why one material is a good conductor and another is an insulator. The conductivity (and its reciprocal, the resistivity) vary enormously from the best insulators to the best conductors. For example, the resistivity of a typical insulator (such as quartz) is of the order of 10^{16} Ω-m, whereas that of a typical conductor (most metals) is of the order of 10^{-8} Ω-m and that of a superconductor is less than 10^{-19} Ω-m.

To understand why some materials conduct and others do not, we must refine the free-electron model and consider the effect of the lattice on the electron energy levels. There are two standard approaches to this problem of determining the energy levels of electrons in a crystal. One is to consider the problem of an electron moving in a periodic potential, and to determine the possible energies by solving the Schrödinger equation. The other is to find the energy levels of the electrons in a solid by following the behavior of the energy levels of individual atoms as they are brought together to form the solid. Both approaches lead to the result that the energy levels are grouped into allowed and forbidden bands. The details of the band structure of a particular material determine whether that material is a conductor, an insulator, or a semiconductor. We shall discuss both of these methods qualitatively.

Consider first the problem of an electron moving in a periodic potential. Figure 9-13a shows a one-dimensional sketch of the potential-energy function for a lattice of positive ions. The most important feature of this potential is not the shape, but the fact that it is periodic. A simpler periodic potential is shown in Figure 9-13b. This potential is called the *Kronig-Penney model*. It has the important feature of periodicity and is easier to treat mathematically; however, even for this model the mathematical solution of the Schrödinger equation is fairly involved, and we shall not present it here. For both potential functions shown in Figure 9-13, the solutions of the Schrödinger equation have the following characteristic: for certain ranges of energy, there exist traveling-wave-type solutions of the Schrödinger equation.

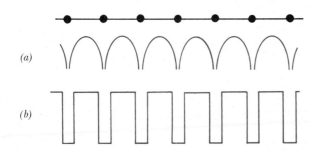

(a)

(b)

Figure 9-13
(a) One-dimensional potential energy of an electron in a crystal. $V(x)$ approaches $-\infty$ at the atom sites. (b) Simplified (Kronig-Penney) model of potential energy of an electron in a crystal.

These energy ranges, called *bands,* are separated by an energy gap E_g in which no traveling wave can exist. Figure 9-14a shows the energy versus the wave number k for a completely free electron. This is, of course, merely a sketch of $E = \hbar^2 k^2 / 2m$. Figure 9-14b shows E versus k for an electron in the periodic potential of Figure 9-13b. The energy gaps occur at

$$ka = \pm n\pi \qquad\qquad 9\text{-}40$$

where a is the lattice spacing. We can understand this result in terms of the Bragg condition

$$n\lambda = 2a \sin \theta$$

In one dimension, $\theta = 90°$ for reflection. Using $k = 2\pi/\lambda$, Equation 9-40 becomes the condition for Bragg reflection. The reason that traveling waves cannot exist for these wave numbers is that the reflection from one atom in the chain is in phase with the reflection from the next, so that standing waves are set up. Figure 9-15 shows a sketch of $|\psi|^2$ for the two types of standing waves for the value $k = \pi/a$:

$$\psi_1 = \sin kx = \sin \frac{\pi x}{a} \qquad \psi_2 = \cos kx = \cos \frac{\pi x}{a}$$

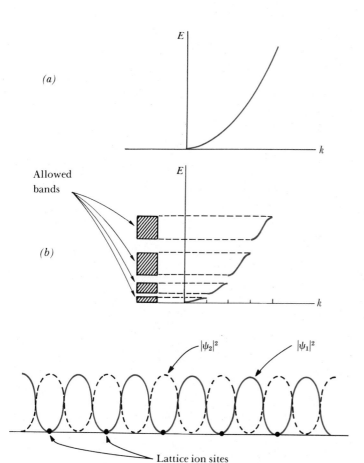

(a)

Allowed
bands

(b)

Lattice ion sites

Figure 9-14
(a) Energy versus k for a free electron. (b) Energy versus k for a nearly free electron in the one-dimensional periodic potential of Figure 9-13. Energy gaps occur at the k values which satisfy the Bragg scattering condition.

Figure 9-15
Probability distribution for standing waves of wave number k in a one-dimensional crystal. The dashed curve $|\psi_2|^2$ is a maximum at the lattice ion sites, and has a lower potential energy than the solid curve $|\psi_1|^2$.

Since ψ_2 gives a concentration of electron-charge density nearer the ion sites than ψ_1, the potential energy is less for ψ_2 than for ψ_1. The energy difference corresponds to the energy gap. Within the allowed energy bands, the energy has a continuous range if the number of atoms in the chain is infinite; for N atoms, there are N allowed energy levels in each band. Since the number of atoms is very large in a macroscopic solid, the energy bands can be considered continuous. Calculations in three dimensions are more difficult, of course, but the results are similar. The allowed ranges of the wave vector **k** are called *Brillouin zones*.

In the second approach to finding the energy levels of electrons in solids, we consider the energy levels of the individual atoms as they are brought together to form a solid. In Section 8-1 we saw that when two H atoms are brought together, the two $1s$ levels (one for each atom) are split into two molecular levels having different energies depending on the space symmetry of the wave functions.

Figure 9-16 shows sketches of one-dimensional s-state wave functions for six atoms. These might represent, for example, six sodium atoms each with a $3s$ electron. In Figure 9-16a, the wave function is symmetric between each pair of atoms. The charge concentration between two atoms is roughly the same as that for

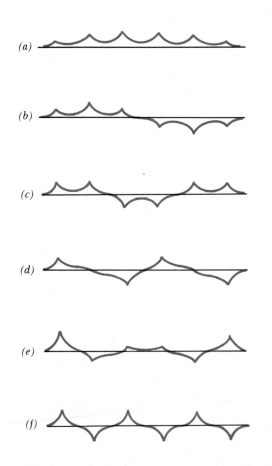

(a)

(b)

(c)

(d)

(e)

(f)

Figure 9-16
Wave functions with different symmetries of six atoms in one dimension. The perfectly symmetric function (a) has the lowest energy, whereas the perfectly antisymmetric function (f) has the highest energy. (*From Shockley, Reference 6.*)

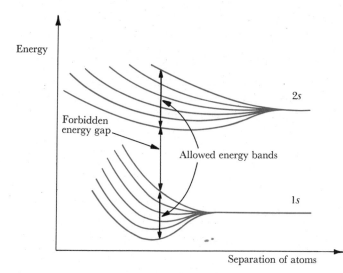

Figure 9-17
Energy splitting of the
1s and 2s states for six
atoms as a function of
separation of the atoms.
(*From Shockley, Reference
6.*)

the symmetric state of two atoms. In Figure 9-16*f*, the wave
function is antisymmetric between any two adjacent atoms and is
similar to that for the antisymmetric state of two atoms. The six
energy levels (twelve, counting spin states), which are all the
same when the atoms are far apart, split into six different energies when they are close. The lowest energy corresponds to
Figure 9-16*a* and the highest to 9-16*f*. The difference in these
two energies depends on the spacing of the atoms but not on the
number of atoms, since the concentration of charge for these extreme cases (perfectly symmetric and perfectly antisymmetric)
does not change when more atoms are added. Figure 9-17 shows
the splitting of the 1s states and 2s states for six atoms as a function of lattice separation. For N atoms, there are N states in a
band, so these bands are nearly continuous in the case of macroscopic solids where $N \sim 10^{23}$. The details of the splitting of the
levels, as in Figure 9-17, depend on the type of atom and the
type of bonding and crystal structure.

We are now ready to understand why some solids are very
good conductors and others are very poor. Consider sodium
first. There is room for two electrons in the 3s state of each
atom, but each separated Na atom has only one 3s electron.
When N atoms are bound in a solid, the 3s energy band is only
half-filled. In addition, the 3p band actually crosses the 3s band.
The allowed energy bands of sodium are shown schematically in
Figure 9-18. The occupied levels are shaded. We see that there
are many allowed energy states available just above the filled
ones, so the valence electrons can easily be raised to a higher-
energy state by an electric field, and accordingly sodium is a
good conductor. Magnesium, on the other hand, has two 3s electrons so the 3s band is filled. However, because the 3p band
crosses the 3s band, just as it does for sodium, magnesium is also
a conductor.

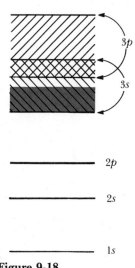

Figure 9-18
Energy-band structure of
sodium. The 3p band
overlaps the half-filled 3s
band. Sodium is a con-
ductor because just above
the filled states are empty
states into which electrons
can be excited by an electric field.

A solid which has only completely-filled bands is an insulator if the energy gap between the last-filled band (called the *valence band*) and the next allowed band (called the *conduction band*) is large. For example, ionic crystals are insulators. The energy bands of a crystal such as NaCl are those from the energy levels of the Na^+ and Cl^- ions. Both of these ions have closed-shell structure, so the highest occupied band in NaCl is completely full and there is a large energy gap between it and the empty conduction band.

If the gap between a filled valence band and an empty conduction band is small, the solid is a semiconductor. Consider carbon, which has two $2s$ electrons and two $2p$ electrons. We might expect carbon to be a conductor because of the four unfilled $2p$ states. However, the $2s$ and $2p$ levels mix when carbon forms covalent bonds.[1] Figure 9-19 shows the splitting of the eight $2s$-$2p$ levels when carbon bonds in the diamond structure. This splitting is due to the nature of the covalent bond and is similar to the splitting of the $1s$ levels in hydrogen discussed in Section 8-1. The energy of the levels corresponding to the four space-symmetric wave functions (one for the $2s$ levels and three for the $2p$ levels) is lowered while the energy of the other four levels (one $2s$ and three $2p$) is raised. The valence band therefore contain four levels per atom which are filled, and the conduction band is empty. At the diamond-lattice spacing of about 1.54 Å, the energy gap between the filled valence band and empty conduction band is about 7 eV, so diamond is an insulator. The band structure is similar for silicon, which has two $3s$ and two $3p$ electrons; and for germanium, which has two $4s$

[1] This mixing, called hybridization, is discussed in Section 8-2.

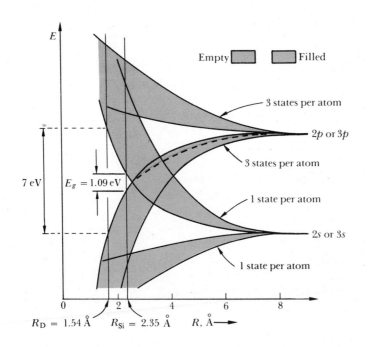

Figure 9-19
Splitting of the $2s$ and $2p$ states of carbon, or the $3s$ and $3p$ states of silicon, versus separation of the atoms. The energy gap between the four filled states in the valence band and the empty states in the conduction band is 7 eV for the diamond-lattice spacing, $R_D = 1.54$ Å. For the silicon spacing $R_{Si} = 2.35$ Å, the energy gap is 1.09 eV. The splitting is similar for the $4s$ and $4p$ levels in germanium, which has an atom spacing of 2.43 Å, giving an energy gap of only 0.7 eV. (*From Sproull, Reference 5.*)

and two $4p$ electrons. At the silicon-lattice spacing of 2.35 Å the energy gap is about 1 eV; at the germanium-lattice spacing of 2.43 Å the energy gap is only about 0.7 eV. These solids are called *intrinsic semiconductors*. Figure 9-20 illustrates the four band structures discussed above.

Intrinsic semiconductors

Even at ordinary room temperature there are some electrons in the conduction band of a semiconductor because of thermal excitation. The number depends on the size of the energy gap between the valence and conduction bands relative to kT. There is an equal number of unoccupied states, or *holes,* in the nearly filled valence band. In the presence of an electric field, the electrons in the conduction band can move because there are many unoccupied states nearby. In addition, electrons in the valence band can move into the holes, thereby creating other holes. In the valence band, in fact, it is easier to speak of the motion of the holes than of the electrons. The holes act like positive charges. An analogy of a two-lane one-way road with one lane full of cars and the other lane empty might help to visualize the conduction of holes. If a car moves out of the full lane into the empty one, it can move freely. As the other cars move up to occupy the space left, the empty space is propagated backwards. Both the forward motion of the car in the nearly empty lane and the backward propagation of the empty space contribute to a net forward propagation of cars.

Hole conduction

The conductivity of germanium at room temperature is about $2/\Omega$-m. As the temperature is increased, more electrons are excited into the conduction band, and correspondingly more holes in the valence band. The conductivity *increases* with temperature because the increase in the number of carriers (electrons and holes) outweighs the increase in the scattering of the lattice due to thermal vibrations. Semiconductors therefore have a negative temperature coefficient of resistivity—as the temperature increases, the resistance decreases.

Question

5. How does the change in the resistivity of copper compare with that of silicon when the temperature increases?

Figure 9-20
Four possible band structures for a solid. In (*a*), the allowed band is only partially full, so electrons can be excited to nearby energy states; (*a*) is a conductor. In (*b*) there is a forbidden band with a large energy gap between the filled band and the next allowed band; thus (*b*) is an insulator. (*c*) is a conductor because the allowed bands overlap. In (*d*), the energy gap between the filled band and the next allowed band is very small so some electrons are excited to the conduction band at normal temperatures, leaving holes in the valence band. (*d*) is a semiconductor.

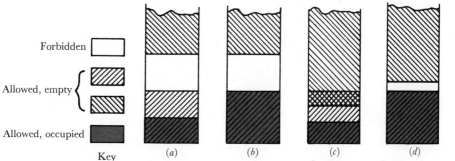

Key   (*a*) Conductor   (*b*) Insulator   (*c*) Conductor   (*d*) Semiconductor

Forbidden
Allowed, empty
Allowed, occupied

9-6   Impurity Semiconductors

In 1949, John Bardeen, Walter Brattain, and William Shockley, all of Bell Telephone Laboratories, initiated a revolution in electronics with the invention of the *transistor*. In 1956 they received the Nobel Prize in physics for their work. The transistor, as well as many other useful devices such as the semiconductor diode, the tunnel diode, etc., make use of *impurity semiconductors*, which result from the controlled addition of certain impurities to intrinsic semiconductors.

Consider, for example, the effect of adding a small amount of arsenic to a germanium crystal. The arsenic atom has five valence electrons in the $n = 4$ shell, rather than four as in germanium. If a germanium atom is replaced with an arsenic atom, four of the electrons participate in covalent bonds in the diamond-structured germanium crystal. The remaining electron of the arsenic atom is not part of the bond and is in fact only loosely bound to the arsenic atom. (A calculation of the "Bohr" orbit of this electron shows it to be larger than the lattice spacing, so this electron is very loosely bound and easily freed from its parent atom.) This extra electron occupies an energy level that is just slightly below the conduction band in the crystal, and is easily excited into the conduction band where it can contribute to the transport of charge in electrical conduction.

The effect on the band structure of doping a germanium crystal with arsenic is shown in Figure 9-21a. The levels just below the conduction band are due to the extra electrons of the arsenic atoms. These levels are called *donor* levels, because the effect of the arsenic atoms is to essentially donate electrons to the conduction band without leaving holes in the valence band. Such a semiconductor is called an *n-type* semiconductor because the major charge carriers are *n*egative electrons. The conductivity of such a doped semiconductor can be controlled by the amount of impurity added. The addition of just one part per million causes a significant change in the conductivity.

Analogous arguments can be made for the replacement of a

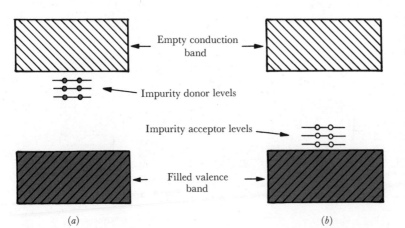

Empty conduction band

Impurity donor levels

Impurity acceptor levels

Filled valence band

(a)   (b)

Figure 9-21
(*a*) Energy bands of an *n*-type semiconductor. Impurity atoms provide filled energy levels just below the empty conduction band and donate electrons to the conduction band. (*b*) Energy bands of a *p*-type semiconductor. Impurity atoms provide empty energy levels just above the filled valence band and accept electrons from it.

germanium atom with an atom with valence 3 such as gallium. The gallium atom accepts electrons from the valence band to complete its four covalent bonds, thus creating a hole in the valence band. The effect on the band structure of doping with gallium is shown in Figure 9-21b. A semiconductor doped with acceptor impurities is called a *p-type* semiconductor because the major charge carriers are holes, which behave like *positive* charges.

In a doped semiconductor the number of donated electrons in an n-type, or holes in a p-type, is typically much greater than the intrinsic number of electron-hole pairs created by thermal excitation electrons from the valence band to the conduction band. In an electric field, the current will therefore consist of both majority carriers (electrons in an n-type and holes in a p-type semiconductor) and minority carriers. The reality of conduction by motion of positive holes is brought out in the Hall effect, illustrated in Figure 9-22. In this figure a thin strip of a doped semiconductor is connected to a battery (not shown), so that there is a current to the right. A uniform magnetic field B is applied perpendicular to the current. For the direction of the current and magnetic field shown, the magnetic force on a moving charged particle $q\mathbf{u} \times \mathbf{B}$ is upward (where $\mathbf{u}$ is the drift velocity) independent of whether the current is due to a positive charge moving to the right or a negative charge moving to the left. Let us assume for the moment that the charge carriers are electrons. The magnetic force will then cause the electrons to drift up to the top of the strip, leaving the bottom of the strip with an excess positive charge. This will continue until the electrostatic field $\mathscr{E}$ caused by the charge separation produces an electric force on the charge carriers just balancing the magnetic force. The condition for balance is $q\mathscr{E} = quB$. If w is the width of the strip, there will be a potential difference called the Hall voltage

$$V_H = \mathscr{E}w = uBw \qquad\qquad 9\text{-}41$$

Hall effect

Hall voltage

between the top and bottom of the strip. This potential difference can be measured with a high-resistance voltmeter. A measurement of the sign of the potential difference (i.e., whether the top of the strip is at higher potential due to a positive charge or lower potential due to a negative charge) deter-

Figure 9-22
The Hall effect. The force on the charge carriers is up whether the carriers are positive charges moving to the right, (a), or negative charges moving to the left, (b). The sign of the charge carriers can be determined by the sign of the potential difference between the top and bottom of the strip, and the drift velocity can be determined by the magnitude of this potential difference.

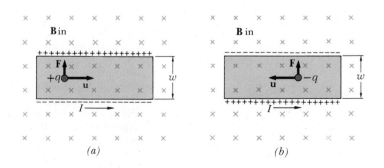

(a)   (b)

mines the sign of the majority carriers. Such measurements reveal that, indeed, the charge carriers are negative in n-type and positive in p-type semiconductors. The value of the Hall voltage provides a measurement of the drift velocity u. Since the current density $j = nqu$ can be easily measured from the total current and cross-sectional area of the strip, measurement of the drift velocity determines n, the number of charge carriers per unit volume.

9-7   Semiconductor Junctions and Devices

Now consider what happens when there is an abrupt change from n- to p-type doping within a semiconductor. A region in which such a change occurs is called a *junction* and is illustrated in Figure 9-23. On one side of the junction (the right side of Figure 9-23) the semiconductor is doped with a donor material so that it is an n type rich in electrons, whereas on the other side it is doped with an acceptor material so that it is rich in holes. The behavior of the electrons and holes near such a junction plays an important role in the operation of semiconductor devices such as diodes and transistors.

pn junction

If we were to join an n-type and p-type semiconductor together, the initially unequal concentrations of electrons and holes would result in the diffusion of electrons across the junction from the n side to the p side and of holes from the p side to the n side, tending to equalize the concentrations. Majority carriers crossing the junction will find a region rich in carriers of the opposite kind and will tend to recombine electrons and holes together to form neutral atoms. The result of this diffusion is a net transport of positive charge from the p side to the n side. This creates a double layer of charge similar to that on a parallel plate capacitor, and there is thus a potential difference across the junction roughly equivalent to a contact potential between two dissimilar metals. In equilibrium the n side with a net positive charge will be at a potential V greater than the p side. This potential difference tends to inhibit further diffusion. Only those electrons or holes at the high-energy end of the distribution will have enough energy to climb the potential barrier that is set up. The current due to these majority carriers will be balanced by that due to the minority carriers (holes in the n side and electrons in the p side) that exist because of the intrinsic

Figure 9-23
A *pn* junction. Because of the difference in concentrations, holes diffuse across the junction to the right and electrons to the left, setting up a potential difference across the junction. In equilibrium, with no bias voltage, the current arising from this diffusion of majority carriers is balanced by that arising from the minority carriers.

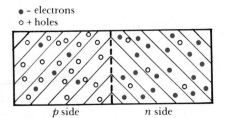

• − electrons
○ + holes

p side n side

thermal excitation and that are attracted across the barrier by the potential difference. In equilibrium these currents will cancel so that there is no net current. In the junction region between the double layer of charge there will be few free charge carriers and therefore a relatively high resistance across the junction.

Let us now consider what happens if we apply an external potential difference across the junction by connecting a battery to the semiconductor. Since the junction is a region of relatively high resistance, most of the external potential difference appears across it. If the positive terminal of the battery is connected to the n side and the negative terminal to the p side (this is called *reverse bias*), the effect will be to increase the potential difference already appearing across the junction. This will further reduce the number of majority carriers near the junction, so the current due to the majority carriers will be slightly reduced. The current due to the minority carriers will be enhanced, but since there aren't very many of these carriers anyway this will be a small effect. The net result is a very small current flow from the n to the p side.

If we now reverse the battery and apply the positive terminal to the p side of the junction (*forward* bias), the effect is to lower the potential across the junction and thereby to increase the flow of majority carriers. As the majority carriers diffuse across the boundary in an attempt to reestablish equilibrium, the battery supplies other charges to take their place so that there is a continuing flow of charge across the junction. We can get an idea of the dependence of the current on applied voltage if we note that the electrons and holes, being at the high-energy end of the distribution, are approximately described by the Maxwell-Boltzmann distribution. Let N_e be the number of conduction electrons in the n region. With no external voltage, only a small fraction proportional to $N_e e^{-eV/kT}$ will have enough energy to diffuse across the contact potential difference. When a forward bias V_b is applied, the number that can cross the barrier is proportional to $N_e e^{-(V-V_b)/kT} = (N_e e^{-V/kT})e^{+V_b/kT}$. The current due to the majority electron carriers in the n region will be

$$I = I_0 e^{+eV_b/kT}$$

where I_0 is the current with no bias. The current due to the minority carriers, the holes from the n side, will be merely I_0, the same as with no bias. (The minority carriers are swept across the junction by the contact potential V with or without a bias voltage.) The net current due to carriers from the n side will therefore be

$$I_{\text{net}} = I_0(e^{+eV_b/kT} - 1) \qquad \text{9-42}$$

Current versus bias voltage across pn junction

If we now consider the current due to the majority and minority carriers from the p side, we obtain the same results. We can use Equation 9-42 for the total current if we interpret I_0 as

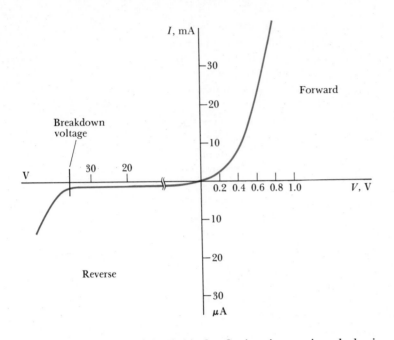

Figure 9-24
Current versus applied
voltage across a *pn* junction. Note the different
scales for the forward
and reverse bias voltages.

the total current due to both kinds of minority carriers, holes in
the *n* region and electrons in the *p* region. For positive V_b the
exponential quickly dominates. For $V = 0$ the current is 0 and
for V less than zero, the current reaches a steady state at $-I_0$
due to the flow of minority carriers. Figure 9-24 shows the
current-voltage characteristic of a typical semiconductor junction. Such a junction can be used as a diode. Note that at extreme values of reverse bias the current suddenly increases in
magnitude. This effect is called *avalanche breakdown*. At such
large electric fields, the electrons are stripped from their atomic
bonds and accelerated across the junction. These in turn cause
others to break loose. While such a process may be disastrous in
a circuit where it is not intended, the fact that breakdown occurs
at such a sharp voltage value makes it of great value in special
voltage reference standards known as "Zener" diodes.

An interesting effect, which we can describe only qualitatively, occurs if both the *n*-type and *p*-type materials are so heavily doped that the donors in the *n*-type material practically fill
the conduction band and the acceptors in the *p*-type material
accept so many electrons that the valence band is nearly empty.
The majority carriers are then so numerous that they must be
treated as a Fermi gas. Because of the detailed structure of the
energy bands (which we shall not consider here) there is a considerable current due to penetration of the potential barrier
across the junction in the presence of a small forward bias. This
"tunneling" current is in addition to the usual current due to diffusion, as already discussed. When the bias voltage is increased
slightly, the tunneling decreases, and although the diffusion
current increases the net current decreases. At large bias voltages the tunneling current is completely negligible and the total
current increases with bias voltage due to diffusion as in an ordinary *np* semiconducting diode. Figure 9-25 shows the current

Tunnel diodes

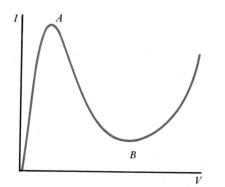

Figure 9-25
Current versus voltage
for a tunnel diode.
Between points A and B
the resistance is negative
because an increase in
voltage inhibits tunneling.
After point B the tunneling is negligible and the
diode behaves like an
ordinary pn junction
diode.

versus voltage curve for such a heavily doped diode called a
tunnel diode.

The basic element of most amplifiers is the transistor, which
contains three distinct regions of semiconductor material. These
regions are called the *emitter,* the *base,* and the *collector.* Figure
9-26*a* is an idealized representation of a *pnp* transistor, where
the three regions are labeled with their common designations.
In a properly designed transistor the emitter doping is much
heavier than in the other two regions, and the base width w is
very narrow.

The device represented in Figure 9-26 contains two pn junctions. In a practical circuit (Figure 9-26*b*) voltages are supplied
so that the emitter-base (EB) junction is forward-biased and the
collector-base (CB) junction is reverse-biased. The difference in
doping levels of the emitter and the base means that practically
all the current across the EB junction consists of positive charges
from the emitter. Because the base width is small and the CB
junction is reverse-biased, most of the positive charges from the
emitter that enter the base diffuse across its narrow width and
enter the collector. The few positive charges that are not collected by the CB junction leave the base through the external
connection. In Figure 9-26*b,* therefore, I_C is almost but not quite
equal to I_E, and I_B is much smaller than either I_C or I_E. It is customary to write.

$$I_C = \beta I_B \qquad\qquad 9\text{-}43$$

where β, the *current gain,* is an important parameter of the transistor. Transistors can be designed to have values of β as low as
10 or as high as several hundred. In a useful transistor, β will be

Transistors

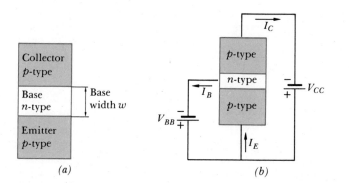

Figure 9-26
(*a*) A *pnp* transistor. (*b*)
The same transistor
biased for normal opera-
tion. The base width w is
very small (of the order
of a hundred angstroms)
so that most of the holes
from the emitter diffuse
across the base and are
collected by the collector.

constant; that is, the device will be linear over a restricted range of base current provided that proper biases are maintained on the two junctions.

Figure 9-27 shows a simple transistor amplifier. A small time-varying voltage, say from the motion of a needle in the groove of a phonograph record, is in series with a bias voltage V_{BB}. The base current is then the sum of a steady current I_B produced by V_{BB} and a varying current i_b produced by the signal voltage v_s. Because v_s may at any instant be either positive or negative, V_{BB} must be included, and must be large enough to ensure that there is always a forward bias on the EB junction. Since the device is linear, the collector current will consist of two parts, a steady current $I_C = \beta I_B$ and a time-varying current $i_c = \beta i_b$. We thus have a *current amplifier* where the time-varying output current i_c is β times the input current i_b. In such an amplifier the steady current I_C and I_B, while essential to proper operation, are usually not of interest. (If an *npn* transistor is used, the bias voltages must be reversed and the currents will all be in the other direction. Otherwise, operation is identical with that described here.) The voltage v_s is related to the current i_b by Ohm's Law,

Transistor amplifier

$$i_b = \frac{v_s}{R_b + r_b}$$

where r_b is the internal resistance of the transistor between base and emitter. Similarly in the collector circuit the current i_c produces a voltage v_L across the resistor R_L:

$$v_L = i_c R_L$$

We already have

$$i_c = \beta i_b = \beta \frac{v_s}{R_b + r_b} \qquad 9\text{-}44$$

and therefore,

$$v_L = \beta \frac{R_L}{R_b + r_b} v_s$$

The *voltage gain* of the amplifier is thus

$$\frac{v_L}{v_s} = \beta \frac{R_L}{R_b + r_b} \qquad 9\text{-}45$$

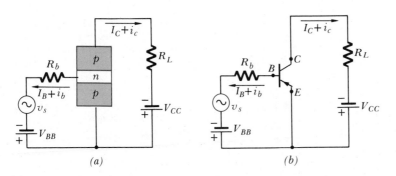

Figure 9-27
(*a*) Simple amplifier circuit using a *pnp* transistor. (*b*) The same circuit with transistor represented by standard symbol.

(*a*)   (*b*)

In a practical case β may be 100, and the ratio $R_L/(R_b + r_b)$ may be $\frac{1}{2}$. Then the voltage gain is 50. A more detailed derivation shows that v_L and v_s are exactly *out of phase;* i.e., when v_s has its most positive value, v_L has its most negative value. For a simple amplifier this phase shift is not important because all input voltages, regardless of frequency, are affected identically. This simple voltage amplifier is the basis of many circuits that are components of communication systems.

The complete amplifier in a record or tape player generally consists of several *stages* (similar to Figure 9-27) connected in *cascade,* so that the output of one becomes the input of the next. Thus the very small voltage produced by the motion of the needle or by the passage of magnetized tape through the player head controls the large amounts of power required to drive a system of loudspeakers. The energy delivered by the speakers is supplied by the sources of direct voltage connected to each transistor.

The resistance of a semiconductor material depends upon the impurity concentration. Reverse-biased diodes have capacitance that can be controlled by controlling the bias voltage. These two facts are exploited in the manufacture of *integrated circuits,* in which several transistors along with associated resistors and capacitors are interconnected on a single tiny piece of silicon. Figure 9-28a shows a 5-W audiofrequency amplifier that uses one integrated circuit containing 14 transistors, 3 diodes, and 10 resistors. Figure 9-28b shows a different integrated circuit containing even more circuit elements.

(a)

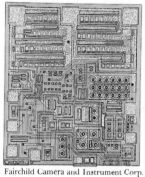

Fairchild Camera and Instrument Corp.

(b)

Figure 9-28
(a) Five-watt audio amplifier using an integrated circuit. The integrated circuit is partly hidden by the piece of metal 3.3 cm long. The five cylinders are capacitors. (b) Photomicrograph of an integrated circuit with more than 50 components on a silicon chip approximately 0.25 cm by 0.2 cm.

Question

6. Why is a semiconductor diode less effective at high temperatures?

9-8 Superconductivity

In 1911, H. Kamerlingh Onnes found that the resistance of mercury dropped suddenly to zero at a temperature of about 4.2 K. Figure 9-29 shows a sketch of his data. Since then this behavior, called *superconductivity,* has been observed in many materials. The critical temperature below which superconductivity sets in varies from material to material, but below this temperature, the resistance is apparently zero. Steady currents have been observed to persist in superconducting rings for several years with no apparent loss (Figure 9-30). Table 9-3 lists some superconducting materials with their critical temperatures. In the presence of a magnetic field, the critical temperature is lower than with no field. As the magnetic field increases, the critical temperature decreases. If the magnetic field is greater than a critical field H_c, superconductivity does not exist at any temperature.

Table 9-3
Critical temperatures for some superconducting elements

Element	T_c (K)
Al	1.2
Hg	4.2
In	3.4
Nb	9.2
Pb	7.2
Sn	3.7
Ta	4.4

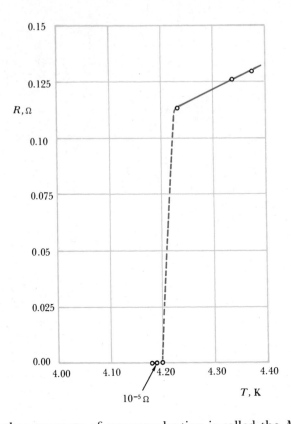

Figure 9-29
Resistance in ohms of a specimen of mercury versus absolute temperature. This plot by Kamerlingh Onnes marked the discovery of superconductivity. (*From Kittel, Reference 3.*)

Another property of superconduction is called the *Meissner effect:* when a superconductor in a magnetic field is cooled below the critical temperature, the magnetic-field lines are expelled from the superconductor (Figure 9-31). A magnetic field cannot exist inside a superconductor.

It has been recognized for some time that superconductivity is due to a collective action of the conducting electrons. In 1957, Bardeen, Cooper, and Schrieffer published a successful theory of superconductivity now known as the *BCS theory.* In this theory, the electrons are coupled in pairs at low temperatures. One electron interacts with the crystal lattice and perturbs it. The perturbed lattice interacts with another electron in such a way that there is an attraction between the two electrons which, at low temperatures, can overcome the Coulomb repulsion between them. The electron-lattice-electron interaction produces an energy gap between the superconducting state, in which the electrons act collectively, and the normal state, in which the electrons act individually. Energy from an electric field can be absorbed by the electrons in this collective state to produce a superconducting current. However, energy cannot be dissipated by individual collisions of electron and lattice unless the temperature is high enough so that the electron bonds can be broken. Materials such as copper or gold that are very good conductors at normal temperatures have weak electron-lattice interactions. In such materials the electron-lattice-electron interaction is correspondingly weak, and these materials do not su-

Helix Technology Corp.

Figure 9-30
Demonstration of persistent currents. Currents are first induced in the superconducting rings. When the lead ball is dropped into the rings, a persistent current is induced on the surface of the ball as it approaches the rings. The current is in the opposite direction to that in the rings, so that the ball is repelled. The ball thus floats at the height at which the weight of the ball is balanced by the magnetic force of repulsion caused by the currents.

(a)

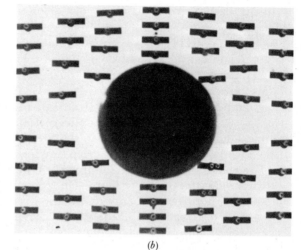

(b)

Figure 9-31
Demonstration of the
Meissner effect. (*a*) A
superconducting tin cyl-
inder is situated with its
axis perpendicular to a
horizontal magnetic field
of 80 G. The direction of
the field lines near the
cylinder is indicated by
the weakly magnetized
compass needles mounted
in a Lucite sandwich so
that they are free to turn.
(The ellipse at the top is
the liquid helium surface
as seen from below.) (*b*)
The same as (*a*) except
that the temperature is
such that the tin cylinder
is normally conducting.
The compass needles now
point in the direction of
the horizontal field except
for small fluctuations due
to their mutual interac-
tions. (*Courtesy of A.
Leitner, Rensselaer Poly-
technic Institute.*)

perconduct. At temperatures as low as 0.05 K, these materials
are still normal conductors. In the superconducting state, the
conduction of heat is *less* than normal because heat conduction
by the electrons is an individual energy-exchange process. This
collective behavior of the electrons in a superconductor is somewhat similar to, but not the same as, the collective behavior of
helium atoms in liquid helium II below 2.17 K, which will be discussed in the next chapter. The superconducting electron gas
and liquid helium II are sometimes called *superfluids*.

Summary

Solids are often found in crystalline form in which a small structure called the unit cell is repeated over and over. The structure
of the unit cell may be face-centered cubic, body-centered cubic,
close-packed hexagonal, or other, depending on the type of
bonding between the atoms, ions, or molecules in the crystal and
on the relative sizes, if there are two kinds as in NaCl.

The attractive part of the potential energy of an ion in an
ionic crystal can be written $V_{att} = -\alpha ke^2/r$, where r is the separation distance between neighboring ions and α is the Madelung constant, which depends on the geometry of the crystal and
is of the order of 1.7. The repulsive part of the potential energy
is due to the exclusion principle. An empirical expression that
works fairly well is $V_{rep} = A/r^n$, where n is about 9.

In the classical free-electron theory of metals, the electrical conductivity is given by $\sigma = ne^2\ell/m\bar{v}$, where v is the mean speed of electrons in the electron gas given by the Maxwell distribution, and ℓ is the mean free path of the electrons, which is inversely proportional to the area of the lattice atoms. The ratio of the electrical to thermal conductivity is proportional to the absolute temperature. This theory accounts for Ohm's law, but it gives the wrong temperature dependence for the conductivity and gives numerical values of σ in disagreement with experiment. The prediction that the heat capacity for metals should be $\frac{3}{2}R$ higher than that for other solids is not observed.

In the quantum-mechanical free-electron theory of metals, the Maxwell-Boltzmann energy distribution is replaced by the Fermi-Dirac distribution, and the wave nature of electron scattering is taken into account. In the Fermi-Dirac distribution at $T = 0$, all the energy states below a certain energy called the Fermi energy E_{F0} are filled and all those above this energy are empty. At higher temperatures, some electrons with energies of the order of kT below the Fermi energy are excited to energy states of the order of kT above that level. The Fermi energy (at $T = 0$) depends on the number density N/V and is given by

$$E_{F0} = \frac{h^2}{8m_e} \left(\frac{3N}{\pi V}\right)^{2/3}$$

The Fermi energy for copper is about 7 eV, much greater than kT at ordinary temperatures. At temperature T the Fermi energy is defined to be that energy E_F for which the probability of occupation of a state is $\frac{1}{2}$. The difference between E_F and E_{F0} is usually negligible.

The classical expression for electrical conductivity can be used if $\bar{v}$ is replaced by the Fermi speed u_F, which is essentially independent of temperature, and ℓ is interpreted as the mean free path for electrons in a lattice of point atoms which are vibrating. The quantum-mechanical theory also gives the ratio of electrical and thermal conductivities to be proportional to the absolute temperature, but the contribution of the electron gas to the heat capacity is very small because there are few unoccupied energy states to which the electrons can be randomly excited.

When many atoms are brought together, the individual energy levels are split into bands of allowed energies separated by gaps of forbidden energies. The splitting depends on the type of bonding and the lattice separation. A material for which the highest occupied band is not full is a conductor. If this band is full, the material is an insulator if there is a gap that is large compared with kT between the filled band and the next allowed band. If the gap is small, as in germanium for which it is about 0.7 eV, the material is an intrinsic semiconductor because there are some electrons in the conduction band due to thermal excitation. Intrinsic semiconductors often have negative temperature coefficients of electrical conduction because the increase in the number of conduction electrons at increased T more than compensates for the increase in the scattering because of the thermal motion of the lattice ions. Impurity semiconductors can be made by doping with an impurity which contributes either filled energy levels just below the conduction band (donor, or n-type) or empty energy levels just above the filled valence band (acceptor, or p-type). The voltage-current characteristics of an impurity semiconductor junction are similar to those of a diode.

In a superconductor, the resistance drops suddenly to zero below a critical temperature T_c that is typically only a few degrees. Superconductivity is due to a collective action of the electrons.

References

1. E. Fermi, *Molecules, Crystals, and Quantum Statistics*, trans. M. Ferro-Luzzi, New York: W. A. Benjamin, Inc., 1966.

2. A. Holden, *The Nature of Solids*, New York: Columbia University Press, 1968. An excellent nonmathematical treatment of the properties of solids.

3. C. Kittel, *Introduction to Solid State Physics*, 5th ed., New York: John Wiley & Sons, Inc., 1976.

4. N. Ashcroft and M. Mermin, *Solid State Physics*, New York: Holt, Rinehart and Winston, 1976.

5. R. Sproull, *Modern Physics*, 2d ed., New York: John Wiley & Sons, Inc., 1963.

6. W. Shockley, *Electrons and Holes in Semiconductors*, Princeton, N. J.: D. Van Nostrand Company, Inc., 1950.

7. A. Leitner, *Introduction to Superconductivity*, East Lansing, Mich.: Michigan State University, 1965. This excellent film, running 48 minutes, is probably the best available introduction to superconductivity.

Exercises

Section 9-1, Structure of Solids

1. Calculate the distance r_0 between the K^+ and Cl^- ions in KCl, assuming that each ion occupies a cube of side r_0. The molecular weight is 74.55 g/mole and the density is 1.984 g/cm³.

2. The crystal structure of KCl is the same as that of NaCl. (*a*) Calculate the electrostatic potential energy of attraction of KCl, assuming that r_0 is 3.14 Å. (*b*) Assuming that $n = 9$ in Equation 9-5, calculate the dissociation energy in eV per ion pair and in kcal/mole. (*c*) The measured dissociation energy is 165.5 kcal/mole. Use this to determine n in Equation 9-5.

3. The distance between the Li^+ and Cl^- ions in LiCl is 2.57 Å. Use this and the molecular weight of 42.4 g/mole to compute the density of LiCl.

4.   Find the value of n that gives the measured dissociation energy of 177 kcal/mole for LiCl, which has the same structure as NaCl, and for which $r_0 = 2.57$ Å.

Section 9-2, Classical Free-Electron Theory of Metals

5.   Find (a) the current density and (b) the drift velocity if there is a current of 1 mA in a No. 14 copper wire. (See Example 9-3.)

6.   A measure of the density of the free-electron gas in a metal is the distance r_s, defined as the radius of the sphere whose volume equals the volume per conduction electron. Show that $r_s = (3/4\pi n)^{1/3}$, where n is the free-electron number density, and calculate r_s in Å for copper.

7.   If the drift velocity in a copper wire is 10^{-2} cm/sec and the cross-sectional area is 1 mm², find (a) the current density, (b) the total current.

8.   Calculate the number density of free electrons in (a) Ag ($\rho = 10.5$ g/cm³) and (b) Au ($\rho = 19.3$ g/cm³), assuming one free electron per atom, and compare your results with the value listed in Table 9-2.

9.   Calculate the number density of free electrons for (a) Mg ($\rho = 1.74$ g/cm³) and (b) Zn ($\rho = 7.1$ g/cm³), assuming two free electrons per atom, and compare your results with the values listed in Table 9-2.

10.   The density of aluminum is 2.7 g/cm³. How many free electrons are there per aluminum atom? (Use Table 9-2 for the number density.)

11.   The density of potassium is 0.851 g/cm³. How many free electrons are there per potassium atom?

12.   The density of tin is 7.3 g/cm³. How many free electrons are there per tin atom?

13.   (a) Using $\ell = 3.7$ Å and $\bar{v} = 1.08 \times 10^5$ m/sec at $T = 300$ K, calculate σ and ρ for copper from Equations 9-13. Using the same value of ℓ, find σ and ρ at (b) $T = 200$ K, (c) $T = 100$ K.

14.   In Equation 9-16 for K/σ, $\bar{v}^2$ should really be $\overline{v^2} = 3kT/m$. Show that this correction leads to a Lorentz number $L = 3k^2/2e^2$ and compute this number in W-Ω/K².

Section 9-3, The Fermi Electron Gas

15.   Calculate the Fermi energies for (a) Al, (b) K, and (c) Sn using the number densities given in Table 9-2.

16.   Calculate the (a) Fermi energy and (b) Fermi temperature for gold at $T = 0$.

17.   Find the average energy of the electrons at $T = 0$ in (a) copper ($E_F = 7.03$ eV), (b) Li ($E_F = 4.74$ eV).

18.   Repeat Exercise 16 for iron.

Section 9-4, Quantum Theory of Conduction

19.   Use Equation 9-31 with $\alpha = \pi^2/4$ to calculate the average energy of an electron in copper at $T = 300$ K and compare your result with the average energy at $T = 0$ and the classical result of $\frac{3}{2}kT$.

20. At what temperature is the heat capacity due to the electron gas in copper equal to 10 percent of that due to lattice vibrations?

21. Find the Fermi speed u_F for (a) Na, for which $E_F = 3.24$ eV, (b) Au, for which $E_F = 5.53$ eV, and (c) Sn, for which $E_F = 10.2$ eV.

22. What is the order of magnitude of the mean time between collisions for the electrons in copper?

23. The resistivities of Na, Au, and Sn at $T = 273$ K are 4.2 $\mu\Omega$-cm, 2.04 $\mu\Omega$-cm, and 10.6 $\mu\Omega$-cm, respectively. Use these values and the Fermi speeds calculated in Exercise 21 to find the mean free paths ℓ for the conduction electrons in these elements.

24. Show that, using the Einstein model, the resistivity can be written

$$\rho = \frac{n_a}{n_e} \frac{m_e u_F}{e^2} \frac{2\pi(\hbar c)^2}{\mathcal{M}c^2(kT_E)} \frac{T}{T_E}$$

where n_a is the number density of atoms, n_e the number density of free electrons (which may or may not equal n_a), $\mathcal{M}$ the molecular weight, and T_E the Einstein temperature.

Section 9-5, Band Theory of Solids

There are no exercises for this section.

Section 9-6, Impurity Semiconductors

25. What kind of semiconductor is obtained if silicon is doped with (a) aluminum, (b) phosphorus? (See Table 7-1 for the electron configurations of all the elements.)

26. What kind of semiconductor is obtained if silicon is doped with (a) indium, (b) antimony?

Section 9-7, Semiconductor Junctions and Devices

27. For what value of bias voltage V_b does the exponential in Equation 9-42 have the value (a) 10, (b) 0.1 for $T = 200$ K?

28. Sketch I_{net} versus V_b for both positive and negative values of V_b from Equation 9-42.

Section 9-8, Superconductivity

There are no exercises for this section.

Problems

1. A one-dimensional model of an ionic crystal consists of a line of alternating positive and negative ions with distance r_0 between each pair. Show that the potential energy of attraction of one ion in such a line is

$$V = -\frac{2ke^2}{r_0} \left(1 - \frac{1}{2} + \frac{1}{3} - \frac{1}{4} + \frac{1}{5} - \cdots\right)$$

Using the result that

$$\ln(1 + x) = x - \frac{x^2}{2} + \frac{x^3}{3} - \frac{x^4}{4} + \cdots$$

show that the Madelung constant for this one-dimensional model is $\alpha = 2 \ln 2 = 1.386$.

2.   (a) Calculate the force $F = -dV/dr$ from Equation 9-4 and show that

$$F = \alpha \frac{ke^2}{r_0^2} \left(\frac{r_0^{n+1}}{r^{n+1}} - \frac{r_0^2}{r^2} \right) \qquad \text{9-46}$$

(b) Note that $F = 0$ at $r = r_0$. Write $r = r_0 + \Delta r = r_0 (1 + \epsilon)$, where $\epsilon = \Delta r/r_0$, and use the binomial expansion $(1 + \epsilon)^n = 1 + n\epsilon + n(n - 1)\epsilon^2/2$ to write F as a power series in Δr and show that, when $r = r_0 + \Delta r$, F is given by

$$F = -C \, \Delta r + C'(\Delta r)^2 + \cdots \qquad \text{9-47}$$

where

$$C = \alpha \frac{(n - 1)ke^2}{r_0^3} \qquad \text{and} \qquad C' = \alpha \frac{(n^2 + 3n - 4)ke^2}{2r_0^4}$$

3.   The quantity C in Equation 9-47 of Problem 2 is the force for a "spring" consisting of a line of alternating positive and negative ions. If these ions are displaced slightly from their equilibrium separation r_0 they will vibrate with frequency $f = (1/2\pi)\sqrt{C/m}$. (a) Use the values of α, n, and r_0 for NaCl and the reduced mass for the NaCl molecule to calculate this frequency. (b) Calculate the wavelength of electromagnetic radiation corresponding to this frequency and compare with the characteristic strong infrared absorption bands in the region of about $\lambda = 61 \ \mu m$ observed for NaCl.

4.   The term $C'(\Delta r)^2$ in Equation 9-47 of Problem 2 is related to the coefficient of thermal expansion of a solid. The time average of the force must be zero since there is no net acceleration averaged over time. From Equation 9-47 we have then

$$\overline{\Delta r} = \frac{C'}{C} \, \overline{(\Delta r)^2}$$

Use the equipartition result that $\frac{1}{2}C\overline{(\Delta r)^2} = \frac{1}{2}kT$ to show that $\overline{\Delta r} = (kC'/C^2)T$, where k is Boltzmann's constant (note that the k in ke^2, in the expressions for C and C', is the Coulomb constant). Evaluate the coefficient of thermal expansion $kC'/C^2 r_0$ for NaCl and compare with the measured value of about $4 \times 10^{-6}/\text{K}$.

5.   The pressure of an ideal gas is related to the average energy of the gas particles by $PV = \frac{2}{3}N\overline{E}$, where N is the number of particles and $\overline{E}$ is the average energy. Use this to calculate the pressure in N/m^2 of the Fermi electron gas in copper, and compare this value with atmospheric pressure, which is about $10^5 \ \text{N/m}^2$. (Note: The units are most easily handled by noting that $1 \ \text{N/m}^2 = 1 \ \text{J/m}^3$ and $1 \ \text{eV} = 1.6 \times 10^{-19} \ \text{J}$.)

6.   The bulk modulus B (which is the reciprocal of the compressibility) of a material is defined by

$$B = -V \frac{\partial P}{\partial V}$$

Use the ideal-gas relation $PV = \frac{2}{3}N\overline{E}$ and Equations 9-24 and 9-27 to show that

$$P = \frac{2}{5} \frac{NE_F}{V} = CV^{-5/3}$$

where C is a constant independent of V and that the bulk modulus of the Fermi electron gas is therefore given by

$$B = \tfrac{5}{3}P = \frac{2}{3} \frac{NE_F}{V}$$

Compute the bulk modulus in $N/m^2 = J/m^3$ for the Fermi gas in copper and compare with the measured value for solid copper of $134 \times 10^9 \ N/m^2$.

7.   The relative binding of the extra electron in the arsenic atom that replaces a silicon or germanium atom can be understood from a calculation of the first Bohr orbit of this electron in these materials. In the Bohr model of hydrogen atom the electron moves in free space at a radius a_0 given by

$$a_0 = \frac{\hbar^2}{m_e k e^2} = \frac{4\pi\epsilon_0 \hbar^2}{m_e e^2}$$

where $k = 1/4\pi\epsilon_0$ is the Coulomb constant. When an electron moves in a crystal we can approximate the effect of the other atoms by replacing ϵ_0 by $\epsilon = K\epsilon_0$ (or equivalently, by replacing k by k/K), where K is the dielectric constant. Calculate the radius of the first Bohr orbit of an electron in (a) silicon ($K = 12$) and (b) germanium ($K = 16$), and compare with the spacing of the lattice ions in these crystals. (A correction for the effective mass of the electron moving in a crystal increases these results by about a factor of 5.) Discuss the possible validity of using the macroscopic concept of the dielectric constant in an atomic calculation.

8.   Replace k by k/K in the expression for the binding energy E_0 of hydrogen to estimate the binding energy of the extra electron of an impurity arsenic atom in (a) silicon ($K = 12$), (b) germanium ($K = 16$). (See Problem 7.)

9.   The number of electrons in the conduction band of an insulator or semiconductor is governed chiefly by the Fermi factor, which in these cases is $e^{-(E-E_F)/kT}$, where E_F is the Fermi-energy level, which is midway between the nearly filled valence band and the nearly empty conduction band. If E_g is the energy gap between the valence and conduction bands, and E is measured from the top of the valence band, $E_F = \tfrac{1}{2}E_g$. The Fermi factor at the bottom of the conduction band $E = E_g$ is then $e^{-E_g/2kT}$. Calculate this Fermi factor for a typical gap energy of (a) 6.0 eV for an insulator and (b) 1.0 eV for a semiconductor at $T = 300$ K. Discuss the significance of these results if there are 10^{22} electrons and $e^{-(E-E_F)/kT}$ is the probability of one electron being in the conduction band.

CHAPTER 10   Quantum Statistics and
Liquid Helium

Objectives

After studying this chapter you should:

1. Know the assumptions that lead to the Maxwell-Boltzmann distribution.

2. Be able to write the mathematical expressions for the Maxwell-Boltzmann, Fermi-Dirac, and Bose-Einstein distributions and discuss the applicability of each.

3. Know the general conditions under which both the Bose-Einstein and Fermi-Dirac distributions can be replaced by the simpler Maxwell-Boltzmann distribution.

4. Know what is meant by the terms boson and fermion.

5. Be able to sketch the heat capacity versus temperature for liquid helium.

6. Know what is meant by the "lambda point" and know its approximate temperature.

7. Be able to describe some of the properties of helium II and discuss how the two-fluid model accounts for these properties.

In our discussion of electrical conduction and heat conduction in metals in Chapter 9, we found that we could treat the free electrons as a gas. The classical Maxwell-Boltzmann distribution could not be used, however, because it implied more than one electron per quantum state, contrary to the Pauli exclusion principle. When we replaced the classical distribution with the quantum Fermi-Dirac distribution, the free-electron gas treatment of metals gave a good account of electrical conduction, heat conduction, and the contribution of the electrons to the heat capacity of metals. We did not need to consider the Fermi-Dirac distribution in detail, because at ordinary temperatures the quantum distribution differs only slightly from that at $T = 0$. In this chapter we shall discuss the Fermi-Dirac distribution in more detail and compare it with the classical Maxwell-

Boltzmann distribution and with another quantum distribution, the Bose-Einstein distribution, which governs indistinguishable identical particles that do not obey the exclusion principle. We shall be particularly interested in investigating the conditions under which the quantum distributions can be replaced by the simpler classical Maxwell-Boltzmann distribution. In Section 10-2 we shall present some of the remarkable properties of liquid helium and discuss qualitatively how the Bose-Einstein distribution can be applied to this system.

10-1 The Quantum Distribution Functions

In the statistical derivation of the Maxwell-Boltzmann distribution function, the following assumptions are made:

1. The equilibrium distribution is the most probable distribution that is consistent with a constant number of particles and constant total energy.

2. The particles are identical but distinguishable.

3. There is no restriction on the number of particles in any state.

(Such a derivation, and derivations of the quantum distributions to be discussed in this section, can be found in many kinetic-theory textbooks such as Reference 6.) To compare the Maxwell-Boltzmann distribution with the quantum distributions, we write it as a product of the number of states having the same energy, g_i (the statistical weight), and the probability F_{MB} that the state is occupied. If n_i is the number of particles with energy E_i, the discrete form of the Maxwell-Boltzmann distribution is

$$n_i = g_i F_{MB} \qquad 10\text{-}1$$

with the Maxwell-Boltzmann factor given by

$$F_{MB} = A e^{-E_i/kT} \qquad 10\text{-}2$$

As usual, the constant A is determined by the normalization condition

$$\Sigma\, n_i = N \qquad 10\text{-}3$$

where N is the total number of particles. (We shall use number distributions in this chapter rather than the fractional distribution function $f_i = n_i/N$ used in Chapter 2.) For the case of continuous energies, or for a very large number of energy states so closely spaced that they can be treated as continuous, we replace g_i by $g(E)\, dE$ (the number of states in the interval dE) and n_i by $n(E)\, dE$ (the number of particles having energies between E and

$E + dE$). The continuous distribution function $n(E)$ is then written

$$n(E) = g(E)\, F_{MB} \qquad\qquad 10\text{-}4$$

Maxwell-Boltzmann distribution

where $g(E)$ is the density of states and

$$F_{MB} = A e^{-E/kT} \qquad\qquad 10\text{-}5$$

Because of the wave nature of all particles, identical particles cannot be distinguished if their wave functions overlap. Assumption 2 is therefore not valid for any type of particle, except in the classical limit in which the wave packets are very narrow and the particles are far enough apart that their wave functions do not overlap. Assumption 2 is important because the most probable distribution is, by definition, the one that can be realized in the greatest number of ways, and the number of ways a particular distribution can be realized depends on whether the particles can be distinguished. We can illustrate this point by considering a simple probability example.

Example 10-1   The Arrangement of Three Objects in Two Boxes   We shall label the boxes H and T since this problem can be done experimentally by flipping three coins. We wish to find the probability that there will be n_H objects in box H and n_T in box T. The *probability* of any arrangement is defined as the number of ways that arrangement can be realized, divided by the total number of combinations possible. Figure 10-1 shows the eight different ways possible if the objects labeled A, B, and C are identical but distinguishable. A ninth way is shown in Figure 10-1i. This is not different from Figure 10-1c if the objects are identical. The probability of two objects in box H and one in box T is $\frac{3}{8}$. The probability of all three being in H is $\frac{1}{8}$. Figure 10-2

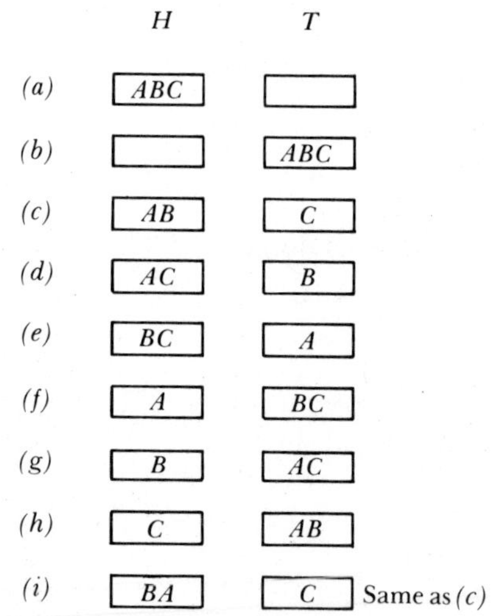

Figure 10-1
The eight different ways of arranging three identical but distinguishable objects in two boxes. A ninth way, shown in (i), is the same as (c). Assuming that each object is equally likely to be in either box, the probability of all three objects being in box H is $\frac{1}{8}$.

shows the situation if the objects are indistinguishable: there are now only four different arrangements. The probability of two in box H and one in T is now $\frac{1}{4}$, and the probability that all three are in box H is also $\frac{1}{4}$.

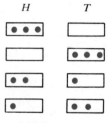

Figure 10-2
The four different ways of arranging three identical and indistinguishable objects in two boxes. For this case, the probability of all three objects being in box H is $\frac{1}{4}$. Compare this with Figure 10-1.

If we keep assumptions 1 and 3 but instead of assumption 2 we assume that the particles in a gas are indistinguishable, we obtain the *Bose-Einstein distribution,* which describes particles with zero or integral spin that do not obey the exclusion principle. Such particles, called *bosons,* include α particles ($S = 0$), deuterons ($S = 1$), photons ($S = 1$), and most atoms and molecules, which have zero spin. If we also give up assumption 3 and introduce the restriction that not more than one particle can occupy a quantum state, we obtain the *Fermi-Dirac distribution,* which applies to spin-$\frac{1}{2}$ particles, such as electrons, called *fermions.* (The derivations of these distribution functions can be found in Reference 6.) These distributions can be written:

$$n(E) = g(E)F_{\text{BE}} \qquad \text{(Bose-Einstein)} \qquad 10\text{-}6$$

and

$$n(E) = g(E)F_{\text{FD}} \qquad \text{(Fermi-Dirac)} \qquad 10\text{-}7$$

where the Bose-Einstein factor is

$$F_{\text{BE}} = \frac{1}{Be^{+E/kT} - 1} \qquad\qquad 10\text{-}8 \qquad \textit{Bose-Einstein distribution}$$

and the Fermi-Dirac factor is

$$F_{\text{FD}} = \frac{1}{Be^{+E/kT} + 1} \qquad\qquad 10\text{-}9 \qquad \textit{Fermi-Dirac distribution}$$

The constant B in each factor is determined by normalization, and depends on the total number of particles. If we compare these distribution functions with the classical distribution function, we note that were it not for the ± 1 in the denominators, all three distribution functions would be the same, with $B = 1/A$. The seemingly small difference of $+1$ or -1 in the denominators of the quantum distribution functions has a profound effect and makes each very different from the other and, in many cases, from the classical Maxwell-Boltzmann distribution function.

Since the Maxwell-Boltzmann distribution is generally easier to use, it is important to know under what conditions it is valid. We see from Equations 10-8 and 10-9 that both quantum distributions approach the Maxwell-Boltzmann distribution when $Be^{E/kT}$ is much greater than 1, which is equivalent to

$$Ae^{-E/kT} \ll 1 \qquad\qquad 10\text{-}10$$

with $A = 1/B$. In the Maxwell-Boltzmann distribution, the

quantity $Ae^{-E/kT}$ is the number of particles per energy state. Thus if the number of particles per state is much less than 1, the quantum distribution differs little from the classical distribution. This is what we would expect. The number of particles will be small if the density of particles is very small, in which case the wave functions should overlap very little and the indistinguishability of particles should have little effect.

For energies of the order of kT, the classical distribution will be a good approximation if A is much less than 1. The simplest procedure for determining whether the classical distribution is valid is to assume it, and calculate A. The classical distribution will then be valid for energies such that $Ae^{-E/kT} \ll 1$. We can compute A from the normalization condition using the density of states found in Section 9-4 for N particles in a cube of side L and volume $V = L^3$. For electrons, the density of states is given by Equation 9-26, which can be written

$$g_e(E) = \frac{4\pi V}{h^3}(2m_e)^{3/2}E^{1/2}$$

10-11

Density of states for electrons

If the particles are He atoms, instead, the density of states is the same except that the mass is now that of the helium atom, and there is no factor of 2 introduced into g because of the two electron spin states. The density of states for bosons of mass M is, therefore,

$$g_{He}(E) = \frac{2\pi V}{h^3}(2M)^{3/2}E^{1/2}$$

10-12

Density of states for He atoms

We can compute A for these two cases from the normalization condition

$$\int_0^\infty n(E)\, dE = N$$

10-13

For electrons we have

$$N = \int_0^\infty n(E)\, dE = A_e \frac{4\pi V}{h^3}(2m_e)^{3/2}\int_0^\infty E^{1/2}e^{-E/kT}\, dE$$

$$= \frac{2V}{h^3}(2\pi m_e kT)^{3/2}A_e$$

or

$$A_e = \frac{N}{V}\frac{h^3}{2(2\pi m_e kT)^{3/2}}$$

10-14

For the case of He, we obtain

$$A_{He} = \frac{N}{V}\frac{h^3}{(2\pi MkT)^{3/2}}$$

10-15

We note that A is proportional to the number density of particles N/V.

Let us now compute these numbers for an electron gas and He gas. Assuming one electron per atom in copper, we have 6.02×10^{23} electrons in 1 mole, which has a volume of 7.09 cm³. This gives a number density N/V of 8.49×10^{22}/cm³. Using $hc = 1.24 \times 10^{-4}$ eV-cm and $kT = 2.59 \times 10^{-2}$ eV at $T = 300$ K, we have

$$A_e = \frac{N}{V} \frac{(hc)^3}{2(2\pi m_e c^2 kT)^{3/2}}$$

$$= \frac{(6.02 \times 10^{23})(1.24 \times 10^{-4} \text{ eV-cm})^3}{(7.09 \text{ cm}^3)2[(2\pi)(5.11 \times 10^5 \text{ eV})(2.59 \times 10^{-2} \text{ eV})]^{3/2}}$$

$$= 3.38 \times 10^3$$

Since $A_e \ll 1$ does *not* hold here, the Maxwell-Boltzmann distribution function is not a good approximation for this electron gas. This is exactly what we found in Chapter 9. If the Maxwell-Boltzmann distribution were applied to this electron gas there would be many electrons per quantum state, in violation of the exclusion principle.

If we compute A_{He} for He gas at standard conditions, we obtain

$$A_{\text{He}} \approx 3.5 \times 10^{-6}$$

This is much smaller because the mass of He is about 8000 times that of the electron. Thus $(M/m_e)^{3/2} \approx 10^6$. Also, the density of He is much smaller. Under standard conditions, a mole occupies 22.4×10^3 cm³, compared with 7.09 cm³ for the electron gas in copper. We see that even for the very light gas, He, only about 1 out of 10^6 states is occupied at $E \approx kT$; so the BE distribution differs little from the MB distribution. The MB distribution therefore gives an excellent approximation to the BE distribution in practically every case of interest; however, it is usually not a good approximation to the FD distribution.

An important application of the Bose-Einstein distribution is to liquid helium, which we shall discuss in the next section. The density of liquid helium is 0.145 g/cm³, so that the volume of 1 mole is $V = (4 \text{ g/mole})/(0.145 \text{ g/cm}^3) = 27.6$ cm³. Using this value rather than 22.4×10^3 cm³ for helium, we obtain for A_{He} at $T = 300$ K, $A_{\text{He}} \approx 2.84 \times 10^{-3}$. At any other temperature T we have

$$A_{\text{He}} = 2.84 \times 10^{-3} \left(\frac{300 \text{ K}}{T}\right)^{3/2}$$

If we set A_{He} equal to 1 and solve for T we obtain $T \approx 6$ K. Although we should be skeptical of applying the density-of-states formula for an ideal gas to liquid helium, this calculation does indicate that, at the high number density found in the liquid state, we should expect the classical distribution function to break down at sufficiently low temperatures.

The constant B in the Fermi-Dirac distribution function is usually written in terms of the Fermi energy level, which we defined in Chapter 9 to be that energy for which the occupation probability is $\frac{1}{2}$. Setting F_{FD} equal to $\frac{1}{2}$ for $E = E_F$ and solving for B we obtain

$$\frac{1}{Be^{E_F/kT} + 1} = \frac{1}{2}$$

or

$$Be^{E_F/kT} = 1$$

$$B = e^{-E_F/kT} \tag{10-16}$$

The Fermi-Dirac factor F_{FD} is then written

$$F_{FD} = \frac{1}{e^{(E-E_F)/kT} + 1} \tag{10-17}$$

Fermi-Dirac factor in terms of Fermi energy

From this expression we can see that, if $E - E_F$ is much greater than kT, we can neglect the 1 in the denominator and the Fermi-Dirac distribution becomes identical with the classical Maxwell-Boltzmann distribution.

Example 10-2   The Fermi energy for a semiconductor is in the energy gap between the valence band and the conduction band. Assuming that it is halfway between these bands, show that the electrons in the conduction band of a semiconductor can be described by the classical distribution. The energy gap E_g between the valence and conduction band is about 1 eV for a semiconductor. Taking $E = 0$ at the bottom of the conduction band, we then have $E - E_F = \frac{1}{2}E_g$. For $E_g = 1$ eV and $kT = 2.59 \times 10^{-2}$ eV at $T = 300$ K, we have

$$\frac{E - E_F}{kT} = \frac{0.5 \text{ eV}}{2.59 \times 10^{-2} \text{ eV}} = 19.3$$

The exponential in the denominator is then $e^{19.3} \approx 2.4 \times 10^8$, which is much greater than 1.

If we examine Equation 10-17 at very low temperatures we note that, as T approaches zero, F_{FD} approaches zero for $E \gg E_F$ and 1 for $E \ll E_F$. This agrees with our discussion of the Fermi electron gas at $T = 0$ in Chapter 9.

The first application of the Bose-Einstein distribution function was by S. Bose in 1924. He showed that the Planck formula for the spectral distribution of blackbody radiation can be derived by considering the radiation to be a "gas" of photons obeying the Bose-Einstein distribution. This calculation will not be presented here, but it is outlined in Problem 4. In the next section we shall discuss another interesting application of the Bose-Einstein distribution—the description of the properties of liquid helium at low temperatures.

Questions

1. What is a boson?

2. What is a fermion?

3. How can identical particles be distinguished classically?

4. Under what conditions does the Maxwell-Boltzmann distribution hold?

10-2 Liquid Helium II

We saw in the previous section that, for ordinary gases, the Bose-Einstein distribution differs very little from the classical Maxwell-Boltzmann distribution, because there are many quantum states per particle due to the low density of gases and the large mass of the molecules. However, for liquid helium, there is approximately one particle per quantum state at very low temperatures, and the classical distribution is invalid. The somewhat daring idea that liquid helium can be treated as an ideal gas obeying the Bose-Einstein distribution was suggested in 1938 by F. London in an attempt to understand the amazing properties of helium at low temperatures. We shall describe some of the properties of liquid helium at low temperatures and give a brief outline of the London theory.

When liquid helium is cooled, a remarkable change in its properties takes place at a temperature of 2.17 K. In 1924, H. Kamerlingh Onnes and J. Boks measured the density of liquid helium as a function of temperature and obtained the curve shown in Figure 10-3. In 1928, W. H. Keesom and M. Wolfke

Figure 10-3
Plot of density of liquid helium versus temperature, by Kamerlingh Onnes and Boks. (*From F. London,* Superfluids, *New York: Dover Publications Inc., 1964. Reprinted by permission of the publisher.*) The photo shows H. Kamerlingh Onnes and J. D. Van der Waals by the helium liquefier in the Kamerlingh Onnes Laboratory in Leiden in 1911. (*Courtesy of the Kamerlingh Onnes Laboratory.*)

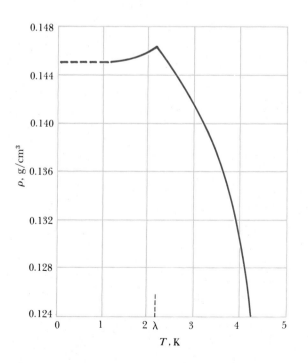

suggested that this discontinuity in the slope of the curve was an indication of a phrase transition. They used the terms "helium I" for the liquid above 2.17 K and "helium II" for the liquid below that temperature. Keesom and K. Clusius in 1932 measured the specific heat as a function of temperature and obtained the curve shown in Figure 10-4. Because of the similarity of this curve to the Greek letter λ, the transition temperature is called the *lambda point*. Figure 10-5 shows this same curve measured with greater resolution. Just above the lambda point, He boils vigorously as it evaporates. The bubbling immediately ceases at the lambda point, although evaporation continues. This effect is due to the sudden increase in the thermal conductivity at the lambda point (Figure 10-6).

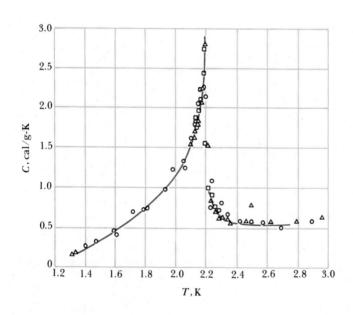

Figure 10-4
Specific heat of liquid helium versus temperature. Because of the resemblance of this curve to the Greek letter λ, the transition point is called the lambda point. (From *F. London,* Superfluids, *New York: Dover Publications, Inc., 1964. Reprinted by permission of the publisher.*)

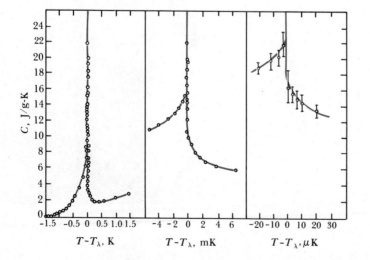

Figure 10-5
The lambda point with high resolution. The specific-heat curve maintains its shape as the scale is expanded. (*From M. J. Buckingham and W. M. Fairbank, "The Nature of the λ-Transition,"* Progress in Low Temperature Physics, *edited by C. J. Gorter, Vol. III, Amsterdam: North-Holland Publishing Company, 1961.*)

Measurements of thermal conductivity show that helium II conducts heat better than helium I by a factor of more than a million; in fact, helium II is a better heat conductor than any metal. This conduction process is different from ordinary heat conduction, for the rate of conduction is not proportional to the temperature difference.

The viscosity of liquid helium II also shows unusual behavior. If it is measured by the method of passing the liquid through a fine capillary, the result depends on the size of the capillary; the measured viscosity approaches zero as the diameter of the capillary is made smaller. However, if the rotating-disk method is used, the measured viscosity is not too different from that of helium I.

Rather than mention the many other interesting properties of liquid helium II, we shall discuss briefly a theory suggested by F. London in 1938 and elaborated by L. Tisza.[1] In this theory, helium II is imagined to consist of two parts, a normal fluid with properties similar to helium I and a superfluid with quite different properties. The density of liquid helium II is the sum of the densities of the normal fluid and the superfluid:

$$\rho = \rho_s + \rho_n \qquad \text{10-18}$$

As the temperature is lowered from the lambda point, the density of the superfluid increases and that of the normal fluid decreases until, at absolute zero, only the superfluid remains. The superfluid is supposed to correspond to the helium molecules being in the lowest possible quantum state, the ground state. These molecules are not excited to higher states, so the superfluid cannot contribute to viscosity. When the viscosity of helium II is measured by the rotating-disk method, only the normal-fluid component exerts a viscous force on the disk. As the temperature is lowered, the fraction of helium in the normal component decreases from 100 percent at the lambda point to 0 percent at $T = 0$ K; thus the viscosity decreases rapidly with temperature in agreement with experiment. On the other hand, when the viscosity is measured by passing the liquid through a fine capillary, the superfluid shows no resistance to flow. As the diameter of the capillary is reduced, the amount of normal fluid getting through the capillary is reduced until, in the limit of very small diameter, only the superfluid gets through and there is no viscosity.

It is not obvious that liquid helium should behave at all like an ideal gas. The atoms do exert forces on each other; however, these are weak van der Waals forces and the fairly low density of liquid helium (0.145 g/cm³) indicates that the atoms are relatively far apart. The ideal-gas model is used mainly because it is relatively simple and because it yields insight into the behavior of this interesting fluid.

[1] Many of these properties are elegantly displayed in the film "Liquid Helium II, the Superfluid," available from the Instructional Media Center, Michigan State University, East Lansing, Michigan 48824.

(a)

(b)

Figure 10-6
(a) Liquid helium being cooled by evaporation just above the lambda point boils vigorously. (b) Below the lambda point the boiling ceases and the superfluid runs out through the fine pores in the bottom of the vessel suspended above the helium bath. (*Courtesy of Clarendon Laboratory. From K. Mendelssohn,* The Quest for Absolute Zero: The Meaning of Low Temperature Physics, *World University Library, New York: McGraw-Hill Book Company, 1966.*)

(a)

(b)

Helix Technology Corp.

(c)

Some remarkable properties of liquid helium II. (a) The Rollin creeping film. The liquid helium in the dish is at a temperature of about 1.6 K. A thin film creeps up the sides of the dish, over the edge, and down the outside to form the drop shown. (b) The thermomechanical effect. A bulb containing liquid helium is in a cold bath of liquid helium II at 1.6 K. When light containing infrared radiation is focused on the bulb, liquid helium rises above the ambient level. The height of the level depends on the narrowness of the tube. If the tube is packed with powder and the top drawn out into a fine capillary, the superfluid spurts out in a jet as shown in (c), hence the name "fountain effect." [(a) and (b), courtesy of A. Leitner, Rensselaer Polytechnic Institute.]

In the Bose-Einstein distribution the number of particles in the energy range dE is given by $n(E) \, dE$, where

$$n(E) = \frac{g(E)}{Be^{E/kT} - 1} = \frac{g(E)}{e^{\alpha}e^{E/kT} - 1} \qquad 10\text{-}19$$

where $g(E)$ is the density of states given by Equation 10-12 and we have written $B = e^{\alpha}$ for convenience. The constant α, which is determined by normalization, cannot be negative for if it were, $n(E)$ would be negative for low values of E. The normalization condition is

$$N = \int n(E) \, dE = \frac{2\pi V}{h^3} (2M)^{3/2} \int_0^{\infty} \frac{E^{1/2} \, dE}{e^{\alpha}e^{E/kT} - 1}$$

$$= \frac{2\pi V}{h^3} (2MkT)^{3/2} \int_0^{\infty} \frac{x^{1/2} \, dx}{e^{\alpha + x} - 1} \qquad 10\text{-}20$$

where $x = E/kT$ and the integral in this equation is a function of α.

The usual justification for using a continuous energy distribution to describe a quantum system with discrete energies is that the energy levels are numerous and closely spaced. In this case, as we have already seen, for a gas of N particles in a macroscopic box of volume V (the container), this condition holds.

However, in replacing the discrete distribution of energy states by a continuous distribution we ignore the ground state. This has little effect for a Fermi gas since there can be only two particles in any single state, and ignoring two particles out of 10^{22} causes no difficulty. In a Bose-Einstein gas, however, there can be any number of particles in a single state. If we ignore the ground state as we have up to now, the normalization condition expressed by Equation 10-20 cannot be satisfied at low temperatures. This implies that at very low temperatures there is a significant number of particles in the ground state.

The integral in Equation 10-20, which can be evaluated numerically, has a maximum value of 2.315 when α has its minimum value of 0. This implies a maximum value of N/V, given by

$$\frac{N}{V} \leq \frac{2\pi}{h^3} (2MkT)^{3/2}(2.315)$$

Since N/V is determined by the density of liquid helium, this implies a minimum temperature, given by

$$T \geq \frac{h^2}{2Mk} \left[\frac{N}{2\pi(2.315)\,V}\right]^{2/3} = T_c \qquad\qquad 10\text{-}21$$

Inserting the known constants and the density of helium, we find for the critical temperature

$$T_c = 3.1 \text{ K} \qquad\qquad 10\text{-}22$$

For temperatures below 3.1 K the normalization Equation 10-20 cannot be satisfied for any value of α. Evidently at these temperatures there is a significant number of particles in the ground state, which we have not included.

If we include the ground state, Equation 10-20 becomes

$$N = N_0 + \frac{2\pi V}{h^3} (2MkT)^{3/2} \int_0^\infty \frac{x^{1/2}\, dx}{e^{\alpha+x} - 1} \qquad\qquad 10\text{-}23$$

where N_0 is the number in the ground state. If we choose $E_0 = 0$ for the energy of the ground state this number is

$$N_0 = \frac{g_0}{e^\alpha\, e^{E_0/kT} - 1} = \frac{1}{e^\alpha - 1} \qquad\qquad 10\text{-}24$$

where g_0 is the statistical weight, which is 1 for a single state. We see that N_0 becomes large as α becomes small. With the inclusion of N_0, which depends on α, the normalization condition (Equation 10-23) can be met and α can be computed numerically for any temperature and density. For temperatures below the critical temperature T_c, it can be shown that α is of the order of N^{-1}, and that the fraction of molecules in the ground state is given approximately by

$$\frac{N_0}{N} \approx 1 - \left(\frac{T}{T_c}\right)^{3/2} \qquad\qquad 10\text{-}25$$

The particles in the ground state constitute the superfluid.

The value $T_c = 3.1$ K is not very different from the observed lambda-point temperature $T = 2.17$ K, especially considering that this calculation is based on the assumption that liquid helium is an ideal gas. The process of molecules dropping into the ground state as the temperature is lowered below T_c is called *Einstein condensation*. Such an occurrence was predicted by Einstein in 1924, before there was any evidence that such a process took place in nature.

Questions

5. Why does the value of the coefficient of viscosity of helium II measured by the capillary method depend on the size of the capillary?

6. What is Einstein condensation?

7. Would you expect a gas or liquid of ^{3}He atoms to be much different from one of ^{4}He atoms? Why or why not?

Summary

The Maxwell-Boltzmann distribution function $n(E)$, which applies to identical but distinguishable particles, can be written

$$n(E) = g(E)F_{MB}$$

where $g(E)$ is the density of states and the Maxwell-Boltzmann factor is given by

$$F_{MB} = Ae^{-E/kT}$$

The normalization constant A is proportional to the number density of particles N/V. The distribution functions for identical and indistinguishable particles can be written

$$n(E) = g(E)F_{FD} \qquad \text{(Fermi-Dirac)}$$

and

$$n(E) = g(E)F_{BE} \qquad \text{(Bose-Einstein)}$$

where again $g(E)$ is the density of states, and the Fermi-Dirac factor F_{FD} and the Bose-Einstein factor F_{BE} are given by

$$F_{FD} = \frac{1}{Be^{E/kT} + 1}$$

and

$$F_{BE} = \frac{1}{Be^{E/kT} - 1}$$

The Fermi-Dirac distribution applies to particles that obey the exclusion principle (electrons, protons, neutrons, etc.). Such particles have intrinsic spin quantum number $\frac{1}{2}$ and are called fermions. The Bose-Einstein distribution applies to particles with zero or integral spin called bosons (α particles, deuterons, photons, etc.), which do not obey the exclusion principle. At high energies ($E \gg kT$), both F_{FD} and F_{BE} approach F_{MB}, with $B = 1/A$. The Maxwell-Boltzmann distribution is a good approximation to either quantum distribution if the normalization constant $A = 1/B$ computed from the Maxwell-Boltzmann distribution function is much less than 1. This is equivalent to the condition that the number of particles is much less than the number of available energy states. It is always met for ordinary gases, because of the large atomic masses and low densities. It is almost never met for electron gases. In the Fermi-Dirac distribution, the constant B is usually written $B = e^{-E_F/kT}$, where E_F is the Fermi energy.

Liquid helium exhibits many unusual and remarkable properties at temperatures below 2.17 K. This temperature is called the lambda point because of the resemblance of the curve of heat capacity versus temperature to the Greek letter λ. The properties of liquid helium can be understood qualitatively by assuming that liquid helium is an ideal gas obeying the Bose-Einstein distribution. Below the lambda point, liquid helium can be described as a mixture of an ordinary fluid and a superfluid in which all the molecules are in a single energy state, the ground state. The process of molecules dropping into the ground state is called Einstein condensation.

References

1. R. Eisberg and R. Resnick, *Quantum Physics of Atoms, Molecules, Solids, Nuclei, and Particles,* New York: John Wiley & Sons, Inc., 1974. An excellent but somewhat more advanced discussion of quantum statistics can be found in Chapter 11 of this book.

2. F. London, *Superfluids. Vol. II: Macroscopic Theory of Superfluid Helium,* 2d rev. ed., New York: Dover Publications, Inc., 1954.

3. C. Lane, *Superfluid Physics,* New York: McGraw-Hill Book Company, 1962.

4. K. Mendelssohn, *The Quest for Absolute Zero: The Meaning of Low Temperature Physics,* World University Library, New York: McGraw-Hill Book Company, 1966.

5. A. Leitner, *Liquid Helium II, the Superfluid,* East Lansing, Mich.: Michigan State University, 1963. This 39-minute film is an excellent introduction to the subject of liquid helium II.

6. J. F. Lee, F. W. Sears, and D. L. Turcotte, *Statistical Thermodynamics,* Reading, Mass.: Addison-Wesley Publishing Co., Inc., 1963.

Exercises

Section 10-1, The Quantum Distribution Functions

1. Construct diagrams like Figure 10-1 for the possible arrangements of two particles in two boxes if the particles are (*a*) identical but distinguishable, and (*b*) identical and indistinguishable. (*c*) Find the probability that both particles will be in the H box for each case.

2.   Find the number density N/V for electrons such that (a) $A_e = 1$, (b) $A_e = 10^{-6}$.

3.   (a) Compute A from Equation 10-15 for O_2 gas at standard conditions. (b) At what temperature is $A = 1$ for O_2?

4.   Compute $e^{(E-E_F)/kT}$ for germanium, at $T = 300$ K for $E = E_g$ and $E_F = \frac{1}{2}E_g$ (as in Example 10-2) where $E_g = 0.07$ eV.

5.   Repeat Exercise 4 for a typical insulator, for which $E_g \approx 6$ eV. Can you use the Maxwell-Boltzmann distribution for the electrons in the conduction band of an insulator? Your answer is not very important. Why not?

Section 10-2, Liquid Helium II

6.   Use the expansion $e^x \approx 1 + x + \cdots$ for $x \ll 1$ to show that $N_0 \approx 1/\alpha$ for small α.

7.   Compute N_0/N from Equation 10-25 for (a) $T = \frac{1}{2}T_c$, (b) $T = \frac{1}{4}T_c$.

Problems

1.   Electrons that escape from a metal in thermionic emission must have energy at least as great as $E = E_F + \phi$, where E_F is the Fermi energy and ϕ is the work function. Show that those which are likely to escape obey the Maxwell-Boltzmann energy distribution. (Assume that $kT \ll \phi$. How good is this assumption?)

2.   Make a plot of F_{FD} versus E for (a) $T = 0.1T_F$, and (b) $T = 0.5T_F$, where $T_F = E_F/k$.

3.   What fraction of electrons in an electron gas have less than the mean energy at $T = 0$?

4.   If the assumptions leading to the Bose-Einstein distribution are modified so that the number of particles is not assumed constant, the resulting distribution has $B = 1$. This distribution can be applied to a "gas" of photons. Consider the photons to be in a cubic box of side L. The momentum components of a photon are quantized by the standing-wave conditions $k_x = n_1\pi/L$, $k_y = n_2\pi/L$, and $k_z = n_3\pi/L$, where $p = \hbar(k_x^2 + k_y^2 + k_z^2)^{1/2}$ is the magnitude of the momentum. (a) Show that the energy of a photon can be written $E = N(\hbar c\pi/L)$, where $N^2 = n_1^2 + n_2^2 + n_3^2$. (b) Assuming two photons per space state because of the two possible polarizations, show that the number of states between N and $N + dN$ is $\pi N^2\, dN$. (c) Find the density of states and show that the number of photons in the energy interval dE is

$$n(E)\, dE = \frac{8\pi(L/hc)^3 E^2\, dE}{e^{E/kT} - 1}$$

(d) The energy density in dE is given by $u(E)\, dE = En(E)\, dE/L^3$. Use this to obtain the Planck blackbody-radiation formula for the energy density in $d\lambda$, where λ is the wavelength:

$$u(\lambda) = \frac{8\pi hc\lambda^{-5}}{e^{hc/\lambda kT} - 1}$$

CHAPTER 11   Nuclear Physics

Objectives

After studying this chapter you should:

1. Know the meaning of isotope, isotone, isobar, and isomer.

2. Know the relation between the nuclear radius and mass number, know the order of magnitude of the nuclear radius, and be able to describe one or more ways of determining it.

3. Know the general shape of the N versus Z curve for stable nuclei.

4. Be able to sketch the nuclear binding-energy curve and discuss the significance of this curve for fission and fusion.

5. Know the exponential law of radioactive decay and be able to work problems using it.

6. Know the meaning of the Q value and the cross section for nuclear reactions.

7. Be able to describe the nuclear-fission chain reaction and discuss the advantages and disadvantages of fission reactors.

8. Know the Lawson criterion for nuclear-fusion reactors.

9. Know the chief mechanisms of energy loss of particles in matter, and be able to explain why some particles have well-defined ranges and others do not.

10. Know the uses of the radiation units roentgen, rad, and rem.

11. Be able to describe various kinds of particle detectors and discuss the advantages and disadvantages of each.

12. Be able to discuss the features of the nuclear shell model.

The first information about the atomic nucleus came from the discovery of radioactivity by Becquerel in 1896. The rays emitted by radioactive nuclei were studied by many physicists in the early decades of the twentieth century. They were first classified by Rutherford as α, β, and γ rays, according to their ability to penetrate matter and ionize air: α radiation penetrates the least and produces the most ionization, and γ penetrates the most with the least ionization. It was soon found that α rays are ⁴He nuclei, β rays are electrons (or *positrons*) and γ rays are very short-wavelength electromagnetic radiation. (A positron is an antielectron, which is identical to an electron except that its charge is positive.) Rutherford's α particle-scattering experiments in 1911 showed that the size of the atomic nucleus is about 10^{-4} or 10^{-5} times that of the atom, about 1 to 10 fm (1 fm $= 10^{-15}$ m). In 1919, Rutherford bombarded nitrogen with α particles and observed scintillations on a zinc sulfide screen due to protons, which have a much longer range in air than α particles. This was the first observation of artificial nuclear disintegration. Such experiments were extended to many other elements in the next few years.

In 1928, the correct theory of α radioactivity as a quantum-mechanical, barrier-penetration phenomenon was given by Gamow, Gurney, and Condon. In 1932, the neutron was discovered by Chadwick, the positron by Anderson, and the first nuclear reaction using artificially accelerated particles was observed by Cockcroft and Walton. It is quite reasonable to mark this year as the beginning of modern nuclear physics. With the discovery of the neutron, it became possible to understand some of the properties of nuclear structure; the advent of nuclear accelerators made many experimental studies possible without the severe limitations on particle type and energy imposed by naturally radioactive sources.

The study of nuclear physics is quite different from that of atomic physics. The simplest atom, the hydrogen atom, can be completely understood by solving the Schrödinger equation using the known potential energy of interaction between the electron and proton, $V(r) = -ke^2/r$ (though, as we have seen, the mathematics needed is fairly complicated). The simplest nucleus (other than a single proton) is the deuteron, consisting of a proton and a neutron. We cannot simply solve the Schrödinger equation for this problem and then compare with experiment, because the potential energy of interaction V is not known. There is no macroscopic way to measure the force between a neutron and a proton. It is clear from the stability of ⁴He that there are other forces much stronger than electromagnetic or gravitational forces between nucleons.[1] The electrostatic potential energy of two protons separated by 1 fm is

$$V = \frac{ke^2}{r} = \frac{1.44 \text{ MeV-fm}}{1 \text{ fm}} = 1.44 \text{ MeV}$$

and the gravitational potential energy is smaller than this by a factor of 10^{-35}. The energy needed to remove a proton or neutron from ⁴He is about 20 MeV. The force responsible for this

---

[1] The word "nucleon" refers to either a proton or a neutron.

large binding energy is called the *nuclear force*. The determination of the characteristics of the nuclear force is one of the central problems of nuclear physics. Information about this force can be obtained from scattering experiments involving protons and neutrons. Although the results of a scattering experiment can be predicted unambiguously from a knowledge of the force law, the force law cannot be completely determined from the results of such an experiment. The results of many scattering experiments do indicate that the nuclear force is the same between any two nucleons—that is, *n-n*, *p-p*, or *n-p*; and that the force is strong when the particles are close together, and drops rapidly to zero when the particles are separated by a few fm. The potential energy of the interaction can be roughly represented by a square well of about 40 MeV depth and a few fm width.

It is not possible to give an adequate discussion of all the many facets of nuclear physics in a short space. In the next two sections we shall describe the discovery of the neutron and discuss some of the properties of nuclear ground states. We shall then consider radioactivity and some nuclear reactions including fission and fusion. After a brief discussion of the interactions of nuclear particles with matter, we shall discuss the detection of nuclear particles, and conclude with a discussion of the nuclear shell model.

11-1 The Discovery of the Neutron

The experiments of Moseley (see Section 4-4) showed that the nuclear charge is Z times the proton charge, where Z is the atomic number, which is about half the atomic weight A (except for hydrogen, for which $Z = A$). Thus the nucleus has a mass about equal to that of A protons, but a charge of only $Z \approx \frac{1}{2}A$ protons. Before the discovery of the neutron, it was difficult to understand this unless there were $A - Z$ electrons in the nucleus to balance the charge without changing the mass very much. The idea that the nucleus contained electrons was supported by observation of β decay, in which electrons are ejected. However, there were serious difficulties with this model. A relatively simple calculation from the uncertainty principle shows that an electron has a minimum kinetic energy of about 100 MeV if it is confined in a region of $r < 10^{-14}$ m. The energies of the electrons in β decay are only of the order of 1 or 2 MeV. There is no evidence for such a strong attractive force between nuclei and electrons as would be implied by a negative potential energy of 50 to 100 MeV inside the nucleus. Furthermore, since the electrostatic potential energy of the electron and nucleus is negative, there is no barrier, as there is in α decay (Figure 11-1). If the electron's total energy were positive, as required for β decay, the electron should leak out im-

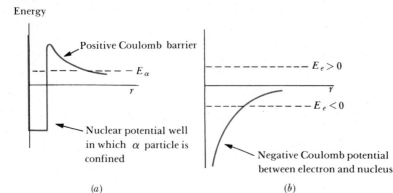

Energy

Positive Coulomb barrier

$-E_\alpha$

r

Nuclear potential well
in which α particle is
confined

(a)

$E_e > 0$

r

$E_e < 0$

Negative Coulomb potential
between electron and nucleus

(b)

Figure 11-1
(a) Potential barrier for
an α particle compared
with (b), potential for a
negative electron. Be-
cause there is no barrier
for the electron, it will
not be bound at all unless
the total energy is nega-
tive, in which case it can
never escape.

mediately. A further difficulty is the observation that the magnetic moments of nuclei are of the order of nuclear magnetons,
$e\hbar/2m_p$, about 2000 times smaller than a Bohr magneton
$e\hbar/2m_e$, which would be expected if there were electrons inside the nucleus.

The most convincing argument against electrons in the nucleus concerns angular momentum. The angular momentum of
the nitrogen nucleus has a quantum number 1, which can be inferred from a very small splitting of atomic spectral lines called
hyperfine structure (see Section 11-2). It is also known (from
molecular spectra) that the nitrogen nucleus obeys Bose-
Einstein rather than Fermi-Dirac statistics. If ^{14}N contained 14
protons and 7 electrons, each with spin $\frac{1}{2}$, the resultant angular
momentum would have to be $\frac{1}{2}$, $\frac{3}{2}$, $\frac{5}{2}$, etc., and the nucleus
would obey Fermi-Dirac statistics.

In 1920 Rutherford suggested that there might be a neutral
particle, possibly a proton and an electron tightly bound
together, which he called a *neutron*. When such a particle was
found by Chadwick in 1932, the idea that electrons were permanent constituents of nuclei was abandoned. Instead, the nucleus
was assumed to contain N neutrons and Z protons, a total of
$A = N + Z$ particles. The notion of the neutron being a proton
and electron bound together has also been abandoned, since the
spin of the neutron is $\frac{1}{2}$.

Optional

In 1930, Bothe and Becker found that when bombarded by α
particles from radioactive polonium, many nuclei became radioactive and emitted what were then thought to be γ rays (see Figure 11-2). Using a much stronger polonium α source, Curie and
Joliot repeated these experiments in 1932 and measured the
penetration of the unknown γ rays in lead. For the most pene-
trating radiation, that from beryllium, they found that it took
4.7 cm of lead to reduce the intensity by $\frac{1}{2}$. Calculating the absorption expected by the Compton scattering process, they found
the energy of the γ rays from Be to be about 15 MeV. This was
much higher than the energy of γ rays emitted by naturally
radioactive substances. They also found that sheets of lead,

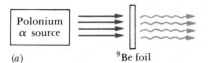

(a)

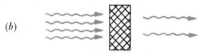

(b)

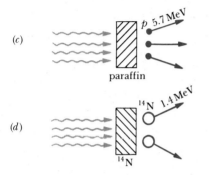

(c)

(d)

(a) An α particle hitting ^{9}Be foil produces unknown radiation.

(b) The intensity of the unknown radiation is reduced to $\frac{1}{2}$ by 4.7 cm of lead. If the radiation is γ radiation, its energy must be about 15 MeV.

(c) Many protons are produced by the unknown radiation incident on paraffin. If the radiation is γ radiation, the energy must be about 50 MeV to produce 5.7-MeV protons by Compton scattering. If the radiation consists of uncharged particles with the same mass as the proton (neutrons), the energy of the neutron need be only 5.7 MeV to produce 5.7-MeV protons by collisions.

(d) The unknown radiation incident on nitrogen gas produces recoil ^{14}N atoms with energy of 1.4 MeV which requires 90-MeV γ rays or 5.7-MeV neutrons.

Figure 11-2
Schematic description of experiments of Bothe and Becker, Curie and Joliot, and Chadwick leading to the discovery of the neutron.

carbon, aluminum, etc., placed between the radiation and their ionization-chamber detector had little effect, but thin sheets of material such as paraffin, which contained hydrogen, produced a secondary radiation that greatly increased the ionization in their detector. They correctly surmised that this secondary radiation consisted of protons. By measuring the range of these protons in air, they found the energy of the protons to be about 5.7 MeV. This result was quite surprising. The ejection of a proton from a substance by a γ ray is analogous to Compton scattering of x rays by electrons. The maximum change in wavelength in Compton scattering by protons is $2h/m_p c$, obtained from Equation 3-37 with $\theta = 180°$. It is easy to calculate that the energy of the γ ray must be at least 50 MeV in order to give the proton 5.7 MeV (see Problem 1). Curie and Joliot found similar results using α particles on boron, though the radiation from boron was weaker.

Chadwick thought it unlikely that the protons were ejected by the Compton scattering process. Not only were rays of very high energy needed, but the number of protons observed was several thousand times greater than predicted for this scattering process. He extended the experiment by exposing many materials, including the gases hydrogen, helium, oxygen, nitrogen, and argon, to the beryllium and boron radiation produced by bombardment with polonium α particles. He found that the beryllium radiation produced recoil atoms of about the same number for all the gases. By measuring the energy of recoil, he could compute the energy of the incident γ rays, just as Curie and Joliot had. He found that the result depended on which gas was used. For example, it required about 50-MeV γ rays to produce recoil hydrogen atoms, but 90-MeV γ rays were needed to produce the recoil nitrogen atoms observed. He then computed the results to be expected if the beryllium radiation consisted, instead, of neutral particles of mass approximately equal to the proton mass. The maximum recoil energy occurs when the neutral particle (which Chadwick called the neutron, after Rutherford) makes a head-on collision with the gas atom. For the case of recoil protons, the neutron will transfer *all* its kinetic energy in a head-on collision, because the two particles have the same mass. Since the maximum energy of the recoil protons was measured to be 5.7 MeV, this must be the energy of the neutrons emitted from beryllium. Assuming this neutron energy, he could calculate the maximum recoil energy of other atoms such as nitrogen. These calculated energies agreed with Chadwick's experimental results. It was not possible to determine the mass of the neutron accurately from the calculations, but it could be determined in principle from the reaction $^4\text{He} + {}^9\text{Be} \rightarrow {}^{12}\text{C} + n$ if the masses were known. The mass of ^9Be had not been measured accurately at the time, but that of boron and nitrogen had. Using the reaction $^4\text{He} + {}^{11}\text{B} \rightarrow {}^{14}\text{N} + n$, Chadwick determined the mass of the neutron from

$$(Mc^2)_{^4\text{He}} + (Mc^2)_{^{11}\text{B}} + E_k\,(^4\text{He})$$
$$= (Mc^2)_{^{14}\text{N}} + (mc^2)_n + E_k(n) \qquad \text{11-1}$$

with the result: $1.005 < m_n < 1.008$ atomic mass units.

Chadwick's paper, "The Existence of a Neutron," makes an excellent introduction to the study of nuclear physics.[1] We quote from the last two paragraphs:

> In conclusion, I may restate briefly the case for supposing that the radiation, the effects of which have been examined in this paper, consists of neutral particles rather than of radiation quanta. Firstly, there is no evidence from electron collisions of the presence of a radiation of such a quantum energy as is necessary to account for the nuclear collisions. Secondly, the quantum hypothesis can be sustained only by relinquishing the conservation of energy and momentum. On the other hand, the neutron hypothesis gives an immediate and simple explanation of the experimental facts; it is consistent in itself and it throws new light on the problem of nuclear structure.
>
> Summary: The properties of the penetrating radiation emitted from beryllium (and boron) when bombarded by the α-particles of polonium have been examined. It is concluded that the radiation consists, not of quanta as hitherto supposed,

---

[1] *Proceedings of the Royal Society*, **A136,** 692 (1932). This paper is reprinted in Reference 1, and in Reference 2 of Chapters 2 and 3.

but of neutrons, particles of mass 1 and charge 0. Evidence is given to show that the mass of the neutron is probably between 1.005 and 1.008. This suggests that the neutron consists of a proton and an electron in close combination, the binding energy being about 1 to 2×10^6 electron-volts. From experiments on the passage of the neutrons through matter, the frequency of their collisions with atomic nuclei and with electrons is discussed.

11-2 Ground-State Properties of Nuclei

In this section we shall discuss some of the properties of nuclei in the ground state. We shall mention only a few methods of determining these properties.[1] We shall use the following standard terminology: the letter N stands for the number of neutrons in a nucleus, and Z for the number of protons (the atomic number); $A = N + Z$ is the total number of nucleons, the *mass number*. The mass number is an integer approximately equal to the atomic weight. A particular nuclear species is called a *nuclide*. Nuclides are denoted by the chemical symbol with a presuperscript giving the value of A, such as ^{16}O or ^{15}O. Sometimes Z is given as a presubscript, such as $^{15}_{8}$O, though this is not necessary because each element (Z number) has a unique chemical symbol. Nuclides with the same Z, such as ^{15}O and ^{16}O, are called *isotopes*. Nuclides with the same N, such as $^{13}_{6}$C and $^{14}_{7}$N, are called *isotones*, while nuclides with the same A, such as ^{14}C and ^{14}N, are called *isobars*. A nucleus in an excited state with a particularly long decay time is called an *isomer,* and the state is called an *isomeric state.* For example, ^{137}Ba has an isomeric state of energy 0.66 MeV above the ground state, with a mean life of about 3.8 min.

Isotopes, isotones, isobars, and isomers

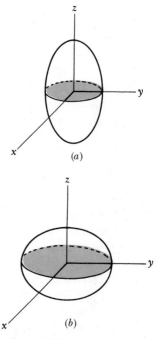

(a)

(b)

Size and Shape of Nuclei

With a few exceptions, nuclei are nearly spherical. Most of the exceptions occur in the rare-earth region (the transition period in the periodic table, $Z = 57$ to $Z = 71$) in which the shape is ellipsoidal, with the major axis differing from the minor axis by about 20 percent or less. In these heavy nuclides, the inner atomic electron wave functions penetrate the nucleus, and deviations from spherical shape show up as small changes in the atomic energy levels. If the nucleus is shaped like a watermelon (Figure 11-3), with the z axis larger than the x and y axes, the average value of z^2 is larger than the average value of x^2 and y^2. In this case the *quadrupole moment,* which is proportional to $3(z^2)_{av} - (x^2 + y^2 + z^2)_{av}$, is positive. This is the most common case for nonspherical nuclei. Nuclei with negative quadrupole moments are shaped more like flattened pumpkins, with the two equal axes longer than the third axis.

Figure 11-3
Nonspherical nuclear shapes. Nuclei with positive quadrupole moments have $(z^2)_{av}$ greater than $(x^2)_{av}$ or $(y^2)_{av}$ and are of watermelon shape, as in (a). Nuclei with negative quadrupole moments have $(z^2)_{av}$ less than $(x^2)_{av}$ or $(y^2)_{av}$ and are shaped like flattened pumpkins, as in (b).

[1] Reference 2 contains a particularly good discussion of the experimental methods used in measuring nuclear properties.

The nuclear radius can be determined by scattering experiments similar to the first ones of Rutherford, or in some cases from measurements of radioactivity. An interesting, nearly classical method of determining the nuclear radius involves the measurement of the energy of positron β decay between *mirror nuclides,* which are nuclides whose Z and N numbers are interchanged (Figure 11-4). For example, ^{15}O, with eight protons and seven neutrons, and ^{15}N, with eight neutrons and seven protons, are mirror nuclides. Assuming that the nuclear force between nucleons is independent of the kind of nucleons, the only difference in energy between ^{15}O and ^{15}N is electrostatic. The electrostatic energy of a ball of uniform charge can be shown to be given by

$$W = \frac{3}{5}\frac{kq^2}{R} \qquad\qquad 11\text{-}2$$

where q is the charge and R is the radius. The energy difference between ^{15}O and ^{15}N is then

$$W = \frac{3}{5}\frac{ke^2}{R}[Z^2 - (Z-1)^2] \qquad\qquad 11\text{-}3$$

with $Z = 8$. A measurement of the energy of the decay, $^{15}O \rightarrow {}^{15}N + \beta^+$ (positron) $+ \nu$ (neutrino), thus gives a measurement of R. Assuming a uniform charge distribution, measurements of the positron decay energies for 18 pairs of mirror nuclides give for the nuclear radius

$$R = R_0 A^{1/3} \qquad \text{with } R_0 = 1.5 \text{ fm} \qquad\qquad 11\text{-}4$$

Radius of nuclear charge distribution

where A is the atomic number. (A quantum-mechanical correction using a charge distribution calculated from the nuclear shell model changes the value of R_0 to 1.2 fm.) The consistency of these results with others is a strong indication that the nuclear energy is the same for these nuclei. The most extensive measurements of nuclear radii have been carried out by Robert Hofstadter et al. in a series of experiments begun in 1953. In these experiments at the Stanford linear accelerator, nuclei were bombarded with electrons having energies of about 200 to 500 MeV. The wavelength of a 500-MeV electron is about 2.5 fm, which is smaller than the radius of heavy nuclei. It is thus possible to learn something about the detailed structure of the charge distribution of nuclei by analyzing the scattering of these electrons.[1] The analysis is fairly complicated because the electrons are relativistic. Figure 11-5a shows some charge distributions obtained from these experiments. The mean electromagnetic radius R_e and the surface thickness t, indicated in Figure 11-5b, are given by

$$t = 2.4 \pm 0.3 \text{ fm}$$
$$\qquad\qquad 11\text{-}5$$
$$R_e = (1.07 \pm 0.02)A^{1/3} \text{ fm}$$

[1] Hofstadter was awarded the Nobel Prize for physics, jointly with R. L. Mössbauer, in 1961.

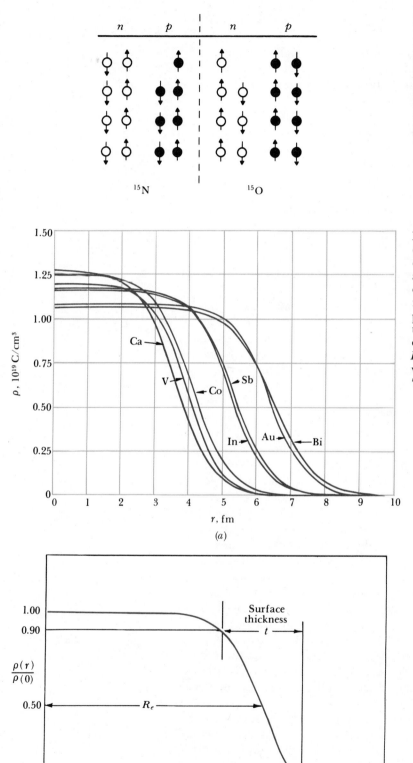

Figure 11-4
Mirror nuclides. If all the neutrons are changed to protons and all the protons are changed to neutrons, ^{15}N becomes its mirror, ^{15}O. The ground-state energy of mirror pairs differs only in the electrostatic energy.

Figure 11-5
(*a*) Charge density versus distance for several nuclei as determined by electron-scattering experiments. (*b*) Definitions of parameters R_e and t used to describe nuclear charge density. [*From R. Hofstadter,* Annual Review of Nuclear Science, **7**, *231 (1957).*]

These results are not very different from those obtained from positron decay.

A different measurement of the nuclear radius can be determined from the attenuation of a beam of fast neutrons. The total cross section for attenuation can be shown to be

$$\sigma = 2\pi(R + \lambdabar)^2 \qquad\qquad 11\text{-}6$$

where $\lambdabar$ is the de Broglie wavelength divided by 2π and R is the nuclear radius. The neutrons must be fast enough so that $\lambdabar < R$ in order to gain information about R from measurement of σ. These experiments do not measure the charge distribution but, instead, measure the "radius" of the nuclear force between a neutron and the nucleus. The results of these measurements are

$$R = R_0 A^{1/3} \qquad \text{with } R_0 = 1.4 \text{ fm} \qquad 11\text{-}7$$

Radius of nuclear force

These different types of experiments thus give compatible but not identical results. The fact that the radius is proportional to $A^{1/3}$ means that the density is the same for all nuclei. The numerical value is easily computed to be about 10^{14} g/cm³. This fantastically high density, compared with ~ 1 g/cm³ for atoms, is a consequence of the fact that nearly all the mass of the atom is concentrated in a region of radius about 10^{-5} that of the atom.

Systematics of N and Z Numbers

There are about 260 stable nuclei. Figure 11-6 shows a plot of the neutron number N versus the proton number Z for stable nuclides. The straight line is $N = Z$. The general shape of this curve can be understood in terms of the exclusion principle and the electrostatic energy of the protons. Consider the kinetic energy of A particles in a one-dimensional square well. The energy is least if $\frac{1}{2}A$ are neutrons and $\frac{1}{2}A$ are protons and greatest if all the particles are of one type (Figure 11-7). There is therefore a tendency, due to the exclusion principle, for N and Z to be equal. If we include the electrostatic energy of repulsion of the protons, the result is changed somewhat. This potential energy is proportional to Z^2. At large A, the energy is increased less by adding two neutrons than by adding one neutron and one proton; so the difference $N - Z$ increases with increasing Z.

There is also a tendency for nucleons to pair. Of the 264 nuclides that are stable, 158 have even Z and even N, 49 have odd Z and even N, 53 have even Z and odd N, and only 4 have both odd N and Z.

Since there are about 100 different elements and about 260 stable nuclides, there is an average of about $2\frac{1}{2}$ isotopes per element. There is an unusually large number of isotopes for nuclei with Z equal to 20, 28, 50, and 82. For example, tin, with $Z = 50$, has 10 stable isotopes. Similarly, nuclides with these same numbers of neutrons have a larger-than-average number of isotones. These numbers, called *magic numbers*, are a manifes-

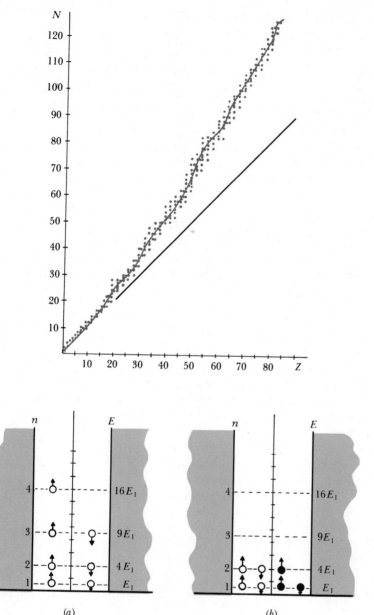

Figure 11-6
Plot of neutron number N versus proton number Z for the stable nuclides. The straight line is $N = Z$.

Figure 11-7
(a) Seven neutrons in an infinite square well. In accordance with the exclusion principle, only two neutrons can be in a given space state. The total energy is $16E_1 + (2 \times 9E_1) + (2 \times 4E_1) + (2 \times 1E_1) = 44E_1$. (b) Four neutrons and three protons in the same infinite square well. Because protons and neutrons are not identical, four particles (two neutrons and two protons) can be in the state $n = 1$. The total energy is $(3 \times 4E_1) + (4 \times 1E_1) = 16E_1$. This is much less than in (a).

tation of shell structure in very much the same way that the atomic magic numbers 2, 10, 18, and 36 correspond to closed-electron-shell structure.

Nuclides that do not fall on the N-versus-Z stability curve are radioactive. We shall discuss radioactivity in Section 11-3.

Masses and Binding Energy

The mass of an atom can be accurately measured in a mass spectrometer, which measures q/M for ions by bending them in a magnetic field. The mass of an atom is not quite equal to the mass of the nucleus plus the mass of the electrons because of the binding energy of the electrons. The binding energy of the electrons is defined by

$$B_{\text{atomic}} = M_N c^2 + Z m_e c^2 - M_A c^2 \qquad 11\text{-}8$$

where M_N is the mass of the nucleus, M_A is the mass of the atom, and m_e is the mass of an electron. Because the binding energies of atoms are only of the order of keV, compared with nuclear binding energies of many MeV, atomic binding energies are usually neglected in nuclear physics. The binding energy of a nucleus with Z protons and N neutrons is defined as

$$B_{\text{nuclear}} = Z m_p c^2 + N m_n c^2 - M_N c^2 \qquad 11\text{-}9 \qquad \textit{Binding energy}$$

where m_p is the mass of a proton and m_n the mass of a neutron. Since the mass of an atom is very nearly equal to the mass of the nucleus plus the mass of the electrons (neglecting the atomic binding energy), the nuclear binding energy can be accurately computed from

$$B_{\text{nuclear}} = Z M_H c^2 + N m_n c^2 - M_A c^2 \qquad 11\text{-}10$$

where M_A is the atomic mass and M_H is the mass of a hydrogen atom. Note that the masses of the Z electrons cancel out. This expression is more convenient to use because it is the mass of the atom that is usually measured in mass spectrometers. The masses of all stable nuclides and of some unstable nuclides are listed in Appendix A.

Once the mass of a nucleus is determined, the binding energy can be computed. It is approximately proportional to the number of nucleons in the nucleus. The binding energy per nucleon B/A is plotted against A in Figure 11-8. The mean value

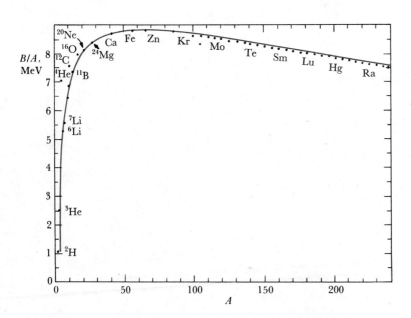

Figure 11-8
The binding energy per particle versus mass number A. The solid curve represents a semi-empirical binding-energy formula, Equation 11-11. (*From R. Leighton, Principles of Modern Physics, New York: McGraw-Hill Book Company, 1959. Used by permission of the publisher.*)

is about 8.3 MeV. The fact that this curve is approximately constant (for $A > 16$) indicates that there is *saturation* of nuclear forces. If each nucleon bonded to every other nucleon, there would be $A - 1$ bonds for each nucleon, and the binding energy per nucleon would be proportional to $A - 1$ rather than constant. Figure 11-8 indicates that, instead, there is a fixed number of bonds per nucleon, as would be the case if each nucleon were attracted only to its nearest neighbors. Such a situation also leads to a constant nuclear density, consistent with the measurements of radius. If the binding energy per nucleon were proportional to the number of nucleons, the radius would be approximately constant, as is the case for atoms.

The fact that the density and the binding energy per nucleon are approximately the same for all nuclei was first noticed in the early 1930s, after a sufficient number of atomic masses had been measured. This led to the comparison of a nucleus with a liquid drop, which also has a constant density, independent of the number of molecules. The energy required to remove molecules from a liquid is the heat of vaporization. This is proportional to the mass or number of molecules in the liquid, just as the binding energy is proportional to the number of nucleons. Using this analogy, Weizsäcker in 1935 developed a formula for the mass of a nucleus (or the binding energy, since the two are related by Equation 11-9) as a function of A and Z, called the *Weizsäcker semiempirical mass formula*. We shall write down one version of this formula and discuss the origin of the terms.

$$B = [+a_1 A - a_2 A^{2/3} - a_3 Z^2 A^{-1/3} - a_4 (A - 2Z)^2 A^{-1}$$
$$\pm a_5 A^{-1}]c^2 \qquad \text{11-11}$$

Weizsäcker semiempirical mass formula

The first term in this equation accounts for the fact that the number of bonds is proportional to A, the number of particles. This gives a constant term to the binding energy per nucleon.

The second term is a correction to the first. The nucleons on the surface of the nucleus have fewer near neighbors, thus fewer bonds, than those in the interior of the nucleus. The effect is analogous to the surface tension of a liquid drop. The surface area is proportional to R^2, which is proportional to $A^{2/3}$. This term is negative because fewer bonds imply a smaller binding energy.

The third term accounts for the positive electrostatic energy of a charged drop, which is proportional to $Z^2/R \propto Z^2 A^{-1/3}$. This positive energy of repulsion decreases the binding energy, so this term is negative.

The fourth term is not analogous to a liquid drop. It is a quantum-mechanical term that accounts for the fact that if $N \neq Z$, the energy of the nucleus increases and the binding energy decreases because of the exclusion principle. The quantity $A - 2Z = N + Z - 2Z = N - Z$ is the number of neutrons in excess of the number of protons. The expression $(A - 2Z)^2/A = (N - Z)^2/A$ is an empirical term that is zero if $N = Z$ and is independent of the sign of $N - Z$.

The last term is an empirical one to account for the pairing tendency of the nucleons. The plus sign is used if Z and N are both even, the negative sign for both Z and N odd. For the case Z or N even and the other odd, the term is taken to be zero. There have been various attempts to fit this expression, or refinements of it, to the binding energies calculated from the measured masses. The solid curve in Figure 11-8 is one such fit, with the following parameters [from A. Green, *Physical Review*, **95**, 1006 (1954)]:

$$a_1c^2 = 15.7 \text{ MeV}$$

$$a_2c^2 = 17.8 \text{ MeV}$$

$$a_3c^2 = 0.712 \text{ MeV}$$

$$a_4c^2 = 23.6 \text{ MeV}$$

$$a_5c^2 = 132 \text{ MeV or } 0$$

Nuclear Angular Momentum and Magnetic Moments

The spin quantum number of the neutron and the proton is $\frac{1}{2}$. The angular momentum of the nucleus is a combination of the spin angular momenta of the nucleons plus any orbital angular momentum due to the motion of the nucleons. This resultant angular momentum is usually called *nuclear spin* and designated by the symbol **I**. Evidence for nuclear spin was first found in atomic spectra. The nuclear spin adds to the angular momentum **J** of the electrons to form a total angular momentum **F**:

$$\mathbf{F} = \mathbf{I} + \mathbf{J}$$

The possible quantum numbers for F are $I + J$, $I + J - 1$, . . . , $|I - J|$, according to the usual rule for combining angular momenta. The number of values of F is $2J + 1$ or $2I + 1$, whichever is the smaller. Because of the energy of interaction of the electronic magnetic moment and the nuclear magnetic moment associated with I, each spectral line is split into $2J + 1$ (or $2I + 1$) components. This splitting, called *hyperfine structure*, is very small, of the order of 10^{-6} eV. It can be observed only with extremely high resolution. For the case $I < J$, the nuclear spin can be determined by counting the number of lines in the hyperfine splitting. The spin of all even-even nuclides (those with even Z and even N) is zero in the ground state. Evidently the nucleons couple together in such a way that their angular momenta add to zero in pairs, as is often the case for electrons in atoms. There is no such simple rule for other nuclides with either odd N or odd Z or both. Some of the successes of the shell model to be discussed in Section 11-8 is the correct prediction of nuclear spins for many nuclei.

The magnetic moment of the nucleus is of the order of the nuclear magneton, $\mu_N = e\hbar/2m_p$, where m_p is the mass of the proton. The exact value is difficult to predict because it depends on the detailed motion of the nucleons. If the proton and neutron obeyed the Dirac relativistic wave equation, as does the electron, the magnetic moment due to spin would be 1 nuclear magneton for the proton and 0 for the neutron because it has no charge. The experimentally determined moments of the nucleons are

$$(\mu_p)_z = +2.79\,\mu_N$$

$$(\mu_n)_z = -1.91\,\mu_N$$

The proton and neutron are more complicated particles than the electron. It is interesting that the deviations of these moments from those predicted by the Dirac equation are about the same magnitude, 1.91 for the neutron and 1.79 for the proton. The reason for the values of the magnetic moments of the nucleons is not yet completely understood.

Optional

Ground-State Properties of the Deuteron

We shall discuss briefly some properties of the deuteron to indicate some of the complexities of the nuclear force. We have already mentioned that the binding energy is 2.22 MeV. The binding energy (1.11 MeV per nucleon) is the lowest found in any nuclide and is much lower than the average of 8 MeV per nucleon. This indicates that saturation has not been reached with only two particles. The angular-momentum quantum number is found to be 1, and the magnetic moment is $0.857\,\mu_N$, which is just $2\frac{1}{2}$ percent smaller than the sum of the moments of the proton and neutron, $2.79 - 1.91 = 0.88\,\mu_N$. This near agreement is consistent with a ground-state orbital angular momentum of 0 and parallel spins. There are no bound excited states of the deuteron. Neutron-proton scattering experiments indicate that the force between n and p in the singlet state (antiparallel spins) is just sufficiently less strong than in the triplet state to make the deuteron unstable if the spins are antiparallel. The quadrupole moment of the deuteron should be zero if the particles are in the spherically symmetric $l = 0$ state. However, there is a small, measurable quadrupole moment. If it is assumed that the orbital quantum number is other than $l = 0$, neither the calculated magnetic moment nor the quadrupole moment is in agreement with experiment. If the force between the neutron and proton is not assumed to be central (along the line joining the particles), orbital angular momentum is not conserved and the ground-state wave function can be a mixture of different l states. The assumption of 96 percent $l = 0$ state and 4 percent $l = 2$ state, with spins parallel, gives the correct quadrupole moment and the correct magnetic moment. Thus the ground-state properties of the deuteron indicate that the nuclear force depends on the orientation of the spins of the particles and has a small noncentral part.

Questions

1. Why is N approximately equal to Z for stable nuclei? Why is N greater than Z for heavy nuclei?

2. What property of the nuclear force is indicated by the fact that all nuclei have about the same density?

3. How does the nuclear force differ from the electromagnetic force?

11-3 Radioactivity

The discovery of natural radioactivity by Becquerel in 1896 marked the beginning of the study of the atomic nucleus. In 1900 Rutherford discovered that the rate of emission of radioactive particles from a substance was not constant but decreased exponentially with time. This exponential time dependence is characteristic of all radioactivity and indicates that it is a statistical process. Because each nucleus is well shielded from others by the atomic electrons, pressure and temperature changes have no effect on nuclear properties.

For a statistical decay (in which the decay of any individual nucleus is a random event), the number of nuclei decaying in a time interval dt is proportional to dt and to the number of nuclei. If $N(t)$ is the number of radioactive nuclei at time t and $-dN$ is the number that decay in dt (the minus sign is necessary because N decreases), we have

$$- dN = \lambda N \ dt \qquad\qquad 11\text{-}12$$

where the constant of proportionality, λ, is called the *decay constant*. λ is the probability per unit time of the decay of any given nucleus. The solution of this equation is

$$N = N_0 e^{-\lambda t} \qquad\qquad 11\text{-}13$$

where N_0 is the number of nuclei at $t = 0$. The decay rate is

$$R = - \frac{dN}{dt} = \lambda N_0 e^{-\lambda t} = R_0 e^{-\lambda t} \qquad\qquad 11\text{-}14$$

Note that both the number of nuclei and the rate of decay decrease exponentially. It is the decrease in the rate of decay, of course, that is determined experimentally. Figure 11-9 shows N versus t. If we multiply the numbers on the N axis by λ, this becomes a graph of R versus t.

We can calculate the mean lifetime from Equation 11-13. The number of nuclei with lifetimes between t and $t + dt$ is the number that decay in dt, which is $\lambda N \ dt$; thus the fraction of lifetimes in dt is

$$f(t) \ dt = \frac{\lambda N \ dt}{N_0} = \lambda e^{-\lambda t} \ dt \qquad\qquad 11\text{-}15$$

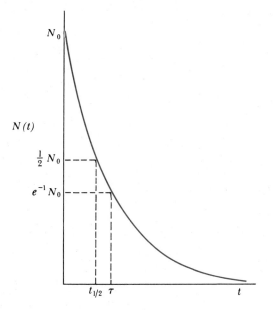

Figure 11-9
Exponential radioactive-
decay law. The number
of nuclei remaining at
time t decreases exponentially with time t. The
half-life $t_{1/2}$ and the mean
life $\tau = 1/\lambda$ are indicated.
The decay rate
$R(t) = \lambda N(t)$ has the
same time dependence.

Using this distribution function, the mean lifetime is

$$\tau = \int_0^\infty t f(t)\, dt$$

$$= \int_0^\infty t \lambda e^{-\lambda t}\, dt = \frac{1}{\lambda} \qquad\qquad 11\text{-}16$$

which is the reciprocal of the decay constant λ. The *half-life* $t_{1/2}$ is defined as the time after which the number of radioactive nuclei has decreased to half its original value. From Equation 11-13

$$\tfrac{1}{2} N_0 = N_0 e^{-\lambda t_{1/2}}$$

or

$$e^{\lambda t_{1/2}} = 2$$

$$t_{1/2} = \frac{\ln 2}{\lambda} = (\ln 2)\tau$$

or, using $\ln 2 = 0.693$,

$$t_{1/2} = \frac{0.693}{\lambda} = 0.693\,\tau \qquad\qquad 11\text{-}17$$

Half-life, mean life, and decay constant

Example 11-1   A radioactive source has a half-life of 1 min. At time $t = 0$ it is placed near a detector and the counting rate is observed to be 2000 counts/sec. (*a*) Find the mean life and the decay constant. (*b*) Find the counting rate at times $t = 1$ min, 2 min, 3 min, 10 min. (*a*) From Equation 11-17 we find for the mean life and the decay constant

$$\tau = \frac{t_{1/2}}{\ln 2} = \frac{1 \text{ min}}{0.693} = 1.44 \text{ min} = 86.6 \text{ sec}$$

and

$$\lambda = \frac{1}{\tau} = 0.693/\text{min} = 1.16 \times 10^{-2}/\text{sec}$$

(b) Since the half-life is 1 min, the counting rate will be half as great at $t = 1$ min as at $t = 0$, so that $R_1 = 1000$ counts/sec at $t = 1$ min and $R_2 = 500$ counts/sec at 2 min and $R_3 = 250$ counts/sec at 3 min. At $t = n$ min $= nt_{1/2}$ the rate will be $R = (\frac{1}{2})^n R_0$, so that at 10 min it will be $R_{10} = (\frac{1}{2})^{10}\, 2000 = 1.95$ counts/sec.

Example 11-2   If the detection efficiency in Example 11-1 is 25 percent, how many radioactive nuclei are there at time $t = 0$? At time $t = 1$ min? How many decay in the first minute? The detection efficiency depends on the distance from the source to the detector and the chance that a radioactive decay particle entering the detector will produce a count. If the source is smaller than the detector and placed very close, about half the emitted particles will enter the detector. (Some detectors are designed so that the source can be placed in a well inside the detector, so that nearly all the emitted particles enter the detector.) If the counting rate at $t = 0$ is 2000 counts/sec and the efficiency is 25 percent, the decay rate at $t = 0$ must be 8000 per sec. The number of radioactive nuclei at $t = 0$ is found from Equation 11-16, using $\lambda = 0.693$/min from Example 11-1.

$$N_0 = \frac{R_0}{\lambda} = \frac{8000/\text{sec}}{0.693/\text{min}} = 6.93 \times 10^5$$

At time $t = 1$ min $= t_{1/2}$ there are half as many nuclei as at $t = 0$, so that $N_1 = \frac{1}{2}(6.93 \times 10^5) = 3.46 \times 10^5$ nuclei. The number that decay in the first minute is therefore 3.46×10^5 nuclei. Note that this is not equal to the initial decay rate of 8000/sec times 60 sec, which would give 4.8×10^5 decays. The reason is that the decay rate is not constant during the whole time of 1 minute, but decreases by half during each minute. Note also that the average decay rate during the first minute is not the mean of 8000 and 4000/sec because the decrease is exponential, not linear, in time.

It often occurs that nucleus A decays into another nucleus B which, in turn, decays into C. For this case, the equation for nucleus B is not as simple as Equation 11-12 because there is a term describing the production of B. Let us first consider a somewhat different situation, that of a radioactive nuclide produced at a constant rate. If this rate is R_0, the equation for $N(t)$ is

$$\frac{dN}{dt} = R_0 - \lambda N \qquad\qquad 11\text{-}18$$

If we start with $N = 0$ at $t = 0$, the solution is

$$N = \frac{R_0}{\lambda}(1 - e^{-\lambda t}) \qquad\qquad 11\text{-}19$$

This function, sketched in Figure 11-10, is the same as that which occurs when charging a capacitor through a resistor with a constant source of emf. After a time of several half-lives, the

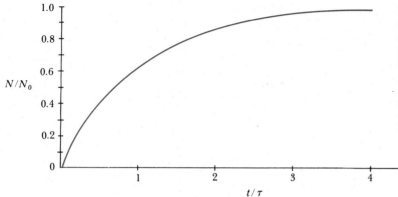

Figure 11-10
Growth of number of radioactive nuclei that are produced at a constant rate. The function N/N_0 is given by $N/N_0 = 1 - e^{-t/\tau}$, where τ is the mean life and N_0 is the production rate R_0 times the mean life.

slope of this curve becomes negligible and the number is nearly equal to its terminal value

$$N \approx \frac{R_0}{\lambda} = R_0\tau \qquad\qquad 11\text{-}20$$

Example 11-3 In a laboratory experiment, silver-foil strips are placed near a neutron source. The capture of neutrons by ^{107}Ag produces ^{108}Ag, which is radioactive and decays by β decay with a half-life of 2.4 min (thus a mean lifetime of about 3.5 min). How long should the foil be left by the neutron source to obtain a large number or radioactive ^{108}Ag nuclei? From Figure 11-10, the number of radioactive ^{108}Ag nuclei will be about 95 percent of the maximum number after three mean lifetimes; thus there is not much point in irradiating the silver foil longer than 10 min.

If the production of the radioactive nuclide is due to the decay of another nuclide the rate of production is not constant and the situation is more complex. If λ_A is the decay constant of A, which decays into B, and λ_B is the decay constant of B, the differential equation for B is

$$\frac{dN_B}{dt} = -\lambda_B N_B + \lambda_A N_A = -\lambda_B N_B + \lambda_A N_{0A}e^{-\lambda_A t} \qquad 11\text{-}21$$

The solution of this equation is not too difficult, but we shall consider only the special case in which nuclide A (called the *parent nuclide*) has a much longer half-life than nuclide B (called the *daughter nuclide*). For this case, $\lambda_A \ll \lambda_B$. For times much less than the half-life of A, the number of A nuclei is approximately constant. Then the rate $\lambda_A N_A \approx \lambda_A N_{0A} = R_0$ and Equation 11-21 is the same as the constant-rate problem just discussed. If we start with $N_B = 0$ at time $t = 0$, the equation for N_B is Equation 11-19 with $R_0 = \lambda_A N_{0A}$. For times satisfying $\tau_B \ll t \ll \tau_A$, the number of B nuclei is approximately constant and equal to the limiting value

$$N_B = \frac{\lambda_A}{\lambda_B} N_{0A} \qquad\qquad 11\text{-}22$$

An example is ^{226}Ra, which decays by α decay into ^{222}Rn, which in turn decays by α decay into ^{218}Po. The half-life of ^{226}Ra is about 1620 years and the half-life of ^{222}Rn is about 3.83 days. For times greater than about 10 days but much less than 1620 years, the number of ^{222}Rn nuclei remains constant, because its rate of formation by decay of ^{226}Ra is equal to the rate at which it decays away. This equilibrium situation is called *secular equilibrium*.

A unit of radioactivity which applies to all types is the curie (Ci), which is defined as

$$1 \text{ Ci} = 3.7 \times 10^{10} \text{ decays/sec} = 3.7 \times 10^{10} \text{ Bq} \qquad 11\text{-}23$$

Curie defined

where the becquerel (1 Bq = 1 decay/sec) is the SI unit of radioactive decay. The curie is the amount of radiation emitted by 1 g of radium. Since the curie is a very large unit, the millicurie (mCi) or microcurie (μCi) are often used.

Alpha Decay

In order for a radioactive substance to be found in nature, either it must have a half-life that is not much smaller than the age of the earth (about 4.5×10^9 years) or it must be continually produced by the decay of another radioactive substance. For a nucleus to be radioactive at all, its mass must be greater than the sum of the masses of the decay products. Many heavy nuclei are unstable to α decay. Because the Coulomb barrier inhibits the decay process (the α particle must "tunnel" through a region in which its energy is less than the potential energy, as shown in Figure 11-1a), the half-life for α decay can be very long if the decay energy is small. When a nucleus emits an α particle, both N and Z decrease by 2, and A decreases by 4. There are four possible α-decay chains, depending on whether A equals $4n$, $4n + 1$, $4n + 2$, or $4n + 3$, where n is an integer. All but one of these are found in nature. The $4n + 1$ series is not, because its longest-lived member (other than the stable end product ^{209}Bi), ^{237}Np, has a half-life of only 2×10^6 years, which is much smaller than the age of the earth.

Figure 11-11 illustrates the thorium series, which has $A = 4n$ and begins with an α decay from ^{232}Th to ^{228}Ra. The daughter nuclide of an α decay is on the left or neutron-rich side of the stability curve (dashed line), so it often decays by β^- decay, in which one neutron changes to a proton by emitting an electron. In Figure 11-11 ^{228}Ra decays by β^- decay to ^{228}Ac, which in turn decays to ^{228}Th. There are then four α decays to ^{212}Pb, which decays by β^- to ^{212}Bi. There is a branch point at ^{212}Bi, which decays either by α decay to ^{208}Tl or by β^- decay to ^{212}Po. The branches meet at the stable lead isotope ^{208}Pb.

The energy released in α decay, Q, is determined by the difference in mass of the parent nucleus and the decay products, which include the daughter nucleus and the α particle. This en-

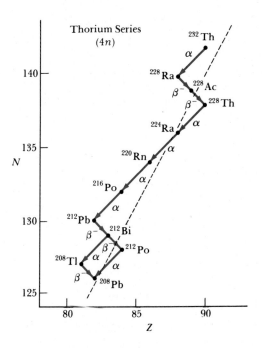

Thorium Series
(4n)

Figure 11-11
The thorium $(4n)$ α-decay
series.

ergy is usually expressed in terms of atomic masses (which include the masses of the electrons) because these are the masses measured in mass spectroscopy. If M_P is the mass of the parent atom, M_D that of the daughter atom, and M_{He} that of the helium atom, the decay energy Q is given by

$$\frac{Q}{c^2} = M_P - (M_D + M_{He}) \qquad 11\text{-}24$$

Note that the mass of the two electrons in the He atom compensates for the fact that the daughter atom has two less electrons than the parent atom.

The kinetic energy of the α particle (for decays to the ground state of the daughter nucleus) is slightly less than the decay energy Q because of the small recoil energy of the daughter nucleus. If the parent nucleus is at rest when it decays, the daughter nucleus and α particle must have equal and opposite momenta. If p is the magnitude of the momentum of either particle, the decay energy is

$$Q = \frac{p^2}{2M_D} + \frac{p^2}{2M_{He}} = \frac{p^2}{2M_{He}}\left(1 + \frac{M_{He}}{M_D}\right) \qquad 11\text{-}25$$

(Since we are not calculating mass *differences*, it doesn't matter whether we use nuclear masses or atomic masses.) Then, writing E_α for $p^2/2M_{He}$ and $M_{He}/M_D = 4/(A - 4)$, where A is the mass number of the parent nucleus, we have

$$E_\alpha = \frac{A - 4}{A} Q \qquad 11\text{-}26$$

Since A is much greater than 4 for most nuclides that decay by α decay, E_α is nearly equal to Q.

If all the α decays proceeded from the ground state of the parent nucleus to the ground state of the daughter nucleus, the emitted α particles would all have the same energy related to total energy available Q by Equation 11-26. When the energies of the emitted α particles are measured with high resolution, a spectrum of energies is observed, as shown in Figure 11-12. The peak in the spectrum labeled α_0 corresponds to α decays to the ground state of the daughter nucleus, with a total energy of $Q = 6.04$ MeV, as calculated from Equation 11-24. The peak labeled α_{30} indicates α particle with energy 30 keV less than those of maximum energy, indicating that the decay is to an excited state of the daughter nucleus at 30 keV above the ground state. (Unless the parent nucleus was recently produced in a reaction or previous decay, it will be in its ground state. The energy spectrum of α particles then indicates the energy levels in the daughter nucleus as in the case shown in Figure 11-12.) This interpretation of the α-particle energy spectrum is confirmed by the observation of a 30-keV γ ray emitted as the daughter nucleus decays to its ground state. Because the half-life for γ decay is very short, this decay follows essentially immediately after the α decay. Figure 11-13 shows an energy-level diagram for ^{223}Ra obtained from the measurement of the α-particle energies in the decay of ^{227}Th and from the observed γ-decay energies. Only the lowest-lying levels and some of the γ-ray transitions are indicated.

Beta Decay

There are three processes in which the mass number A remains unchanged, while Z and N change by ± 1. These are β^- decay, in which a neutron inside a nucleus changes into a proton with the emission of an electron, β^+ decay, in which a proton inside a nucleus changes into a neutron with the emission of a positron, and electron capture, in which a proton in a nucleus captures an

Figure 11-12
Alpha-particle spectrum from ^{227}Th. The highest-energy α particles correspond to decay to the ground state of ^{223}Ra with a transition energy of $Q = 6.04$ MeV. The next highest-energy particles, α_{30}, result from transitions to the first excited state of ^{223}Ra, 30 keV above the ground state. The energy levels of the daughter nucleus, ^{223}Ra, can be determined by measurement of the α-particle energies. [*Data taken from R. C. Pilger, Ph.D. Thesis, University of California Radiation Laboratory Report UCRL-3877 (1957).*]

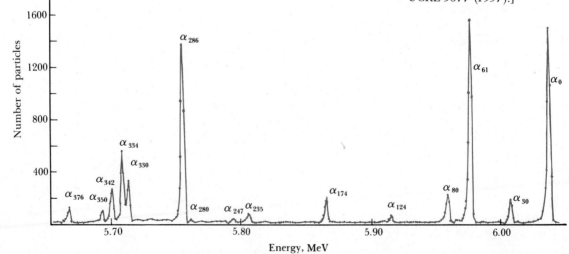

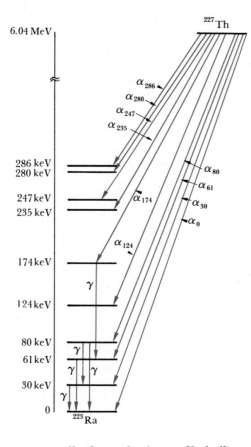

Figure 11-13
Energy levels of ^{227}Ra determined by measurement of α-particle energies from ^{227}Th, as shown in Figure 11-12. Only the lowest-lying levels and some of the γ-ray transitions are shown.

atomic electron (usually from the inner K shell) and changes into a neutron. We shall discuss each of these processes briefly.

The simplest example of β decay is that of the free neutron, which decays into a proton plus an electron with a half-life of about 10.8 min. The energy of decay is 0.78 MeV, which is the difference between the rest energy of the neutron (939.57 MeV) and that of the proton plus electron (938.28 + 0.511 MeV). More generally, in β^- decay, a nucleus of mass number A and atomic number Z changes into one with mass number A and atomic number $Z' = Z + 1$ with the emission of an electron. The energy of decay Q is c^2 times the difference between the mass of the parent nucleus and that of the decay products. If we add the mass of Z electrons to both the parent nucleus and the decay products, we can write Q in terms of the *atomic* masses:

$$\frac{Q}{c^2} = M_P - M_D$$

Another way of understanding this result is to note that in β^- decay an electron of mass m_e leaves the system, which is now a daughter *ion* of nuclear charge $Z + 1$ and Z atomic electrons. To obtain the mass of the neutral daughter atom, we must add the mass of an electron m_e so the total mass change is just the difference in mass between the parent and daughter *atoms*. If the decay energy Q is shared by the daughter atom and the emitted electron, the energy of the electron is uniquely determined by conservation of energy and momentum, just as in α decay. Experimentally, however, the energies of the electrons emitted in β decay are observed to vary from zero to the maximum energy available. A typical energy spectrum is shown in Figure 11-14; compare this with the discrete spectrum of α-particle energies of Figure 11-12.

To explain the apparent nonconservation of energy in β decay, Pauli in 1930 suggested that a third particle is emitted which he called the neutrino. The mass of the neutrino was assumed to be very much less than that of the electron, because the maximum energy of the electrons emitted is nearly equal to the total available for the decay. The neutrino was observed experimentally in 1957. It is now known that the neutrino is massless, and that there are two kinds, one (ν_e) associated with electrons and one (ν_μ) with muons. Moreover, each has an antiparticle, written $\bar{\nu}_e$ and $\bar{\nu}_\mu$. (These particles will be discussed more fully in Chapter 12.) It is the antineutrino that is emitted in the decay of the neutron, which is written

$$n \longrightarrow p + \beta^- + \bar{\nu}_e$$

In β^+ decay, a proton changes into a neutron with the emission of a positron (and a neutrino). A free proton cannot decay by positron emission because of conservation of energy (the rest mass of the neutron plus positron is greater than that of the proton), but because of binding-energy effects, a proton inside a nucleus can. A typical β^+ decay is

$$^{13}_{7}\text{N} \longrightarrow {}^{13}_{6}\text{C} + \beta^+ + \nu_e \qquad \qquad 11\text{-}27$$

As in all nuclear transformations, the decay energy Q is related to the difference in mass between the parent nucleus and the decay products. Note that if we add the mass of Z electrons to the nuclear masses ($Z = 7$ in the case of Equation 11-27), we obtain the mass of the daughter atom plus two extra electron masses (the positron and electron have identical mass). The decay energy for β^+ decay is related to the atomic mass of the parent and daughter atoms by

$$\frac{Q}{c^2} = M_P - (M_D + 2m_e) \qquad \qquad 11\text{-}28$$

Again, we can understand this by noting that in β^+ decay a positron of mass m_e leaves the system, which is now a negative daughter ion of nuclear charge $Z - 1$ and Z atomic electrons. To obtain the mass of the neutral daughter atom, we must subtract the mass of another electron, giving a net change of $2m_e$ in addition to the difference in mass of the parent and daughter atoms.

As we have mentioned, electrons (or positrons) do not exist inside the nucleus. They are created in the process of decay, just as photons are created when an atom makes a transition from a higher- to a lower-energy state.

In electron capture, a proton inside a nucleus captures an atomic electron, and changes into a neutron with the emission of

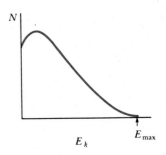

Figure 11-14
Energy spectrum of electrons emitted in β decay. The number of electrons per unit energy interval is plotted versus energy. The fact that all the electrons do not have the same energy E_{max} suggests that there is another particle emitted which shares the energy available for decay.

a neutrino. The energy available for this process is given by

$$\frac{Q}{c^2} = M_P - M_D \qquad\qquad 11\text{-}29$$

Whenever the mass of an atom of atomic number Z is greater than that of the adjacent atom with atomic number $Z - 1$, electron capture is possible. If the mass difference is greater than $2m_e$, β^+ decay is also possible, and these two processes compete. The probability of electron capture is negligible unless the atomic electron is in the immediate vicinity of the nucleus. This probability is proportional to the square of the electron wave function integrated over the volume of the nucleus. It is significant only for the inner K-shell or L-shell electrons.

An important example of β decay is that of ^{14}C, which is used in radioactive dating:

$$^{14}C \longrightarrow {}^{14}N + \beta^- + \bar{\nu}_e$$

Carbon dating

The half-life for this decay is 5730 years. The radioactive isotope ^{14}C is produced in the upper atmosphere from nuclear reactions caused by cosmic rays. The chemical behavior of carbon atoms with ^{14}C nuclei is the same as those with ordinary ^{12}C nuclei, e.g., atoms with these nuclei combine with oxygen to form CO_2 molecules. Since living organisms continually exchange CO_2 with the atmosphere, the ratio of ^{14}C to ^{12}C in a living organism is the same as the equilibrium ratio in the atmosphere, which is about 1.3×10^{-12}. When an organism dies, it no longer absorbs ^{14}C from the atmosphere, so that the ratio $^{14}C/^{12}C$ continually decreases due to the radioactive decay of ^{14}C. A measurement of the decay rate per gram of carbon thus allows the calculation of the time of death of the organism.

Example 11-4 Calculate the decay rate per gram of carbon of a living organism, assuming the ratio $^{14}C/^{12}C = 1.3 \times 10^{-12}$. From Equations 11-12 and 11-17 the decay rate is related to the number of radioactive nuclei as follows:

$$R = -\frac{dN}{dt} = \lambda N = \frac{0.693}{t_{1/2}} N$$

The number of ^{12}C nuclei in 1 g of carbon is

$$N_{^{12}C} = \frac{6.02 \times 10^{23} \text{ nuclei/mole}}{12 \text{ g/mole}} = 5.02 \times 10^{22} \text{ nuclei/g}$$

The number of radioactive ^{14}C nuclei is 1.3×10^{-12} times this number. The decay rate is therefore

$$R = \frac{0.693}{5730 \text{ years}} (1.3 \times 10^{-12}) \left(\frac{5.02 \times 10^{22}}{g}\right) \times$$

$$\left(\frac{1 \text{ year}}{3.16 \times 10^7 \text{ sec}}\right)\left(\frac{60 \text{ sec}}{\text{min}}\right)$$

$$= 15 \text{ decays/min-g}$$

The decay rate for a living organism is therefore 15 decays per minute per gram of carbon.

Decay rate per gram of carbon

Example 11-5   A bone containing 200 g of carbon has a β decay rate of 400 decays/min. How old is the bone? We first obtain a rough estimate. If the bone were from a living organism, we expect the decay rate to be (15 decays/min-g) $\times$ 200 g = 3000 decays/min. Since 400/3000 is roughly $\frac{1}{8}$, (actually 1/7.5) the sample must be about 3 half-lives old, or about 3 $\times$ 5730 years. To find the age more accurately, we note that after n half-lives, the decay rate decreases by a factor of $(\frac{1}{2})^n$. We therefore find n from

$$(\tfrac{1}{2})^n = \tfrac{400}{3000}$$

or

$$2^n = \tfrac{3000}{400} = 7.5$$

$$n \ln 2 = \ln 7.5$$

$$n = \frac{\ln 7.5}{\ln 2} = 2.91$$

The   age   is   therefore $t = nt_{1/2} = 2.91(5730 \text{ years}) =$ 16,700 years.

Gamma Decay and the Mössbauer Effect

In γ decay a nucleus in an excited state decays to a lower-energy state by the emission of a photon. This decay is the nuclear analogue of the emission of light by atoms. Since the spacing of the nuclear energy levels is of the order of MeV (as compared with eV in atoms), the wavelengths of the emitted photons are of the order of

$$\lambda = \frac{hc}{E} \sim \frac{1240 \text{ MeV-fm}}{1 \text{ MeV}} = 1240 \text{ fm}$$

Gamma-ray emission usually follows beta decay or alpha decay. For example, if a radioactive parent nucleus decays by beta decay to an excited state of the daughter nucleus, the daughter nucleus often decays down to its ground state by γ emission. The mean life for γ decay is often very short. Direct measurements of mean lives down to about 1 μsec are possible, and sophisticated coincidence techniques can be used to measure mean lives as short as 10^{-11} sec. Measurements of lifetimes smaller than 10^{-11} sec are difficult but can sometimes be accomplished by determining the natural line width Γ and using the uncertainty relation $\tau = \hbar/\Gamma$.

Optional

An interesting feature of γ decay is the Mössbauer effect. The photons emitted by nuclei making a transition from an excited state to the ground state are not monoenergetic, but have a distribution of energies because of the energy width of the excited state. The natural width of the energy distribution of photons, Γ, is related to the mean lifetime of the excited state by $\Gamma = \hbar/\tau$, consistent with the uncertainty principle. The width and thus the lifetime can, in principle, be determined by the technique of *resonance fluorescence,* the absorption and reemission of a photon emitted by a nucleus of the same type. If the excited state has energy centered at E_0 with width Γ, the cross section for the absorption of photons has a sharp maximum at the excitation energy E_0 and drops to half the maximum value at $E_0 \pm \frac{1}{2}\Gamma$. The integral of the cross section over energy is proportional to Γ, so a measurement of the absorption cross section versus energy can be used to determine Γ and therefore the lifetime τ. Resonance fluorescence is observed for atomic transitions, but generally not for nuclear transitions because of the nuclear recoil. This recoil is negligible for atomic transitions because of their low energy (~ 1 eV), whereas for nuclear transitions of energy of the order of 1 MeV, the recoil shifts the energy of the emitted photon completely off resonance. A nucleus or atom emitting a photon of energy E will recoil with momentum approximately equal to E/c. The energy of the photon will be reduced by the nuclear recoil energy $E_r = p^2/2M \approx E^2/2Mc^2$. If the energy of the excited state is E_0, the energy of the emitted photon will then be $E_0 - E_r$ with $E_r \approx E_0^2/2Mc^2$. Similarly, if the nucleus is to absorb a photon and make a transition of energy E_0, the photon must have energy $E_0 + E_r$. The center of the distribution of energies of the emitted photons is therefore displaced from that of the distribution of absorbed photons by $2E_r = E_0^2/Mc^2$ (Figure 11-15). If the width of the excited state is less than this amount, no photons will be emitted by a nucleus with energy great enough to be absorbed by another nucleus of the same kind, and resonance fluorescence cannot take place.

For a typical mean lifetime of 10^{-8} sec, the width is about $\Gamma = \hbar/\tau = 10^{-7}$ eV. For a typical atom of atomic number $A = 100$, the nuclear recoil energy is about 10^{-11} eV for $E_0 = 1$ eV and about 10^{-1} eV for $E_0 = 1$ MeV. For atomic transitions, therefore, the recoil energy is negligible compared with the natural width, Γ, but for nuclear transitions, it is much greater than Γ, and there is no overlap of the emission and absorption lines unless these lines are broadened greatly by the Doppler effect.

The thermal motion of atoms gives a Doppler broadening of the lines, but does not shift the central energy because the motion is random and the chances of increase or decrease of energy are equal. At ordinary room temperatures, the Doppler energy width is of the order of $D = 10^{-6}E_0$, which is about 10 times the natural width for atomic transitions and much larger than the natural width for nuclear transitions. Because the Doppler width D is of the order of magnitude of the recoil energy E_r for nuclear transitions, the emission and absorption lines do have

Figure 11-15
(a) Nuclear energy level has a half width Γ. (b) Shift in energy of emitted photon due to recoil of nucleus. (c) Shift in energy of absorptive transition because of the need of the absorbing nucleus to recoil. If the shifted levels (b) and (c) overlap, resonance fluorescence can occur.

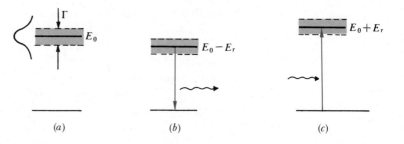

(a) (b) (c)

Rudolf L. Mössbauer, winner of the Nobel Prize (1961) for his discovery of recoilless resonance absorption of γ rays— the Mössbauer effect.

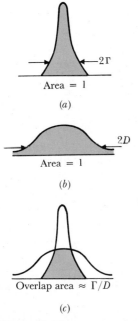

Figure 11-16
(*a*) Unbroadened line with natural width Γ. (*b*) Doppler-broadened line with the same area as in (*a*). (*c*) Absorption in resonance fluorescence is proportional to the overlap area, which is approximately Γ/D.

some overlap, and some resonance fluorescence is possible (Figure 11-16).

In 1950, P. Moon successfully observed nuclear resonance fluorescence by placing a source on the rim of an ultracentrifuge rotor that revolved such that the tip had a speed of about 800 m/sec, introducing an external Doppler shift compensating for the recoil loss. By varying the speed, he was able to measure the absorption cross section as a function of energy and determine a mean life of the order of 10^{-11} sec for his source.

In 1958, Mössbauer used an ^{191}Ir source of 129-keV photons, for which the Doppler broadening at room temperature is about twice the recoil shift, so that resonance fluorescence could be observed. When he cooled the source and absorber, he expected to see the absorption decrease because of the decrease in the Doppler width and thus in the overlap. Instead, however, he observed an increase in the absorption. The absorption, in fact, was as much as was to be expected if there had been no recoil and no Doppler broadening. The explanation of this effect is, qualitatively, that at sufficiently low temperatures an atom in a solid cannot recoil individually because of the quantization of energy states in the lattice. The recoil momentum is absorbed by the crystal as a whole. The effective mass is thus the mass of the crystal, which is so much larger than that of the atom that the recoil energy is completely negligible. The emitted photon, therefore, has energy E_0 that can be absorbed by another nucleus

Equipment for measuring the resonance absorption of γ rays in Mössbauer-effect experiments. A ^{57}Fe source and absorber are maintained at low temperatures in the vertical cylinders. The source (left cylinder) is mounted on the carriage of a lathe to control its velocity. The absorber (right cylinder) is fixed. The horizontal cylinder is a sodium iodide detector for γ rays. At the far right is a 256-channel analyzer used to read the γ spectrum from the detector. (A magnet used in other experiments to measure hyperfine splitting of the absorption line is shown at right center.) (*Courtesy of Argonne National Laboratory.*)

without recoil. Mössbauer was able to destroy the resonance by moving the source or absorber, thereby introducing an external Doppler shift. However, this shift need be only of the order of Γ, which is 4.6×10^{-6} eV for ^{191}Ir. The velocity needed to obtain a Doppler shift of this energy is only a few centimeters per second. In 1961, Mössbauer received the Nobel Prize for this experiment and for his theoretical explanation of the results.

Questions

4. Why do extreme changes in temperature and pressure of a radioactive sample have little or no effect on the radioactivity?

5. Why is the decay series $A = 4n + 1$ not found in nature?

6. A decay by α emission is often followed by β decay. When this occurs, it is by β^- and not β^+ decay. Why?

11-4 Nuclear Reactions

Most of the information about nuclei is obtained by bombarding them with various particles and observing the results. Although the first experiments were limited by the need to use radiation from naturally occurring sources, they produced many important discoveries. In 1932, Cockcroft and Walton succeeded in producing the reaction $p + {}^7\text{Li} \rightarrow {}^8\text{Be} \rightarrow {}^4\text{He} + {}^4\text{He}$, using artificially accelerated protons. At about the same time, the van de Graaff electrostatic generator was built (by R. van de Graaff in 1931), as was the first cyclotron (by E. O. Lawrence and M. S. Livingston in 1932). Since then, an enormous technology has been developed for accelerating and detecting particles, and many nuclear reactions have been studied.

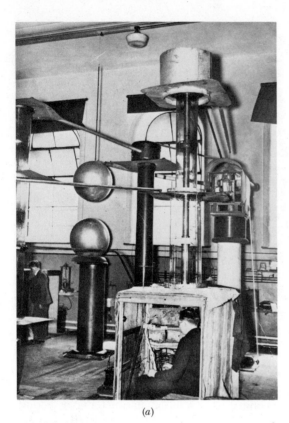

(a)

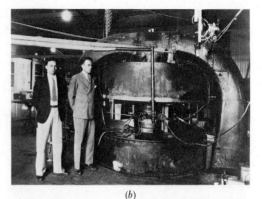

(b)

(c)

(a) The Cockcroft-Walton accelerator. Walton is sitting in the foreground. J. D. Cockcroft and E. T. S. Walton produced the first transmutation of nuclei with artificially accelerated particles in 1932, for which they received the Nobel Prize (1951). (b) M. S. Livingston and E. O. Lawrence standing in front of their 27-in cyclotron in 1934. Lawrence won the Nobel Prize (1939) for the invention of the cyclotron. (c) A modern 83-in cyclotron at the University of Michigan. (A deflecting magnet is shown in the foreground.) [(a) *Courtesy of Cavendish Laboratory. (b) Courtesy of Lawrence Radiation Laboratory, University of California, Berkeley. (c) Courtesy of University of Michigan.*]

When a particle is incident on a nucleus, several different things can happen. The particle may be scattered elastically or inelastically (in which case the nucleus is left in an excited state and decays by emitting photons or other particles) or the original particle may be absorbed and another particle or particles emitted.

Figure 11-17 illustrates the several possible stages of a nuclear reaction according to the theory of V. Weisskopf and H. Feshbach. *Shape elastic scattering* refers to the reflection of the incident wave at the edge of the nuclear potential well. If the incident particle interacts with a single nucleon in the nucleus so that the nucleon leaves the nucleus, the reaction is called a *direct reaction.* Direct reactions are more probable at high energies. If the nucleon does not leave the nucleus but interacts with several other nucleons, complicated excited states can be formed in the nucleus. If the energy is shared by many particles, the excited nucleus is called a *compound nucleus.* The compound nucleus can decay by emission of one or more particles (including photons). The decay of the compound nucleus can be treated as a statistical process independent of the detailed manner of formation.

In this section we shall study some of the systematics of nuclear reactions and some typical reactions produced by incident neutrons, protons, photons, or deuterons. We shall limit the discussion to energies of less than 140 MeV. (At higher energies,

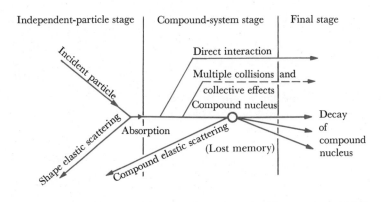

Figure 11-17
Stages in a nuclear reaction, according to Weisskopf. [*From V. Weisskopf, Reviews of Modern Physics,* **29,** *174 (1957).*]

mesons and other particles can be created. The study of high-energy reactions is generally undertaken to reveal the properties of elementary particles and of the nuclear force rather than the structure of the nucleus.)

Energetics

Consider a general reaction of particle x on nucleus X resulting in nucleus Y and particle y. The reaction is usually written

$$x + X \longrightarrow Y + y + Q$$

or, for short, $X(x,y)Y$. The quantity Q, defined by

$$Q = (m_x + m_X - m_y - m_Y)c^2 \qquad \text{11-30}$$

Q value of reaction

is the energy released in the reaction. The Q value for the reaction $n + {}^1\text{H} \to {}^2\text{H} + \gamma + Q$ is $+2.22$ MeV, whereas for the inverse reaction $\gamma + {}^2\text{H} \to n + {}^1\text{H} + Q$, it is -2.22 MeV. If Q is positive, the reaction is said to be *exothermic*. It is energetically possible even if the particles are at rest. If Q is negative, the reaction is *endothermic* and cannot occur below a threshold energy. In the reference frame in which the total momentum is zero (the center-of-mass frame), the threshold energy for the two particles is just $|Q|$. However, most reactions occur with one particle X at rest. In this frame, called the *laboratory frame,* the incident particle must have energy greater than $|Q|$ because, by conservation of momentum, the kinetic energy of y and Y cannot be zero. Consider the nonrelativistic case of x, of mass m, incident on X, of mass M (Figure 11-18). In the center-of-mass frame, both

Figure 11-18
Energetics of nuclear reaction in center-of-mass system and laboratory system. The energies are related by $E_{\text{lab}} = [(M + m)/M]E_{\text{cm}}$.

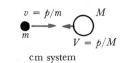

cm system

$$p = mv = MV$$

$$E_{\text{cm}} = p^2/2m + p^2/2M = (m + M)p^2/2mM$$

lab system

$$p_{\text{lab}} = m(v + V) = mv(1 + m/M) = \frac{M + m}{M}p$$

$$E_{\text{lab}} = \frac{p_L^2}{2m} = \left(\frac{p^2}{2m}\right)\left(\frac{M + m}{M}\right)^2 = \frac{M + m}{M}E_{\text{cm}}$$

particles have momentum of equal magnitude, and the total kinetic energy is

$$E_{cm} = \frac{p^2}{2m} + \frac{p^2}{2M} = \frac{1}{2} p^2 \left(\frac{m + M}{mM} \right) \qquad \text{11-31}$$

where $p = mv = MV$. We transform to the lab frame by adding V to each velocity so that M is at rest and m has velocity $v + V$. The momentum of m in the lab frame is then

$$p_{lab} = m(v + V) = mv \left(1 + \frac{m}{M} \right) = p \left(\frac{m + M}{M} \right)$$

and its energy is

$$E_{lab} = \frac{p_{lab}^2}{2m} = \frac{p^2}{2m} \left(\frac{m + M}{M} \right)^2 = \frac{m + M}{M} E_{cm} \qquad \text{11-32}$$

The threshold for an endothermic reaction in the lab frame is thus

$$E_{th} = \frac{m + M}{M} |Q| \qquad \text{11-33} \qquad \textit{Threshold energy}$$

[If the incident particle is a photon, the Lorentz transformation must be used. For low energies, the momentum of a photon is small and approximate methods can be used. For a photon, $pc = E$, whereas for a proton or neutron, $pc = (2mc^2E)^{1/2} \gg E$ for $E \ll 940$ MeV.]

Cross Section

The cross section for a nuclear reaction is defined as the number of reactions per unit time per nucleus divided by the incident intensity (number of incident particles per unit time per unit area). Consider, for example, the bombardment of ^{13}C by protons. A number of reactions might occur. Elastic scattering is written ^{13}C$(p,p)^{13}$C; the first p indicates an incident proton, the second indicates that the particle that leaves is also a proton. If the scattering is inelastic, the outgoing proton is indicated by p' and the nucleus in an excited state by ^{13}C* and one writes ^{13}C$(p,p')^{13}$C*. Some other possible reactions are

(p,n)	^{13}C$(p,n)^{13}$N
capture	^{13}C$(p,\gamma)^{14}$N
(p,α)	^{13}C$(p,\alpha)^{10}$B

The total cross section is the sum of the partial cross sections:

$$\sigma = \sigma_{p,p} + \sigma_{p,p'} + \sigma_{p,n} + \sigma_{p,\gamma} + \sigma_{p,\alpha} + \cdots$$

Cross sections have the dimensions of area. Since nuclear cross sections are of the order of the square of the nuclear radius, a convenient unit for them is the *barn*, defined by

$$1 \text{ barn} = 10^{-28} \text{ m}^2 \qquad \text{11-34} \qquad \textit{Barn defined}$$

The cross section for a particular reaction is a function of energy. For an endothermic reaction, it is zero for energies below the threshold.

The Compound Nucleus

In 1936, Niels Bohr pointed out that many low-energy reactions could be described as two-stage processes—the formation of a compound nucleus and its subsequent decay. In this description, the incident particle is absorbed by the target nucleus and the energy is shared by all the nucleons of the compound nucleus. After a time that is long compared with the time necessary for the incident particle to cross the nucleus, enough of the excitation energy of the compound nucleus becomes concentrated in one particle and that particle escapes. The emission of a particle is a statistical process which depends only on the state of the compound nucleus and not on how it was produced. An incident 1-MeV proton has a speed of about 10^9 cm/sec, so that it takes time $R/v \approx 10^{-13}/10^9 = 10^{-22}$ sec to cross a nuclear distance. The lifetime of a compound nucleus can be inferred to be about 10^{-16} sec. This is too short to be measured directly, but it is so long compared with 10^{-22} sec that it is reasonable to assume that the decay is independent of the formation.

The compound nucleus for the reactions listed above is $^{14}N^*$. This nucleus can be formed by many other reactions, such as

$$11\text{-}35$$

Since the decay of $^{14}N^*$ is independent of the formation, we can write the cross section for a particular reaction such as $^{13}C(p,n)^{13}N$ as the product of the cross section for the formation of the compound nucleus, σ_c, and the relative probability of decay by neutron emission, P_n

$$\sigma_{p,n} = \sigma_c P_n \qquad\qquad 11\text{-}36$$

An illustration of the statistical decay of the compound nucleus is afforded by the energy distribution of neutrons from the reaction (Figure 11-19)

$$\gamma + {}^{209}\text{Bi} \longrightarrow {}^{208}\text{Bi} + n$$

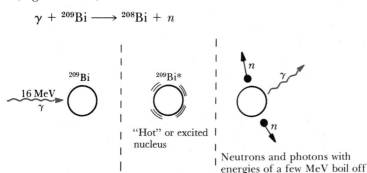

Figure 11-19
Nuclear reaction via formation of compound nucleus. The 16-MeV photon is absorbed by the ^{209}Bi nucleus, producing an excited nucleus which lives so long that excitation energy is shared by many nucleons. The excited nucleus then decays by emitting neutrons and photons of energy of the order of a few MeV.

The cross section for this reaction and other photonuclear reactions, such as (γ,γ) or (γ,p), shows a broad resonance with a width of several MeV and a maximum at $E_m \approx 14$ MeV (for Bi and other heavy elements) to 20 MeV (for light elements such as C). Consider an incident 16-MeV photon. The maximum energy of an emitted neutron is $16 - 7.4 = 8.6$ MeV since the binding energy of the last neutron is about 7.4 MeV. Some neutrons are observed to have this energy, corresponding to a direct reaction leaving ^{208}Bi in the ground state, and some have about 7.1 MeV, leaving ^{208}Bi in an excited state. About 90 percent of the neutron yield from this reaction has a Maxwell-Boltzmann distribution in energy with a peak of about 2 MeV; this is to be expected if the compound nucleus ^{209}Bi* has a long lifetime (compared with 10^{-22} sec), and the neutrons "evaporate," just as molecules evaporate from a liquid that is heated. After emitting a neutron of about 2 MeV, the ^{209}Bi* nucleus is still in an excited state and emits more neutrons or photons as it decays to the ground state.

Determination of Excited States from Nuclear Reactions

The excited states of a nucleus can be determined in two ways from nuclear reactions. A peak in the cross section $\sigma(E)$ as a function of energy indicates an excited state of the compound nucleus. Information about the lifetime of the excited states of the compound nucleus is obtained by measuring the width of these *resonances*. Figure 11-20 shows the cross section for formation of ^{14}N by the reaction ^{10}B $+ \alpha \rightarrow ^{14}$N* as a function of the α-particle energy. The peaks in this curve indicate energy levels in the ^{14}N nucleus. The Q value for this reaction is $M(^{10}$B$)c^2 + M(\alpha)c^2 - M(^{14}N)c^2 = 11.61$ MeV.

The kinetic energy in the center-of-mass frame is related to the lab energy of the α particle by

$$E_{cm} = \frac{M}{M + m} E_{lab} = \frac{10}{14} E_{lab}$$

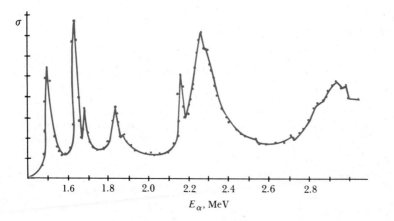

E_α, MeV

Figure 11-20
Cross section for the reaction ^{10}B $+ \alpha \rightarrow ^{14}$N* versus energy. The resonances indicate energy levels in the compound nucleus ^{14}N*. [*From F. L. Talbott and N. P. Heydenburg,* Physical Review, **90**, *186 (1953).*]

The peak in Figure 11-20 at $E_{\text{lab}} = 1.63$ MeV corresponds to an excited state in ^{14}N of energy $E = 11.61 + (\frac{10}{14})(1.63) = 12.77$ MeV. The same level can be excited by the reaction ^{12}C $+ ^{2}$H $\rightarrow ^{14}$N*. For this case, the Q value is 10.26 MeV. Thus the deuteron energy in the lab must be

$$E_{^2\text{H}} = (\tfrac{14}{12})(12.77 - 10.26) = 2.93 \text{ MeV}$$

Another way to determine the energy levels in a nucleus is to observe the energies of particles scattered inelastically. In this case, the energy levels of the product nucleus are determined. Figure 11-21 shows the energy spectrum of protons from the reaction $p + ^{14}$N $\rightarrow ^{14}$N* $+ p'$ using 6.92-MeV protons. (The horizontal scale in this figure is proportional to the momentum of the protons, since this is what is measured experimentally.) The two peaks in the curve correspond to energy losses of 2.31 and 3.75 MeV, which indicate energy levels in ^{14}N of 2.31 and 3.75 MeV. The excited product nucleus decays from these states by γ emission. A measurement of the γ-ray spectrum is an alternative method of determining these energy levels in ^{14}N. (Note that the compound nucleus for this reaction is ^{15}O, not ^{14}N.) The method of inelastic scattering can determine energy levels lying just above the ground state, whereas the levels excited in the compound nucleus must be much higher because of the Q values for formation of the compound nucleus. The Q value is the binding energy of the incident particle in the compound nucleus, which is always of the order of 6 to 10 MeV; thus levels of energy less than about 6 MeV cannot be reached in the compound nucleus.

Neutron Capture

The most likely reaction with a nucleus for a neutron of energy of more than about 1 MeV is scattering. However, even if the scattering is elastic, the neutron loses some energy to the nucleus. If a neutron is scattered many times in a material, its energy decreases until it is of the order of kT. The neutron is then equally likely to gain or lose energy from a nucleus when it is

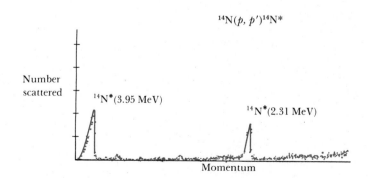

^{14}N$(p, p')^{14}$N*

Number scattered

^{14}N*(3.95 MeV)

^{14}N*(2.31 MeV)

Momentum

Figure 11-21
Spectrum of protons scattered from ^{14}N, indicating energy levels in ^{14}N. [*From C. R. Bockelman et al.,* Physical Review, **92**, *665 (1953)*.]

elastically scattered. A neutron with energy of the order of kT is called a *thermal neutron*. At low energies, a neutron is likely to be captured, with the emission of a γ ray from the excited nucleus:

$$n + {}^{A}_{Z}M \longrightarrow {}^{A+1}_{Z}M + \gamma$$

Neutron capture

Since the binding energy of a neutron is of the order of 6 to 10 MeV and the kinetic energy of the neutron is negligible compared with this, the excitation energy of the compound nucleus is from 6 to 10 MeV, and γ rays of this energy are emitted. If there are no resonances, the cross section $\sigma(n, \gamma)$ varies smoothly with energy, decreasing with increasing energy roughly as $E^{-1/2} \propto 1/v$, where v is the speed of the neutron. This energy dependence is easily understood, because the time spent by a neutron near a nucleus is inversely proportional to the speed of the neutron—so the capture cross section is proportional to this time. Superimposed on the $1/v$ dependence are large fluctuations in the capture cross section due to resonances. Figure

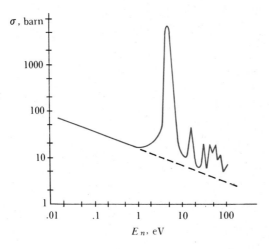

Figure 11-22
Neutron-capture cross section for Ag versus energy. The dashed-line extension is expected if there were no resonances and the cross section were merely proportional to the time spent near the nucleus, i.e., proportional to $1/v$. The resonance widths of a few eV indicate states with lifetimes of the order of $h/\Gamma \approx 10^{-16}$ sec. (*From R. Evans*, The Atomic Nucleus, *New York: McGraw-Hill Book Company, 1955. Used by permission of the publisher.*)

11-22 shows the neutron-capture cross section for silver as a function of energy. The dashed line indicates the $1/v$ dependence. At the maximum of the resonance, the value of the cross section is very large—greater than 5000 barns—compared with the value of about 10 barns just past the resonance. The fact that this resonance has a width of only a few eV shows that the compound nucleus has a state with a lifetime of the order of $\hbar/(1\ \text{eV}) \approx 10^{-16}$ sec. Many elements show similar resonances in the neutron-capture cross section. The maximum cross section for ^{113}Cd is about 57,000 barns, so this material is useful for shielding against low-energy neutrons.

Questions

7. What is meant by the cross section for a nuclear reaction?

8. Why is the neutron-capture cross section (excluding resonances) proportional to $E^{-1/2}$?

9. What is meant by the Q value of a reaction? Why is the reaction threshold not equal to Q?

10. Why can't low-lying energy levels (1 to 2 MeV above the ground state) be studied using neutron capture?

11-5 Fission, Fusion, and Nuclear Reactors

Two nuclear reactions, fission and fusion, are of particular importance. In the fission of ^{235}U, for example, the uranium nucleus is excited by the capture of a neutron and splits into two nuclei, each of about half the total mass. The Coulomb force of repulsion drives the fission fragments apart, with the energy eventually showing up as thermal energy. In fusion, two light nuclei such as those of deuterium and tritium (^{2}H and ^{3}H) fuse together to form a heavy nucleus (in this case ^{4}He plus a neutron). A typical fusion reaction is

$$^2\text{H} + {}^3\text{H} \longrightarrow {}^4\text{He} + n + 17.6 \text{ Mev} \qquad 11\text{-}37$$

Figure 11-23 shows a plot of the mass difference per nucleon $(M - Zm_p - Nm_n)/A$ versus A in units of MeV/c^2. This is just the negative of the binding-energy curve of Figure 11-7. From this figure we see that the rest mass per particle of both very

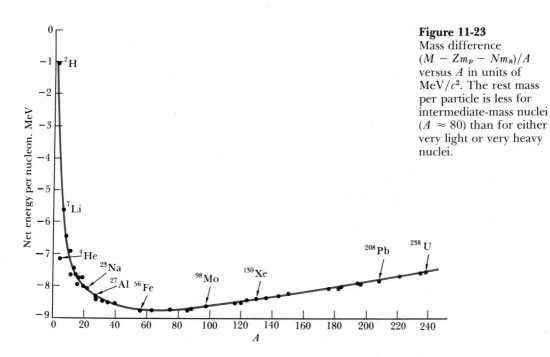

Figure 11-23
Mass difference
$(M - Zm_p - Nm_n)/A$
versus A in units of
MeV/c^2. The rest mass
per particle is less for
intermediate-mass nuclei
($A \approx 80$) than for either
very light or very heavy
nuclei.

heavy nuclides ($A \approx 200$) and for very light nuclides ($A \approx 2$) is more than that for nuclides of intermediate mass. Thus in both fission and fusion the total rest mass decreases and energy is released. Since for $A = 200$ the rest energy is about 1 MeV per nucleon greater than for $A = 100$, about 200 MeV is released in the fission of a heavy nucleus. This is much more energy than is released in atomic or chemical reactions such as combustion (about 4 eV). The energy release in fusion depends on the particular reaction. For the ^{2}H + ^{3}H reaction, above, 17.6 MeV is released. Although this is less than the energy released in fission, it is a greater amount of energy per unit mass.

Example 11-6   Calculate the total energy released in kilowatt-hours in the fission of 1 g of ^{235}U, assuming 200 MeV is released per fission. The number of ^{235}U nuclei in 1 g is

$$N = \frac{6.02 \times 10^{23}}{235} = 2.56 \times 10^{21}$$

The energy released per gram is then

$$\frac{200 \text{ MeV}}{\text{Nucleus}} \times \frac{2.56 \times 10^{21} \text{ nuclei}}{\text{g}} \times \frac{1.6 \times 10^{-19} \text{ J}}{\text{eV}}$$

$$\times \frac{1 \text{ h}}{3600 \text{ sec}} \frac{1 \text{ kW}}{1000 \text{ J/sec}} = 2.28 \times 10^4 \text{ kW-h}$$

Example 11-7   Compare the energy release per gram in the fusion of deuterium and tritium with that of fission. We can compare the energy per gram by first comparing the energy per nucleon. In fission this is about 1 MeV/nucleon. In the fusion of ^{2}H + ^{3}H it is 17.6 MeV/5 = 3.52, or about $3\frac{1}{2}$ times as great. The energy released per gram in this fusion reaction is therefore about $3\frac{1}{2}$ times that in fission.

The application of both fission and fusion to the development of nuclear weapons has had a profound effect on our lives in the past 40 years. The peaceful application of these reactions to the development of our energy resources may have an even greater effect on the future. In this section we shall look at some of the features of fission and fusion that are important for their application in reactors to generate power.

The fission of uranium was discovered in 1939 by Hahn and Strassmann, who found, by careful chemical analysis, that medium-mass elements such as barium and lanthanum were produced in the bombardment of uranium with neutrons. The discovery that several neutrons were emitted in the fission process led to speculation concerning the possibility of using these neutrons to cause further fissions, thereby producing a chain reaction. Just three years later, in 1942, a group led by Fermi at the University of Chicago constructed the first nuclear

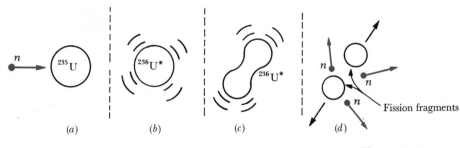

(a)   (b)   (c)   (d)

Figure 11-24
Schematic description of nuclear fission. (*a*) Absorption of a neutron by ^{235}U leads to ^{236}U* in excited state (*b*). In (*c*), the oscillation of ^{236}U* has become unstable. (*d*) The nucleus splits apart, emitting several neutrons which can produce fission in other nuclei.

reactor utilizing such a chain reaction. When ^{235}U captures a neutron, the nucleus emits γ rays as it de-excites to the ground state about 15 percent of the time, and fissions about 85 percent of the time. The fission process is analogous to the oscillation of a liquid drop. If the oscillations are violent enough, the drop splits in two, as shown in Figure 11-24. Using a liquid-drop model, Bohr and Wheeler calculated the critical energy E_c needed by the compound nucleus ^{236}U to fission. For this nucleus the energy is 5.3 MeV, which is less than the binding energy of the last neutron in ^{236}U, 6.4 MeV. The addition of a neutron to ^{235}U therefore produces an excited state of the ^{236}U nucleus with more than enough energy to break apart. On the other hand, the critical energy for the ^{239}U nucleus is 5.9 MeV while the binding energy of the last neutron is only 5.2 MeV, so that when a neutron is captured by ^{238}U to form a compound nucleus ^{239}U, the excitation energy is not great enough for fission, and the nucleus de-excites by γ emission.

A fissioning nucleus can break into two medium-weight fragments in many different ways. A typical breakup is

$$n + {}^{235}\text{U} \rightarrow ({}^{236}\text{U*}) \rightarrow {}^{141}_{56}\text{Ba} + {}^{92}_{36}\text{Kr} + 3n + Q \qquad 11\text{-}38$$

For this reaction Q is 175 MeV. Other fission reactions induced by neutron capture of ^{235}U lead to other end products.

Figure 11-25 shows the distribution of fission products versus mass number. The average number of neutrons emitted in the fission of ^{235}U is about 2.5. In order to sustain a chain reaction, one of these neutrons (on the average) must be captured by another ^{235}U nucleus and cause it to fission. The *reproduction constant* of a reactor, k, is defined as the average number of neutrons from each fission that cause a further fission. The maximum possible value of k is 2.5, but it is less than this for two important reasons: (1) neutrons may escape from the region containing fissionable nuclei, and (2) neutrons may be captured by other nuclei in the reactor not leading to fission. If k is exactly 1, the reaction will be self-sustaining. If k is less than 1 it will die out. If k is significantly greater than 1, the reaction rate will increase rapidly and "run away." In the design of nuclear bombs, such a runaway reaction is desired; in power reactors, the value of k must be kept very nearly equal to 1.

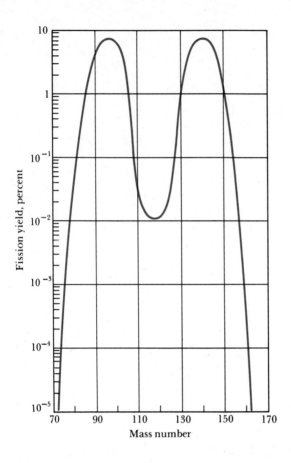

Figure 11-25
Distribution of fission fragments of ^{235}U. Symmetric fission, in which the uranium nucleus splits into two nuclei of equal mass, is less likely than asymmetric fission, in which the fragments have unequal masses.

Since the neutrons emitted in fission have energies of the order of 1 MeV whereas the cross section for capture leading to fission in ^{235}U is largest at small energies, the chain reaction is most easily sustained if the neutrons are slowed down before they escape from the reactor. At high energies (1 to 2 MeV) neutrons lose energy rapidly by inelastic scattering from ^{238}U, leaving the nucleus in an excited state that decays by γ emission. Once the neutron energy is below the excitation energies of the reactor elements (about 1 MeV), the main process of energy loss is by elastic scattering, in which a fast neutron collides with a nucleus essentially at rest and transfers some of its kinetic energy. Such energy transfers by elastic scattering are efficient only if the masses of the two bodies are comparable, e.g., a neutron will not transfer much energy in an elastic collision with a heavy ^{238}U nucleus. (Natural uranium contains 99.3 percent ^{238}U and only 0.7 percent fissionable ^{235}U.) A *moderator* consisting of material, such as water or carbon, containing light nuclei is therefore placed around the fissionable material in the core of a reactor to slow down the neutrons. Neutrons are slowed down by elastic collisions with the nuclei of the moderator until the neutrons are in thermal equilibrium with the moderator, at which time they have approximately a Maxwell-Boltzmann energy distribution with average energy of $\frac{3}{2}kT$. Because of the relatively large neutron-capture cross section of the hydrogen in water, reactors using ordinary water as a moderator must use enriched uranium, in which the ^{235}U content is increased from 0.7 percent to 1 to 4 percent. Natural uranium can be used if heavy water (D_2O) replaces ordinary (light) water (H_2O) as the moderator. Although heavy water is expensive, most Canadian reactors use it for a moderator to avoid the cost of constructing uranium-enrichment facilities.

The ability to control the reproduction factor k is important if a power reactor is to operate with any degree of safety. There are both natural negative-feedback mechanisms and mechanical methods of control. If k is greater than 1 and the reaction rate increases, the temperature of the reactor increases. If water is used as a moderator its density decreases with increasing temperature, and it becomes a less effective moderator. A second important method is the use of control rods made of a material, such as cadmium, which has a very large neutron-capture cross section. When a reactor is started up, the control rods are inserted, so that k is less than 1. As they are gradually withdrawn from the reactor, the neutron capture decreases and k increases to 1. If k becomes greater than 1, the rods are again inserted.

Control of the reaction rate of a nuclear reactor with mechanical control rods is possible only because some of the neutrons emitted in the fission process are *delayed*. The time needed to slow down a neutron from 1 or 2 MeV to thermal energy is only of the order of a millisecond. If all the neutrons emitted in fission were prompt neutrons, i.e., emitted immediately in the fission process, mechanical control would not be possible because the reactor would run away before the rods could be inserted. However, about 0.65 percent of the neutrons emitted are delayed by an average time of about 14 sec. These neutrons are emitted not in the fission process itself, but in the decay of the fission fragments. A typical decay is

$$^{87}\text{Br} \longrightarrow {}^{87}\text{Kr*} + \beta^-$$
$$\phantom{^{87}\text{Br} \longrightarrow} \rule[2pt]{0.5pt}{12pt}$$
$$\phantom{^{87}\text{Br} \longrightarrow} \longrightarrow {}^{86}\text{Kr} + n$$

In this decay, which has a 56-sec half-life, the $^{87}\text{Kr*}$ nucleus has enough excitation energy to emit a neutron. This neutron is thus delayed by 56 sec on the average. The effect of the delayed neutrons can be seen in the following examples.

Example 11-8 If the average time between fission generations (i.e., the time for a neutron emitted in one fission to cause another) is 1 msec and the reproduction factor is 1.001, how long would it take for the reaction rate to double? If $k = 1.001$, the rate after N generations is $(1.001)^N$. Setting this rate equal to 2 and solving for N we obtain $N = 700$ generations for the rate to double. The time is then 700×1 msec = 0.70 sec. This is not enough time for the mechanical control of such an excursion.

Example 11-9   Assuming 0.65 percent of the neutrons emitted are delayed by 14 sec, find the average generation time and the doubling time if $k = 1.001$. Since 99.35 percent of the generations are 10^{-3} sec and 0.65 percent are 14 sec, the average time is

$$t_{av} = 0.9935(10^{-3}) + 0.0065(14) = 0.092 \text{ sec}$$

The time for 700 generations is then

$$700 \times 0.092 \text{ sec} = 64.4 \text{ sec}$$

Because of the limited supply of the small fraction of ^{235}U in natural uranium and the limited capacity of enrichment facilities, reactors using the fission of ^{235}U cannot meet our energy needs for very long. Two possibilities hold much promise for the future; breeder reactors and controlled nuclear fusion reactors.

The breeder reactor makes use of the fact that when the relatively plentiful but nonfissionable ^{238}U nucleus captures a neutron, it decays by β decay (half-life 20 min) to ^{239}Np, which in turn decays by β decay (half-life 2 days) to the fissionable nuclide ^{239}Pu. Since plutonium fissions with fast neutrons, no moderator is needed. A reactor with a mixture of ^{238}U and ^{239}Pu will breed as much or more fuel than it uses if one or more of the neutrons emitted in the fission of ^{239}Pu is captured by ^{238}U. Practical studies indicate that a typical breeder reactor can be expected to double its fuel supply in from 7 to 10 years.

There are several safety problems inherent with breeder reactors. Since fast neutrons are used, the time between generations is essentially determined by the fraction of delayed neutrons, which is only 0.3 percent for the fission of ^{239}Pu. Mechanical control is therefore much more difficult. Also, since the operating temperature of a breeder reactor is relatively high, and a moderator is not desired, a heat exchanger such as liquid sodium metal is used rather than water (which is the moderator as well as the heat transfer material in an ordinary reactor). When the temperature of the reactor increases, the decreased density of the heat exchanger now leads to positive feedback since it now absorbs fewer neutrons than before. There is also the general safety problem with the large-scale production of plutonium, which is extremely poisonous and is the material used in nuclear bombs. In addition, there is the problem of storage of the radioactive waste products produced in any reactor. For example, a single 1000-MW ordinary reactor produces in 1 year about 3 million Ci of the radioactive ^{90}Sr (half-life 28.8 years), enough to contaminate Lake Michigan above the legal limit if it were mixed uniformly in the lake. Despite elaborate storage methods of this and other long-lived radioactive waste products, the long-term safety of large-scale production of these wastes is always open to question.

The production of power from the fusion of light nuclei holds great promise because of the relative abundance of the fuel and the lack of some of the dangers presented in fission reactors. Unfortunately, at the present time the technology has not been developed sufficiently to use this plentiful energy source. We shall consider the $^2H + {}^3H$ reaction (Equation 11-37); other reactions present similar problems.

Because of the Coulomb repulsion between the ^2H and ^3H nuclei, kinetic energies of the order of 10 keV or greater are needed to get the nuclei close enough for the attractive nuclear forces to become effective to cause fusion. Such energies can be obtained in an accelerator, but since the scattering of one nucleus by the other is much more probable than fusion, the bombardment of one nucleus by another from an accelerator requires more energy input than is recovered. To obtain energy from fusion, the particles must be heated to a temperature great enough so that the fusion reaction will occur in random thermal collisions. The temperature corresponding to $kT = 10$ keV is of the order of $T \sim 10^8$ K. (This is roughly the temperature in the interiors of stars, where such reactions do indeed take place.) At such temperatures a gas consists of positive ions and negative electrons; such a gas is called a *plasma*. One of the problems arising in the attempt to produce controlled fusion reactions is that of confining such a plasma long enough for the reactions to take place. The energy required to heat a plasma is proportional to the density of the ions, n, whereas the collision rate is proportional to the square of the density, n^2. If τ is the confinement time, the output energy is proportional to $n^2\tau$. If the output energy is to exceed the input energy we have

$$C_1 n^2 \tau > C_2 n$$

where C_1 and C_2 are constants. In 1957 the British physicist J. D. Lawson evaluated these constants from estimates of efficiencies of various hypothetical fusion reactors and derived the following relation between density and confinement time, known as *Lawson's criterion:*

$$n\tau > 10^{14} \text{ sec-particles/cm}^3 \qquad \text{11-39}$$

Lawson's criterion

There are two schemes currently under investigation for achieving the Lawson criterion. (At present, the product of density and confinement time attainable by either scheme is several orders of magnitude too small.) In one scheme a magnetic field is used to confine a hot plasma.[1] Densities of 3×10^{13} particles/cm^3 at temperatures corresponding to $kT \approx 1$ keV have been achieved for times up to about 0.03 sec, each of which is about a factor of 10 too low. In a second scheme, called *inertial confinement,* a pellet of solid deuterium and tritium is bombarded from all sides by intense pulsed laser beams of energy of the order of 10^4 J and duration of about 10^{-8} sec. Computer simulation studies indicate the pellet should be compressed to about

[1] An elementary discussion of a magnetic bottle can be found in P. A. Tipler, *Physics,* New York: Worth Publishers, Inc., 1976, p. 843. The principal schemes for magnetic confinement of plasmas are shown in Reference 16.

10^4 times its normal density and heated to a temperature greater than 10^8 K, producing about 10^6 J of fusion energy in a time of 10^{-11} sec, which is so brief that confinement is achieved by inertia alone.

Questions

11. Why is a moderator needed in an ordinary nuclear fission reactor?

12. What happens to neutrons produced in fission that do not produce another fission?

13. What is the advantage of using ^{238}U as a fuel to breed ^{239}Pu rather than using ^{235}U? What are the disadvantages?

14. Why does fusion occur spontaneously in the sun but not on earth?

11-6 The Interaction of Particles with Matter

In this section we shall discuss briefly the main interactions of charged particles, neutrons, and photons with matter. An understanding of these interactions is important for the study of nuclear detectors, shielding, and the effects of radiation. We shall not attempt to give a detailed theory, but instead indicate the principal factors involved in stopping or attenuating a beam of particles.

Charged Particles

When a charged particle traverses matter, it loses energy mainly by excitation and ionization of electrons in the matter. If the particle energy is large compared with the ionization energies of the atoms, the energy loss in each encounter with an electron will be only a small fraction of the particle energy. (A heavy particle cannot lose a large fraction of its energy to a free electron because of conservation of momentum, as discussed in Chapter 4 for α particles.) Since the number of electrons in matter is so large, we can treat the problem as that of a continuous loss of energy. After a fairly well-defined distance, called the *range,* the particle has lost all its kinetic energy and stops. Near the end of the range, the continuous picture of energy loss is not valid because individual encounters are important. The statistical variation of the path length for monoenergetic particles is called *straggling.* For electrons, this can be quite important; however, for heavy particles of several MeV or more, the path lengths vary by only a few percent or less.

We can get an idea of the important factors in the stopping of a heavy charged particle by considering a simple model. Let ze be the charge and M the mass of a particle moving with speed v past an electron of mass m_e and charge e. Let b be the impact

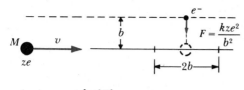

$$\text{Impulse} \approx Ft = \frac{kze^2}{b^2} \frac{2b}{v}$$

Figure 11-26
Model for calculating the
energy lost by a charged
particle in a collision with
an electron. The impulse
given to the electron is of
the order Ft, where
$F = kze^2/b^2$ is the maximum force and $t = 2b/v$
is the time for the particle
to pass the electron.

parameter. We can estimate the momentum imparted to the electron by assuming that the force has the constant value $F = kze^2/b^2$ for the time it takes the particle to pass the electron, which is of the order of $t \approx 2b/v$ (Figure 11-26). The momentum given to the electron is equal to the impulse, which is of the order of magnitude

$$p \approx Ft = \frac{kze^2}{b^2} \frac{2b}{v} = \frac{2kze^2}{bv}$$

(The same result is obtained by integration of the variable impulse, assuming the particle moves in a straight line and the electron remains at rest.) The energy given to the electron is then

$$E_e = \frac{p^2}{2m_e} = \frac{2k^2z^2e^4}{m_ev^2b^2} \qquad \text{11-40}$$

This is thus the energy lost by the particle in one encounter.

In a cylindrical shell of thickness db and length dx (see Figure 11-27), there are $Z(N_A/A)\,\rho 2\pi b\, db\, dx$ electrons, where Z is the atomic number, A the atomic weight, N_A Avogadro's number, and ρ the mass density. The energy lost to these electrons is then

$$-dE = \frac{2k^2z^2e^4}{m_ev^2b^2} Z \frac{N_A}{A} 2\pi b\rho\, db\, dx$$

If we integrate from some minimum b to some maximum b, we obtain

$$-\frac{dE}{dx} = \frac{4\pi k^2z^2e^4(Z/A)N_A\rho}{m_ev^2} L \qquad \text{11-41}$$

where

$$L = \ln\left(\frac{b_{max}}{b_{min}}\right) \qquad \text{11-42}$$

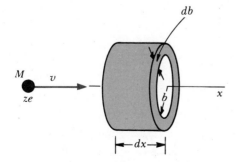

Volume of shell is $2\pi b\, db\, dx$
Number in shell is $n2\pi b\, db\, dx$

Figure 11-27
In path length dx, the
charged particle collides
with $n2\pi b\, db\, dx$ electrons
with impact parameters in
db, where $n = Z(N_A/A)\rho$
is the number of elec-
trons per unit volume in
the material.

The maximum and minimum values of b can be estimated from general considerations. For example, this model is certainly not valid if the collision time is longer than the period of the electron in its orbit. The requirement that $2b/v$ be less than this time sets an upper limit on b. The lower limit on b can be obtained from the requirement that the maximum velocity the electron can receive from a collision is $2v$ (obtained from the classical mechanics of collisions of a heavy particle with a light particle). In any case, L is a slowly varying function, and the main dependence of the energy loss per unit length is given by the other factors in Equation 11-41. We see that $-dE/dx$ varies inversely with the square of the velocity of the particle and is proportional to the square of the charge of the particle. Since $Z/A \approx \frac{1}{2}$ for all matter, the energy loss is roughly proportional to the density of the material.

The variation of $-dE/dx$ with energy is shown in Figure 11-28. From points B to C on this curve, the energy loss is proportional to $1/v^2$. For relativistic particles, the speed does not vary much with energy and the curve varies only because of the term L. The low-energy portion of the curve from A to B is not given by this theory. At very low energies, the energy-loss function is not simple. Particles with kinetic energy greater than their rest energy are called *minimum ionizing particles*. The energy loss per unit path length is approximately constant for these particles, and the maximum path length or range is proportional to the energy. (The range-energy relation for electrons is more complicated because of the statistical fluctuations due to the small mass of the electron.) A sketch of the range versus energy is given in Figure 11-29. The range of 4-MeV α particles is about 2.5 cm in air. Since the range is inversely proportional to the density of the stopping material, the range is much smaller in dense material.

It is convenient to divide out the density dependence by defining a thickness parameter:

$$\ell = \rho x \qquad \text{in g/cm}^2 \qquad \qquad 11\text{-}43$$

If we then express the energy loss as $dE/d\ell$, it does not vary much from one material to another.

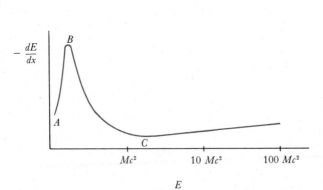

Figure 11-28
Energy loss $-dE/dx$ versus energy for a charged particle. The energy loss is approximately proportional to v^{-2}, where v is the speed of the particle. Thus, in the nonrelativistic region B to C, $-dE/dx$ is proportional to E^{-1}, and in the relativistic region above C, $-dE/dx$ is roughly independent of E. At low energies in the region A to B, the theory is complicated because the charge of the particle varies due to capture and loss of electrons.

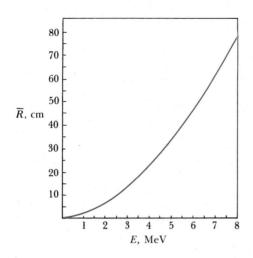

Figure 11-29
Mean range versus energy for protons in dry air. Except at low energies, the range is approximately linear with energy.

If the energy of the charged particle is large compared with its rest energy, the energy loss due to radiation as the particle slows down is important. This radiation is called *Bremsstrahlung*. The ratio of the energy lost by radiation and that lost through ionization is proportional to the energy of the particle and to the atomic number Z of the stopping material. This ratio equals 1 for electrons of about 10 MeV in lead.

The fact that the rate of energy loss for heavy charged particles is very great at very low energies (as seen from the low-energy peak in Figure 11-28) has important applications in nuclear radiation therapy. Figure 11-30 shows the energy loss versus penetration distance of charged particles in water. Most of the energy is deposited near the end of the range. The peak in this curve is called the *Bragg peak*. A beam of heavy charged particles can be used to destroy cancer cells at a given depth in the body without destroying other, healthy cells if the energy is carefully chosen so that most of the energy loss occurs at the proper depth.

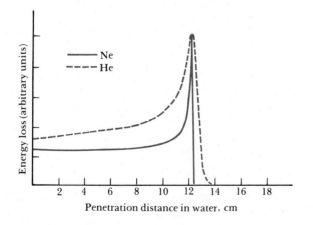

Figure 11-30
Energy loss of helium ions and neon ions in water versus depth of penetration. Most of the energy loss occurs near the end of the path in the Bragg peak. In general, the heavier the ion the narrower the peak.

Neutrons

Since neutrons are uncharged, they do not interact with electrons in matter. Neutrons lose energy by nuclear scattering, or they may be captured. For energies large compared with kT, the most important processes are elastic scattering and inelastic scattering. If we have a collimated beam of intensity I, any scattering or absorption will remove neutrons from the beam. If the sum of the cross sections for all such processes is σ, the number removed from the beam is σI per nucleus. If n is the number of nuclei per cubic centimeter, the number of neutrons removed in a distance dx is

$$-dN = \sigma InA\ dx \qquad\qquad 11\text{-}44$$

where A is the area of the neutron beam and $N = IA$ is the number of neutrons. Thus

$$-dN = \sigma nN\ dx$$

which gives

$$N = N_0 e^{-\sigma nx} \qquad\qquad 11\text{-}45$$

or

$$I = I_0 e^{-\sigma nx} \qquad\qquad 11\text{-}46$$

The number of neutrons in the beam decreases exponentially with distance, and there is no range.

The main source of energy loss for a neutron is usually elastic scattering. (In materials of intermediate weight, such as iron and silicon, inelastic scattering is also important. We shall neglect inelastic scattering here.) The maximum energy loss possible in one elastic collision occurs when the collision is head on. This can be calculated by considering a neutron of mass m with speed v_L making a head-on collision with a nucleus of mass M at rest in the laboratory frame (see Problem 6). The result is that the fractional energy lost by a neutron in one such collision is

$$-\frac{\Delta E}{E} = \frac{4mM}{(M + m)^2} = \frac{4(m/M)}{[1 + (m/M)]^2} \qquad\qquad 11\text{-}47$$

This fraction has a maximum value of 1 when $M = m$ and approaches $4(m/M)$ for $M \gg m$.

Photons

The intensity of a photon beam, like that of a neutron beam, decreases exponentially with distance through an absorbing material. The intensity is given by Equation 11-46, where σ is the cross section for absorption per atom. The important processes that remove photons from a beam are the photoelectric effect at low energies, Compton scattering at intermediate energies, and pair production at high energies. The total cross section for absorption σ is the sum of the partial cross sections σ_{pe}, σ_{cs}, σ_{pp} for these three processes. The photonuclear cross sections, such as $\sigma(\gamma,n)$, are very small compared with these atomic cross sections and can usually be neglected. The cross section for the photoelectric effect dominates at very low energy, but decreases rapidly with increasing energy. The photoelectric effect cannot occur unless the electron is bound in an atom, which can recoil to conserve momentum.

If the photon energy is large compared with the binding energy of the electrons, the electrons can be considered to be free, and Compton scattering is the principal mechanism for the removal of photons from the beam. If the photon energy is greater than $2m_ec^2 = 1.02$ MeV, the photon can disappear, with the creation of an electron-positron pair. This process is called *pair production*.

The cross section for pair production increases rapidly with the photon energy and is the dominant term in σ at high energies. Like the photoelectric effect, pair production cannot occur in free space. If we consider the reaction $\gamma \rightarrow e^+ + e^-$, there is some reference frame in which the total momentum of the electron-positron pair is zero, but there is no reference frame in which the photon's momentum is zero. Thus a nucleus is needed nearby to absorb the momentum by recoil. The cross section for pair production is proportional to Z^2 of the absorbing matter. The three partial cross sections, σ_{pe}, σ_{cs}, and σ_{pp}, are shown with the total cross section as functions of energy in Figure 11-31.

Dosage

The biological effects of radiation are principally due to the ionization produced. There are three units used to measure these effects. The roentgen (R) is defined as the amount of radiation that produces 1 statcoulomb of electric charge (either posi-

The roentgen, rad, and rem

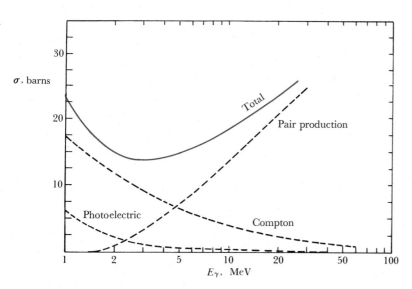

Figure 11-31
Photon attenuation cross section versus energy for lead. The cross sections for the photoelectric effect, Compton scattering, and pair production are shown by the dashed curves. The total cross section (solid curve) is the sum of these. [*From C. Davisson and R. Evans,* Reviews of Modern Physics, **24,** *79 (1952).*]

tive ions or electrons) in 1 cm³ of dry air at standard conditions. [A statcoulomb is a unit of charge in the esu system and is equal to $1/(3 \times 10^9)$ C.] The R has been largely replaced by the rad (radiation *absorbed dose*), which is defined in terms of the energy absorbed. One rad is the amount of radiation that deposits 10^{-2} J/kg of energy in any material.[1] Since 1 R is equivalent to the deposit of about 8.7×10^{-3} J/kg, these units are roughly equal. The amount of biological damage depends not only on the energy absorbed, which is equivalent to dependence on the number of ion pairs formed, but also on the spacing of the ion pairs; if the ion pairs are closely spaced as in the case of ionization by α particles, the biological effect is enhanced. The unit rem (*roentgen equivalent in man*) is the dose that has the same biological effect as 1 rad of β or γ radiation.

$$1 \text{ rem} = 1 \text{ rad} \times \text{RBE} \qquad\qquad 11\text{-}48$$

where the factor RBE (*Relative Biological Effectiveness* factor) is tabulated for α particles, protons, and neutrons of various energies. For example, RBE is 1 for β and γ rays, about 4 or 5 for slow neutrons, about 10 for fast neutrons, and between 10 and 20 for α particles of energies from 5 to 10 MeV. Table 11-1 compares the various radiation units we have discussed. Some typical human radiation exposures are listed in Table 11-2 and some dose-limit recommendations are listed in Table 11-3.

11-7 The Detection of Particles

The detection of charged particles and photons is based on the fact that these particles produce ionization when traversing matter. There are three methods for using this ionization to detect the particle: (1) the electron-ion pairs can be separated in an electric field and collected to produce an electric signal (ionization chambers, proportional counters, Geiger counters, and semiconductor detectors[2]); (2) light emitted when the electron-ion pairs recombine can be observed, or collected and converted to an electric signal by a photomultiplier (scintillators); and (3) the path of the particle can be determined by making the ionization observable, so that it can be seen or photographed (photographic emulsions, cloud chamber, bubble chamber, spark chamber), or by recording the location of the electron-ion pairs electrically (wire spark chamber).

The detection of neutrons is somewhat different. They can be detected by detecting the γ rays emitted in neutron capture, or by detecting the ionization produced by protons recoiling from an *n-p* collision in hydrogen-containing material. In this section we will describe some of the commonly used detectors briefly and discuss their advantages and disadvantages.

---

[1] The SI unit J/kg is now called a gray (Gy). Then 1 rad = 10^{-2} Gy.

[2] Semiconductor detectors are somewhat different in that it is electron-hole pairs that are separated and detected, rather than electron-ion pairs.

Table 11-1
Radiation and dose units

Quantity	Customary unit		SI unit	
	Name	Symbol	Name	Symbol
Energy	electron volt	eV	joule	J
Exposure	roentgen	R	coulomb per kilogram	C/kg
Absorbed dose	rad	rad or rd	gray	Gy = J/kg
Dose equivalent	rem	rem	rem	rem
Activity	curie	Ci	becquerel	Bq = 1/sec

1 MeV = 1.602×10^{-13} J

1 R = 2.58×10^{-4} C/kg

1 rad = 10^{-2} J/kg = 10^{-2} Gy

1 Ci = 3.7×10^{10} decays/sec = 3.7×10^{10} Bq

From S. C. Bushong, *The Physics Teacher*, **15**, no. 3, 135 (1977).

Table 11-2
Sources and average intensities of human radiation exposure

	mrad/y
Natural sources	
Cosmic rays	45
Terrestrial, external exposure	60
Terrestrial, internally deposited radionuclides	25
	130
Man made sources	
Diagnostic x rays	70
Weapons testing	<1
Power generation	<1
Occupational	≤1
	70
Total	200

From S. C. Bushong, *The Physics Teacher*, **15**, no. 3, 135 (1977).

Table 11-3
Recommended dose limits

Maximum permissible dose equivalent for occupational exposure

Combined whole-body occupational exposure	
Prospective annual limit	5 rems in any one year
Retrospective annual limit	10–15 rems in any one year
Long-term accumulation to age N years	$(N - 18) \times 5$ rems
Skin	15 rems in any one year
Hands	75 rems in any one year (25/qtr)
Forearms	30 rems in any one year (10/qtr)
Other organs, tissue, and organ systems	15 rems in any one year (5/qtr)
Fertile women (with respect to fetus)	0.5 rem in gestation period
Dose limits for nonoccupationally exposed	
Population average	0.17 rem in any one year
An individual in the population	0.5 rem in any one year
Students	0.1 rem in any one year

Adapted from NCRP Report #39, 1971, as given in S. C. Bushong, *The Physics Teacher*, **15**, no. 3, 135 (1977).

Gas-Filled Detectors

Figure 11-32 shows a simple gas-filled detector that can be an ionization chamber, a proportional counter, or a Geiger-Müller counter (often called a Geiger or GM counter). The thin wire along the cylindrical axis is maintained at a positive potential V relative to the outer cylindrical conductor. In Figure 11-33, the charge collected at either electrode after an ionizing particle passes through is sketched as a function of the potential difference V. In region A the potential is so low that many electron-ion pairs recombine before they can be collected. In region B, all the primary electrons and ions are collected. A gas-filled detector operating in this region is called an *ionization chamber*. It is usually used to measure the total amount of ionization in some time interval rather than to detect individual particles. For example, a small chamber can be clipped to the pocket of someone working in a radiation laboratory to determine the total dosage received during the day. The chamber is originally charged to potential $V = Q/C$ where Q is the charge on either electrode (the inner wire or outer cylinder) and C is the capacitance. Ionizing radiation passing through the chamber reduces the charge and therefore the potential because the negative electrons drift to the positively charged inner wire and the positive ions drift to the negative outer conductor. The total radiation dose received during the day is proportional to the change in the potential difference between the inner and outer conductors, which can be easily measured and calibrated. Ionization chambers can also be used for a continuous measure of radiation level by using a large chamber and measuring the current output (after amplification, if necessary). Since the current output is proportional to the rate of production of electron-ion pairs, the output can be easily calibrated in rad/min.

If the potential difference V is greater, as in region C of Figure 11-33, the electric field near the inner wire is so great that the electrons are accelerated enough to produce secondary ionization. The charge collected depends strongly on the potential difference, as indicated by the steepness of the curve. In this region the collected charge resulting from ionization produced by a single particle is great enough to produce a voltage pulse that can be amplified and detected. The energy loss of the particle can also be determined because the output pulse is proportional to the original ionization—hence the name *proportional counter*.

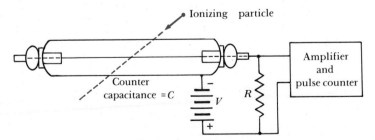

Figure 11-32
Gas-filled particle detector. The inner wire and outer cylinder are charged to a potential difference V, with the inner wire positive.

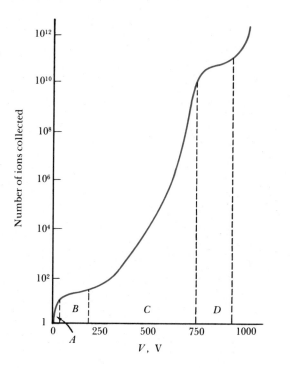

Figure 11-33
Charge collected as a function of potential difference V for the gas-filled detector of Figure 11-32. In region A the potential is so low that the electrons and ions recombine before they can be collected. In the ionization-chamber region B all the original electrons and ions are collected. In the proportional region C, the collected charge is greater because of secondary ionization and is proportional to the energy loss of the particle. In the Geiger-Müller region D, the charge collected is much greater because of secondary ionization, but it is no longer proportional to the energy lost by the particle.

In region D, called the Geiger-Müller region, a single electron from an electron-ion pair will trigger an avalanche of electron-ion pairs. This produces a large voltage pulse that is easily amplified and detected but is not proportional to the energy loss of the original particle. After such an avalanche, the tube is insensitive to other particles for about 200 μsec, called the *dead time*. A gas-filled detector operated in this region is called a Geiger counter. Despite the drawbacks of a relatively long dead time, and the inability to distinguish the type of particle or its energy, Geiger counters are still used today because they are relatively simple, inexpensive, rugged, and reliable.

Scintillators and Semiconductor Detectors

When ionizing particles traverse certain materials, the electrons in the material are excited and give off visible light. The first scintillation detector, built in 1903 by Sir William Crookes, consisted of a ZnS screen and a microscope for viewing the scintillations given off when α particles struck the screen. Such a detector was used by Geiger and Marsden to detect the α particles in the Rutherford scattering experiment. Various other materials such as thallium-activated sodium iodide, NaI(Tl), and some plastics have the important property of being transparent to the light they emit.

In a modern detector, the light is detected by a photosensitive cathode which is part of a photomultiplier, as shown in Figure 11-34. When an ionizing particle passes through the crystal, the light produced strikes the photosensitive cathode and electrons

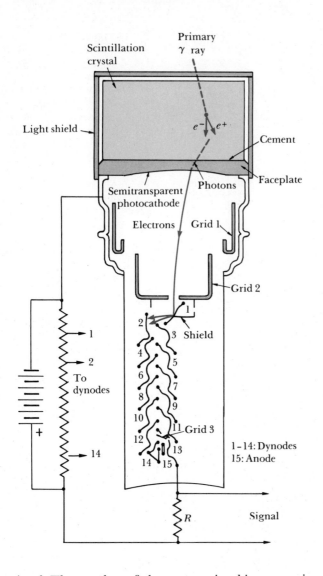

Scintillation crystal

Primary
γ ray

Light shield

Cement

e^- e^+

Photons

Faceplate

Semitransparent
photocathode

Electrons

Grid 1

Grid 2

1
2
3 Shield
4
5
6
7
8
9
10
11 Grid 3
12
13
14 15

1
2
To
dynodes

14

+

R

Signal

1–14: Dynodes
15: Anode

Figure 11-34
Scintillation crystal for detecting γ rays and photomultiplier tube. The incident photon excites the crystal lattice and produces visible light that strikes the photocathode, causing it to emit electrons. The resistor chain and voltage source establishes a voltage gradient between each pair of dynodes to accelerate the electrons. The number of electrons is multiplied at each dynode. The current at the anode produces a voltage pulse across the output resistor R. (*Redrawn from H. Enge,* Introduction to Nuclear Physics, *Reading, Mass.: Addison-Wesley Publishing Company, 1966.*)

are emitted. The number of electrons emitted is proportional to the energy deposited in the crystal by the ionizing particle. Each electron from the cathode is accelerated by an electric field and strikes a second electrode (called a dynode) causing several secondary electrons to be emitted. These electrons strike a second dynode, and electron multiplication continues. The electric current arriving at the anode produces a voltage pulse across the output resistor. This voltage pulse is proportional to the number of electrons striking the cathode, and therefore to the energy deposited in the crystal. The pulses are then amplified and counted electronically. Often they are analyzed and stored in a multichannel pulse height analyzer according to their magnitude, with the larger pulses stored in higher numbered channels. Figure 11-35 shows a pulse height spectrum from a ^{24}Na source, which emits γ rays of energy 2.75 and 1.37 MeV as measured with a NaI(Tl) scintillation detector. In this figure, the number of pulses is plotted against the channel number, which

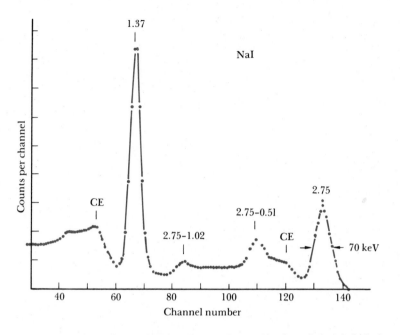

Figure 11-35
Pulse height spectrum of
^{24}Na measured with a NaI
crystal and pulse height
analyzer. The channel
number measures the en-
ergy deposited in the
crystal. The peaks labeled
1.37 and 2.75 are due to
the two γ rays emitted by
^{24}Na of energy 1.37 MeV
and 2.75 MeV. Pulses
corresponding to lower
energy occur when an in-
cident photon undergoes
Compton scattering and
leaves the crystal. The
peaks labeled CE (Comp-
ton edge) are from
maximum-energy elec-
trons produced by Comp-
ton scattering. If the
2.75-MeV photon pro-
duces an electron-
positron pair and the
positron later is anni-
hilated, producing two
0.51-MeV photons, the
escape of one or both
photons produces pulses
at 2.75 − 0.51 and
2.75 − 1.02 MeV. (*Data
from R. L. Heath, "Scintillation Spectrometry,"
USAEC.*)

measures the pulse height. The peak at the far right, labeled
2.75, is called the photopeak. It is due to pulses of maximum
height, which arise when the 2.75-MeV photon deposits all of
its energy in the crystal. The continuum of pulses below the photopeak result when the photon undergoes Compton scattering
and escapes the crystal with an energy that depends on the scattering angle. Only the energy given to the recoil electron is then
deposited in the crystal. The Compton-scattered electrons with
maximum energy give rise to pulses labeled CE (for Compton
edge). If an electron-positron pair are produced, both particles
come to rest and the positron annihilates with an electron,
usually producing two photons each of energy 0.51 MeV. The
pulses labeled 2.75–0.51 and 2.75–1.02 are due to cases in which
one or both of these photons escape from the crystal. The peak
labeled 1.37 is the photopeak for the 1.37-MeV γ ray, and that
labeled CE is its Compton edge.

Sodium iodide crystals have a high efficiency for γ rays (because of their density, and the large Z value of iodine). The
relatively short decay time for their pulses (about 0.25 μsec)
permits much higher counting rates than with a Geiger counter,
but this decay time is long compared with plastic scintillators.
Large NaI(Tl) crystals have the disadvantage of being relatively
expensive. On the other hand, plastics such as polystyrene are
inexpensive, can be easily machined into any desired shape, and
produce pulses of very short duration (several nsec). They have
low efficiencies for γ rays and are used mainly for electrons
or other charged particles, or for neutrons, in which case the
ionization is produced by a recoil nucleus in the plastic. Plastic
scintillators also have the disadvantage that the light output
is not linearly related to the energy loss in the crystal, as it is
in an inorganic crystal such as sodium iodide.

In a semiconductor detector, an ionizing particle produces electron-hole pairs in the depletion layer of a *pn* junction in a semiconductor.[1] The electrons and holes are collected at opposite terminals because of a voltage across the junction, just as electrons and ions are collected in a gas-filled counter. Semiconductor detectors have the following advantages: (1) Small size (silicon wafers are typically 1 inch in diameter and a few millimeters thick), (2) high efficiency (compared with gas detectors, because of their greater density), (3) fast pulses (typically a few nanoseconds), and (4) pulse height proportional to energy loss, with excellent resolution. Gamma rays can be detected with germanium detectors, which have lithium ions drifted through them to enlarge their depletion layer [these detectors are sometimes called "jelly" detectors because of their designation Ge(Li)]. Because of noise due to thermal excitation, these detectors must be operated at liquid nitrogen temperatures, which is a disadvantage. Furthermore, Ge(Li) detectors must be stored at liquid nitrogen temperature to prevent further, unwanted drifting of the Li into the Ge wafer. Figure 11-36 shows the pulse height spectrum of γ rays from ^{24}Na produced by a Ge(Li) detector and a pulse height analyzer. Note the excellent resolution compared with that from NaI(Tl) in Figure 11-35. [The resolution of NaI(Tl) is limited by the statistical fluctuations in the number of visible photons produced in the scintillator.] For the detection of electrons and other charged particles, silicon detectors can be used. These detectors can be stored at room temperatures because the mobility of Li is less in Si then it is in Ge.

---

[1] Semiconductors and *pn* junctions are discussed in Sections 9-5, 9-6, and 9-7.

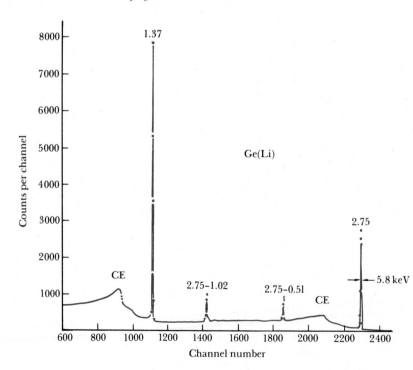

Figure 11-36
Pulse height spectrum of ^{24}Na measured with a Ge(Li) semiconductor detector. This spectrum is the same as that of Figure 11-35 except the resolution is much better. (*Data from SAIP, Malakoff, France.*)

Track Recording Devices

The first nuclear detector was an ordinary photographic plate that Becquerel found to be exposed when left in his desk drawer next to a uranium salt crystal. (The subsequent investigation and discovery of radioactivity by Becquerel in 1896 marked the beginning of nuclear physics.) Ionizing radiation activates silver halides in the emulsion so that upon development, grains of silver mark the track of the particle. (This process is the same as in ordinary photography.) Nuclear emulsions are thicker than ordinary film so that the part of the track not parallel to the emulsion surface can be seen, and are specially designed to have more silver grains, smaller and more uniform grain size, etc. After development the emulsions are viewed with a microscope.

The cloud chamber, invented by C. T. R. Wilson in 1912, makes use of the fact that in a supersaturated vapor, ion pairs act as centers for the formation of liquid drops just as dust particles act as centers for the formation of raindrops. Thus when a ionizing particle passes through such a vapor, the path is marked by a series of visible liquid drops.

The cloud chamber has largely been replaced by the bubble chamber, invented by D. A. Glaser in 1952, in which an ionizing particle causes a superheated transparent liquid to boil with the bubbles forming along its path. Figure 11-37 shows a bubble chamber photograph of the paths of a μ^-, π^+, and proton resulting from the interaction of a neutrino (invisible) and a proton in a liquid-hydrogen bubble chamber. Bubble chambers are used extensively in elementary-particle physics experiments (see Chapter 12). They have excellent track resolution and good efficiency because their density is greater than that of the gas in a cloud chamber (or spark chamber, to be discussed below), but

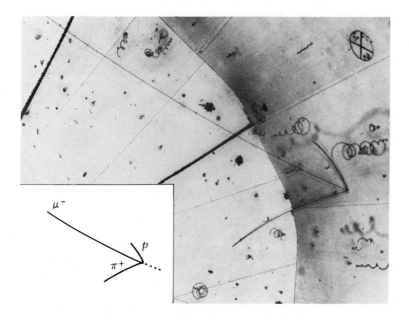

Figure 11-37
Bubble chamber photograph of tracks of a proton, π^+, and μ^- produced by the interaction of a neutrino and a proton in the liquid-hydrogen chamber. (*Courtesy Argonne National Laboratory.*)

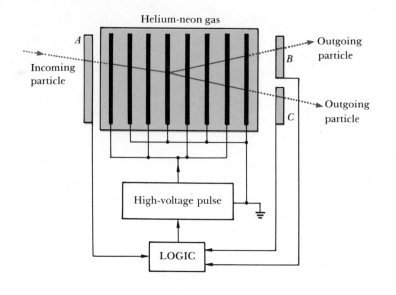

Helium-neon gas

Incoming particle

Outgoing particle

Outgoing particle

High-voltage pulse

LOGIC

Figure 11-38
Spark chamber arrange-
ment to photograph
paths of particles when
one particle is incident
from the left and two
particles exit to the right.
The path is visible be-
cause of sparks produced
at sites of ionization
caused by the particles.
Unless the plastic scintillators A, B, and C send
signals with the proper
time characteristics to the
LOGIC circuit, voltage is
not applied to the plates
and the sparks do not
occur. (*Redrawn from
H. Frauenfelder and
E. Henley,* Subatomic
Physics, *Englewood
Cliffs, N.J.: Prentice-Hall,
Inc., 1974.*)

they have the disadvantage that the liquid must be superheated
before the ionizing particle passes through it so the chamber
cannot be triggered.

This disadvantage is not present in the spark chamber, which
consists of a series of metal plates separated by an inert gas such
as neon. If the voltage between two successive plates is high
enough, the formation of an ion in the gas will cause electrical
breakdown in the form of a spark. The path of an ionizing particle is thus marked by a series of sparks which can be observed
and photographed. Figure 11-38 illustrates the feature of
triggering. Boxes marked A, B, and C are plastic scintillators.
This experiment is designed so that sparks will be produced and
photographed only if there is a charged particle incident from
the left and two charged particles exit to the right essentially
simultaneously. Unless there is a signal from A followed by
signals from B and C a short time later, the voltage on the plates
is too low for breakdown to occur. If a signal arrives from A followed by signals from B and C within the correct time interval,
the electronic circuitry in the box marked "logic" sends a
high-voltage pulse to the plates. Since the electron-ion pairs persist for several microseconds before recombination, and the
triggering can be accomplished in a fraction of a microsecond,
the paths of the particles will be marked by the sparks, which can
be photographed. (The bubble chamber cannot be similarly
triggered because of the relatively long time needed to bring the
liquid to its superheated condition.)

In a variation on the spark chamber, the metal plates are replaced by a grid of wires which forms a coordinate system. The
location of the spark can then be recorded electrically, and the
information fed directly to a computer for analysis.

Optional

11-8 The Shell Model

Although the gross features of the binding energy of nuclei are well accounted for by the semiempirical mass formula, the binding energy and other properties do not vary with perfect smoothness from nucleus to nucleus. It is not surprising that the smooth curve predicted by Equation 11-11 does not fit the data for very small A, for which the addition of a single proton or a neutron makes a drastic difference. However, even for large A there are some large fluctuations of nuclear properties in neighboring nuclei. Consider the binding energy of the last neutron in a nucleus. (Note that this is not the same as the average binding energy per nucleon.) We can calculate this from the semiempirical mass formula by computing the difference in mass $M[(A - 1), Z] + m_n - M(A,Z)$. Figure 11-39 shows a plot of the difference between the experimentally measured binding energy of the last neutron and that calculated from Equation 11-11 as a function of the neutron number N. There are large fluctuations near $N = 20, 28, 50, 82,$ and 126. These are also the neutron numbers of the nuclei that have an unusually large number of isotones. Also, nuclei with these proton numbers (except that there is no element with $Z = 126$) have an unusually large number of isotopes.

Magic numbers

These numbers are called "magic numbers." In the region between these magic numbers, the binding energy of the last neutron is predicted quite accurately by the semiempirical mass formula. Figure 11-39 should be compared with Figure 7-18, which shows the binding energy of the last electron in an atom as a function of atomic number Z. The similarity of these two figures suggests a shell structure of the nucleus analogous to the shell structure of atoms. (There is other evidence for these magic numbers, such as neutron capture and cross sections, which we shall not describe here. See Reference 3 for a more detailed discussion of the evidence for shell structure.)

Although the unusual stability of the nuclei with N or Z equal to one of the magic numbers was noticed in the 1930s, there was no successful explanation in terms of shell structure until 1949.

Figure 11-39
Difference in the measured binding energy of the last neutron and that calculated from mass formula versus neutron number. Note the similarity of this curve and the ionization potential of atoms versus Z (Figure 7-18). The neutron numbers 50, 82, and 126 correspond to closed shells. [*From J. Harvey, Physical Review*, **81**, *353* (*1951*).]

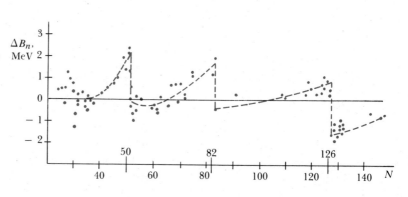

In the discussion of atoms in Chapter 7, we started with a fixed positive charge $+Ze$ and computed the energies of individual electrons, assuming at first that each electron was independent of the others as long as the exclusion principle was not violated. The interaction of the outer electrons with the inner core could be taken care of by assuming an effective nuclear charge which is less than Z because of the screening of the inner electrons. This works quite well since the electrons are fairly far from each other in an atom. We could therefore use the individual electron quantum states of the hydrogen atom described by n, l, m_l, and m_s for the electrons in complex atoms as a first approximation. The atomic magic numbers come about naturally due to the large energy difference between one shell or subshell and the next. The actual calculations of atomic wave functions and atomic energies require powerful approximation or numerical techniques, but they can be done reliably because the forces involved are well known.

The situation is not the same for the nuclear shell model. In the first place, there is no originally given central potential analogous to the fixed positive charge of the atom. The interaction of the nucleons with each other is the only interaction present. The situation is complicated by the fact that we know little about the force between nucleons except that it is strong and has a short range. At first sight, it is difficult to imagine a neutron or proton moving almost freely in a well-defined orbit when there are $A - 1$ particles nearby exerting very strong forces on it. Despite these difficulties, the observed properties, such as are illustrated in Figure 11-39, give a strong motivation to try a model in which each nucleon moves about more or less freely in an average potential field produced by the other nucleons. The assumption that the nucleon can move in an orbit without making many collisions can be rationalized by using the exclusion principle. Consider N neutrons in some potential well. In the ground state, the N lowest-energy levels will be filled. A collision between two neutrons that does not result in their merely exchanging states is forbidden by the exclusion principle if there are no unfilled states. A collision involving the exchange of identical particles has no effect. Thus, only those nucleons in the highest filled levels, where there are empty states available nearby, can collide with each other. This is analogous to the result that most of the free electrons in a metal cannot absorb energy in random collisions with the lattice because all the nearby energy levels are full.

The first shell-model calculations failed to produce the correct magic numbers. In 1949, Mayer and Jensen independently showed that, with a modification in these calculations, the magic numbers do follow directly from a relatively simple shell model. We shall consider only some of the qualitative aspects of the nuclear shell model. Detailed calculations of energies and wave functions require many approximations, the understanding of which is a major continuing problem of nuclear physics.

Let us consider one nucleon of mass m moving in a spherically symmetric potential, $V(r)$. The Schrödinger equation for this problem in three dimensions is the same as Equation 7-4. Since we are assuming $V(r)$ to be independent of θ and ϕ, the angular part of the equation can be separated and solved, as discussed in Chapter 7. The result is that the square of the angular momentum is restricted to the values $l(l + 1)\hbar^2$, and the z component to the values $m_l\hbar$. The radial equation is then Equation 7-13:

$$-\frac{\hbar^2}{2mr^2}\frac{d}{dr}(r^2R') + \left[\frac{l(l + 1)\hbar^2}{2mr^2} + V(r)\right]R = ER$$

11-49

The solution of this equation, of course, depends on the form of the potential energy $V(r)$. Though $V(r)$ is not known, we certainly expect it to be quite different from the $1/r$ potential used in Chapter 7 for atoms. Because the nuclear force is so strong and is negligible beyond a few fermis of the nuclear surface, it does not matter too much what the exact form of $V(r)$ is. Various guesses have been made. The simplest is the finite square well,

$$V(r) = -V_0 \qquad \text{for } r < r_N$$

$$V(r) = 0 \qquad \text{for } r > r_N$$

where r_N is the nuclear radius. This corresponds to an infinite attractive force at the nuclear surface [the force is related to $V(r)$ by $F_r = -dV/dr$] and no force for other r. A more reasonable potential is obtained by rounding the corners of the square well shown in Figure 11-40; however, this makes the problem more difficult to solve analytically.

Figure 11-41a shows the energy levels obtained for an infinite square well and Figure 11-41b shows the level for the finite, rounded well. The levels are labeled with the radial quantum number n and the code letter s for $l = 0$, p for $l = 1$, etc. (Unlike the case of the $1/r$ potential, there is no restriction $l < n$ for these potentials.) The first number after the code letter is the number of identical particles that can exist in that level [this

Figure 11-40
(a) Nuclear-potential well with rounded corners. (b) Finite-square-well approximation.

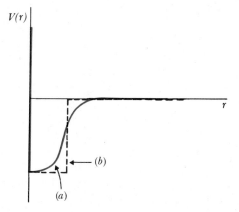

Figure 11-41
Energy levels for a single
particle in (a) an infinite
square well and (b) a
finite square well with
rounded corners. The
number of particles in
each level is given in
parentheses, followed by
the total number through
that level in brackets.

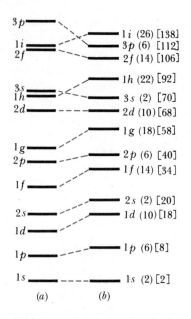

number is $2(2l + 1)$]. The second number is the total number of
particles up to and including that level. From this calculation we
should expect the magic numbers to be 2, 8, 20, 40, 70, 92, and
138, since there are relatively large energy differences after
these numbers. Though the first three numbers, 2, 8, and 20, do
agree with the observed stability of ^{4}He ($N = 2$ and $Z = 2$), ^{16}O
($Z = N = 8$), and ^{40}Ca($Z = N = 20$), the rest of the numbers
are not the magic numbers observed. For example, there is no
evidence from this figure for the magic number 50. Calculations
using various other potential wells give about the same ordering
and spacing of the energy levels.

In 1949 Maria G. Mayer and J. Hans Jensen independently
pointed out that, if the nuclear force is strongly dependent on
the orientation of the spin of the nucleon relative to the orbital
angular momentum, the magic numbers could be obtained. We
have seen that in atoms there is such a "spin-orbit" force, since
the energy of the atom does depend on whether j is $l + \frac{1}{2}$ or
$l - \frac{1}{2}$; however, this fine-structure splitting of the energy levels
is very small compared with the energy difference between the
shells or subshells, and can be neglected in the first approximation to atomic energies. The situation is different for nuclei.
The spin-orbit force is very strong and leads to a large splitting
of the energy level if l is large. Figure 11-42 shows the energy
levels with the assumption of a strong spin-orbit force. The
levels with **L** parallel to **S** have a lower energy than those with
L antiparallel to **S**, i.e., the higher j values have lower energies,
unlike the situation for atoms. (The fact that the force of attraction between a nucleon and a nucleus is greater when spin and
orbital angular momentum of the nucleon are parallel can also
be shown from scattering experiments.) The splitting is large for
large l values. The splitting of the $1g$ level is so great that a large
energy difference occurs between the $1g_{9/2}$ ($n = 1$, $l = 4$,

Figure 11-42
Energy levels for a single particle in a nuclear well, including spin-orbit splitting. The number of particles in each level is given at the right, followed by the total number through that level in brackets. The total numbers just before the large energy gaps are the magic numbers. The spacing shown here is for protons; the spacing for neutrons is slightly different. [*From P. F. A. Klinkenberg,* Reviews of Modern Physics, **24,** *63 (1952).*]

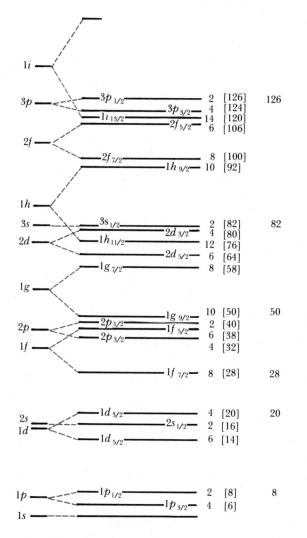

$j = l + s = 4\frac{1}{2}$) and the $1g_{7/2}$ ($n = 1$, $l = 4$, $j = 1 - s = 3\frac{1}{2}$) levels. Since there are $2j + 1$ values of m_j, there can be 10 neutrons in the $1g_{9/2}$ level, making a total of 50 up through this level. The next large energy difference occurs because of the splitting of the $1h$ level, and this accounts for the magic number 82.

From qualitative considerations alone it is not possible to decide the exact order of the energy levels, for example, whether the $2s_{1/2}$ is higher or lower than the $1d_{3/2}$ level. Questions of this nature can usually be answered from empirical evidence. As an example of the predictions of the shell model, we shall consider a nuclide with one neutron or one proton outside a closed shell. These are the simplest nuclides (except for those with closed shells of both neutrons and protons) and are somewhat analogous to the alkali atoms with one outer electron. Many of the energy levels of these nuclides can be understood in terms of the excitation of one odd nucleon. Table 11-4 lists several nuclei with ±1 nucleon outside a closed shell, along with the predicted state of this nucleon and the measured spin of the

Table 11-4

Angular momenta of nuclei with closed shells $\pm$ 1 nucleon

Element	Number of odd particles	Predicted	Measured nuclear spin
$^{11}_{5}$B	5	$p_{3/2}$	$\frac{3}{2}$
$^{13}_{6}$C	7	$p_{1/2}$	$\frac{1}{2}$
$^{15}_{7}$N	7	$p_{1/2}$	$\frac{1}{2}$
$^{17}_{8}$O	9	$d_{5/2}$	$\frac{5}{2}$
$^{27}_{13}$Al	13	$d_{5/2}$	$\frac{5}{2}$
$^{39}_{19}$K	19	$d_{3/2}$	$\frac{3}{2}$
$^{207}_{82}$Pb	125	$p_{1/2}$	$\frac{1}{2}$
$^{209}_{83}$Bi	83	$h_{9/2}$	$\frac{9}{2}$

From Mayer and Jensen, Reference 3.

nucleus. (The neutron number is given as a subscript.) In all these cases, the spin prediction is correct. This simple shell model is also quite successful in predicting the magnetic moments for these nuclei. For example, the magnetic moment of ^{17}O is observed to be -1.89, which is quite close to that of a single neutron. This is predicted by the shell model for this nucleus, since the other eight neutrons and the eight protons form a closed shell. For a more complete discussion of the success of this shell model, the reader is referred to Mayer and Jensen's excellent book, Reference 3. (This model is sometimes called the *extreme individual-particle model* to distinguish it from various, more sophisticated extensions in which the nucleon interactions are taken into account.)

The most serious deficiency of the simple shell model is in the region of the rare-earth nuclei. The quadrupole moments predicted from the orbital motion of the individual protons are much smaller than those observed. Many of the excited states of these nuclei can be most simply understood as being due to the rotation or vibration of the nucleus as a whole, considering it to be a deformed liquid drop. From the shell-model point of view, the rare-earth nuclei lie about midway between the neutron magic numbers of 82 and 126. This is just the region for which shell-model calculations are the most difficult since there are many particles outside a closed shell. There are several extensions of the shell model that have been fairly successful in the understanding of these nonspherical nuclei. In one of these, called the *collective model,* the closed-shell core nucleons are treated as a liquid drop deformed by the interaction with the outer nucleons that orbit about the core and drag it with them. In another model, called the *unified model,* the Schrödinger equation is solved for individual particles in a nonspherically symmetric potential corresponding to an ellipsoidal nucleus. Much of the work with these models was done by J. Rainwater, A. Bohr (son of Niels Bohr), and B. Mottleson in the years 1950–1960. These physicists received the Nobel Prize for this work in 1975.

Summary

All nuclei are made up of N neutrons and Z protons. The nuclear radius is proportional to $A^{1/3}$, so that the volume is proportional to the number of particles A and the density is independent of A. The radius varies from about 1 fm for the proton to about 10 fm for the heaviest nuclei.

The nuclear binding energy is given by

$$B_{\text{nuclear}} = (Nm_n + ZM_H - M_A)c^2$$

where M_A is the mass of the corresponding atom, M_H that of the hydrogen atom, and m_n the mass of the neutron. The binding energy per nucleon is about 8.3 MeV for most nuclei. The general variation of the binding energy from nucleus to nucleus can be understood by analogy with a liquid drop, which also has constant density. There are important deviations from the average behavior of nuclear properties due to shell structure. The nuclear shell model, assuming a strong spin-orbit force, accounts for most of these deviations.

Stable nuclei have $N = Z$ for light nuclei and N slightly greater than Z for heavy nuclei. Radioactive nuclei decay by the emission of ^{4}He nuclei (α decay), electrons or positrons (β decay), or photons (γ decay). Both the number of nuclei and the decay rate follow the exponential law of radioactivity

$$R = \lambda N = \lambda N_0 e^{-\lambda t}$$

where R is the decay rate, N the number of nuclei, N_0 the number at time $t = 0$, and λ the decay constant, which is related to the mean life τ and half-life $t_{1/2}$ by $\lambda = 1/\tau = 0.693/t_{1/2}$. A unit of radioactivity, the curie (1 Ci $= 3.7 \times 10^{10}$ decays/sec), is the activity of 1 g of radium.

Information concerning excited states of nuclei can be obtained from radioactive decay and from nuclear reactions. Many low-energy reactions can be described as proceeding through an intermediate compound nucleus whose decay depends on the energy and angular momentum of the state but not on the manner of its formation.

In nuclear fission a neutron is captured by a fissionable nucleus such as ^{235}U which then breaks into two parts, emitting two or three neutrons. A chain reaction occurs if, on the average, one or more of the emitted neutrons is captured by another fissionable nucleus, thereby causing another fission. Because the neutrons emitted in fission are fast (1 to 2 MeV) whereas those causing fission are usually slow (< 1 eV), a moderator is used in a fission reactor to slow down the neutrons. In a breeder reactor, neutron capture by nonfissionable ^{238}U leads to the production of fissionable ^{239}Pu. A breeder reactor can be designed so that it produces more fuel than it consumes.

In nuclear fusion, energy is released in the fusion of light nuclei such as ^{2}H and ^{3}H. A controlled fusion reactor is not yet a practical reality because of the extremely high temperatures ($\sim 10^8$ K) needed for the ions to overcome their Coulomb repulsion and come close enough to fuse. The Lawson criterion on the particle density n and confinement time τ needed for a practical fusion reactor is $n\tau > 10^{14}$ sec-particles/cm^3. The two schemes currently under investigation for achieving the Lawson criterion are magnetic confinement of a plasma (gas of charged ions) and laser-induced implosion of solid fuel pellets.

Charged particles lose energy almost continuously in matter by ionization (and at high energies by radiation) and have a well-defined range. The range increases with increasing energy of the particle. Neutrons and photons do not have a range; the intensity of a neutron or photon beam decreases exponentially with distance through matter. Neutrons lose energy in a discrete way by elastic or inelastic collisions with nuclei, and are absorbed at low energies. Photons lose energy by Compton scattering or are absorbed at low energies in the photoelectric effect or at high energies in pair production.

A unit of radiation dose is the rad, defined as the amount of radiation that deposits 10^{-2} J/kg in any material. The rad is approximately equal to an older unit, the roentgen (R), defined in terms of the amount of charge produced by ionization in 1 cm^3 of air. The biological effect of radiation is measured in rems, where the rem is defined as the rad times the Relative Biological Effectiveness factor (RBE), which equals 1 for photons and electrons, and varies from about 5 to 20 for α particles, neutrons, and protons.

The detection of charged particles and photons is based on the fact that these particles produce electron-ion pairs in ordinary matter, electron-hole pairs in semiconductors, and visible light in certain materials used as scintillators. In gas-filled counters (ionization chambers, proportional counters, and Geiger counters) and semiconductor detectors the electron-ion or electron-hole pairs are separated by an electric field and collected. In scintillation detectors the light produced is converted to an electric current and amplified by a photomultiplier. In track-recording devices (photographic emulsions, cloud chambers, bubble chambers, and spark chambers) the tracks are made visible by the formation of silver grains, liquid drops, bubbles, or sparks at the electron-ion pair sites.

References

1.   R. Beyer (ed.), *Foundations of Nuclear Physics,* New York: Dover Publications Inc., 1949. This paperback contains 13 original papers, eight in English—by Anderson, Chadwick, Cockroft and Walton, Fermi, Lawrence and Livingston, Rutherford (two papers), and Yukawa—the others are in German or French.

2.   R. Evans, *The Atomic Nucleus,* New York: McGraw-Hill Book Company, 1955.

3.   M. Mayer and J. H. D. Jensen, *Elementary Theory of Nuclear Shell Structure,* New York: John Wiley & Sons, Inc., 1955.

4. American Association of Physics Teachers, *Mössbauer Effect: Selected Reprints,* New York: American Institute of Physics, 1963.

5. H. Frauenfelder, *The Mössbauer Effect,* New York: W. A. Benjamin, Inc., 1962.

6. B. Harvey, *Introduction to Nuclear Physics and Chemistry,* 2d ed., Englewood Cliffs, N.J.: Prentice-Hall, Inc., 1969.

7. I. Kaplan, *Nuclear Physics,* 2d ed., Reading, Mass.: Addison-Wesley Publishing Company, Inc., 1963.

8. E. Fermi, *Nuclear Physics,* rev. ed., Chicago: University of Chicago Press, 1974.

9. W. Price, *Nuclear Radiation Detection,* 2d ed., New York: McGraw-Hill Book Company, 1964.

10. H. Semat and J. Albright, *Introduction to Atomic and Nuclear Physics,* 5th ed., New York: Holt, Rinehart and Winston, Inc., 1972.

11. E. H. Thorndike, *Energy and Environment: A Primer for Scientists and Engineers,* Reading, Mass.: Addison-Wesley Publishing Company, 1976. This is an excellent paperback, filled with information and references, worthy of serious study by anyone interested in energy resources and environmental concerns.

12. F. L. Culler, Jr., and W. O. Harms, "Energy from Breeder Reactors," *Physics Today,* **25,** no. 5, 28 (1972).

13. G. T. Seaborg and J. L. Bloom, "Fast Breeder Reactors," *Scientific American,* **223,** no. 5, 13 (1970).

14. D. J. Rose, "Controlled Nuclear Fusion: Status and Outlook," *Science,* **172,** 797 (1971).

15. J. Nuckolls, J. Emmett, and L. Wood, "Laser-Induced Thermonuclear Fusion," *Physics Today,* **26,** no. 8, 46 (1973).

16. W. C. Gough and B. J. Eastlund, "The Prospects of Fusion Power," *Scientific American,* **224,** no. 2, 50 (1971).

17. M. J. Lubin and A. P. Fraas, "Fusion by Laser," *Scientific American,* **224,** no. 6, 13 (1971).

Exercises

Section 11-1, The Discovery of the Neutron

There are no exercises for this section.

Section 11-2, Ground-State Properties of Nuclei

1. Give at least one isotope and one isotone of each of the following nuclides: (*a*) ^{16}O, (*b*) ^{208}Pb, (*c*) ^{120}Sn.

2. Give at least one isobar and one isotope of each of the following nuclides: (*a*) ^{14}N, (*b*) ^{63}Cu, (*c*) ^{238}U.

3. Use the masses in the table in Appendix A to compute the total binding energy and the binding energy per nucleon of the following nuclides: (*a*) ^{4}He, (*b*) ^{12}C, (*c*) ^{56}Fe.

4. Use the masses in the table in Appendix A to compute the total binding energy and the binding energy per nucleon of the following nuclides: (*a*) ^{2}H, (*b*) ^{16}O, (*c*) ^{58}Ni.

5. Compute the "charge distribution radius" from Equation 11-5 and the "nuclear force radius" from Equation 11-7 for the following nuclides: (a) ^{16}O, (b) ^{63}Cu, (c) ^{208}Pb.

Section 11-3, Radioactivity

6. The half-life of radium is 1620 years. (a) Calculate the number of disintegrations per second of 1 g of radium and show that the disintegration rate is approximately 1 Ci. (b) Calculate the approximate energy of the α particle in the decay ^{226}Ra $\rightarrow$ ^{222}Rn $+ \alpha$, assuming the energy of recoil of the Rn nucleus is negligible. (Use the mass table of Appendix A.)

7. The stable isotope of sodium is ^{23}Na. What kind of radioactivity would you expect of (a) ^{22}Na, (b) ^{24}Na?

8. A radioactive silver foil ($t_{1/2} = 2.4$ min) is placed near a Geiger counter and 1000 counts/sec are observed at time $t = 0$. (a) What is the counting rate at $t = 2.4$ min? at 4.8 min? (b) If the counting efficiency is 20 percent, how many radioactive nuclei are there at time $t = 0$? at time $t = 2.4$ min? (c) At what time will the counting rate be about 30 counts/sec?

9. The counting rate from a radioactive source is 8000 counts/sec at time $t = 0$. Ten min later the rate is 1000 counts/sec. (a) What is the half-life? (b) What is the decay constant? (c) What is the counting rate after 1 min?

10. Show that if the decay rate is R_0 at $t = 0$ and R_1 at some later time t_1, the decay constant λ and half-life are given by

$$\lambda = t_1^{-1} \ln \frac{R_0}{R_1}$$

and

$$t_{1/2} = \frac{0.693 \, t_1}{\ln (R_0/R_1)}$$

11. Use the results of Exercise 10 to find the decay constant and half-life if the counting rate is 1000 counts/sec at time $t = 0$ and 800 counts/sec 1 min later.

12. The counting rate from a radioactive source is measured every minute. The resulting number of counts per second are 1000, 820, 673, 552, 453, 371, 305, 250, ... (a) Plot the counting rate versus time and use your graph to estimate the half-life. (b) Use the results of Exercise 10 to calculate the half-life.

13. ^{62}Cu is produced at a constant rate [e.g., by the (γ,n) reaction on ^{63}Cu placed in a high-energy x-ray beam] and decays by β^+ decay with a half-life of about 10 min. How long does it take to produce 90 percent of the equilibrium value of ^{62}Cu?

14. If the rate of production of ^{62}Cu is 100/sec and the half-life for decay of ^{62}Cu is 10 min, make a plot of the number of ^{62}Cu nuclei versus time.

15. A sample of wood contains 10 g of carbon and shows a ^{14}C decay rate of 100 counts/min. What is the age of the sample?

16. A bone claimed to be 10,000 years old contains 15 g of carbon. What should the decay rate of ^{14}C be from this bone?

17. Suppose you could observe a single radioactive nucleus. Explain why it is more likely for that nucleus to decay during the first second than during the 10th second of observation.

Section 11-4, Nuclear Reactions

18. What is the compound nucleus for the reaction of α particles on ^{16}O? What are some of the possible product nuclei and particles for this reaction?

19. If the width of a resonance for the formation of a compound nucleus is 10 eV, what is the lifetime of this state?

20. (a) Find the Q value for the reaction ^{3}H $+ \, ^1$H $\rightarrow \, ^3$He $+ \, n + Q$. (b) Find the threshold for this reaction if stationary ^{1}H nuclei are bombarded with ^{3}H nuclei from an accelerator. (c) Find the threshold for this reaction if stationary ^{3}H nuclei are bombarded with ^{1}H nuclei from an accelerator.

21. Find the Q values for the following fusion reactions:

(a) ^{2}H $+ \, ^2$H $\rightarrow \, ^3$He $+ \, n + Q$
(b) ^{2}H $+ \, ^2$H $\rightarrow \, ^3$H $+ \, ^1$H $+ Q$
(c) ^{2}H $+ \, ^3$He $\rightarrow \, ^4$He $+ \, ^1$H $+ Q$
(d) ^{6}Li $+ \, n \rightarrow \, ^3$H $+ \, ^4$He $+ Q$

Section 11-5, Fission, Fusion, and Nuclear Reactors

22. The radius of ^{141}Ba ($Z = 56$) computed from Equation 11-4 is 7.8 fm and that of ^{92}Kr ($Z = 36$) is 6.8 fm. Compute the electrostatic potential energy of two point charges $+Z_1e$ and $+Z_2e$ separated by the sum of these radii.

23. Assuming an average energy of 200 MeV/fission, calculate the number of fissions per second needed for a 500-MW reactor.

24. If the reproduction factor in a reactor is $k = 1.1$, find the number of generations needed for the power level of the reactor to (a) double, (b) increase by a factor of 10, (c) increase by a factor of 100. (d) In each case find the time needed if there were no delayed neutrons so that the generation time were 10^{-3} sec. (e) Find the time needed in each case if the delayed neutrons make the average generation time 0.1 sec.

Section 11-6, The Interaction of Particles with Matter

25. The total absorption cross section for γ rays in lead is about 20 barns at 15 MeV. (a) What thickness of lead will reduce the intensity of 15-MeV γ rays by a factor of e? (b) What thickness will reduce the intensity by a factor of 100?

26. The range of 4-MeV α particles in air is 2.5 cm ($\rho_{air} = 1.29 \times 10^{-3}$ g/cm^3). Assuming the range to be inversely proportional to the density, find the range of 4-MeV α particles in (a) water, (b) lead ($\rho = 11.3$ g/cm^3).

27. The range of 6-MeV protons in air is approximately 45 cm. Find the approximate range of 6-MeV protons in (a) water, (b) lead (see Exercise 26).

28. If the absorption cross section for neutrons in iron is 2.5 barns, find the thickness of iron that will reduce the intensity of a neutron beam by a factor of 2 through absorption.

Section 11-7, The Detection of Particles

There are no exercises for this section.

Section 11-8, The Shell Model

29. The nuclei listed below have filled j shells plus or minus one nucleon. (For example, $^{29}_{14}$Si has the $1d_{5/2}$ shell filled for both neutrons and protons, plus one neutron in the $2s_{1/2}$ shell.) Use the shell model to predict the orbital and total angular momentum of these nuclei: $^{29}_{14}$Si, $^{37}_{17}$Cl, $^{71}_{31}$Ga, $^{59}_{27}$Co, $^{73}_{32}$Ge, $^{33}_{16}$S, and $^{87}_{38}$Sr.

Problems

1. (a) Using the Compton-scattering result that the maximum change in wavelength is $\Delta\lambda = 2hc/Mc^2$ and the approximation $\Delta E \approx hc\,\Delta\lambda/\lambda^2$, show that for a photon to lose an amount of energy E_p to a proton, the energy of the photon must be at least $E = (\frac{1}{2}Mc^2E_p)^{1/2}$. (b) Calculate the photon energy needed to produce a 5.7-MeV proton by Compton scattering. (c) Calculate the energy given a ^{14}N nucleus in a head-on collision with a 5.7-MeV neutron. (d) Calculate the photon energy needed to give a ^{14}N nucleus this energy by Compton scattering.

2. A photon of energy E is incident on a deuteron at rest. In the center-of-mass reference frame, both the photon and deuteron have momentum p. Prove that the approximation $p \approx E/c$ is good by showing that the deuteron with this momentum has energy much less than E. If the binding energy of the deuteron is 2.22 MeV, what is the threshold energy in the lab for photodisintegration?

3. Use Equations 11-11 and 11-9 with $N = A - Z$ to find an expression for the nuclear mass $M_N(A, Z)$. For constant odd A, M_N is a parabolic function of Z. The minimum value of M_N should occur near the Z value for the stable isobar for that value of A. Using the constants given in Section 11-2, find the Z for which $dM_N/dZ = 0$ for the following A values: $A = 27, A = 65$, and $A = 139$. Does this calculation give the correct stable isobars ^{27}Al, ^{65}Cu, and ^{139}La?

4. ^{228}Ac decays into ^{228}Th with a half-life of 6.13 h. ^{228}Th decays into ^{224}Ra with a half-life of 1.9 years. Sketch the number of ^{228}Th nuclei present as a function of time, assuming that at $t = 0$ there are N_0 nuclei of ^{228}Ac and no nuclei of ^{228}Th.

5. An empirical expression for the range of α particles in air is $R(\text{cm}) = 0.31E^{3/2}$ for E in MeV and $4 < E < 7$ MeV. (a) What is the range in air of a 5-MeV α particle? (b) Express this range in g/cm^2, using $\rho = 1.29 \times 10^{-3}$ g/cm^3 for air. (c) Assuming the range in g/cm^2 is the same in aluminum ($\rho = 2.70$ g/cm^3), find the range in aluminum in cm for a 5-MeV α particle.

6. Consider a neutron of mass m moving with speed v_L and colliding head on with a nucleus of mass M. (a) Show that the speed of the center of mass in the lab frame is $V = mv_L/(m + M)$. (b) What is the speed of the nucleus in the center-of-mass frame before the collision? after the collision? (c) What is the speed of the nucleus in the original lab frame after the collision? (d) Show that the energy of the nucleus after the collision is $\frac{1}{2}M(2V)^2 = [4mM/(m + M)^2]\frac{1}{2}mv_L^2$, and use this to obtain Equation 11-47.

7.   The cross section for the reaction ^{63}Cu (p,n) ^{63}Zn is about 0.5 barn for 11-MeV protons. Copper has a density of 8.9 g/cm^3 and is about 69 percent ^{63}Cu. (*a*) What is the rate of production of ^{63}Zn nuclei, using a 10^{-6}-A proton beam on a copper target of thickness 0.1 mm? (*b*) The half-life for ^{63}Zn is 38 min. How much ^{63}Zn is there after 38 min of activation? (*c*) How much is there after 10 days of activation? What is the decay rate of ^{63}Zn at this time?

8.   (*a*) Show that after N head-on collisions of a neutron with a carbon nucleus at rest, the energy of a neutron is approximately $(0.72)^N E_0$, where E_0 is the original energy. (*b*) How many head-on collisions are required to reduce the energy of a neutron from 2 MeV to 0.02 eV, assuming stationary carbon nuclei?

9.   On the average, a neutron loses 63 percent of its energy in a collision with a hydrogen atom and 11 percent of its energy in a collision with a carbon atom. Calculate the number of collisions needed to reduce the energy of a neutron from 2 MeV to 0.02 MeV if the neutron collides with (*a*) hydrogen atoms, (*b*) carbon atoms (see Problem 8).

10. Energy is generated in the sun and other stars by fusion. One of the fusion cycles, the proton-proton cycle, consists of the following reactions:

$$^1\text{H} + {}^1\text{H} \longrightarrow {}^2\text{H} + \beta^+ + \nu$$

$$^1\text{H} + {}^2\text{H} \longrightarrow {}^3\text{He} + \gamma$$

followed by either

$$^3\text{He} + {}^3\text{He} \longrightarrow {}^4\text{He} + {}^1\text{H} + {}^1\text{H}$$

or

$$^1\text{H} + {}^3\text{He} \longrightarrow {}^4\text{He} + \beta^+ + \nu$$

(*a*) Show that the net effect of these reactions is

$$4\,{}^1\text{H} \longrightarrow {}^4\text{He} + 2\beta^+ + 2\nu + \gamma$$

(*b*) Show that the rest-mass energy of 24.7 MeV is released in this cycle, not counting the 2×0.511 MeV released when each positron meets an electron and is annihilated according to $e^+ + e^- \rightarrow 2\gamma$. (*c*) The sun radiates energy at the rate of about 4×10^{26} W. Assuming that this is due to the conversion of four protons into helium plus γ rays and neutrinos, which releases 26.7 MeV, what is the rate of proton consumption in the sun? How long will the sun last if it continues to radiate at its present level? (Assume that protons constitute about half the total mass of the sun, which is about 2×10^{30} kg.)

11. The intensity of a neutrino beam decreases exponentially with distance according to Equation 11-46, just like that of a beam of neutrons or photons. The absorption cross section of iron for neutrinos is of the order of 10^{-20} barn. Find the thickness of iron needed to reduce the intensity of a neutrino beam by a factor of e. Compare this thickness with the distance from the earth to the sun (about 1.5×10^{11} m).

CHAPTER 12   Elementary Particles

Objectives

After studying this chapter you should:

1. Be able to estimate the pion rest mass from the uncertainty principle and the range of the nuclear force.

2. Be able to list the four basic interactions, some of the particles that participate in each interaction, and the mediating particle for each interaction.

3. Know the meaning of the terms hadron, lepton, baryon, meson, resonance particle, isospin, charge multiplet, strangeness, hypercharge, parity, eightfold way, quark.

4. Be able to list the conservation laws that are obeyed in all interactions, and those that are obeyed in hadronic but not weak interactions.

5. Be able to describe two methods for detecting the existence of very-short-lived particles.

6. Know the relationship between the strangeness or hypercharge of a hadron multiplet, and the average charge of the multiplet.

7. Know the requirements for a particle to have a distinct antiparticle.

8. Be able to discuss the quark model of hadrons.

In Dalton's atomic theory of matter (1808), the atom was considered to be the smallest indivisible constituent of matter, i.e., an elementary particle. But with the discovery of the electron by Thomson (1897), the Bohr-Rutherford theory of the nuclear atom, and the discovery of the neutron (1932), it was clear that atoms and even nuclei have considerable structure. At that time the structure of matter and its interactions could be described with just four "elementary" particles: the proton, neutron, electron, and the massless photon, which can be considered to have been discovered by Einstein in 1905. However, in 1932, the *antielectron,* or *positron,* was discovered; and shortly thereafter, many other particles were predicted and discovered.

At present there are several hundred particles that are considered by some to be as elementary as any other. Some of these have such short lifetimes (10^{-23} sec) that their detection is very indirect. However, stability is not a good criterion for elementarity. For example, the deuteron, consisting of a proton and a neutron, is not considered an elementary particle even though it is stable, whereas the (elementary) free neutron decays with a mean lifetime of about 15 min. In fact, the only stable particles (besides the massless particles) are the proton and the electron (and their antiparticles). The stability of these particles is merely a consequence of their being the least massive particles of their class. At present the term "elementary particle" is applied to any of the vast array of subnuclear particles that have been discovered or predicted; it does not necessarily imply that these particles are ultimate building blocks of all matter.

12-1 The Positron and Other Antiparticles

In 1932, the same year as the discovery of the neutron, the positron was discovered by Carl Anderson. This particle is identical to the electron but has positive charge and therefore its intrinsic magnetic moment is parallel, rather than antiparallel, to its spin. It is called the antielectron and written e^+. The existence of such an antielectron had been predicted by Dirac from his relativistic wave equation, though there was some difficulty about the interpretation of this prediction.

The energy of a relativistic particle is related to its rest mass and momentum by $E^2 = p^2c^2 + (mc^2)^2$, or $E = \pm (p^2c^2 + m^2c^4)^{1/2}$. Though we can usually choose the plus sign and ignore the negative-energy solution with a "physical argument," the mathematics of the Dirac equation require the existence of wave func-

An aerial view of the principal components of the accelerator system at the Fermi National Accelerator Laboratory, Batavia, Illinois. The largest circle is the Main Accelerator, 4 miles in circumference. The smaller circle, with cooling water pond in the center, is the Booster Accelerator. Experimental areas lie at a tangent to the left. (*Courtesy of Fermi National Accelerator Laboratory.*)

tions corresponding to these negative-energy states. Dirac got around this difficulty by postulating that all the negative-energy states were filled and thus would not be observable. Only holes in this "infinite sea" of negative-energy states would be observable. The discovery of a particle with mass identical to that of the electron but with positive charge indicated that the interpretation was reasonable. Figure 12-1 shows Anderson's original cloud-chamber track and Figure 12-2 shows tracks of electron-positron pairs created by high-energy photons in a lead plate.

The process of pair production can be thought of as the interaction of a photon with a negative-energy electron, raising the electron to a positive-energy state (where it appears as a normal electron) and leaving a hole, which appears as a positron (Figure 12-3). If the Dirac equation can be applied to other spin-$\frac{1}{2}$ particles, such as protons or neutrons, these particles should also have antiparticles. (It is not obvious that the Dirac equation should apply to these particles, because of their anomalous magnetic moments.) The creation of a proton-antiproton pair requires at least $2m_pc^2 = 1896$ MeV, which was not available except in cosmic rays until the development of high-energy accelerators in the 1950s. The antiproton (designated $\bar{p}$) was discovered by Segrè and his coworkers at Berkeley in 1955 using a beam of protons of 6.2-GeV kinetic energy from the Bevatron particle accelerator (Figure 12-4).[1] The antineutron ($\bar{n}$) (a particle with the same mass as the neutron but with a positive magnetic moment) was discovered two years later.

Particles other than those obeying the Dirac equation can also have antiparticles. In general, an antiparticle must have exactly

Figure 12-1
First cloud-chamber photograph of a positron track. The particle came in from the bottom, and followed a circular path in the magnetic field (directed into the page). It was slowed down in traversing the 6-mm lead plate. The direction of the particle is known because of the greater curvature of the track above the plate. (*Courtesy of C. D. Anderson*).

---

[1] The antiproton was first created in the reaction $p + p \rightarrow p + p + p + \bar{p}$, which requires an initial kinetic energy of $2m_pc^2$ in the center-of-mass reference frame. A relativistic calculation shows that in the laboratory frame, in which stationary protons are bombarded with protons of kinetic energy E_k, the laboratory energy $E_k = 6m_pc^2 \approx 5.6$ GeV is needed to achieve kinetic energy of $2m_pc^2$ in the center-of-mass frame.

Figure 12-2
Tracks of electron-positron pairs produced by 300-MeV synchrotron x rays. (*Courtesy of Lawrence Radiation Laboratory, University of California, Berkeley.*)

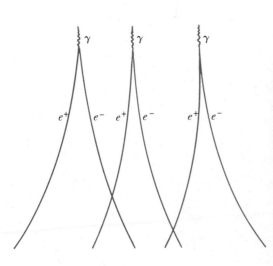

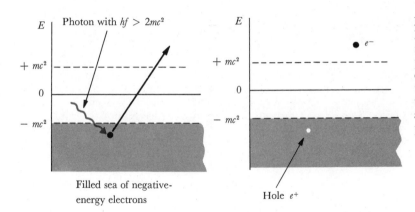

Figure 12-3
Pair production resulting from the collision of a photon with a negative-energy electron. The electron is excited to a positive energy state, leaving a hole that appears as a positron.

the same mass as the particle, but with electric charge, baryon number, and strangeness of sign opposite that of the particle. (Baryon number and strangeness will be defined in Section 12-5.) In some cases, such as the photon and the π^0 meson, the antiparticle is identical to the particle itself.

Although the positron is stable, it has only a short-term existence in our universe because of the large supply of electrons in matter. The fate of the positron is annihilation according to the reaction

$$e^+ + e^- \longrightarrow \gamma + \gamma \qquad\qquad 12\text{-}1$$

Positron annihilation

The probability of this reaction is large only if the positron is at rest or nearly at rest. Two photons are needed to conserve linear momentum. The fact that we call electrons *particles* and positrons *antiparticles* does not imply that positrons are less fundamental than electrons, but merely reflects the nature of our part of the universe. If our matter were made up of negative protons, positive electrons and neutrons with positive magnetic moments, then particles such as positive protons, negative electrons, and neutrons with negative magnetic moments would suffer quick annihilation and would be called antiparticles.

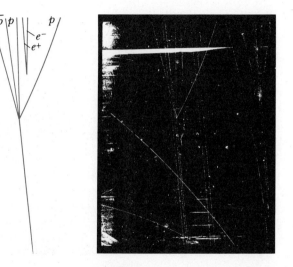

Figure 12-4
Bubble-chamber tracks showing creation of proton-antiproton pair in the collision of an incident 25-BeV proton from the Brookhaven Alternating Gradient Synchrotron with a liquid hydrogen nucleus (stationary proton). The reaction is $p + p \rightarrow p + p + p + \bar{p}$. (*Courtesy of R. Ehrlich.*)

Bevatron at the University of California Lawrence Radiation Laboratory, Berkeley, with which the antiproton was discovered. Protons are injected from a linear accelerator (shown at lower right) and accelerated to 6.2 GeV. (1 GeV = 10^9 eV was formerly called 1 BeV, hence the name bevatron.) (*Courtesy of Lawrence Radiation Laboratory, University of California, Berkeley.*)

12-2 The Discovery of the Neutrino

The history of the discovery of the neutrino is interesting because of the long period of time that elapsed between its prediction (1930) and its direct detection (1956). As discussed in Chapter 11, if a neutron or any nucleus decays by β decay into just two particles, the energy of each particle is completely determined by conservation of energy and momentum. We consider, for example, the decay of the neutron

$$n \longrightarrow p + e^- \qquad\qquad 12\text{-}2$$

Experimentally, we observe a spectrum of electron energies from zero to that predicted by energy and momentum conservation. This result led to the prediction of a third unseen particle named the neutrino ("the little neutral one"). The fact that some of the electrons have the maximum energy allowed by conservation of energy and momentum implies that the rest mass of the neutrino is zero. We can also see that Equation 12-2 does not conserve angular momentum unless there is another particle. The spin of the proton and that of the electron is $\frac{1}{2}$. The combination of two particles of spin $\frac{1}{2}$ can result in a total spin of either 0 or 1, but not of $\frac{1}{2}$ as required to equal the spin of the neutron. The decay of the neutron is therefore written[1]

$$n \longrightarrow p + e^- + \bar{\nu}_e \qquad\qquad 12\text{-}3$$

There were many attempts to observe the neutrino directly, but until 1956 none was successful due to the very weak interaction of neutrinos with matter. (For example, the cross section for absorption in matter is of the order of 10^{-20} barn.) In 1948, C. Sherwin showed that linear momentum was also not conserved in β decay unless a third particle was emitted. He measured the momentum of the β rays from the decay of ^{32}P into

---

[1] The third particle in this reaction is actually an antineutrino, and is therefore written with a bar over it. This is required by the conservation of leptons, discussed in Section 12-5.

^{32}S and simultaneously measured the recoil momentum of ^{32}S. He found the magnitude of the additional momentum needed for conservation of momentum to be E/c, where E is the additional energy needed for conservation of energy. With this experiment, the existence of the neutrino as a massless particle carrying energy E, momentum E/c, and angular momentum with quantum number $\frac{1}{2}$ was firmly established despite a lack of direct detection. In 1956, Reines and Cowen observed the neutrino · "directly" by observing the inverse reaction $\bar{\nu} + {}^{1}\text{H} \rightarrow n + e^{+}$ using an intense flux of (anti)neutrinos from the decay of fission products in a reactor.

12-3 Mesons

In 1935, H. Yukawa proposed a theory of nuclear forces that implied the existence of a particle with about 200 times the mass of the electron. This particle, first called the meson, is now known as the π *meson* or the *pion*. Yukawa's theory was based on an analogy with electrodynamics. The force exerted by a charge q_1 on another charge q_2 at a distance r is given by Coulomb's law kq_1q_2/r^2, where the constant k depends on the units used. To avoid the idea of action at a distance, we invoke the idea of an electromagnetic field which carries the force from one particle to the other. In an alternate and less familiar description based on quantum electrodynamics, the electromagnetic field is described in terms of photons that are emitted by one particle and absorbed by the other. Since these photons are not observed, they are called *virtual* photons. The ability to emit and absorb a virtual photon is proportional to the charge q of the particle. The factor $1/r^2$ in Coulomb's law is purely geometric. We can use this model to calculate the range of the electric force from the uncertainty principle. If a charged particle emits a (virtual) photon, the energy of the two-particle system increases by an amount $\Delta E = hf$, where f is the frequency of the photon. This violates the law of conservation of energy; but *such a violation is not forbidden if it cannot be observed*, i.e., if it occurs within a time T less than $\hbar/\Delta E$, which is the time needed to measure the energy within an uncertainty of ΔE. The maximum distance the photon can travel is then

$$R = cT = \frac{c\hbar}{\Delta E} = \frac{c\hbar}{hf} = \frac{c}{2\pi f} = \frac{\lambda}{2\pi} \qquad 12\text{-}4$$

where λ is the wavelength of the photon. Since photons can have any wavelength, the range of the electric force is infinite.

Yukawa suggested that the strong nuclear force such as that between a proton and a neutron can be similarly described in terms of the exchange of a particle of rest mass m_π that is related to the range of the force. We can use a simplified argument to estimate the rest mass of the particle from the experimental observation that the range of the strong nuclear force is of the order of a few fermis (1 fm = 10^{-15} m). If a neutron or proton emits a particle of rest mass m_π, the energy of the system increases by at least the rest energy $m_\pi c^2$. This is allowed if the time is less than $\hbar/\Delta E = \hbar/m_\pi c^2$. Since the speed of the particle must be less than that of light, the greatest distance the particle can travel in this time is

$$R = cT = \frac{c\hbar}{\Delta E} = \frac{c\hbar}{m_\pi c^2} = \frac{\hbar}{m_\pi c} = \lambdabar_C \qquad \text{12-5}$$

where $\lambdabar_C$ is the Compton wavelength of the particle divided by 2π. The rest energy of the particle is therefore related to the range by

$$m_\pi c^2 = \frac{\hbar c}{R} = \frac{197 \text{ MeV-fm}}{R} \qquad \text{12-6} \qquad \textit{Meson mass estimate}$$

If we take 1.5 fm for the range of the strong nuclear force, we obtain a rest energy for the meson of about 130 MeV. (Yukawa estimated the rest energy to be 100 MeV, about 200 times that of the electron.)

Such a particle of intermediate mass, the *muon*, with mass 105 MeV/c^2, was discovered in 1937 and thought to be Yukawa's meson. However, in 1946 it was shown that the muon did not interact strongly with nucleons and therefore could not be connected with the strong nuclear force. Shortly after this, the π meson, or pion, of mass about 140 MeV/c^2 was discovered. The pion does interact strongly with nucleons and is now considered to be the mediator of the strong nuclear force just as the photon is the mediator of the electromagnetic force.

Question

1. The force of gravity varies as $1/r^2$ and has an infinite range like that of electricity. It is thought to be mediated by a particle called the *graviton*. What is the mass of the graviton?

12-4 The Basic Interactions and Classification of Particles

All the different forces observed in nature can be understood in terms of four basic interactions occurring among elementary particles. In order of decreasing strength these are the strong nuclear (or *hadronic*) interaction, the electromagnetic interaction, the weak nuclear (or Fermi) interaction, and the gravitational interaction. Molecular forces and most of the everyday forces we observe between macroscopic objects, e.g., contact forces, friction, and forces exerted by springs and strings, are complicated manifestations of the electromagnetic interaction. Although gravity plays an important role in our life, it is so weak compared with the other forces that it plays essentially no role in the interactions between elementary particles. The four interactions provide a convenient structure for the classification of elementary particles.

Some particles participate in all four interactions, while others participate in only one, two, or three of them. For example, the neutron interacts with other neutrons or protons via the strong nuclear interaction, it decays by β decay via the weak interaction, it has a magnetic moment and therefore exerts electromagnetic forces on moving charged particles via the electromagnetic interaction, and it attracts other particles via gravity because of its mass. On the other hand, the *graviton* (as yet undetected, but thought to be the mediator of the gravitational interaction just as photons and mesons are mediators of the electromagnetic and hadronic interactions respectively) participates only in the gravitational interaction. Between these extremes are the photon, which participates in electromagnetic and gravitational interaction but not the strong or weak nuclear interactions, and the electron, which participates in all but the strong nuclear interaction. In general, as we move from strong to weak we add to the list of particles that participate. In this section we shall discuss some of the features of the interactions, such as relative strengths and typical interaction (or decay) times, and describe some of the characteristics of the particles that participate in each interaction. The relative strengths listed are only approximate since there is no unambiguous method of comparison, particularly for the weak interaction. Similarly, the lifetime of a particle depends on many factors, so the interaction or decay times listed are typical order-of-magnitude times associated with that particular interaction.

1. Strong Nuclear, or Hadronic, Interaction

> *Relative strength* 1 (chosen arbitrarily)
> *Interaction time* 10^{-23} sec

The strongest of the interactions is responsible for holding nuclei together. Its range is of the order of a few fermis. The particles that participate in this interaction are called *hadrons*. There are two distinct kinds of hadrons, *baryons* and *mesons*. The baryons have spin $\frac{1}{2}$ (or $\frac{3}{2}$, $\frac{5}{2}$, . . .) and are therefore fermions. They are generally the most massive of the elementary particles. The mesons have integral spin $(0, 1, 2, \ldots)$, and are therefore bosons. Particles that decay via the hadronic interaction have such short lifetimes ($\sim 10^{-23}$ sec) that their detection is quite indirect. They are usually observed as resonances in the cross section for scattering of one hadron on another and are therefore also called *resonance particles*. We shall say more about resonance particles later.

Hadrons, baryons, mesons

Table 12-1 lists all the elementary particles which are hadronically stable, i.e., which do not decay via the hadronic interaction. There are nine such baryons, each with a distinct antiparticle. Note that the baryons cluster into "charge multiplets" of about the same mass: the nucleons (n and p) of mass about 939 MeV, the lambda of mass about 1116 MeV, the sigma particles of mass about 1190 MeV, the xi particles of mass about 1315 MeV, and

Table 12-1

Particles stable against hadronic decay

Name	Symbol	Mass (MeV/c^2)	Spin ($\hbar$)	Charge (e)	Antiparticle	Mean lifetime (sec)	Typical decay products†
Baryons							
Nucleon	p^+ (proton)‡	938.3	$\frac{1}{2}$	+1	$\bar{p}^{-*}$	infinite	
	n (neutron)	939.6	$\frac{1}{2}$	0	$\bar{n}$	930	$p^+ + e^- + \bar{\nu}_e$
Lambda	Λ^0	1116	$\frac{1}{2}$	0		2.5×10^{-10}	$p^+ + \pi^-$
Sigma	Σ^+	1189	$\frac{1}{2}$	+1	$\bar{\Sigma}^-$	0.8×10^{-10}	$n + \pi^+$
	Σ^0	1193	$\frac{1}{2}$	0	$\bar{\Sigma}^0$	10^{-20}	$\Lambda^0 + \gamma$
	Σ^-	1197	$\frac{1}{2}$	-1	$\bar{\Sigma}^+$	1.7×10^{-10}	$n + \pi^-$
Xi	Ξ^0	1315	$\frac{1}{2}$	0	$\bar{\Xi}^0$	3.0×10^{-10}	$\Lambda^0 + \pi^0$
	Ξ^-	1321	$\frac{1}{2}$	-1	$\bar{\Xi}^+$	1.7×10^{-10}	$\Lambda^0 + \pi^-$
Omega	Ω^-	1672	$\frac{3}{2}$	-1	$\bar{\Omega}^+$	1.3×10^{-10}	$\Xi^0 + \pi^-$
Mesons							
Pion	π^+	139.6	0	+1	π^-	2.6×10^{-8}	$\mu^+ + \nu_\mu$
	π^0	135	0	0	π^0	0.8×10^{-16}	$\gamma + \gamma$
	π^-	139.6	0	-1	π^+	2.6×10^{-8}	$\mu^- + \bar{\nu}_\mu$
Kaon	K^+	493.7	0	+1	$\bar{K}^-$	1.24×10^{-8}	$\pi^+ + \pi^0$
	K^0	497.7	0	0	$\bar{K}^0$	0.88×10^{-10} and	$\pi^+ + \pi^-$
						5.2×10^{-8}§	$\pi^+ + e^- + \bar{\nu}_e$
Eta	η^0	549	0	0		2×10^{-19}	$\gamma + \gamma$
Leptons							
Electron	e^-	0.511	$\frac{1}{2}$	-1	e^+ (positron)	infinite	
Muon	μ^-	105.7	$\frac{1}{2}$	-1	μ^+	2.2×10^{-6}	$e^- + \bar{\nu}_e + \nu_\mu$
Electron's neutrino	ν_e	0	$\frac{1}{2}$	0	$\bar{\nu}_e$	infinite	
Muon's neutrino	ν_μ	0	$\frac{1}{2}$	0	$\bar{\nu}_\mu$	infinite	
Photon	γ	0	1	0	γ	infinite	
Graviton¶		0	2	0		infinite	

† Other decay modes also occur for most particles.
‡ The symbol p without the + sign is often used. The nucleons are also denoted by the symbols N^+ and N^0.
* The sign given here is that of the antiparticle. The notation with the bar over the sign also is sometimes used to indicate the negative of the sign.
§ The K^0 has two distinct lifetimes, sometimes referred to as K^0_{short} and K^0_{long}. All other particles have a unique lifetime.
¶ The graviton has not yet been discovered, and there is no commonly accepted symbol to denote it.

the omega of mass 1672 MeV. The differences in masses within the multiplets (such as between the neutron and proton) are thought to be due solely to the electric charge of the particles. There are six mesons in Table 12-1: three pions, two kaons, and the eta particle. The kaons have distinct antiparticles, but the pions and the eta do not. The mesons also cluster into charge multiplets. As with the baryons, the mass differences are due to their electric charge. Note that the mass of the π^+ is exactly equal to that of the π^-, as it must if these particles are antiparticles of each other. We shall discuss this grouping of the hadrons into mass multiplets more fully in Section 12-7.

Because the hadronic interaction is so strong, hadrons are complex particles. Each hadron is partially a mixture of other hadrons. For example, a single proton continually emits and reabsorbs (virtual) mesons. It is therefore sometimes a mixture of a p and a π^0 or a n and a π^+ (however, a proton cannot emit a virtual π^- because of strict conservation of charge). Similarly, a neutron is sometimes a mixture of a proton and π^- or a neutron and a π^0. This complex structure of nucleons is responsible for the magnetic moment of the neutron and the anomalous magnetic moment of the proton. The virtual mesons can of course be absorbed by other hadrons if they are close enough (as discussed in Section 12-3); this emission and absorption accounts for the short-range strong hadronic force. The mesons are the mediators of this force. Because the masses of the kaons and the eta are much greater than that of the pions, the exchange of these particles contributes much less to the hadronic force, and then only at extremely close range. The size of the hadrons is of the order of a few fermis, as is the range of the force.

2. Electromagnetic Interaction

Relative strength 10^{-2} $(\frac{1}{137})$
Interaction time 10^{-16}–10^{-21} sec

All particles with electric charge or magnetic moment, plus the photon, participate in the electromagnetic interaction. This includes all the hadrons plus the electron and muon (and their antiparticles), but not the neutrinos or graviton listed in Table 12-1. The mediator of the electromagnetic interaction is the photon. Decays via the electromagnetic interaction always result in the emission of one or more photons. In Table 12-1 we see that the Σ^0, η^0, and π^0 decay via the electromagnetic interaction.

3. Weak Interaction

Relative strength 10^{-13} (depends on definition)
Interaction time 10^{-10}–10^{-16} sec

Particles that participate in the weak interaction include all the hadrons plus the electron e^-, the muon μ^-, the neutrinos ν_e and ν_μ, and their antiparticles (e^+, μ^+, $\bar{\nu}_e$, and $\bar{\nu}_\mu$). These particles, called *leptons,* are less massive than the lightest hadron. *Leptons*
The leptons all have spin $\frac{1}{2}$ and are therefore fermions. The mediator of the weak interaction is thought to be a particle called the *intermediate boson W* (for weak). The W particle has *Intermediate boson*
not been detected, and its existence is not yet established. It is thought to carry electric charge and to be very massive, but its mass is not known. The weak interaction is probably the least understood of the four interactions, but in many ways the leptons are less complex particles than the hadrons. The range

of the weak interaction and the size of the leptons is less than 0.1 fm, and all indications are that leptons are in fact point particles.

The existence of the muon is a puzzle. Except for its mass, which is about 200 times that of the electron, it is identical to the electron. Decays proceeding via the weak interaction usually (but not always) result in the emission of neutrinos. The neutrinos associated with β decay (ν_e and $\bar{\nu}_e$) are apparently distinct from those associated with the emission of muons. This can be seen from experiments in which neutrons (in nuclei) are bombarded by neutrinos from the decay of pions such as $\pi^+ \rightarrow \mu^+ + \nu_\mu$. The reaction $\nu_\mu + n \rightarrow e^- + p$ is never observed, whereas the reaction $\nu_\mu + n \rightarrow \mu^- + p$ is.

4. Gravitational Interaction

> *Relative strength* 10^{-43}
> *Interaction time*   ?

All particles participate in the gravitational interaction, but this interaction is so weak as to be unimportant in the discussion of elementary particles. The interaction is long range, with the force decreasing as $1/r^2$, as does the electrostatic force. The mediating particle for this force is the *graviton*, which should be massless and have spin 2, but this particle has not been observed and probably will not be in the near future because of the extreme weakness of gravitational interactions.

Graviton

Questions

2. How are baryons and mesons similar? How are they different?

3. What properties do all leptons have in common?

4. The mass of the muon is nearly equal to that of the pion. How do these particles differ?

12-5 The Conservation Laws

One of the maxims of nature is "anything that can happen, does." If a conceivable decay or reaction does *not* occur, there must be a reason. The reason is usually expressed in terms of a conservation law. For example, we do not expect to observe the decay $n \rightarrow p + e^+ + \nu_e$ because of conservation of charge. The total electric charge after a decay or reaction must equal that before. Similarly, the decay of a free photon into a neutron, a positron, and a neutrino cannot occur because of conservation of energy. In addition to the familiar laws of conservation of energy, charge, linear momentum, and angular momentum, there are three other conservation laws that are always obeyed. These are the conservation of baryon number, conservation of lepton number, and conservation of muon number (or simply the conservation of baryons, leptons, and muons).

Consider the possible decay $p \rightarrow \pi^0 + e^+$. This decay would conserve charge, angular momentum, and energy but neither baryons nor leptons. The conservation of leptons and baryons means that whenever a lepton or baryon particle is created, an antiparticle of the same type is also created. We describe this further by assigning a lepton number $L = +1$ to all leptons, $L = -1$ to all antileptons, and $L = 0$ to all other particles. Similarly, a baryon number $B = +1$ is assigned to baryons, $B = -1$ to antibaryons, and $B = 0$ for all other particles. The conservation of baryon number rules out a reaction such as $n + n \rightarrow n + \bar{n}$, because the baryon number is 2 on the left side and 0 on the right side. The conservation of lepton number implies that the neutrino emitted in the decay of the free neutron is an antineutrino, $n \rightarrow p + e^- + \bar{\nu}_e$. The fact that neutrinos and antineutrinos are different is illustrated by an experiment sensitive to the detection of the reaction $^{37}\text{Cl} + \bar{\nu}_e \rightarrow {}^{37}\text{Ar} + e^-$, using an intense antineutrino beam from the decay of reactor neutrons. This reaction is not observed, though the reaction $p + \bar{\nu}_e \rightarrow n + e^+$ is.

As we have already seen, the neutrinos from decays involving electrons or positrons are apparently different from those involving muons. The neutrinos from pion decays such as $\pi^- \rightarrow \mu^- + \bar{\nu}_\mu$ do not produce such reactions as $\bar{\nu}_\mu + p \rightarrow n + e^+$ or $\nu_\mu + n \rightarrow p + e^-$. We describe this by assigning a muon family number $+1$ to μ^- and ν_μ, -1 to the antiparticles μ^+ and $\bar{\nu}_\mu$, and 0 for all other particles. We then require the muon family number to be the same before and after a reaction. Since there are only two other known leptons (e^- and ν_e) and their antiparticles (e^+ and $\bar{\nu}_e$), the conservation of leptons plus that of muon family number implies that electron family number is also conserved, if $+1$ is assigned to e^- and ν_e and -1 to e^+ and $\bar{\nu}_e$, and 0 to all other particles. Alternatively, we could list the conservation of muon family number and the conservation of electron family number and omit the conservation of lepton family number, which is then implied by these two.

There are also conservation laws that are not as universal as those just discussed. Some quantities are conserved in reactions or decays that proceed by some interactions, but not if they proceed by others. This is somewhat analogous to the selection rules discussed in atomic transitions. The selection rule $\Delta l = \pm 1$ holds for electric dipole transitions from one atomic state to another. An atom in a state with $l = 2$ cannot decay to a lower energy state with $l = 0$ via electric dipole radiation because of this selection rule, but it can decay via an electric quadrupole transition, which is generally much slower than electric dipole transitions.

There are four quantities that are conserved in hadronic interactions but not in weak interactions. These are called *isotopic spin* (or *isospin*), *strangeness*, *hypercharge*, and *parity*. (All these quantities except isospin are also conserved in electromagnetic interactions.) We shall begin by considering isospin.

Isospin

As previously mentioned, the hadrons are grouped into charge multiplets of one, two, or three particles of nearly identical mass. Were it not for the electromagnetic field, the particles within a multiplet would be the same. The nuclear force between two protons, two neutrons, or a neutron and a proton is the same. The "splitting" of particle states because of the electromagnetic field is analogous to the splitting of atomic energy states in a magnetic field. Because of the analogy with isotopes (similar atoms with slightly different masses) and with the splitting of different spin states, the term "isospin" is used to describe this multiplicity. The isospin T of the nucleon is $\frac{1}{2}$, with the two states $T_3 = +\frac{1}{2}$ for the proton and $T_3 = -\frac{1}{2}$ for the neutron. The isospin is also $\frac{1}{2}$ for the xi-particle doublet, and 0 for the lambda and omega hyperon singlets. It is 1 for the Σ triplet (with $T_3 = +1$ for Σ^+, 0 for Σ^0, and -1 for Σ^-). In the case of the mesons, the pion isospin triplet has $T = 1$, the kaon doublet $T = \frac{1}{2}$, and the eta singlet $T = 0$. The rules of combining isospin are the same as those for combining real spin or angular momentum. Decays and reactions in which the total isospin of the system is not conserved do not proceed via the hadronic interaction.

The property of strangeness was introduced to explain the nonoccurrence, or unusually long half-lives, of certain seemingly possible reactions and decays. Consider the reaction (Figure 12-5)

Strangeness

$$p + \pi^- \longrightarrow \Lambda^0 + K^0 \qquad\qquad 12\text{-}7$$

The cross section for this reaction at sufficiently high energies is large (of the order of millibarns), as would be expected if it takes place via the hadronic interaction. However, the decay times for

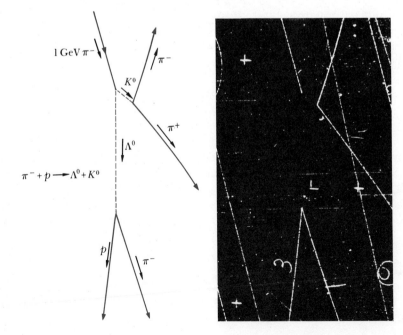

Figure 12-5
Bubble-chamber tracks showing the creation and decay of two strange particles, the K^0 and Λ^0. These neutral particles are identified by the tracks of their decay products. The lambda particle is so named because of the similarity of the tracks of its decay products and the Greek letter Λ. (*Courtesy of Lawrence Radiation Laboratory, University of California, Berkeley.*)

the Λ^0 and K^0 are of the order of 10^{-10} sec, characteristic of weak interactions, and not 10^{-23} sec as would be expected for hadronic decay. Other particles showing similar behavior were called *strange particles*. For example, the Ξ^- particle decays into a proton and two negative pions in a two-step process:

$$\Xi^- \longrightarrow \Lambda^0 + \pi^-$$
$$ \longrightarrow \pi^- + p \qquad\qquad 12\text{-}8$$

Each step has a half-life of the order of 10^{-10} sec.

It was suggested by A. Pais in 1952 that these strange particles Λ^0, K^0, and others were always produced in pairs and never singly, even though all other conservation conditions are met. This suggestion led M. Gell-Mann and K. Nishijima to postulate a new property, which was labeled strangeness by Gell-Mann. The strangeness number S was arbitrarily chosen to be zero for the "ordinary" particles, i.e., the nucleons (n and p) and the pions (π^+, π^-, and π^0). The strangeness of the K^0 was arbitrarily chosen to be $+1$. Then by conservation of strangeness in strong interactions, the strangeness of Λ^0 must be $S = -1$, since the total strangeness of the particles on the left side of Equation 12-7 is zero. The strangeness of other particles could be assigned using various reactions and decays. For example, the reaction $p + K^- \to n + \Lambda^0$ proceeds via the strong interaction, and conservation of strangeness gives $S = -1$ for the K^- meson. Similarly the reaction $p + \pi^- \to n + K^- + K^+$ implies $S = +1$ for the K^+, and so on. The fact that there seemed to be two kaons with $S = +1$ (K^0 and K^+), but only one with $S = -1$ (K^-), led Gell-Mann to suggest the existence of a fourth kaon, $\overline{K}^0$ with $S = -1$, which was later found experimentally.

The strangeness number of a hadron multiplet is simply related to the position of the center of charge $\langle Q \rangle$, which is the average charge of the multiplet. Figure 12-6 shows the hadronically stable hadrons arranged according to mass and charge. The center of charge of the nucleon doublet (n and p) is at $\langle Q \rangle = +\frac{1}{2}e$. This doublet has strangeness $S = 0$. The Λ^0 singlet with $S = -1$ has its center of charge at $\langle Q \rangle = 0$, one step (of size $\frac{1}{2}e$) to the left of the nucleons. The Σ triplet (Σ^+, Σ^0, Σ^-) also has $S = -1$ and its center of charge at 0, one step to the left of the nucleons. The Ξ doublet has its center of charge at $\langle Q \rangle = -\frac{1}{2}e$, two steps to the left of the nucleons, and its strangeness is $S = -2$; whereas the singlet Ω^- is three steps to the left of the nucleons and has strangeness -3. The strangeness of the mesons is similarly related to the position of the center of charge of the meson multiplet relative to that of the pions, which is at $\langle Q \rangle = 0$. The strangeness of an antiparticle is the negative of that of the corresponding particle. In hadronic interactions the strangeness number is conserved, but in weak interactions it can change by ± 1 or 0.

The conservation laws and the properties of charge (Q), lepton number (L), baryon number (B), and strangeness (S) give

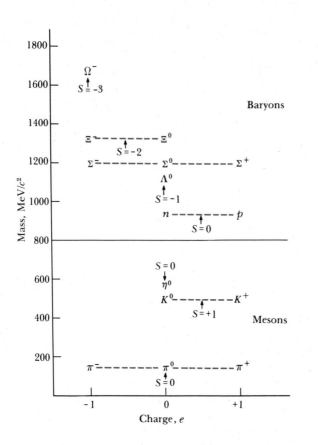

Figure 12-6
Plot of hadrons versus
rest energy (vertical) and
charge (horizontal). The
strangeness of a multiplet
is defined for baryons by
the number of places the
center of charge of the
multiplet is displaced
from that of the nucleon
doublet. For mesons,
strangeness is the number
of places the center of
charge is displaced from
that of the pion triplet.
The hypercharge Y is
twice the average charge
of the multiplet. It is
the same as the
strangeness for mesons
and 1 more than the
strangeness for the
baryons.

us some insight into the relation between particles and their antiparticles. A particle and its antiparticle must have opposite signs
for the values of each of these properties. Any particle that has a
nonzero value for any of these properties will therefore have a
distinct antiparticle. The photon, graviton, and the π^0 have
$Q = 0$, $L = 0$, $B = 0$, and $S = 0$ and are therefore in some
sense their own antiparticles. The π^+ and π^- mesons are somewhat special because they have charge but have zero values for
L, B, and S. They are therefore antiparticles of each other, but
since there is no conservation law for mesons, it is impossible to
say which is the particle and which the antiparticle.

*Relation between particles
and antiparticles*

A property of hadrons that is related to strangeness but somewhat simpler is *hypercharge,* which is defined as the number of
steps by which the average charge of a multiplet differs from
zero. Since each step is $\frac{1}{2}e$, the hypercharge quantum number Y
is defined by

Hypercharge

$$Y = 2\frac{\langle Q \rangle}{e} \qquad 12\text{-}9$$

The hypercharge of the mesons is identical to their strangeness.
The hypercharge of the baryons is 1 greater than their
strangeness. In general we can write

$$Y = S + B \qquad 12\text{-}10$$

Table 12-2
Some properties of the hadronically stable hadrons†

Particle	T	T_3	B	S	Y
p	$\frac{1}{2}$	$+\frac{1}{2}$	1	0	1
n	$\frac{1}{2}$	$-\frac{1}{2}$	1	0	1
Λ^0	0	0	1	-1	0
Σ^+	1	$+1$	1	-1	0
Σ^0	1	0	1	-1	0
Σ^-	1	-1	1	-1	0
Ξ^0	$\frac{1}{2}$	$+\frac{1}{2}$	1	-2	-1
Ξ^-	$\frac{1}{2}$	$-\frac{1}{2}$	1	-2	-1
Ω^-	0	0	1	-3	-2
π^+	1	$+1$	0	0	0
π^0	1	0	0	0	0
π^-	1	-1	0	0	0
K^+	$\frac{1}{2}$	$+\frac{1}{2}$	0	$+1$	$+1$
K^0	$\frac{1}{2}$	$-\frac{1}{2}$	0	$+1$	$+1$
η^0	0	0	0	0	0

† The isospin T is the same for all the particles in a multiplet and for their antiparticles. The values of $T_3, B, S,$ and Y for the antiparticles are the negatives of those for the particles.

where B is the baryon number. Since the baryon number is strictly conserved, the hypercharge number obeys the same rules as the strangeness number, namely, $\Delta Y = 0$ in strong interactions and $\Delta Y = \pm 1$ or 0 in weak interactions. Note that, if it were not for the conservation of strangeness or hypercharge, all the baryons except the nucleons would decay via the hadronic interaction and live only for about 10^{-23} sec. Table 12-2 gives the values of $T, T_3, B, S,$ and Y for the hadrons listed in Table 12-1.

Optional

Parity

As our final example of a conservation law, we consider *parity*. The parity of a nucleus or particle is defined in the same way as for an atom. If the wave function changes sign upon reflection of the coordinates, the parity is said to be odd, or -1. If the wave function does not change sign, the parity is even, or $+1$. The parity quantum number P is different from the other quantum numbers we have been considering in that it can have only the values $+1$ or -1. If the value of the parity of a system changes, the new value is -1 times the old value. Parity is therefore a multiplicative property rather than an additive property like baryon number, strangeness, hypercharge, etc. The parity of an atomic wave function is related to the orbital angular mo-

T. D. Lee (left) and C. N. Yang, cowinners of the Nobel Prize (1957) for their suggestion of the nonconservation of parity and their work in elementary-particle theory. (*Courtesy of H. Schrader, Princeton University.*)

mentum by $P = (-1)^l$. The parity is odd or even depending on whether l is odd or even. In our discussion of radiation from atoms, we saw that the parity of an atom can change just as the angular momentum of the atom changes when the atom emits light. In fact, for electric dipole transitions, $\Delta l = \pm 1$, so the parity and angular momentum quantum numbers always change. However, if the complete system including the photon is considered, the total angular momentum and the total parity do not change in atomic transitions.

Until 1956 it was assumed that parity is conserved in all nuclear reactions and radioactive decays. In that year, T. D. Lee and C. N. Yang suggested that parity might not be conserved in weak interactions. This suggestion grew out of attempts to understand the peculiar behavior of what were then known as the τ and θ mesons. These particles are identical in every way except that the θ meson decays into two pions with positive parity, whereas the τ decays into three pions with negative parity. (Each elementary particle can be assigned an intrinsic parity. That of the pion is negative.) The τ-θ puzzle was—are there two different particles with all properties identical except parity, or is it possible that parity is not conserved in some reactions? After careful study Lee and Yang found that all the experimental evidence for parity conservation pertained to strong or electromagnetic interactions and not to weak interactions. They suggested that the nonconservation of parity could be observed experimentally by measuring the angular distribution of electrons emitted in β decay of nuclei that have their spins aligned in some direction. Such an experiment was performed in December 1956 by a group led by C. S. Wu and E. Ambler. The results confirmed their predictions.[1] The τ and θ mesons are a single particle now known as the K^0 meson, which has two distinct modes of decay.

The conservation of parity essentially means that a process described by the coordinates x, y, z appears the same if described by the coordinates $x' = -x$, $y' = -y$, and $z' = -z$. The system x, y, z is called a *right-handed coordinate system* because $\mathbf{x} \times \mathbf{y}$ is in the $+\mathbf{z}$ direction. Similarly, the system x', y', z' is called a *left-handed coordinate system* because $\mathbf{x}' \times \mathbf{y}'$ is the negative $\mathbf{z}'$ direction. No rotation can change a right-handed coordinate system into a left-handed one; but reflection in a mirror does, as shown in Figure 12-7. We can thus state the law of conservation of parity in more physical terms: if parity is conserved, the mirror image of a process cannot be distinguished from the process itself. Figure 12-8 shows a spinning nucleus emitting an electron in the direction of its spin. In the mirror, the nucleus appears to be emitting the electron in the direction opposite to that of its spin. If parity is conserved in β decay, the chance of emission in the direction of the nuclear spin must equal the chance of emission in the opposite direction, i.e., there can be no preferred direction. Whether or not one direction is actually preferred in β decay is usually not observable because the nuclear spins are randomly oriented. Wu et al. aligned the nuclei in ^{60}Co by placing their sample in a magnetic field at a very low temperature (about 0.01 K). They found that more particles were emitted opposite to the spin of the nucleus than in the direction of the spin, indicating that parity is not conserved in weak interactions.

[1] See Philip Morrison, "The Overthrow of Parity," *Scientific American*, **196**, no. 4, 45 (1957).

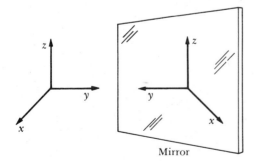

Figure 12-7
The mirror image of a right-handed coordinate system ($\mathbf{x} \times \mathbf{y}$ in the $\mathbf{z}$ direction) is a left-handed coordinate system ($\mathbf{x} \times \mathbf{y}$ in the $-\mathbf{z}$ direction). No combination of translation and rotation can change a right-handed coordinate system into a left-handed system.

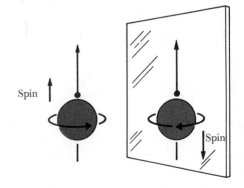

Figure 12-8
Spinning nucleus emitting an electron in the direction of its spin. In the mirror, the image nucleus is emitting the electron in the direction opposite to its spin.

Table 12-3
Conserved quantities in elementary-particle interactions

Quantity	When conserved
Charge	Always
Energy	Always
Linear momentum	Always
Angular momentum	Always
Baryon number	Always
Lepton number†	Always
Muon family number†	Always
Electron family number†	Always
Parity	In hadronic and electromagnetic interactions; not in weak interactions
Isospin	In hadronic interactions
Strangeness and hypercharge	In hadronic and electromagnetic interactions, changes by ± 1 or 0 in weak interactions

† These are not independent. Conservation of two of these implies conservation of the third.

Table 12-3 summarizes the conservation laws discussed in this section.

Questions

5. Can the strangeness or hypercharge of a new particle be determined even if the number of particles in the multiplet is unknown? How, or why not?

6. Suppose a new uncharged meson is discovered. What condition is necessary for it to have a distinct antiparticle?

7. How might Table 12-1 be different if strangeness were not conserved in hadronic interactions?

12-6 Resonance Particles

Particles that are unstable against decay by the hadronic interaction have mean lives of the order of 10^{-23} sec, and therefore cannot be detected by ordinary means. For example, if such a particle moves with nearly the speed of light, it can travel a distance of only about $r = cT = (3 \times 10^8 \text{ m/sec})(10^{-23} \text{ sec}) = 3 \times 10^{-15}$ m = 3 fm (about the size of a nucleus), far too short to leave a track in a bubble chamber. The existence of such particles is inferred from resonances in the scattering cross sections of one hadron on another, or from the energy distribution of nuclear reaction products.

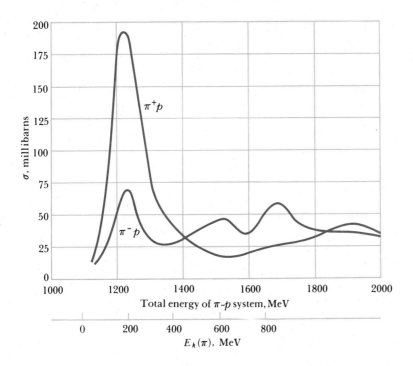

Figure 12-9
Cross section for scattering of π^+ and π^- mesons by protons. The resonance at a pion energy of 195 MeV, corresponding to a total center-of-mass energy (including rest energy) of 1236 MeV, indicates the existence of a new particle called the Δ particle. Other resonances in the $\pi^- + p$ scattering indicate other particles of greater rest energy.

Figure 12-9 shows the cross section σ versus energy for the scattering of π^+ mesons by protons. There is a strong resonance at a pion kinetic energy of 195 MeV (in the laboratory frame). This corresponds to a total energy in the center-of-mass frame (including the rest energies of the π^+ and p) of 1236 MeV. The width of this resonance in the cm frame is about 100 MeV, which correspond to a lifetime of the state of the order of $T = \hbar/\Delta E \approx 10^{-23}$ sec. Despite the brief lifetime of this resonance state, such a state is now considered to be a particle that is in many ways as fundamental as those in Table 12-1 which are stable against hadronic decay. The particle is designated as $\Delta^{++}(1236)$. It has zero strangeness, since both p and π^+ have zero strangeness. Furthermore, the isospin is $\frac{3}{2}$, since $T = \frac{1}{2}$ and $T_3 = +\frac{1}{2}$ for the proton, and $T = 1$ and $T_3 = +1$ for the π^+. The spin and parity can be inferred from angular distribution measurements.

Figure 12-9 also shows the cross section for the scattering of π^- on protons. The resonance at total cm energy of 1236 MeV also shows in this experiment, but its cross section is not nearly as great as that of $\pi^+ + p$ scattering. There are also additional resonances in the $\pi^- + p$ cross section not seen in the $\pi^+ + p$ scattering. This is because the $\pi^- + p$ system is a mixture of isospin states. Since $T = 1$ for the pion and $\frac{1}{2}$ for the nucleon, a system of pion plus nucleon can have either $T = \frac{3}{2}$ or $T = \frac{1}{2}$. Since $T_3 = +1$ for the π^+ and $T_3 = \frac{1}{2}$ for the p, then $\pi^+ + p$ have $T = \frac{3}{2}$. However, $T_3 = -1$ for the π^- so that $\pi^- + p$ is a mixture of both $T = \frac{3}{2}$ and $T = \frac{1}{2}$. The resonances at total cm energy of 1520 MeV and 1688 MeV are designated N because they have the same isospin (and strangeness) as the nucleon and are therefore considered to be in some sense excited states of the nucleon.

The detection of short-lived particles by scattering resonances is limited because particles such as Λ^0 and Σ are not available as targets. Another method of detection is based on the dependence of the energy distribution of decay particles upon the number of particles. Consider the decay $A \rightarrow B + C$. Suppose that we detect only the particle C. As we have previously discussed, conservation of energy and momentum imply a unique energy for particle C. On the other hand, if this decay involves three particles in the final state such as $A \rightarrow B + C + D$, we expect a distribution of energies for particle C as shown in Figure 12-10a, which can be calculated from statistics. If, however, some of the decays proceed via a two-step process such as

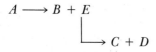

the energy distribution of particle C will look like Figure 12-10c. Figure 12-11 shows the energy distribution of the π^+ observed in the reaction

$$K^- + p \longrightarrow \pi^+ + \pi^- + \Lambda^0$$

(This is just part of a very complicated analysis of this reaction.) The solid curve is that expected if there are three particles in the final state. The large peak at kinetic energy of about 300 MeV implies the existence of an unseen particle with a very short lifetime. The reaction observed is actually a mixture of two reactions, each of which proceeds in two steps. These reactions are

$$K^- + p \longrightarrow \pi^+ + \Sigma^-$$
$$\qquad\qquad \longmapsto \pi^- + \Lambda^0$$

and

$$K^- + p \longrightarrow \pi^- + \Sigma^+$$
$$\qquad\qquad \longmapsto \pi^+ + \Lambda^0$$

Table 12-4 lists some of the hadronically unstable mesons and baryons. Many more have been discovered, some of whose properties are not yet completely established.

Question

8. If a track as short as 1 mm can be seen in a bubble chamber, how long must a particle live to make such a track if it travels at approximately the speed of light?

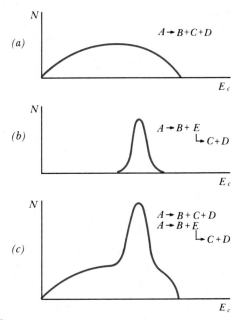

Figure 12-10
Energy distribution of particle C if (a) the reaction yields three particles $A \rightarrow B + C + D$, (b) the reaction yields two particles, one of which then decays, as in $A \rightarrow B + E$, $E \rightarrow C + D$, or (c) the reaction sometimes yields three particles, as in (a), and sometimes two particles, as in (b). The existence of an unseen particle E can be inferred from a peak in the energy distribution as in (b) or (c).

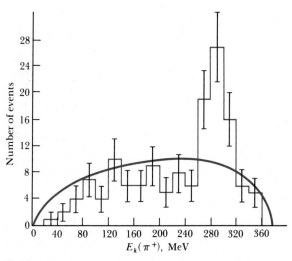

Figure 12-11
Kinetic energy distribution of the π^+ mesons in the reaction $K^- + p \rightarrow \pi^- + \pi^+ + \Lambda^0$. The solid curve is that expected from statistics if there are three particles in the final state. The peak indicates that the reaction proceeds in two steps, either $K^- + p \rightarrow \pi^+ + \Sigma^-$ followed by $\Sigma^- \rightarrow \pi^- + \Lambda^0$ or $K^- + p \rightarrow \pi^- + \Sigma^+$; $\Sigma^+ \rightarrow \pi^+ + \Lambda^0$. The rest energy of the $\Sigma^\pm$ resonance particles is 1385 MeV.

Table 12-4
A partial list of resonance particles

Particle	Mass (MeV/c^2)	Width (MeV)	T	B	S	J^P†
ρ (765)	765	120	1	0	0	1^-
K^* (890)	891	50	$\frac{1}{2}$	0	$+1$	1^-
$\bar{K}^*$ (890)	891	50	$\frac{1}{2}$	0	-1	1^-
Δ (1236)	1236	120	$\frac{3}{2}$	1	0	$\frac{3}{2}^+$
Σ (1385)	1383	36	1	1	-1	$\frac{3}{2}^+$
Λ (1405)	1405	40	0	1	-1	$\frac{1}{2}^-$
N (1470)	1470	210	$\frac{1}{2}$	1	0	$\frac{1}{2}^+$
N (1520)	~1520	105–150	$\frac{1}{2}$	1	0	$\frac{3}{2}^-$
Λ (1520)	~1520	16	0	1	-1	$\frac{3}{2}^-$
Ξ (1530)	1529	7	$\frac{1}{2}$	1	-2	$\frac{3}{2}^+$
N (1535)	1500–1600	50–160	$\frac{1}{2}$	1	0	$\frac{1}{2}^-$
Δ (1650)	1615–1695	130–200	$\frac{3}{2}$	1	0	$\frac{1}{2}^-$
Δ (1670)	1650–1720	175–300	$\frac{3}{2}$	1	0	$\frac{3}{2}^-$
Σ (1670)	1670	50	1	1	-1	$\frac{3}{2}^-$
Λ (1670)	1670	15–38	0	1	-1	$\frac{1}{2}^-$
N (1670)	1655–1680	105–175	$\frac{1}{2}$	1	0	$\frac{5}{2}^-$
N (1688)	1680–1692	105–180	$\frac{1}{2}$	1	0	$\frac{5}{2}^+$
Λ (1690)	1690	27–85	0	1	-1	$\frac{3}{2}^-$
Σ (1750)	1750	50–80	1	1	-1	$\frac{1}{2}^-$

†J^P stands for the spin and parity of the particle.

12-7 The Eightfold Way and Quarks

Among the many attempts at understanding and classifying the jumble of hadrons, the most successful scheme is known as the "eightfold way."[1] It was suggested by Gell-Mann and Y. Ne'eman in 1961.

In this scheme, hadrons of the same intrinsic angular momentum and parity (J^P, where J is the spin-angular-momentum quantum number and P is the parity) are considered to be a *supermultiplet* of particles. These particles would have the same mass were it not for the strong interaction, just as the isospin multiplets (such as the *np* doublet) are considered to have the same mass except for the splitting due to the electromagnetic interaction. Figure 12-12 shows an energy-level diagram for the $J^P = \frac{1}{2}^+$ baryons. The energy splittings between the isospin multiplets (from 78 to 125 MeV) are about 20 times the splitting within the multiplets.

The eightfold way is based on a mathematical theory known as *group theory*, developed by the Norwegian mathematician

UPI

Murray Gell-Mann, who proposed the existence of strangeness, developed the classification system for hadrons [SU(3)], and postulated the existence of fractionally charged particles, which he called quarks. He won the Nobel prize in 1969. (*Courtesy of Niels Bohr Library, American Institute of Physics.*)

[1] From a saying attributed to the Buddha: "Now this, O monks, is the noble truth of the way that leads to the cessation of pain; this is the noble *Eightfold Way:* namely, right views, right intention, right speech, right action, right living, right effort, right mindfulness, right concentration."

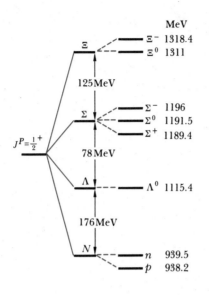

MeV

Ξ⁻ 1318.4
Ξ⁰ 1311

Ξ

125 MeV

Σ⁻ 1196
Σ⁰ 1191.5
Σ⁺ 1189.4

Σ

$J^P = \frac{1}{2}^+$

78 MeV

Λ⁰ 1115.4

Λ

176 MeV

N

n 939.5
p 938.2

Figure 12-12
Supermultiplet of the hadronically stable $J^P = \frac{1}{2}^+$ baryons. In the absence of any interactions, all these particles should have the same mass. The hadronic or strong nuclear interaction splits the mass states into four states, corresponding to the nucleon (N), lambda (Λ), sigma (Σ), and xi (Ξ) particles. The weaker electromagnetic interaction further splits the particles into the N doublet, Σ triplet, and Ξ doublet.

S. Lie. The simplest Lie group is known as SU(2), for special unitary group of 2×2 matrices. A special condition on the 2×2 arrays reduces the number of components from 4 to 3. The three independent components of these arrays correspond to the three components of angular momentum (or isospin). As we have seen previously, the various possible values of angular momentum J have corresponding states which occur in multiplets having 1, 2, 3, 4, . . . , $(2J + 1)$, . . . elements which we describe as having angular momentum of 0, $\frac{1}{2}$, 1, $\frac{3}{2}$, . . . units. The next higher Lie group is known as SU(3), for special unitary group of 3×3 arrays. Again, a special condition reduces the number of components from 9 to 8 (hence the name eightfold way). The eight quantities in the application of SU(3) group theory to hadrons consist of the three components of isospin, the hypercharge, and four that are yet to be named. Without going into any details of group theory, we shall merely state that the SU(3) group leads to multiplets of 1, 3, 8, 10, . . . elements. Rather than assigning a single number to these multiplets analogous to the angular momentum quantum number of SU(2), it is more useful to make two-dimensional diagrams called *weight diagrams,* which are the geometric patterns of points, triangles, and hexagons shown in Figure 12-13.

Figure 12-13
Weight diagrams occurring in SU(3) group theory. The circle and dot at the origin in the hexagon indicate two particles at the origin, making this pattern an octet.

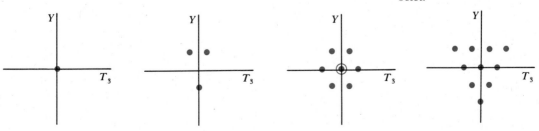

In these plots, the axes are labeled Y and T_3, since these variables seem to work in the application of SU(3) theory to hadron supermultiplets. The dot within the circle in the hexagon indicates two elements at the origin. Note that only the three-element pattern has no element at the origin. Figure 12-14 shows a plot of the hypercharge Y versus T_3 for the $J^P = \frac{1}{2}^+$ baryons, and Figure 12-15 shows the same plot for the $J^P = 0^-$ mesons. A plot of Y versus T_3 for the $J^P = \frac{3}{2}^+$ baryons is shown in Figure 12-16. Neither the Ξ nor the Ω^- had been discovered prior to 1961. Note that the difference in rest energy between each line of the decuplet of Figure 12-16 is about 140 MeV. A constant energy difference between successive multiplets in the decuplet is predicted by SU(3) theory. The prediction of the Ω^- particle by Gell-Mann in 1961 and its discovery in 1964 was one of the first spectacular successes of the eightfold way. Note that the Ω^- is the only particle in Figure 12-16 that is not a resonance particle. The mass of the Ω^- is just small enough that energy conservation prevents it from decaying via a strangeness-conserving strong interaction such as $\Omega^- \rightarrow \Xi^0 + K^-$.

Other patterns of singlets, octets, and decuplets can be formed from the many hadronically unstable baryons and mesons, but there are no observed patterns of three particles corresponding to that allowed by the SU(3) theory. The triplet multiplet in SU(3) theory is one of the fundamental multiplets because all other multiplets can be formed from it. Gell-Mann in 1964 and Zweig independently in 1965 postulated the existence of a triplet of hypothetical particles from which all hadrons can be constructed. These particles are called *quarks* (proposed by Gell-Mann from a quotation from *Finnegans Wake* by James Joyce). The three quarks are labeled *u*, *d*, and *s* (for *u*p, *d*own, and *s*ideways or *s*trange; somewhat older labels no longer in use are *p*, *n*, and *λ*). The proposed properties of these quarks are listed in Table 12-5. They are plotted on a Y versus T_3 diagram in Figure 12-17. The up and down quarks are considered to be a

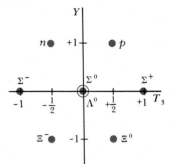

Figure 12-14
Plot of hypercharge Y versus T_3 for the $J^P = \frac{1}{2}^+$ baryons.

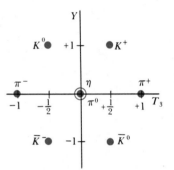

Figure 12-15
Plot of Y versus T_3 for the $J^P = 0^-$ mesons.

Quarks

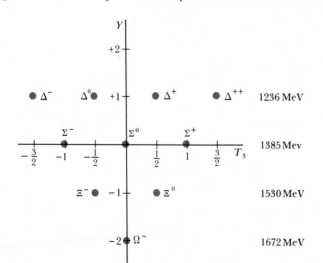

Figure 12-16
Plot of Y versus T_3 for the $J^P = \frac{3}{2}^+$ baryons. According to the SU(3) theory, there should be a constant mass difference between each set of multiplets. This plot was used to predict accurately the existence of the Ω^- baryon with mass 1672 MeV, hypercharge -2, and spin $\frac{3}{2}$ by Gell-Mann three years before it was found experimentally. All the particles in this decuplet except the Ω^- are resonance particles.

Table 12-5
Properties of quarks and antiquarks†

Symbol	Mass (GeV/c^2)	Q (e)	T	T_3	B	S	Y
u	0.34	$+\frac{2}{3}$	$\frac{1}{2}$	$+\frac{1}{2}$	$\frac{1}{3}$	0	$\frac{1}{3}$
d	0.34	$-\frac{1}{3}$	$\frac{1}{2}$	$-\frac{1}{2}$	$\frac{1}{3}$	0	$\frac{1}{3}$
s	0.54	$-\frac{1}{3}$	0	0	$\frac{1}{3}$	-1	$-\frac{2}{3}$
$\bar{u}$	0.34	$-\frac{2}{3}$	$\frac{1}{2}$	$-\frac{1}{2}$	$-\frac{1}{3}$	0	$-\frac{1}{3}$
$\bar{d}$	0.34	$+\frac{1}{3}$	$\frac{1}{2}$	$+\frac{1}{2}$	$-\frac{1}{3}$	0	$-\frac{1}{3}$
$\bar{s}$	0.54	$+\frac{1}{3}$	0	0	$-\frac{1}{3}$	$+1$	$+\frac{2}{3}$

† A fourth quark, the charmed quark, is not listed. It is not needed to construct any of the particles in Tables 12-1 or 12-4.

doublet of approximately equal mass (unknown, but thought to be about 0.34 GeV/c^2), whereas the strange quark is a singlet with different and probably greater mass (~ 0.54 GeV/c^2) than the u and d quarks. Note that the quarks, unlike any known particles, have fractional electric charge and baryon numbers. Each quark has an antiquark with charge, baryon number, and strangeness equal to the negative of that of the quark.

Baryons are considered to be bound states of three quarks. For example, the proton is a combination of two up quarks plus a down quark, whereas the neutron consists of two down quarks and an up quark. Examination of the properties listed in Table 12-5 shows that these combinations give the proper spin, baryon number, and electric charge, as well as strangeness 0. All strange baryons must have one or more s quarks. For example, the Λ^0 consists of one of each kind of quark, giving it a strangeness of -1, baryon number 1, and electric charge 0. Mesons are thought to be bound states of a quark and antiquark. For example, the π^+ meson is thought to be a bound state of an up quark and an antidown quark.

The great strength of the quark model is that all the allowed combinations of three quarks or quark-antiquark pairs result in known hadrons. Furthermore, all known hadrons can be constructed out of either three quarks or a quark-antiquark pair. For example, there are nine possible quark-antiquark combinations, which can be shown to group into a singlet and an octet, whereas the 27 possible combinations of three quarks group into a singlet, two octets, and a decuplet. These are exactly the groupings observed for the hadrons.

In 1967, a fourth kind of quark was proposed to explain some discrepancies in certain decay rates between theory based on the three-quark model and experiment. The fourth quark is known as the *charmed* quark. One of the results of the introduction of this fourth quark was the prediction of a new meson consisting of the charmed quark and its antiquark. Such a particle called the *J* particle (or the ψ particle) was discovered in September 1975 simultaneously by a group led by S. Ting at the Brook-

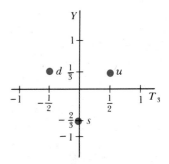

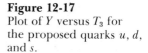

Figure 12-17
Plot of Y versus T_3 for the proposed quarks u, d, and s.

haven National Laboratory and one led by B. Richter at the Stanford Linear Accelerator (SLAC). Despite the greatest interest and effort, no isolated quark has ever been observed. Whether an isolated quark will ever be found in nature is still an open question.

Summary

The forces found in nature can be classified into four basic interactions: hadronic (also called strong), electromagnetic, weak, and gravitational. Each interaction is considered to be mediated by the exchange of a particle: the hadronic by the pion, the electromagnetic by the photon, the weak by the intermediate boson (W particle), and the gravitational by the graviton. The W particle and the graviton have not yet been discovered.

Elementary particles can be classified by their spin (fermions and bosons), by their mass, and by the interactions in which they participate. Two important classes of particles are the hadrons and the leptons. Hadrons participate in the hadronic interaction and include baryons (neutrons, protons, etc.), which are fermions, and mesons (pions, kaons, etc.), which are bosons. The leptons (electrons, muons, and neutrinos) do not participate in hadronic interactions. There are four leptons (e^-, μ^-, ν_e, and ν_μ) and their antiparticles. The heaviest lepton (the muon) is lighter than the lightest hadron (the pion).

Particles that are unstable to decay via the hadronic interaction have mean lives of the order of 10^{-23} sec and cannot be detected directly. The existence of such resonance particles can be inferred from resonances in certain scattering cross sections, and from the energy distribution of nuclear reaction products.

Hadrons of a given spin and parity (J^P) occur in charge multiplets: particles of nearly equal mass but different charge. The small mass difference within a charge multiplet is the result of the electromagnetic interaction. These charge multiplets can be grouped into supermultiplets of 1, 8, or 10 particles. The mass difference between charge multiplets in a supermultiplet is thought to be the result of the hadronic interaction.

All interactions and decays conserve energy, linear momentum, angular momentum, charge, baryon number, lepton number, electron family number, and muon family number. In addition, hadronic interactions conserve isospin, strangeness, hypercharge, and parity, whereas weak interactions do not conserve these quantities. The strangeness and hypercharge quantum numbers of a hadron charge multiplet are simply related to the average charge of the multiplet.

Every hadron can be thought of as made up of as yet undetected particles called quarks. These particles differ from all other particles in that they have fractional baryon numbers, and electric charge that is not an integral multiple of e. Baryons are considered to be made up of three quarks, whereas mesons are considered to be made up of a quark-antiquark pair.

References

An excellent introduction to the subject of elementary particles can be found under "Particles, Subatomic" in the 1975 edition of the *Encyclopedia Britannica*. Other sources of information are:

1. C. Swartz, *The Fundamental Particles,* Reading, Mass.: Addison-Wesley Publishing Co., Inc., 1965.

2. M. Longo, *Fundamentals of Elementary Particle Physics,* New York: McGraw-Hill Book Company, 1973.

3. L. J. Tassie, *The Physics of Elementary Particles,* London: Longmans Group, Ltd., 1973.

4. M. Gell-Mann and E. Rosenbaum, "Elementary Particles," *Scientific American,* **197,** no. 1, 72 (1957).

5. G. Chew, M. Gell-Mann, and A. Rosenfeld, "Strongly Interacting Particles," *Scientific American,* **210,** no. 2, 74 (1964).

6. R. D. Hill, "Resonance Particles," *Scientific American,* **208,** no. 1, 38 (1963).

7. S. Glashow, "Quarks with Color and Flavor," *Scientific American,* **233,** no. 4, 38 (1975).

8. D. B. Cline, A. K. Mann, and C. Rubbia, "The Search for New Families of Elementary Particles," *Scientific American,* **234,** no. 1, 44 (1976).

9. Y. Nambu, "The Confinement of Quarks," *Scientific American,* **235,** no. 5, 48 (1976).

Exercises

Section 12-1, The Positron and Other Antiparticles

1. The fate of an antiproton is usually annihilation via the reaction $p + \bar{p} \rightarrow \gamma + \gamma$. Assume that the proton and antiproton annihilate at rest. (*a*) Why must there be two photons rather than just one? (*b*) What is the energy of each photon? (*c*) What is the wavelength of each photon?

2. A high-energy photon in the vicinity of a nucleus can create an electron-positron pair via pair production: $\gamma \rightarrow e^- + e^+$ (the nucleus is needed to absorb momentum). What energy photon is needed for pair production?

3. Find the energy of the photon needed for the following pair-production reactions: (*a*) $\gamma \rightarrow \pi^+ + \pi^-$, (*b*) $\gamma \rightarrow p + \bar{p}$.

4. Write the symbol for the antiparticle of each of the following: (*a*) e^-, (*b*) n, (*c*) p, (*d*) π^+, (*e*) K^0.

Section 12-2, The Discovery of the Neutrino

5. (*a*) What is the total energy shared by the proton, electron, and neutrino in the β decay of a free neutron? (*b*) If the decay process went via $n \rightarrow p + e^-$ with no neutrino, find the energy of the proton and that of the electron? (Hint: The electron is relativistic.)

6. Find (*a*) the energy of the electron, (*b*) the energy of the ^{32}S nucleus, and (*c*) the momentum of each, in the decay $^{32}\text{P} \rightarrow {}^{32}\text{S} + e^-$, assuming no neutrino in the final state (see Exercise 5). (The rest mass of ^{32}P is 31.973909 u.)

Section 12-3, Mesons

7. Find the approximate range of the strong nuclear force if the mediating particle were the K^+ meson rather than the pion.

8. A hadron can sometimes emit a virtual proton-antiproton pair. (a) Find the time of existence of such a pair allowed by the uncertainty principle. (b) How far could such particles travel in this time if they moved with speed $0.9c$?

Section 12-4, The Basic Interactions and Classification of Particles

9. Name the interaction responsible for each of the following decays:
(a) $n \rightarrow p + e^- + \bar{\nu}_e$, (b) $\pi^0 \rightarrow \gamma + \gamma$, (c) $\Delta^+ \rightarrow \pi^0 + p$,
(d) $\pi^+ \rightarrow \mu^+ + \nu_\mu$.

10. Which of the following decays—$\pi^0 \rightarrow \gamma + \gamma$ or $\pi^- \rightarrow \mu^- + \bar{\nu}_\mu$—would you expect to have the longer lifetime? Why?

11. Compute the ratio of the gravitational force and the electrostatic force between two protons separated by a distance r.

Section 12-5, The Conservation Laws

12. Indicate which of the reactions listed below violate one or more conservation laws, and name the law or laws in each case.

(a) $p^+ \longrightarrow \pi^+ + e^+ + e^-$

(b) $p + \bar{p} \longrightarrow \gamma + \gamma$

(c) $n \longrightarrow p^+ + \pi^-$

(d) $\mu^- \longrightarrow e^- + \nu_\mu + \bar{\nu}_e$

(e) $e^+ + e^- \longrightarrow \gamma$

(f) $\nu_e + p \longrightarrow n + e^+$

(g) $\Omega^- \longrightarrow p + \pi^- + \mu^- + \bar{\nu}_\mu$

13. Compute the change in strangeness in each reaction listed and state whether the reaction can proceed via the hadronic interaction, the weak interaction, or must proceed via two or more steps.

(a) $\Omega^- \longrightarrow \Xi^0 + \pi^-$

(b) $\Xi^0 \longrightarrow p + \pi^- + \pi^0$

(c) $\Delta^+ \longrightarrow p + \pi^0$ $(S = 0$ for the Δ particles)

14. The rules for determining the isospin of two or more particles are the same as those for combining angular momentum. For example, since $T = \frac{1}{2}$ for nucleons, the combination of two nucleons can have either $T = 1$ or $T = 0$, or may be a mixture of these isospin states. Since $T_3 = +\frac{1}{2}$ for the proton, the combination $p + p$ has $T_3 = +1$ and therefore must have $T = 1$. Find T_3 and the possible values of T for the following: (a) $n + n$, (b) $n + p$, (c) $\pi^+ + p$, (d) $\pi^- + n$, (e) $\pi^+ + n$.

Section 12-6, Resonance Particles

There are no exercises for this section.

Section 12-7, The Eightfold Way and Quarks

15. Find the baryon number, charge, isospin, and strangeness for the following quark combinations and identify the corresponding hadron: (a) *uud*, (b) *udd*, (c) *uuu*, (d) *uss*, (e) *dss*, (f) *suu*, (g) *sdd*.

16. Find the baryon number, charge, isospin, and strangeness of the following quark combinations and identify the corresponding hadron (the charge and strangeness of the antiquarks are the negatives of those of the corresponding quarks, as with any other particle-antiparticle pair): (a) $u\bar{d}$, (b) $\bar{u}d$, (c) $u\bar{s}$, (d) $s\bar{s}$, (e) $\bar{d}s$.

17. Some quark combinations can exist in two or more isospin states, with each state corresponding to a different hadron. One such combination is *uds*. (a) What is the value of T_3 for this combination? (b) What are the possible values of total isospin T for this combination? (Use the rules for combination of ordinary spin.) (c) Find the baryon number, charge, and strangeness of this combination, and identify the hadron corresponding to each isospin state.

Problems

There are no problems for this chapter.

Properties of Nuclei

The masses in column 6 are atomic masses, which include the mass of Z electrons. They are taken from J. H. E. Mattauch, W. Thiele, and A. H. Wapstra, "1964 Atomic Mass Table," *Nuclear Physics,* **67,** 1 (1965). Other data are from *The Chart of the Nuclides,* 9th ed., revised to July 1966, by D. T. Goldman and J. R. Roesser; distributed by Educational Relations, General Electric Company, Schenectady, N.Y.

(1)	(2)	(3)	(4)	(5)	(6)	(7)	(8)	(9)
Z	Element	Symbol	Chemical Atomic Weight	Mass Number (* Indicates Radioactive) A	Atomic Mass	Spin and Parity	Percent Abundance	Half-Life (if Radioactive) $t_{1/2}$
0	(Neutron)	n		1*	1.008 665	$\frac{1}{2}$ +		10.8 min
1	Hydrogen	H	1.0079	1	1.007 825	$\frac{1}{2}$ +	99.985	
	Deuterium	D		2	2.014 102	1 +	0.015	
	Tritium	T		3*	3.016 050	$\frac{1}{2}$ +		12.26 y
2	Helium	He	4.0026	3	3.016 030	$\frac{1}{2}$ +	0.00013	
				4	4.002 603	0 +	≈100	
				6*	6.018 892	0 +		0.81 sec
3	Lithium	Li	6.939	6	6.015 125	1 +	7.42	
				7	7.016 004	$\frac{3}{2}$ +	92.58	
				8*	8.022 487	2 +		0.85 sec
4	Beryllium	Be	9.0122	7*	7.016 929	$\frac{3}{2}$ −		53 d
				9	9.012 186	$\frac{3}{2}$ −	100	
				10*	10.013 534	0 +		2.7×10^6 y
5	Boron	B	10.811	10	10.012 939	3 +	19.78	
				11	11.009 305	$\frac{3}{2}$ −	80.22	
				12*	12.014 353	1 +		0.020 sec
6	Carbon	C	12.01115	10*	10.016 810	0 +		19 sec
				11*	11.011 431	$\frac{3}{2}$ −		20.5 min
				12	12.000 000	0 +	98.89	
				13	13.003 354	$\frac{1}{2}$ +	1.11	
				14*	14.003 242	0 +		5730 y
				15*	15.010 599	$\frac{1}{2}$ +		2.25 sec
7	Nitrogen	N	14.0067	12*	12.018 641	1 +		0.011 sec
				13*	13.005 738	$\frac{1}{2}$ −		9.96 min
				14	14.003 074	1 +	99.63	
				15	15.000 108	$\frac{1}{2}$ −	0.37	
				16*	16.006 103	2 −		7.35 sec
				17*	17.008 450			4.14 sec
8	Oxygen	O	15.9994	14*	14.008 597	0 +		71 sec
				15*	15.003 070	$\frac{1}{2}$ +		124 sec
				16	15.994 915	0 +	99.759	
				17	16.999 133	$\frac{5}{2}$ +	0.037	
				18	17.999 160	0 +	0.204	
				19*	19.003 578	$\frac{5}{2}$ +		29 sec
9	Fluorine	F	18.9984	17*	17.002 096			66 sec
				18*	18.000 936	1 +		110 min
				19	18.998 405	$\frac{1}{2}$ +	100	
				20*	19.999 987			11 sec
				21*	20.999 951			4.4 sec

(1)	(2)	(3)	(4)	(5)	(6)	(7)	(8)	(9)
			Chemical Atomic Weight	Mass Number (* Indicates Radioactive) A	Atomic Mass	Spin and Parity	Percent Abundance	Half-Life (if Radioactive) $t_{1/2}$
Z	Element	Symbol						
10	Neon	Ne	20.183	18*	18.005 710	$0 +$		1.46 sec
				19*	19.001 881	$\frac{1}{2} +$		18 sec
				20	19.992 440	$0 +$	90.92	
				21	20.993 849	$\frac{3}{2} +$	0.257	
				22	21.991 385	$0 +$	8.82	
				23*	22.994 473			38 sec
11	Sodium	Na	22.9898	21*	20.997 654	$\frac{3}{2} +$		23 sec
				22*	21.994 437	$3 +$		2.58 y
				23	22.989 771	$\frac{3}{2} +$	100	
				24*	23.990 963	$4 +$		15.0 h
12	Magnesium	Mg	24.312	23*	22.994 125	$\frac{3}{2} +$		12 sec
				24	23.985 042	$0 +$	78.70	
				25	24.986 809	$\frac{5}{2} +$	10.13	
				26	25.982 593	$0 +$	11.17	
				27*	26.984 344	$\frac{1}{2} +$		9.5 min
13	Aluminum	Al	26.9815	26*	25.986 892	$5 +$		7.4×10^5 y
				27	26.981 539	$\frac{5}{2} +$	100	
				28*	27.981 905			2.3 min
14	Silicon	Si	28.086	28	27.976 929	$0 +$	92.21	
				29	28.976 496	$\frac{1}{2} +$	4.70	
				30	29.973 763	$0 +$	3.09	
				31*	30.975 349	$\frac{3}{2} +$		2.62 h
				32*	31.974 020	$0 +$		≈ 700 y
15	Phosphorus	P	30.9738	30*	29.978 317	$1 +$		2.5 min
				31	30.973 765	$\frac{1}{2} +$	100	
				32*	31.973 909	$1 +$		14.3 d
				33*	32.971 728			25 d
16	Sulfur	S	32.064	32	31.972 074	$0 +$	95.0	
				33	32.971 462	$\frac{3}{2} +$	0.76	
				34	33.967 865	$0 +$	4.22	
				35*	34.969 031	$\frac{3}{2} +$		86.7 d
				36	35.967 089	$0 +$	0.014	
17	Chlorine	Cl	35.453	35	34.968 851	$\frac{3}{2} +$	75.53	
				36*	35.968 309	$2 +$		3×10^5 y
				37	36.965 898	$\frac{3}{2} +$	24.47	
18	Argon	Ar	39.948	36	35.967 544	$0 +$	0.337	
				37*	36.966 772	$\frac{3}{2} +$		35.1 d
				38	37.962 728	$0 +$	0.063	
				39*	38.964 317	$\frac{7}{2} -$		270 y
				40	39.962 384	$0 +$	99.60	
				42*	41.963 048	$0 +$		33 y
19	Potassium	K	39.102	39	38.963 710	$\frac{3}{2} +$	93.10	
				40*	39.964 000	$4 -$	0.0118	1.3×10^9 y
				41	40.961 832	$\frac{3}{2} +$	6.88	
20	Calcium	Ca	40.08	40	39.962 589	$0 +$	96.97	
				41*	40.962 275	$\frac{7}{2} -$		7.7×10^4 y
				42	41.958 625	$0 +$	0.64	
				43	42.958 780	$\frac{7}{2} -$	0.145	
				44	43.955 492	$0 +$	2.06	
				46	45.953 689	$0 +$	0.0033	

(1) Z	(2) Element	(3) Symbol	(4) Chemical Atomic Weight	(5) Mass Number (* Indicates Radioactive) A	(6) Atomic Mass	(7) Spin and Parity	(8) Percent Abundance	(9) Half-Life (if Radioactive) $t_{1/2}$
(20)	(Calcium)			48	47.952 531	0 +	0.18	
21	Scandium	Sc	44.956	41*	40.969 248	$\frac{7}{2}$ −		0.55 sec
				45	44.955 920	$\frac{7}{2}$ −	100	
22	Titanium	Ti	47.90	44*	43.959 572	0 +		47 y
				46	45.952 632	0 +	7.93	
				47	46.951 768	$\frac{5}{2}$ −	7.28	
				48	47.947 950	0 +	73.94	
				49	48.947 870	$\frac{7}{2}$ −	5.51	
				50	49.944 786	0 +	5.34	
23	Vanadium	V	50.942	48*	47.952 259			16.1 d
				50*	49.947 164	6 +	0.24	$\approx 6 \times 10^{15}$ y
				51	50.943 961	$\frac{7}{2}$ −	99.76	
24	Chromium	Cr	51.996	48*	47.953 762	0 +		23 h
				50	49.946 054	0 +	4.31	
				52	51.940 513	0 +	83.76	
				53	52.940 653	$\frac{3}{2}$ −	9.55	
				54	53.938 882	0 +	2.38	
25	Manganese	Mn	54.9380	54*	53.940 362	3 +		303 d
				55	54.938 050	$\frac{5}{2}$ −	100	
26	Iron	Fe	55.847	54	53.939 616	0 +	5.82	
				55*	54.938 299	$\frac{3}{2}$ −		2.4 y
				56	55.939 395	0 +	91.66	
				57	56.935 398	$\frac{1}{2}$ −	2.19	
				58	57.933 282	0 +	0.33	
				60*	59.933 964	0 +		$\approx 10^{5}$ y
27	Cobalt	Co	58.9332	59	58.933 189	$\frac{7}{2}$ −	100	
				60*	59.933 813	5 +		5.24 y
28	Nickel	Ni	58.71	58	57.935 342	0 +	67.88	
				59*	58.934 342	$\frac{3}{2}$ −		8×10^{4} y
				60	59.930 787	0 +	26.23	
				61	60.931 056	$\frac{3}{2}$ −	1.19	
				62	61.928 342	0 +	3.66	
				63*	62.929 664			92 y
				64	63.927 958	0 +	1.08	
29	Copper	Cu	63.54	63	62.929 592	$\frac{3}{2}$ −	69.09	
				65	64.927 786	$\frac{3}{2}$ −	30.91	
30	Zinc	Zn	65.37	64	63.929 145	0 +	48.89	
				66	65.926 052	0 +	27.81	
				67	66.927 145	$\frac{5}{2}$ −	4.11	
				68	67.924 857	0 +	18.57	
				70	69.925 334	0 +	0.62	
31	Gallium	Ga	69.72	69	68.925 574	$\frac{3}{2}$ −	60.4	
				71	70.924 706	$\frac{3}{2}$ −	39.6	
32	Germanium	Ge	72.59	70	69.924 252	0 +	20.52	
				72	71.922 082	0 +	27.43	
				73	72.923 462	$\frac{9}{2}$ +	7.76	
				74	73.921 181	0 +	36.54	
				76	75.921 405	0 +	7.76	
33	Arsenic	As	74.9216	75	74.921 596	$\frac{3}{2}$ −	100	

(1)	(2)	(3)	(4)	(5)	(6)	(7)	(8)	(9)
			Chemical Atomic Weight	Mass Number (* Indicates Radioactive) A	Atomic Mass	Spin and Parity	Percent Abundance	Half-Life (if Radioactive) $t_{1/2}$
Z	Element	Symbol						
34	Selenium	Se	78.96	74	73.922 476	$0+$	0.87	
				76	75.919 207	$0+$	9.02	
				77	76.919 911	$\frac{1}{2}-$	7.58	
				78	77.917 314	$0+$	23.52	
				79*	78.918 494	$\frac{7}{2}+$		7×10^4 y
				80	79.916 527	$0+$	49.82	
				82	81.916 707	$0+$	9.19	
35	Bromine	Br	79.909	79	78.918.329	$\frac{3}{2}-$	50.54	
				81	80.916 292	$\frac{3}{2}-$	49.46	
36	Krypton	Kr	83.80	78	77.920 403	$0+$	0.35	
				80	79.916 380	$0+$	2.27	
				81*	80.916 610	$\frac{7}{2}+$		2.1×10^5 y
				82	81.913 482	$0+$	11.56	
				83	82.914 131	$\frac{9}{2}+$	11.55	
				84	83.911 503	$0+$	56.90	
				85*	84.912 523	$\frac{9}{2}+$		10.76 y
				86	85.910 616	$0+$	17.37	
37	Rubidium	Rb	85.47	85	84.911 800	$\frac{5}{2}-$	72.15	
				87*	86.909 186	$\frac{3}{2}-$	27.85	5.2×10^{10} y
38	Strontium	Sr	87.62	84	83.913 430	$0+$	0.56	
				86	85.909 285	$0+$	9.86	
				87	86.908 892	$\frac{9}{2}+$	7.02	
				88	87.905 641	$0+$	82.56	
				90*	89.907 747	$0+$		28.8 y
39	Yttrium	Y	88.905	89	88.905 872	$\frac{1}{2}-$	100	
40	Zirconium	Zr	91.22	90	89.904 700	$0+$	51.46	
				91	90.905 642	$\frac{5}{2}+$	11.23	
				92	91.905 031	$0+$	17.11	
				93*	92.906 450	$\frac{5}{2}+$		9.5×10^5 y
				94	93.906 313	$0+$	17.40	
				96	95.908 286	$0+$	2.80	
41	Niobium	Nb	92.906	91*	90.906 860			(long)
				92*	91.907 211			$\approx 10^7$ y
				93	92.906 382	$\frac{9}{2}+$	100	
				94*	93.907 303	$6+$		2×10^4 y
42	Molybdenum	Mo	95.94	92	91.906 810	$0+$	15.84	
				93*	92.906 830			$\approx 10^4$ y
				94	93.905 090	$0+$	9.04	
				95	94.905 839	$\frac{5}{2}$ (?)	15.72	
				96	95.904 674	$0+$	16.53	
				97	96.906 021	$\frac{5}{2}$ (?)	9.46	
				98	97.905 409	$0+$	23.78	
				100	99.907 475	$0+$	9.63	
43	Technetium	Tc		97*	96.906 340			2.6×10^6 y
				98*	97.907 110			1.5×10^6 y
				99*	98.906 249	$\frac{9}{2}+$		2.1×10^5 y
44	Ruthenium	Ru	101.07	96	95.907 598	$0+$	5.51	
				98	97.905 289	$0+$	1.87	
				99	98.905 936	$\frac{5}{2}+$	12.72	

(1)	(2)	(3)	(4)	(5)	(6)	(7)	(8)	(9)
Z	Element	Symbol	Chemical Atomic Weight	Mass Number (* Indicates Radioactive) A	Atomic Mass	Spin and Parity	Percent Abundance	Half-Life (if Radioactive) $t_{1/2}$
(44)	(Ruthenium)			100	99.904 218	0 +	12.62	
				101	100.905 577	$\frac{5}{2}$ +	17.07	
				102	101.904 348	0 +	31.61	
				104	103.905 430	0 +	18.58	
45	Rhodium	Rh	102.905	103	102.905 511	$\frac{1}{2}$ −	100	
46	Palladium	Pd	106.4	102	101.905 609	0 +	0.96	
				104	103.904 011	0 +	10.97	
				105	104.905 064	$\frac{5}{2}$ +	22.23	
				106	105.903 479	0 +	27.33	
				107*	106.905 132			7×10^6 y
				108	107.903 891	0 +	26.71	
				110	109.905 164	0 +	11.81	
47	Silver	Ag	107.870	107	106.905 094	$\frac{1}{2}$ −.	51.82	
				109	108.904 756	$\frac{1}{2}$ −	48.18	
48	Cadmium	Cd	112.40	106	105.906 463	0 +	1.22	
				108	107.904 187	0 +	0.88	
				109*	108.904 928	$\frac{5}{2}$ +		453 d
				110	109.903 012	0 +	12.39	
				111	110.904 188	$\frac{1}{2}$ +	12.75	
				112	111.902 762	0 +	24.07	
				113	112.904 408	$\frac{1}{2}$ +	12.26	
				114	113.903 360	0 +	28.86	
				116	115.904 762	0 +	7.58	
49	Indium	In	114.82	113	112.904 089	$\frac{9}{2}$ +	4.28	
				115	114.903 871	$\frac{9}{2}$ +	95.72	
50	Tin	Sn	118.69	112	111.904 835	0 +	0.96	
				114	113.902 773	0 +	0.66	
				115	114.903 346	$\frac{1}{2}$ +	0.35	
				116	115.901 745	0 +	14.30	
				117	116.902 958	$\frac{1}{2}$ +	7.61	
				118	117.901 606	0 +	24.03	
				119	118.903 313	$\frac{1}{2}$ +	8.58	
				120	119.902 198	0 +	32.85	
				121*	120.904 227			25 y
				122	121.903 441	0 +	4.72	
				124	123.905 272	0 +	5.94	
51	Antimony	Sb	121.75	121	120.903 816	$\frac{5}{2}$ +	57.25	
				123	122.904 213	$\frac{7}{2}$ +	42.75	
				125*	124.905 232	$\frac{7}{2}$ +		2.7 y
52	Tellurium	Te	127.60	120	119.904 023	0 +	0.089	
				122	121.903 064	0 +	2.46	
				123*	122.904 277	$\frac{1}{2}$ +	0.87	1.2×10^{13} y
				124	123.902 842	0 +	4.61	
				125	124.904 418	$\frac{1}{2}$ +	6.99	
				126	125.903 322	0 +	18.71	
				128	127.904 476	0 +	31.79	
				130	129.906 238	0 +	34.48	
53	Iodine	I	126.904	127	126.904 070	$\frac{5}{2}$ +	100	
				129*	128.904 987	$\frac{7}{2}$ +		1.6×10^7 y

(1) Z	(2) Element	(3) Symbol	(4) Chemical Atomic Weight	(5) Mass Number (* Indicates Radioactive) A	(6) Atomic Mass	(7) Spin and Parity	(8) Percent Abundance	(9) Half-Life (if Radioactive) $t_{1/2}$
54	Xenon	Xe	131.30	124	123.906 120	0 +	0.096	
				126	125.904 288	0 +	0.090	
				128	127.903 540	0 +	1.92	
				129	128.904 784	$\frac{1}{2}$ +	26.44	
				130	129.903 509	0 +	4.08	
				131	130.905 085	$\frac{3}{2}$ +	21.18	
				132	131.904 161	0 +	26.89	
				134	133.905 815	0 +	10.44	
				136	135.907 221	0 +	8.87	
55	Cesium	Cs	132.905	133	132.905 355	$\frac{7}{2}$ +	100	
				134*	133.906 823	4 +		2.1 y
				135*	134.905 770	$\frac{7}{2}$ +		2×10^6 y
				137*	136.906 770	$\frac{7}{2}$ +		30 y
56	Barium	Ba	137.34	130	129.906 245	0 +	0.101	
				132	131.905 120	0 +	0.097	
				133*	132.905 879			7.2 y
				134	133.904 612	0 +	2.42	
				135	134.905 550	$\frac{3}{2}$ +	6.59	
				136	135.904 300	0 +	7.81	
				137	136.905 500	$\frac{3}{2}$ +	11.32	
				138	137.905 000	0 +	71.66	
57	Lanthanum	La	138.91	137*	136.906 040			6×10^4 y
				138*	137.906 910	$\frac{5}{2}$ −	0.089	$\approx 10^{11}$ y
				139	138.906 140	$\frac{7}{2}$ +	99.911	
58	Cerium	Ce	140.12	136	135.907 100	0 +	0.193	
				138	137.905 830	0 +	0.250	
				140	139.905 392	0 +	88.48	
				142*	141.909 140	0 +	11.07	5×10^{15} y
59	Praseodymium	Pr	140.907	141	140.907 596	$\frac{5}{2}$ +	100	
60	Neodymium	Nd	144.24	142	141.907 663	0 +	27.11	
				143	142.909 779	$\frac{7}{2}$ −	12.17	
				144*	143.910 039	0 +	23.85	2.1×10^{15} y
				145	144.912 538	$\frac{7}{2}$ −	8.30	
				146	145.913 086	0 +	17.22	
				148	147.916 869	0 +	5.73	
				150	149.920 960	0 +	5.62	
61	Promethium	Pm		143*	142.910 991			265 d
				145*	144.912 691			18 y
				146*	145.914 632			1600 d
				147*	146.915 108	$\frac{7}{2}$ +		2.6 y
62	Samarium	Sm	150.35	144	143.911 989	0 +	3.09	
				146*	145.912 992	0 +		1.2×10^8 y
				147*	146.914 867	$\frac{7}{2}$ −	14.97	1.08×10^{11} y
				148*	147.914 791	0 +	11.24	1.2×10^{13} y
				149*	148.917 180	$\frac{7}{2}$ −	13.83	4×10^{14} y
				150	149.917 276	0 +	7.44	
				151*	150.919 919	$\frac{7}{2}$ −		90 y
				152	151.919 756	0 +	26.72	
				154	153.922 282	0 +	22.71	

(1) Z	(2) Element	(3) Symbol	(4) Chemical Atomic Weight	(5) Mass Number (* Indicates Radioactive) A	(6) Atomic Mass	(7) Spin and Parity	(8) Percent Abundance	(9) Half-Life (if Radioactive) $t_{1/2}$
63	Europium	Eu	151.96	151	150.919 838	$\frac{5}{2}$ +	47.82	
				152*	151.921 749	3 −		12.4 y
				153	152.921 242	$\frac{5}{2}$ +	52.18	
				154*	153.923 053	3 −		16 y
				155*	154.922 930	$\frac{5}{2}$ +		1.8 y
64	Gadolinium	Gd	157.25	148*	147.918 101	0 +		85 y
				150*	149.918 605	0 +		1.8×10^6 y
				152*	151.919 794	0 +	0.20	1.1×10^{14} y
				154	153.920 929	0 +	2.15	
				155	154.922 664	$\frac{3}{2}$ −	14.73	
				156	155.922 175	0 +	20.47	
				157	156.924 025	$\frac{3}{2}$ −	15.68	
				158	157.924 178	0 +	24.87	
				160	159.927 115	0 +	21.90	
65	Terbium	Tb	158.925	159	158.925 351	$\frac{3}{2}$ +	100	
66	Dysprosium	Dy	162.50	156*	155.923 930	0 +	0.052	2×10^{14} y
				158	157.924 449	0 +	0.090	
				160	159.925 202	0 +	2.29	
				161	160.926 945	$\frac{5}{2}$ +	18.88	
				162	161.926 803	0 +	25.53	
				163	162.928 755	$\frac{5}{2}$ +	24.97	
				164	163.929 200	0 +	28.18	
67	Holmium	Ho	164.930	165	164.930 421	$\frac{7}{2}$ −	100	
				166*	165.932 289	0 −		1.2×10^3 y
68	Erbium	Er	167.26	162	161.928 740	0 +	0.136	
				164	163.929 287	0 +	1.56	
				166	165.930 307	0 +	33.41	
				167	166.932 060	$\frac{7}{2}$ +	22.94	
				168	167.932 383	0 +	27.07	
				170	169.935 560	0 +	14.88	
69	Thulium	Tm	168.934	169	168.934 245	$\frac{1}{2}$ +	100	
				171*	170.936 530	$\frac{1}{2}$ +		1.9 y
70	Ytterbium	Yb	173.04	168	167.934 160	0 +	0.135	
				170	169.935 020	0 +	3.03	
				171	170.936 430	$\frac{1}{2}$ −	14.31	
				172	171.936 360	0 +	21.82	
				173	172.938 060	$\frac{5}{2}$ −	16.13	
				174	173.938 740	0 +	31.84	
				176	175.942 680	0 +	12.73	
71	Lutecium	Lu	174.97	173*	172.938 800	$\frac{7}{2}$ +		1.4 y
				175	174.940 640	$\frac{7}{2}$ +	97.41	
				176*	175.942 660		2.59	2.2×10^{10} y
72	Hafnium	Hf	178.49	174*	173.940 360	0 +	0.18	2.0×10^{15} y
				176	175.941 570	0 +	5.20	
				177	176.943 400	$\frac{7}{2}$ −	18.50	
				178	177.943 880	0 +	27.14	
				179	178.946 030	$\frac{9}{2}$ +	13.75	
				180	179.946 820	0 +	35.24	
73	Tantalum	Ta	180.948	180	179.947 544		0.0123	

(1) Z	(2) Element	(3) Symbol	(4) Chemical Atomic Weight	(5) Mass Number (* Indicates Radioactive) A	(6) Atomic Mass	(7) Spin and Parity	(8) Percent Abundance	(9) Half-Life (if Radioactive) $t_{1/2}$
(73)	(Tantalum)			181	180.948 007	$\frac{7}{2}$ +	99.988	
74	Tungsten	W	183.85	180	179.947 000	0 +	0.14	
	(Wolfram)			182	181.948 301	0 +	26.41	
				183	182 950 324	$\frac{1}{2}$ −	14.40	
				184	183.951 025	0 +	30.64	
				186	185.954 440	0 +	28.41	
75	Rhenium	Re	186.2	185	184.953 059	$\frac{5}{2}$ +	37.07	
				187*	186.955 833	$\frac{5}{2}$ +	62.93	5×10^{10} y
76	Osmium	Os	190.2	184	183.952 750	0 +	0.018	
				186	185.953 870	0 +	1.59	
				187	186.955 832	$\frac{1}{2}$ −	1.64	
				188	187.956 081	0 +	13.3	
				189	188.958 300	$\frac{3}{2}$ −	16.1	
				190	189.958 630	0 +	26.4	
				192	191.961 450	0 +	41.0	
				194*	193.965 229	0 +		6.0 y
77	Iridium	Ir	192.2	191	190.960 640	$\frac{3}{2}$ +	37.3	
				193	192.963 012	$\frac{3}{2}$ +	62.7	
78	Platinum	Pt	195.09	190*	189.959 950	0 +	0.0127	7×10^{11} y
				192	191.961 150	0 +	0.78	
				194	193.962 725	0 +	32.9	
				195	194.964 813	$\frac{1}{2}$ −	33.8	
				196	195.964 967	0 +	25.3	
				198	197.967 895	0 +	7.21	
79	Gold	Au	196.967	197	196.966 541	$\frac{3}{2}$ +	100	
80	Mercury	Hg	200.59	196	195.965 820	0 +	0.146	
				198	197.966 756	0 +	10.02	
				199	198.968 279	$\frac{1}{2}$ −	16.84	
				200	199.968 327	0 +	23.13	
				201	200.970 308	$\frac{3}{2}$ −	13.22	
				202	201.970 642	0 +	29.80	
				204	203.973 495	0 +	6.85	
81	Thallium	Tl	204.19	203	202.972 353	$\frac{1}{2}$ +	29.50	
				204*	203.973 865	2 −		3.75 y
				205	204.974 442	$\frac{1}{2}$ +	70.50	
		(Ra E″)		206*	205.976 104			4.3 min
		(Ac C″)		207*	206.977 450			4.78 min
		(Th C″)		208*	207.982 013	5 +		3.1 min
		(Ra C″)		210*	209.990 054			1.3 min
82	Lead	Pb	207.19	202*	201.927 997	0 +		3×10^5 y
				204*	203.973 044	0 +	1.48	1.4×10^{17} y
				205*	204.974 480			3×10^7 y
				206	205.974 468	0 +	23.6	
				207	206.975 903	$\frac{1}{2}$ −	22.6	
				208	207.976 650	0 +	52.3	
		(Ra D)		210*	209.984 187	0 +		22 y
		(Ac B)		211*	210.988 742			36.1 min
		(Th B)		212*	211.991 905	0 +		10.64 h
		(Ra B)		214*	213.999 764	0 +		26.8 min

(1) Z	(2) Element	(3) Symbol	(4) Chemical Atomic Weight	(5) Mass Number (* Indicates Radioactive) A	(6) Atomic Mass	(7) Spin and Parity	(8) Percent Abundance	(9) Half-Life (if Radioactive) $t_{1/2}$
83	Bismuth	Bi	209.980	207*	206.978 438			30 y
				208*	207.979 731			3.7×10^5 y
				209	208.980 394	$\frac{9}{2}$ −	100	
		(Ra E)		210*	209.984 121	1 −		5.1 d
		(Th C)		211*	210.987 300			2.15 min
				212*	211.991 876	1 −		60.6 min
		(Ra C)		214*	213.998 686			19.7 min
				215*	215.001 830			8 min
84	Polonium	Po		209*	208.982 426	$\frac{1}{2}$ −		103 y
		(Ra F)		210*	209.982 876	0 +		138.4 d
		(Ac C′)		211*	210.986 657			0.52 sec
		(Th C′)		212*	211.989 629	0 +		0.30 μsec
		(Ra C′)		214*	213.995 201	0 +		164 μsec
		(Ac A)		215*	214.999 423			0.0018 sec
		(Th A)		216*	216.001 790	0 +		0.15 sec
		(Ra A)		218*	218.008 930	0 +		3.05 min
85	Astatine	At		215*	214.998 663			≈ 100 μsec
				218*	218.008 607			1.3 sec
				219*	219.011 290			0.9 min
86	Radon	Rn						
		(An)		219*	219.009 481			4.0 sec
		(Tn)		220*	220.011 401	0 +		56 sec
		(Rn)		222*	222.017 531	0 +		3.823 d
87	Francium	Fr						
		(Ac K)		223*	223.019 736			22 min
88	Radium	Ra	226.05					
		(Ac X)		223*	223.018 501	$\frac{1}{2}$ +		11.4 d
		(Th X)		224*	224.020 218	0 +		3.64 d
		(Ra)		226*	226.025 360	0 +		1620 y
		(Ms Th$_1$)		228*	228.031 139	0 +		5.7 y
89	Actinium	Ac		227*	227.027 753	$\frac{3}{2}$ +		21.2 y
		(Ms Th$_2$)		228*	228.031 080			6.13 h
90	Thorium	Th	232.038					
		(Rd Ac)		227*	227.027 706			18.17 d
		(Rd Th)		228*	228.028 750	0 +		1.91 y
				229*	229.031 652	$\frac{3}{2}$ +		7300 y
		(Io)		230*	230.033 087	0 +		76,000 y
		(UY)		231*	231.036 291			25.6 h
		(Th)		232*	232.038 124	0 +		1.39×10^{10} y
		(UX$_1$)		234*	234.043 583	0 +		24.1 d
91	Protactinium	Pa		231*	231.035 877	$\frac{3}{2}$ −		32,480 y
		(Uz)		234*	234.043 298			6.66 h
92	Uranium	U	238.03	232*	232.037 168	0 +		72 y
				233*	233.039 522	$\frac{5}{2}$ +		1.62×10^5 y
				234*	234.040 904	0 +	0.0057	2.48×10^5 y
		(Ac U)		235*	235.043 915	$\frac{7}{2}$ −	0.72	7.13×10^8 y
				236*	236.045 637	0 +		2.39×10^7 y
		(UI)		238*	238.048 608	0 +	99.27	4.51×10^9 y

(1)	(2)	(3)	(4) Chemical Atomic Weight	(5) Mass Number (* Indicates Radioactive) A	(6) Atomic Mass	(7) Spin and Parity	(8) Percent Abundance	(9) Half-Life (if Radioactive) $t_{1/2}$
Z	Element	Symbol						
93	Neptunium	Np		235*	235.044 049			410 d
				236*	236.046 624			5000 y
				237*	237.048 056	$\frac{5}{2}$ +		2.14×10^6 y
94	Plutonium	Pu		236*	236.046 071	0 +		2.85 y
				238*	238.049 511	0 +		89 y
				239*	239.052 146	$\frac{1}{2}$ +		24,360 y
				240*	240.053 882	0 +		6700 y
				241*	241.056 737	$\frac{5}{2}$ +		13 y
				242*	242.058 725	0 +		3.79×10^5 y
				244*	244.064 100	0 +		7.6×10^7 y

APPENDIX B   Probability Integrals

When calculating various average values using the Maxwell-Boltzmann distribution, integrals of the following type occur:

$$I_n = \int_0^\infty x^n e^{-\lambda x^2}\, dx$$

where n is an integer. These can be obtained from I_0 and I_1 by differentiation. Consider I_n to be a function of λ and take the derivative with respect to λ:

$$\frac{dI_n}{d\lambda} = \int_0^\infty - x^2 x^n e^{-\lambda x^2}\, dx = - I_{n+2} \qquad\qquad \text{B-1}$$

Thus if I_0 is known, all the I_n for even n can be obtained, and if I_1 is known, all the I_n for odd n can be obtained from Equation B-1. I_1 can easily be evaluated, using the substitution $u = \lambda x^2$. Then $du = 2\lambda x\, dx$ and

$$I_1 = \int_0^\infty x e^{-\lambda x^2}\, dx = \tfrac{1}{2}\lambda^{-1} \int_0^\infty e^{-u}\, du = \tfrac{1}{2}\lambda^{-1}$$

Then I_3 and I_5 are

$$I_3 = - \frac{d(\tfrac{1}{2}\lambda^{-1})}{d\lambda} = \tfrac{1}{2}\lambda^{-2}$$

and

$$I_5 = - \frac{dI_3}{d\lambda} = \lambda^{-3}$$

The evaluation of I_0 is more difficult, but it can be done using a trick. We evaluate $I_0{}^2$:

$$I_0{}^2 = \int_0^\infty e^{-\lambda x^2}\, dx \int_0^\infty e^{-\lambda y^2}\, dy = \int_0^\infty \int_0^\infty e^{-\lambda(x^2+y^2)}\, dx\, dy$$

where we have used y as the dummy variable of integration in the second integral. If we now consider this to be an integration over the xy plane, we can change to polar coordinates $r^2 = x^2 + y^2$ and $\tan \phi = y/x$. The element of area $dx\, dy$ becomes $r\, dr\, d\phi$ and the integration over positive x and y becomes integration from $r = 0$ to $r = \infty$ and from $\phi = 0$ to $\phi = \pi/2$. Then we have

$$I_0{}^2 = \int_0^\infty \int_0^{\pi/2} e^{-\lambda r^2} r\, dr\, d\phi = \frac{\pi}{2} I_1 = \frac{\pi}{4}\, \lambda^{-1}$$

and

$$I_0 = \tfrac{1}{2} \sqrt{\pi}\, \lambda^{-1/2}$$

We then obtain $I_2, I_4, \ldots$ by differentiation. For example,

$$I_2 = - \frac{dI_0}{d\lambda} = \tfrac{1}{4} \sqrt{\pi}\, \lambda^{-3/2}$$

APPENDIX C   Separation of the Schrödinger Equation in Spherical Coordinates

When Equation 7-5, $\psi(r,\theta,\phi) = R(r)f(\theta)g(\phi)$, is substituted into Equation 7-4 we obtain

$$-\frac{\hbar^2}{2\mu}\frac{1}{r^2}\frac{d}{dr}(r^2fgR') - \frac{\hbar^2}{2\mu r^2}\left[\frac{1}{\sin\theta}\frac{d}{d\theta}(f'gR\sin\theta)\right.$$

$$\left.+ \frac{1}{\sin^2\theta}fg''R\right] + VfgR = EfgR \qquad \text{C-1}$$

where $R' = dR/dr$, $f' = df/d\theta$, and $g'' = d^2g/d\phi^2$. If we multiply each term in this equation by $2\mu r^2/(\hbar^2 fgR)$ and rearrange slightly, we obtain an equation that on one side contains only the variable r:

$$\frac{1}{R(r)}\frac{d}{dr}[r^2R'(r)] + \frac{2\mu r^2}{\hbar^2}[E \quad V(r)]$$

$$= -\frac{1}{f(\theta)\sin\theta}\frac{d}{d\theta}[\sin\theta f'(\theta)]$$

$$-\frac{1}{\sin^2\theta}\frac{g''(\phi)}{g(\phi)} \qquad \text{C-2}$$

No matter how θ and ϕ vary, the left side of Equation C-2 remains constant, because it depends only on r. Similarly, the right side does not change no matter how r varies. Thus neither side can depend on r, θ, or ϕ, so that both sides must equal some constant, which we call α. The equation for r can then be written

$$-\frac{\hbar^2}{2\mu r^2}\frac{d}{dr}(r^2R') + \left[\frac{\alpha\hbar^2}{2\mu r^2} + V(r)\right]R = ER \qquad \text{C-3}$$

This is Equation 7-13 with $\alpha = l(l+1)$.

Setting the right side of Equation C-2 equal to α and rearranging, we obtain

$$\frac{g''(\phi)}{g(\phi)} = -\alpha\sin^2\theta - \frac{\sin\theta}{f(\theta)}\frac{d}{d\theta}[\sin\theta f'(\theta)] \qquad \text{C-4}$$

Again, since the left side depends only on ϕ and the right on θ, both sides must be a constant, which we call K. The equation for $g(\phi)$ is then

$$g''(\phi) = Kg(\phi) \qquad \text{C-5}$$

The solution of this equation is an exponential function $e^{-\sqrt{K}\phi}$ or $e^{+\sqrt{K}\phi}$. It does not satisfy the continuity equation $g(\phi + 2\pi) = g(\phi)$ unless K is a negative number. If we let $K = -m^2$, the solution of Equation C-5 is $g(\phi) = e^{im\phi}$. As stated in Chapter 7, the requirement that $g(\phi + 2\pi) = g(\phi)$ implies that m is a positive or negative integer, or zero.

The equation for $f(\theta)$ is obtained from Equation C-4 by setting the right side equal to $-m^2$. It can be put into a standard form called Legendre's equation by changing the variable to $u = \cos\theta$. Then

$df/d\theta = (df/du)(du/d\theta) = -\sin\theta(df/du)$. If we now write f' for df/du we obtain

$$(1 - u^2)f'' - 2uf' + \left(\alpha - \frac{m^2}{1 - u^2}\right)f = 0 \qquad\qquad \text{C-6}$$

This equation is solved by standard methods of differential equations. The solution becomes infinite at $\theta = 0$ or $\theta = 180°$ unless α is restricted to the values

$$\alpha = l(l + 1)$$

The well-behaved functions that satisfy Equation C-6 are called associated Legendre functions. They depend on the values of l and $|m|$ and are written $f_{l,|m|}$. Some of the associated Legendre functions (with normalization constants omitted) are:

$$f_{0,0} = 1 \qquad\qquad\qquad f_{1,0} = \cos\theta$$
$$f_{1,1} = \sin\theta$$

$$f_{2,0} = 3\cos^2\theta - 1 \qquad f_{3,0} = 5\cos^3\theta - 3\cos\theta$$
$$f_{2,1} = \sin\theta\cos\theta \qquad f_{3,1} = (5\cos^2\theta - 1)\sin\theta$$
$$f_{2,2} = \sin^2\theta \qquad\qquad f_{3,2} = \cos\theta\sin^2\theta$$
$$f_{3,3} = \sin^3\theta$$

For the special case $m = 0$, Equation C-6 becomes

$$(1 - u^2)f'' - 2uf' + \alpha f = 0 \qquad\qquad \text{C-7}$$

The well-behaved solutions of this equation are called Legendre polynomials. The first few such polynomials, normalized such that $f = 1$ when $u = 1$, are

$$f_0 = 1$$
$$f_1 = u$$
$$f_2 = \tfrac{1}{2}(3u^2 - 1)$$
$$f_3 = \tfrac{1}{2}(5u^3 - 3u)$$

By direct substitution you can show that f_0 satisfies Equation C-7 if $\alpha = 0$, f_1 satisfies the equation if $\alpha = 2 = 1(1 + 1)$, f_2 satisfies the equation if $\alpha = 6 = 2(2 + 1)$, and so forth.

APPENDIX D — # General Physical Constants, Useful Combinations, Conversion Factors, and Numerical Data

General Physical Constants

Constant	Symbol	Value	Unit
Speed of light	c	$2.997\ 924\ 58 \times 10^8$	m/sec
Elementary charge	e	$1.602\ 189 \times 10^{-19}$	C
Avogadro constant	N_A	$6.022\ 04 \times 10^{23}$	particles/mole
Faraday constant	F	$96,484.6$	C/mole
Planck constant	h	$6.626\ 18 \times 10^{-34}$	J-sec
		$4.135\ 70 \times 10^{-15}$	eV-sec
	$\hbar = h/2\pi$	$1.054\ 589 \times 10^{-34}$ ⌉	J-sec
		$6.582\ 17 \times 10^{-16}$	eV-sec
Gas constant	R	$8.314\ 4$	J/mole-K
		$1.987\ 2$	cal/mole-K
Boltzmann constant	k	$1.380\ 66 \times 10^{-23}$	J/K
		$8.617\ 4 \times 10^{-5}$	eV/K
Stefan-Boltzmann constant	σ	$5.670\ 3 \times 10^{-8}$	W/m²-K⁴
Gravitational constant	G	6.672×10^{-11}	N-m²/kg²
Coulomb constant	$k = 1/4\pi\epsilon_0$	$8.987\ 551\ 79 \times 10^9$	N-m²/C²
Permittivity of vacuum	ϵ_0	$8.854\ 187\ 82 \times 10^{-12}$	C²/J-m
Permeability of vacuum	μ_0	$4\pi \times 10^{-7}$	N/A²
Fine-structure constant	α	$7.297\ 351 \times 10^{-3}$	
	α^{-1}	$137.036\ 0$	
Rydberg constant	R_∞	$1.097\ 373\ 18 \times 10^7$	m⁻¹
Rydberg frequency	cR_∞	$3.289\ 842\ 0 \times 10^{15}$	Hz
Rydberg energy	hcR_∞	$13.605\ 80$	eV
Bohr radius	a_0	$5.291\ 771 \times 10^{-11}$	m
Electron rest mass	m_e	$9.109\ 53 \times 10^{-31}$	kg
		$5.485\ 802 \times 10^{-4}$	u
		$0.511\ 003$	MeV/c^2
Proton rest mass	m_p	$1.672\ 648 \times 10^{-27}$	kg
		$1.007\ 276\ 47$	u
		938.280	MeV/c^2
Neutron rest mass	m_n	$1.674\ 954 \times 10^{-27}$	kg
		$1.008\ 665\ 01$	u
		939.573	MeV/c^2
Unified mass unit	u	$1.660\ 566 \times 10^{-27}$	kg
		931.502	MeV/c^2
Bohr magneton	μ_B	$9.274\ 08 \times 10^{-24}$	J/T
		$5.788\ 378 \times 10^{-9}$	eV/G
Nuclear magneton	μ_N	$5.050\ 82 \times 10^{-27}$	J/T
		$3.152\ 452 \times 10^{-12}$	eV/G
Electron Compton wavelength	$h/m_e c$	$2.426\ 309 \times 10^{-12}$	m
Proton Compton wavelength	$h/m_p c$	$1.321\ 410 \times 10^{-15}$	m
Ratio, proton mass to electron mass	m_p/m_e	$1836.151\ 5$	

These values (taken from *Physics Today,* September 1974) are recommended by the Committee on Data for Science and Technology of the International Council of Scientific Unions. They have been rounded so that the uncertainty is in the last digit only.

Useful Combinations

hc = 1.2399×10^{-6} eV-m
   = 12,399 eV-Å = 1239.9 MeV-fm
$\hbar c$ = 1.9733×10^{-7} eV-m
   = 1973.3 eV-Å = 197.33 MeV-fm
ke^2 = 1.4400×10^{-9} eV-m
   = 14.400 eV-Å = 1.4400 MeV-fm
kT = 2.353×10^{-2} eV at T = 273 K
   = 2.585×10^{-2} eV at T = 300 K

Conversion Factors

1 Å = 0.1 nm = 10^{-10} m
1 nm = 10 Å = 10^{-9} m
1 fm = 10^{-6} nm = 10^{-15} m
1 barn = 10^{-24} cm^2

1 T = 1 weber/m^2 = 10^4 G
1 atm = 1.013×10^5 N/m^2

1 eV = 1.602 189 $\times 10^{-19}$ J
1 J = 10^{-7} erg = 6.241 461 $\times 10^{18}$ eV
1 cal = 4.1840 J
1 u = 931.502 MeV/c^2
1 g = 6.022 04 $\times 10^{23}$ u = 5.609 56 $\times 10^{32}$ eV/c^2
1 eV/particle = 23.060 36 kcal/mole
1 kcal/mole = 4.336 446 $\times 10^{-2}$ eV/particle

1 roentgen = 258 μC/kg
1 rad = 10^{-2} J/kg = 10^{-2} Gy
1 Ci = 3.7×10^{10} decays/sec = 3.7×10^{10} Bq

For additional data see the front and back endpapers and the following tables in the text:

Table 1-1   Rest energies of some elementary particles and light nuclei

Table 2-1   Molar heat capacities of some gases at 15°C and 1 atm

Table 2-2   Values of the integral $I_n = \int_0^\infty x^n e^{-\lambda x^2}\,dx$ for n = 0 to 5

Table 2-3   Some values of the molecular radius computed from viscosity measurements

Table 3-1   Results of Wilson's determination of e

Table 3-2   Rise and fall times of a single oil drop with calculated number of elementary charges on drop

Table 3-3   Heat capacities in cal/mole-K for Au, Diamond, Al, and Be

Table 7-1   Electron configurations of the atoms in their ground states

APPENDIX E — Periodic Table of Elements

The values listed are based on $^{12}_{6}C = 12$ u exactly. For artificially produced elements, the approximate atomic weight of the most stable isotope is given in brackets.

PERIOD	SERIES	GROUP I	II	III	IV	V	VI	VII	VIII			0
1	1	1 **H** 1.00797										2 **He** 4.003
2	2	3 **Li** 6.942	4 **Be** 9.012	5 **B** 10.81	6 **C** 12.011	7 **N** 14.007	8 **O** 15.9994	9 **F** 19.00				10 **Ne** 20.183
3	3	11 **Na** 22.990	12 **Mg** 24.31	13 **Al** 26.98	14 **Si** 28.09	15 **P** 30.974	16 **S** 32.064	17 **Cl** 35.453				18 **Ar** 39.948
4	4	19 **K** 39.102	20 **Ca** 40.08	21 **Sc** 44.96	22 **Ti** 47.90	23 **V** 50.94	24 **Cr** 52.00	25 **Mn** 54.94	26 **Fe** 55.85	27 **Co** 58.93	28 **Ni** 58.71	
4	5	29 **Cu** 63.54	30 **Zn** 65.37	31 **Ga** 69.72	32 **Ge** 72.59	33 **As** 74.92	34 **Se** 78.96	35 **Br** 70.909				36 **Kr** 83.80
5	6	37 **Rb** 85.47	38 **Sr** 87.62	39 **Y** 88.905	40 **Zr** 91.22	41 **Nb** 92.91	42 **Mo** 95.94	43 **Tc** [98]	44 **Ru** 101.1	45 **Rh** 102.905	46 **Pd** 106.4	
5	7	47 **Ag** 107.870	48 **Cd** 112.40	49 **In** 114.82	50 **Sn** 118.69	51 **Sb** 121.75	52 **Te** 127.60	53 **I** 126.90				54 **Xe** 131.30
6	8	55 **Cs** 132.905	56 **Ba** 137.34	57–71 Lanthanide series*	72 **Hf** 178.49	73 **Ta** 180.95	74 **W** 183.85	75 **Re** 186.2	76 **Os** 190.2	77 **Ir** 192.2	78 **Pt** 195.09	
6	9	79 **Au** 196.97	80 **Hg** 200.59	81 **Tl** 204.37	82 **Pb** 207.19	83 **Bi** 208.98	84 **Po** [210]	85 **At** [210]				86 **Rn** [222]
7	10	87 **Fr** [223]	88 **Ra** [226]	89–103 Actinide series†								

* Lanthanide Series	57 **La** 138.91	58 **Ce** 140.12	59 **Pr** 140.91	60 **Nd** 144.24	61 **Pm** [147]	62 **Sm** 150.35	63 **Eu** 152.0	64 **Gd** 157.25	65 **Tb** 158.92	66 **Dy** 162.50	67 **Ho** 164.93	68 **Er** 167.26	69 **Tm** 168.93	70 **Yb** 173.04	71 **Lu** 174.97

† Actinide Series	89 **Ac** [227]	90 **Th** 232.04	91 **Pa** [231]	92 **U** 238.03	93 **Np** [237]	94 **Pu** [242]	95 **Am** [243]	96 **Cm** [247]	97 **Bk** [247]	98 **Cf** [251]	99 **Es** [254]	100 **Fm** [253]	101 **Md** [256]	102 **No** [254]	103 **Lw** [257]

Answers

These results are usually rounded to three significant figures. Differences in the third figure may result from rounding the input data and are not important.

Chapter 1 **Exercises** 1. (a) 236 μsec (b) 2.36×10^{-12} sec (c) no

5. (a) 5.96×10^{-8} sec (b) 16.1 m (c) 7.02 m

7. (a) 0.8 m (b) 4.44 nsec

9. 80 c-min

13. 60 min, the same as the result of Exercise 12

15. about $0.21c = 39,000$ mi/sec

17. 0.6 c

19. (a) $x_2' - x_1' = \gamma(x_2 - x_1)$ (b) times t_1 and t_2 in Equations 1-19 are equal to t_0, but times t_1' and t_2' in Equations 1-18 are not equal and not known

21. (a) $u_x = v$, $u_y = c/\gamma$ (b) $u_x^2 + u_y^2 = c^2$

25. (a) 1.25 (b) 0.383 MeV/c (c) 0.639 MeV (d) 0.128 MeV

27. (a) 7.09 (b) 3.59 MeV/c (c) 3.62 MeV (d) 3.11 MeV

29. 1.11×10^{-16} kg; increase

31. 6.26 MeV

33. (a) 4.5×10^{-9} u (b) about 8×10^{-9} %

35. 4×10^{-6} kg/h

37. (a) 4.97 MeV/c (b) 0.995

39. (a) 20.505 MeV/c (b) 20.511 MeV/c

41. (a) 300 MeV/c (b) 46.8 MeV

Problems 1. 9.01×10^{11} sec $\approx$ 28,500 years

3. (a) 0.75% (b) 68.7%

9. (a) 120 min (b) 240 min (c) 60 min; 120 min; 240 min; observer in S' makes the same table

Chapter 2 **Exercises** 1. 5.31×10^{-23} g

5. (a) 1.93 km/sec (b) 1.01×10^4 K

7. (a) 1.67 (b) 1.4 (c) 1.29

9. $C_v = R$; $C_p = 2R$; $\gamma = 2$

11. $s_{av} = 6.0$; $\sigma = 1.89$

13. $C = 1/2a$; $x = 0$; $x_{rms} = a/\sqrt{3}$; $\sigma = a/\sqrt{3}$

15. $v = 2.51$ km/sec; $v_m = 2.22$ km/sec

17. (b) 2.40×10^{21} (c) 1.46×10^{21} (d) 3.25×10^{20} (e) 3.04×10^{7}

19. (b) 1.08×10^{17} (c) 5.38×10^{15}

21. (b) 0 (c) 5.00×10^{21} (d) 3.66×10^{20} (e) 5.13×10^{-5} (i.e., zero molecules most of the time)

23. 0.155

25. (a) 2.55 Å (b) 3.13 Å (c) 2.26 Å

27. (a) 400 times (b) 0

Problems   1. $A = 2/L$; $\overline{x} = L/2$; $x_m = L/2$; $x_{\mathrm{rms}} = \left(\dfrac{1}{3} - \dfrac{1}{2\pi^2}\right)^{1/2} L = 0.532\,L$

3. $(2kT/\pi m)^{1/2}$

5. $\Delta f/f \approx 2 \times 10^{-6}$, or two parts per million. The classical Doppler effect formula is adequate for this problem.

13. (a) $N(x) = N_0 e^{-\alpha x}$ (d) 0

Chapter 3   **Exercises**   1. 1.602×10^{-19} C

3. (a) 1.759×10^{11} C/kg (b) 9.579×10^{9} C/kg

5. (a) 1.04 mm (b) 1.35 cm (c) 1.25 G

7. $\sigma = 0.25 \times 10^{-19}$ C; $\sigma/e = 25\%$

9. (a) 8.594×10^{-4} m/sec (b) 2.76×10^{-6} m

11. (a) 9.66×10^{5} nm (b) 9.66×10^{3} nm (c) 966 nm

13. (a) 4.14×10^{3} K (b) 0.0966 K (c) 9.66×10^{-4} K

15. 16

17. (a) $0.951\,kT$, approximately the equipartition-theorem result of kT (b) $4.54 \times 10^{-4}\,kT$, much less than kT

19. (a) 12.4 keV (b) 1240 MeV = 1.24 GeV (c) 4.14×10^{-7} eV

21. (a) 4.13 eV (b) 2.22 eV (c) 1.32 V

23. 683.4 nm; 1.82 eV

25. 1.51×10^{13} photons/sec

27. 2.17×10^{-17} J

29. 10,000 V

31. 2.43 Å

33. (a) $0.951\,kT \approx kT$ (b) $0.582\,kT \approx \frac{1}{2}kT$ (c) $4.54 \times 10^{-4}\,kT \ll kT$

35. $2.76\,R = 5.48$ cal/mole-K

Problems   1. (a) $\sigma = 22.0 \times 10^{-6}$; $\sigma/\overline{x} = 0.41\%$ (b) using $n' = 9\frac{1}{3}$ you get 7σ, compared with 1.45σ using $n' = 9$

3. (a) 4.98×10^{17} eV/sec-m² (b) taking 1Å² for the area, the energy/sec is 4.98×10^{-3} eV/sec (c) 402 sec = 6.7 min

5. 2.55 keV ($\lambda = 4.86$ Å); $\Delta\lambda = 1.21 \times 10^{-2}$ Å for any λ, $\Delta\lambda \approx \frac{1}{400}\lambda$ if $\lambda = 4.86$ Å; recoil energy is 6.35 eV

7. (a) $C = \frac{1}{18}$; $\bar{x} = 2$ (b) $C = \frac{1}{21}$; $\bar{x} = 1.67$ (c) $C = \frac{1}{12}$; $\bar{x} = 1.33$

9. T_E is about 130 K for Au; 1060 K for Diamond; 225 K for Al; 560 K for Be

11. $f = 2.08 \times 10^{10}$ $T_E = 2.7 \times 10^{12}$ Hz for Au; 22×10^{12} Hz for Diamond; 4.7×10^{12} Hz for Al; 11.6×10^{12} Hz for Be

13. (a) $(1 + e^{-\epsilon/kT})^{-1}$ (b) $\epsilon(1 + e^{\epsilon/kT})^{-1}$

15. $a/\lambda = 4.965$; $\lambda/a = 0.2014$ gives $\lambda_m T = 0.2014$ $hc/k = 2.898 \times 10^{-3}$ m-K, which agrees with Equation 3-17

Chapter 4   **Exercises**   1. 10^6 collisions; 10^4 layers

3. 32.5 MeV

5. 9.36 MeV

7. (a) 260 fm (b) 0.013

9. (a) 1.01×10^{-4} (b) 1.29×10^{-6}

11. 121.6 nm, 102.6 nm, 97.25 nm, series limit is 91.18 nm. None are in the visible spectrum.

13. Paschen: 1.88×10^3 nm, 1.28×10^3 nm, 1.09×10^3 nm; series limit is 821 nm; Brackett: 4.05×10^3 nm, 2.63×10^3 nm, 2.17×10^3 nm; series limit is 1.46×10^3 nm

15. (a) 285 fm (b) 2.27 keV (c) 0.546 nm

17. (a) 0.610 Å; 0.578 Å (b) 0.542 Å

19. Calcium ($Z = 20$)

21. Tin ($Z = 50$)

23. (a) $E_1 = 37.6$ eV; $E_2 = 150$ eV (b) 110.3 Å; this is about a factor of 10 smaller than those emitted by the hydrogen atom

Problems   1. (a) -108.8 eV (b) $+27.2$ eV (c) 27.2 eV

3. (a) 1.054 mA (b) 9.27×10^{-24} A-m²

5. 1.79 Å

7. 1.19×10^{-5}

9. (a) 2.17×10^{-5} (b) 1.66×10^{-7}

11. 38.7 MeV

13. (a) 1.04×10^{14} Hz (b) 2.88×10^{-6} m

Chapter 5   **Exercises**   1. (a) 14.36 fm (b) 6.42 fm (c) 5.08 fm

3. (a) 150 eV (b) 6.01×10^{-6} eV (c) 1.5×10^{-14} eV

5. (a) 2.06×10^{-2} eV (b) 8.18×10^{-2} eV

7. 12.4 fm

9. (a) 41.2° (b) 34.8°

11. (a) 2.10 Å (b) 35.7°

13. (a) 4.29×10^{14} to 7.5×10^{14} Hz (b) 1.7 cm to 17 m (c) 3 m (d) 10^{10} Hz

15. (b) 0.12 mm

17. (a) 0.0344 mm (b) 0.034°

19. (a) $y = 0.004 \cos(0.2\,x - 10\,t) \cos(7.8x - 390\,t)$ (b) $v_p = 50$ m/sec (c) $v_g = 50$ m/sec (d) nondispersive (e) $\Delta x = 15.7$ m $= 2\pi/\Delta k$ using $\Delta k = k_2 - k_1 = 0.4$

21. (a) 10 μsec (b) 10^5 Hz

23. 10^{-21} m

25. 6.6×10^{-16} sec using $\Delta E\,\Delta t = \hbar$

27. (a) 2×10^{-2} eV (b) 0.83 MeV

Problems 5. (b) 0.82 unit (c) $k_1 = \pi$; $k_3 = \pi/3$; $\Delta k = 2\pi/3$; $\Delta k\,\Delta x = 3.42$ using $\Delta x = 2x_1$ (d) 12 units (e) 60 units

7. (b) 0.05 mm (c) 100 m

9. (b) $x_1 = 1.7$ (d) 0.16 (e) 0.54

11. 5.27×10^{-34} J

13. $\Delta E = 10^{-6}$ eV for $hf_0 = 1$ eV, about 10 times the natural line width; $\Delta E = 1$ eV for $hf_0 = 1$ MeV

15. Using $\ell = 700$ Å and $\bar{v} = 400$ m/sec gives $\tau_c = 1.75 \times 10^{-10}$ sec between collisions and $\Gamma_c = \hbar/\tau_c = 4 \times 10^{-6}$ eV. Since τ_c is proportional to ℓ, which is inversely proportional to the pressure P, τ_c will be 10^{-8} sec at $P \sim 57$ atm.

Chapter 6 **Exercises** 3. (a) 0.02 (b) 0.01 (c) 0

5. (a) 0.02 (b) 0.01 (c) 0

7. 3×10^{19}

9. (a) 10^{-21} J-sec (b) 10^{13}

13. $0.283\,L^2$

15. $\sigma_x = 0.181\,L$; $\sigma_p = \pi\hbar/L$; $\sigma_x\sigma_p = 0.568\,\hbar$

17. (a) $C_0 = (m\omega/\pi\hbar)^{1/4}$ (b) $\hbar/2m\omega$ (c) $\tfrac{1}{4}\hbar\omega$

21. $E_{311} = 11(\hbar^2\pi^2/2mL^2)$; $E_{222} = 12(\hbar^2\pi^2/2mL^2)$; $E_{321} = 14(\hbar^2\pi^2/2mL^2)$

Problems 5. (b) $T = 0.823$, $R = 0.177$ (c) $T = 0.971$, $R = 0.029$ (d) $T = 0.9993, R = 0.0007$

7. $P \approx 0.15$

Chapter 7 **Exercises** 1. $\sqrt{2}\,\hbar$; 45°, 90°, and 135°

3. (a) $2\hbar^2$ (b) $6\hbar^2$ (c) $5\hbar^2$; no

5. (a) 3.49×10^{-3} J-sec; $l \approx 3.32 \times 10^{31}$ (b) about 10^{-30} rev/min

7. (a) 4 (b) $(n, l, m_l) = (2,0,0); (2,1,+1); (2,1,0); (2,1,-1)$

11. (a) 5.05×10^{-27} J/T (b) 3.15×10^{-12} eV/G

13. l could be either 0 or 1.

15. (a) $j = 2\frac{1}{2}$ or $1\frac{1}{2}$ (b) $2.96\hbar$ for $j = 2\frac{1}{2}$; $1.94\hbar$ for $j = 1\frac{1}{2}$ (c) $2\frac{1}{2}\hbar$, $1\frac{1}{2}\hbar$, $\frac{1}{2}\hbar$, $-\frac{1}{2}\hbar$, $-1\frac{1}{2}\hbar$, $-2\frac{1}{2}\hbar$. All of these values except $\pm 2\frac{1}{2}\hbar$ occur in two different ways, e.g., $+1\frac{1}{2}\hbar$ arises from $m_l = +1$ and $m_s = +\frac{1}{2}$, or from $m_l = +2$ and $m_s = -\frac{1}{2}$.

17. (a) 2, 1, or 0 (b) 1 or 0 (c) 3, 2, 1, 0 (d) $1\frac{1}{2}$ or $\frac{1}{2}$ (e) 3, 2, 1, 0 the same as in (c)

19. (a) 2.10509 eV for 589 nm and 2.10295 eV for 589.6 nm (using $hc = 1239.9$ eV-nm) (b) 2.14×10^{-3} eV (c) 18.5 T

21. 0.48 Å

23. B, Al, and Ga

25. 1.84

27. The $3D_{5/2} \rightarrow 3P_{1/2}$ transition violates $\Delta j = \pm 1$ or 0.

29. (a) -4.32 eV for $4s$; -2.71 eV for $4p$; -0.93 eV for $4d$ (b) 2.25 for $4s$; 1.78 for $4p$; 1.05 for $4d$

31. $4S_{1/2} \rightarrow 3S_{1/2}$ violates $\Delta l = \pm 1$; $4D_{5/2} \rightarrow 3P_{1/2}$ violates $\Delta j = \pm 1$ or 0; $4D_{3/2} \rightarrow 3S_{1/2}$ violates $\Delta l = \pm 1$.

33. He and Mg

Problems   1. (c) 1.52×10^{-2} eV (d) 8.18×10^5 Å

5. 9×10^{-15}

7. (a) 1.9×10^{-2} T (b) 2.2×10^{-6} eV (c) 56 cm

9. (a) $g_{P_{1/2}} = \frac{2}{3}$; $g_{S_{1/2}} = 2$ (b) $g_{P_{3/2}} = \frac{4}{3}$

Chapter 8   **Exercises**   1. (a) -5.16 eV (b) 4.64 eV (c) 0.20 eV

3. (a) 1 eV/molecule = 23.06 kcal/mole (b) 97.5 kcal/mole

5. Si and Ge (and possibly also Sn and Pb, though these elements have empty shells)

7. 1.79×10^{-4} eV

9. (a) 18.6 u (b) 2.80 Å

11. half; no

Problems   1. (a) -6 eV (b) $6 - (4.23 + 1.3) = 0.47$ eV (c) $n \approx 5.6$; $C \approx 67.6$

3. (a) 0.179 eV (b) 2.8×10^{-47} kg-m^2 = 1.68 u-Å^2 (c) 0.980 u; $r_0 = 1.33$ Å

5. (a) 29.8 u (b) 9.01×10^{-6} eV (c) 4.36×10^9 Hz

7. (a) for $H^{35}Cl$, $\mu = 0.979593$ u; for $H^{37}Cl$, $\mu = 0.981077$ u; $\Delta\mu/\mu = 0.0015$ (b) Δf is very difficult to measure from the figure, but $\Delta f/f$ seems to be of the order of 0.001.

Chapter 9   **Exercises**   1. 3.15 Å

3. 2.07 g/cm³

5. (a) 479 A/m² = 47.9 mA/cm² (b) 3.53×10^{-6} cm/sec

7. (a) 136 A/cm² (b) 1.36 A

9. (a) 8.62×10^{22}/cm³ (b) 1.31×10^{23}/cm³

11. 1.07 free electron per atom

13. (a) $\sigma = 8.194 \times 10^6/\Omega$-m; $\rho = 1.22 \times 10^{-7}$ Ω-m = 12.2 $\mu\Omega$-m (b) $\sigma = 1.00 \times 10^7/\Omega$-m; $\rho = 9.96 \times 10^{-8}$ Ω-m (c) $\sigma = 14.2 \times 10^6/\Omega$-m; $\rho = 7.05 \times 10^{-8}$ Ω-m

15. (a) 11.7 eV (b) 2.12 eV (c) 10.2 eV

17. (a) 4.22 eV (b) 2.82 eV

19. Using $E_{av} = \frac{3}{5}E_F$, with $E_F = 7.03$ eV, gives $E_{av} = 4.2180$ eV. The contribution of the electron gas is about 0.0002 eV at $T = 300$ K so the total average energy is 4.2182 eV. (This is inaccurate because the 7.03 eV has been rounded.) The classical result $\frac{3}{2}kT$ is 3.88×10^{-2} eV at $T = 300$ K.

21. (a) 1.07×10^6 m/sec (b) 1.40×10^6 m/sec (c) 1.90×10^6 m/sec

23. 342 Å for Na; 414 Å for Au; 43.1 Å for Sn

25. (a) p-type (b) n-type

27. (a) 39.7 mV (b) -39.7 mV

Problems   3. (a) 1.25×10^{13} Hz (b) 23.9 μm, which is the correct order of magnitude

5. 3.81×10^{10} N/m² = 3.76×10^5 atm

7. (a) 6.35 Å (b) 8.46 Å; use of macroscopic quantity K is somewhat reasonable because these radii are large compared with crystal spacing

9. (a) 4×10^{-51} (b) 4×10^{-9}; both these probabilities are very small, but if there are 10^{22} electrons available, there will be a significant number in the conduction band in the semiconductor ($4 \times 10^{-9} \times 10^{22} = 4 \times 10^{13}$), whereas the probability of even one electron in the conduction band of the insulator is still small ($4 \times 10^{-51} \times 10^{22} = 4 \times 10^{-29}$)

Chapter 10   **Exercises**   1. (c) $\frac{1}{4}$ for distinguishable particles; $\frac{1}{3}$ for indistinguishable particles

3. (a) 1.75×10^{-7} (b) 0.54 K

5. about 4×10^{-51}; the Maxwell-Boltzmann distribution can be used, but this is not very important as there are essentially no electrons in the conduction band

7. (a) 0.646 (b) 0.875

Problems   3. 0.465

Chapter 11   Exercises   1. (a) isotopes of ^{16}O are ^{15}O, ^{17}O; isotones, ^{15}N, ^{14}C (b) isotopes of ^{208}Pb are ^{206}Pb, ^{204}Pb; isotones ^{207}Tl, ^{209}Bi; (c) isotopes of ^{120}Sn are ^{118}Sn, ^{116}Sn; isotones ^{121}Sb; ^{122}Te

3. (a) 28.3 MeV and 7.07 MeV/nucleon (b) 92.16 MeV and 7.68 MeV/nucleon (c) 488.15 MeV and 8.72 MeV/nucleon

5. (a) 2.70 and 3.53 fm (b) 4.26 and 5.57 fm (c) 6.34 and 8.29 fm

7. (a) β^+ (b) β^-

9. (a) 3.33 min (b) 0.208/min (c) 6500 counts/sec

11. $\lambda = 3.72 \times 10^{-3}$/sec, $t_{1/2} = 3.11$ min

13. 33.2 min

15. 3.35×10^3 y

17. To decay in the 2d second it must survive the 1st second. To decay in the 10th second it must survive the first 9 seconds. If p is the probability of decay per second, the probability of *not* decaying in a second is $1 - p$. The probability of decaying in the 1st second is then p, that of decaying during the 2d second is $(1 - p)p$, and that of decaying during the 10th second is $(1 - p)^9 p$, which is less than p.

19. 6.6×10^{-17} sec using $\tau = \hbar/\Gamma$

21. (a) 3.27 MeV (b) 4.03 MeV (c) 18.35 MeV (d) 4.785 MeV

23. 1.56×10^{19} fissions/sec

25. (a) 1.54 cm (b) 7.1 cm

27. (a) 0.58 mm (b) 5.14×10^{-2} mm

29. $l = 0$, $j = \frac{1}{2}$ for ^{29}Si; $l = 2$, $j = \frac{3}{2}$ for ^{37}Cl; $l = 1$, $j = \frac{3}{2}$ for ^{71}Ga; $l = 3$, $j = \frac{7}{2}$ for ^{59}Co; $l = 4$, $j = \frac{9}{2}$ for ^{73}Ge; $l = 2$, $j = \frac{3}{2}$ for ^{33}S; $l = 4$, $j = \frac{9}{2}$ for ^{87}Sr

Problems   1. (b) 51.7 MeV (c) 1.42 MeV (d) 96 MeV

3. For $A = 27$, $Z = 12.8 \approx 13$ corresponds to stable ^{27}Al; for $A = 65$, $Z = 29.4 \approx 29$ corresponds to stable ^{65}Cu; for $A = 139$, $Z = 58.7 \approx 59$ corresponds to $^{139}_{59}$Pr, which is not stable. The stable element with $A = 139$ is $^{139}_{57}$La which has $N = 82$, a magic number. The disagreement is a result of shell structure.

5. (a) 3.47 cm (b) 4.47×10^{-3} g/cm^2 (c) 1.66×10^{-3} cm

7. (a) 1.82×10^9 per sec (b) 3×10^{12} (c) 6×10^{12} nuclei; 1.82×10^9 per sec

9. (a) about 19 collisions (calculation gives $N = 18.5$) (b) about 158 collisions

11. 1.18×10^{19} m, which is about 7.9×10^7 times the distance from the earth to the sun.

Chapter 12   Exercises   1. (a) to conserve momentum (b) 938 MeV (c) 1.32 fm

3. (a) 279.2 MeV (b) 1876.6 MeV

5. (a) 0.782 MeV (b) 752 eV; 0.7812 MeV

7. 0.40 fm

9. (a) weak (b) electromagnetic (c) hadronic (d) weak

11. 8.1×10^{-37}

13. (a) $\Delta S = +1$, weak (b) $\Delta S = +2$, two step (c) $\Delta S = 0$, hadronic

15. $B = 1$ for each of the combinations. (a) $q = +1$, $T = \frac{1}{2}$, $S = 0$, proton (b) $q = 0$, $T = \frac{1}{2}$, $S = 0$, neutron (c) $q = +2$, $T = \frac{3}{2}$, $S = 0$, Δ^{++} (d) $q = 0$, $T = \frac{1}{2}$, $S = -2$, Ξ^0 (e) $q = -1$, $T = \frac{1}{2}$, $S = -2$, Ξ^- (f) $q = +1$, $T = 1$, $S = -1$, Σ^+ (g) $q = -1$, $T = 1$, $S = -1$, Σ^-

17. (a) $T_3 = 0$ (b) $T = 1$ or $T = 0$ (c) $B = 1$, $q = 0$, $S = -1$; this quark combination is a linear combination of a Λ^0 ($T = 0$) and a Σ^0 ($T = 1$)

Index